Life Cycle Assessment in the Agri-food Sector

Bruno Notarnicola • Roberta Salomone
Luigia Petti • Pietro A. Renzulli
Rocco Roma • Alessandro K. Cerutti
Editors

Life Cycle Assessment in the Agri-food Sector

Case Studies, Methodological Issues and Best Practices

Rete Italiana LCA

Springer

Editors
Bruno Notarnicola
Ionian Department of Law, Economics and Environment
University of Bari
Bari
Italy

Pietro A. Renzulli
Ionian Department of Law, Economics and Environment
University of Bari
Bari
Italy

Roberta Salomone
Department SEA
University of Messina
Messina
Italy

Rocco Roma
Department of Agro-Environmental Science
University of Bari
Bari
Italy

Luigia Petti
Department of Economics
University of Chieti-Pescara
Pescara
Italy

Alessandro K. Cerutti
Department Agricultural Sciences Forestry and Food
University of Turin
Grugliasco
Italy

ISBN 978-3-319-11939-7 ISBN 978-3-319-11940-3 (eBook)
DOI 10.1007/978-3-319-11940-3

Library of Congress Control Number: 2014954801

Springer Cham Heidelberg New York Dordrecht London

Printed on acid-free paper

Springer is part of Springer Science+Business Media (www.springer.com)

Foreword

The economic, social and environmental importance of the agri-food sector is well known. The present challenge is how to provide a growing population with good, safe, healthy food while decreasing the pressure and impacts on ecosystems, resource and human health.

LCA is the appropriate method to identify, with high degree of detail, environmental hotspots, compare techniques and crops and inform with scientific data the decision makers both at firm and political level. However, LCA application in the agri-food sector is a complex and challenging endeavour. This book represents a major step forward, identifying complex methodological issues, presenting good examples of case studies and best practices.

The main methodological and data availability issues, involving biological processes and technical systems are exhaustively described in the first chapter of this book, together with a complete survey of major international initiatives and of the (too) many labels used to inform customers about the environmental quality of products.

In the subsequent chapters, case studies on five important product groups, i.e. wine, oil, cereals, fruit, livestock, are thoroughly addressed in terms of best practices, data sources, major environmental impacts, and mitigation strategies.

The book is a very valuable source of data and information for many people, with a primary focus on the LCA practitioner community that will use it as a state-of-the-art reference anytime they have to model agri-food products decision makers and consumers will enjoy the deep and exhaustive description of environmental problems and mitigation strategies, finding scientific basis for informed decisions.

This book is really timely: next year on May 2015 the World Expo titled "Feeding the planet, Energy for life", will present the most advanced solutions of the agri-food sector as source of food, materials and energy. In the meantime, the second wave of pilot Environmental Footprint promoted by the European Commission is starting with, besides others, case studies on wine, oil and meat.

Let me conclude this short foreword with some proud words. Editors and authors of this book are all members of the Associazione Rete Italiana LCA[1], the Italian

---

[1] www.reteitalianalca.it

LCA Network I am honoured to chair. The "Rete" is a scientific, not-for-profit association whose mission is to foster environmental protection through the wide application of the Life Cycle Assessment. So far it has been capable to convene major Italian LCA experts, organise several scientific conferences, award young scientists, and, now, promote and support the writing of this book. Not bad for a 3 year old association.

ENEA, Chair of the Italian LCA Network Paolo Masoni
Bologna
25 May 2014

Preface

This book stems from a joint effort of some members of the Italian Network of LCA, who are particularly interested in Life Cycle Assessment (LCA) of agri-food product systems, with the aim of thoroughly and critically evaluating the state of the art of food LCA and its application to some particular food chains.

The Italian Network of LCA was launched in 2006 as an initiative of the "Italian National Agency for New Technologies, Energy and Sustainable Economic Development" (ENEA) with the aim of creating a network useful for the exchange, in Italy, of information, methodologies and good practices on the state-of-the art and on the prospects of the LCA methodology.

Six years later, during the VI Conference of the Italian Network of LCA hosted in Bari (Italy) in June 2012, a step forward was taken and the Network became a scientific Association founded by ENEA, the Politecnico di Milano, the Universities of Bari, Palermo, Chieti-Pescara, and Padova, and the "Interuniversity Consortium Chemical Reactivity and Catalysis (CIRCC)". Nowadays, membership association is open to all physical persons interested in the promotion of the development of the LCA methodology within the Italian territory.

The main aims of the Association of the Italian Network of LCA are: promoting the exchange of information and good practices on the state-of-the art and prospects of LCA studies in Italy; promoting the dissemination of the LCA methodology at national level; stimulating the interaction between the parties that deal with LCA and encouraging the process of networking among various stakeholders for the implementation of projects at national and international level; finally, the Association supports the life-cycle approach and the LCA methodology among the institutional bodies. Among the Association's activities, apart from the annual conference, are those of the information services (website, newsletter and mailing list) and of the "Working Groups" (WG).

In particular, along the years, nine WGs have been created: Food and Agri-industrial; Energy and Sustainable Technologies; Construction sector; Chemical products and processes; Tourist services; Management and waste treatment; Wood furniture; Automotive & Electric-Electronic; Development and Improvement of LCA methodology: Research and Exchange Board of experience (DIRE). Some of these WGs have been (and still are) involved in the definition of databases and methodological approaches mostly applicable to the specificities of the Italian territory and economy.

As far as the "Food and Agri-industrial" WG is concerned, it was constituted in 2008 with the aim of increasing the specific knowledge regarding the application of the LCA methodology to the Italian food and agro-industrial sector and also with the aim of spreading its use for the improvement of the environmental performances of the involved supply chains. The WG is made up of five sub-working groups which study some of the most important food supply chains, namely wine, olive oil, cereals and derived products, meat and fruit.

Among the past activities developed by this WG, the LCAFood 2010– VII international conference on "Life Cycle Assessment in the Agri-food sector" (hosted in Bari in September 2010) is worthy of mention.

This book represents another challenge undertaken by the "Food and Agro-industrial" WG aiming at highlighting, in an as much as possible exhausting manner, environmental hotspots, methodological issues and best practices for the agri-food sector from a life cycle perspective. Its writing has involved several Life Cycle Assessment (LCA) researchers and practitioners (from both private and public Italian organisations) with the aim of creating some practical guidelines for the LCA community and the main actors of the agro-food chains (e.g. farmers, manufacturing companies, consumers, etc.). The book is focussed, in particular, on some of the most relevant and productive agri-food supply chains within the European context, namely: olive oil, wine, cereal and derived products, livestock and derived edible products, and fruit.

In fact, since the end of the 1990's, researchers have scientifically highlighted the fact that most food chains are not sustainable from an environmental perspective due to the impacts occurring in different phases of their life cycle. So, in order to address these relevant issues, several European policies related to sustainable production and consumption began to be promoted with the aim, among others, of quantifying environmental performances of agri-food products. In particular, in 2003, the so-called Strategy for Sustainable Production and Consumption (SPC) was launched aiming at reducing the environmental, social, and economic impact of products and services throughout their entire life cycle. The concept of SPC can be applied to all the existing products sold and bought on the market and hence also to food and drink products. These products in particular play a fundamental role in the everyday life of consumers, whose demand for high quality food has increased in the last few years. In a similar manner to other products, their production and consumption (from farm to fork and end of life) have environmental implications; nevertheless, because of specific aspects related to health, nutrition, well-being, cultural identity, and lifestyle, they need to be considered and treated differently from all other products. In the same period, the Directorate General Joint Research Centre/Institute for Prospective Technology Studies (DG JRC/IPTS) launched a project called Environmental Impacts of Products (EIPRO) in an attempt to identify those products with the most relevant environmental impacts throughout their life cycle, from cradle to grave, taking into account the full food production and distribution chain from farm to fork. Among its results, the report, published in 2006, highlighted that the food and drink sector accounts for 20–30 % of the environmental impact of private consumption. Subsequently, in 2007, the Strategic Research Agenda (SRA) (2007–2020) of the European Technology Platform (ETP) Food for Life was published defining sustainable food production as the most important challenge that facing the European food industry.

Sustainability tools and, in particular, LCA have been applied for more than 20 years to agricultural and food systems for finding more sustainable ways of food production and consumption and as a means of supporting environmental decision-making via the identification of the environmental impacts throughout the systems' life cycles.

One of the reasons for the growing consideration of the academic community for aspects regarding food LCA is the fact that methodological issues (for example, the definition of the functional unit, difficulties in data collection, pesticides and their exposure, fertiliser dispersion models, impact categories of land use and water use) are different from the typical ones arising from industrial product LCAs. Until now such topics have been tackled with many different approaches that do not represent standardised methods, hence much has to be done to build a consistent, practical and life cycle science based approach to product level sustainability information reporting for all food, beverage, and agriculture products.

This book has been written with the intention of contributing to the identification of practical recommendations to these still open key issues, adding value to the international discourse. It consists of six chapters.

The first chapter has been designed to be propaedeutic to the subsequent ones, providing the reader an as exhaustive as possible overview of the key concerns, applications, and methodological uncertainties of agri-food life cycle assessment (LCA). It comprises: a review of the main international initiatives, eco-labels and declarations, and footprints together with some of the most important LCA initiatives developed by the main stakeholders of the agri-food chains; a general synopsis of the main methodological issues strictly linked to the application of the LCA methodology to the agri-food sector; a state of the art of the major existing international LCI databases and of the national and international initiatives currently under development; finally, an overview of the main dietary issues in the sense that in the context of food sustainability the importance of consumer behaviour and, in particular, dietary behaviour is becoming increasingly recognised, together with the product and its production chain.

On the contrary, each of the other five chapters focuses deeply and critically on one of the chosen agri-food supply chains. Even if each one is developed in its own different way, they are built on a common framework consisting of:

- an as comprehensive as possible state-of-the art of all the international LCA case studies developed on a specific agri-food sector, which represents a building block and a starting point for the subsequent steps;
- the identification of the main environmental hotspots and of the still open methodological issues specifically related to each sector;

- a critical analysis of these key points for identifying and developing practical guidelines to overcome these issues.

These "lessons learnt" are intended to be a support for LCA practitioners and for all the involved stakeholders when developing an LCA study in the agri-food sector.

Specifically, regarding Chaps. 2–6:

- the second chapter focuses on the olive oil industry, one of the most significant sectors within the European Union. The related production process is characterised by a variety of different practices and techniques for both the agricultural and processing phases, causing several adverse effects on the environment. After a description of the international state of the art of LCA implementation in this specific sector, a brief description of other life cycle thinking methodologies and tools (such as simplified LCA, footprint labels and Environmental Product Declarations) is given by the authors. Then, the methodological problems connected with the application of LCA in the olive oil production sector are analysed in depth, starting from a critical comparative analysis of the applicative LCA case studies in the olive oil production supply chain. Finally, guidelines for the application of LCA in the olive oil production sector are proposed.
- the third chapter regards the wine sector; a critical review of LCA case studies is presented by the authors in order to compile a list of scientifically-sound environmental improvements suggested by published LCAs. Next it identifies: the critical environmental issues of wine production and the essential elements that an LCA case study in the sector should consider; optimal sets of indicators and methodologies for the evaluation of the environmental impacts of wine; finally, best practices for environmental improvement in the wine sector are presented;
- the fourth chapter is focussed on cereal and derived products, vital for the production of commodities of worldwide importance that entail particular environmental hot spots originating from their widespread use and from their particular nature. After a brief introduction to the sector and supply chain, the chapter reviews some of the current cereal-based life cycle thinking literature, with a particular emphasis on LCA. Next, an analysis of the LCA methodological issues emerging from the literature review is carried out. The following section discusses some practices and approaches that should be considered when performing cereal-based LCAs in order to achieve the best possible results. Conclusions are drawn in the final part of the chapter and some indications are given of the main hot spots in the cereal supply chain.
- the fifth chapter regards livestock and derived edible products; like the olive oil industry, it is one of the most significant sectors from an economic perspective in Europe representing more than 40 % of the economic value of EU primary productions. This sector consists of a huge diversity of processes and techniques depending on the animal species and the final products. Because of these differences, livestock productions are associated with several adverse effects on the environment, especially in the breeding phases and feeding composition and management. In this chapter, after an overview of the structural and economic characteristics of the most significant livestock supply chain and its main

environmental problems, a description of the international state of the art of LCA implementations for livestock is given. Methodological problems connected with the application of LCA are investigated, starting with the critical analysis of international papers and the few Italian papers in the scientific literature. Finally, the best practices regarding LCA methodology implementation are proposed, in order to improve results and manage the methodological problems identified.
- finally, the sixth chapter focusses on fruit products, generally considered to be some of the less environmentally damaging foods in western diets. In fact studies investigating the carbon footprint of different food choices have reported that fruit is the category with the least environmental impact. However, these studies use data from environmental assessments of generic fruit production, which take no account of specific issues within orchard systems and fruit supply chains. Indeed, modern food production is very diverse, with high levels of specialisation and complexity. This chapter starts with an overview about the fruit industry in Europe and the main environmental burdens related to fruit production. Then, life cycle thinking methodologies and approaches in the sector are presented reporting a state of the art of international LCA practices and other life cycle methodologies and tools for product environmental assessment. Finally, based on the results of the critical analysis of international experiences, methodological problems concerned with the application of LCA to the sector are described and lessons learnt and practical guidelines are proposed.

The authors of this book would like to thank the Italian Network of LCA for its financial support.

The views expressed in this book are those of the authors and do not necessarily represent the views of the European Commission, THE UN Food and Agriculture Organisation or any other organisation cited in the text.

Bari
6 June 2014

The Editors

Contents

1 **Life Cycle Assessment in the agri-food sector: an overview of its key aspects, international initiatives, certification, labelling schemes and methodological issues** 1
Bruno Notarnicola, Giuseppe Tassielli, Pietro A. Renzulli and Agata Lo Giudice

2 **Life Cycle Assessment in the Olive Oil Sector** 57
Roberta Salomone, Giulio Mario Cappelletti, Ornella Malandrino, Marina Mistretta, Elena Neri, Giuseppe Martino Nicoletti, Bruno Notarnicola, Claudio Pattara, Carlo Russo and Giuseppe Saija

3 **Life Cycle Assessment in the Wine Sector** 123
Luigia Petti, Ioannis Arzoumanidis, Graziella Benedetto, Simona Bosco, Maurizio Cellura, Camillo De Camillis, Valentina Fantin, Paola Masotti, Claudio Pattara, Andrea Raggi, Benedetto Rugani, Giuseppe Tassielli and Manfredi Vale

4 **Life Cycle Assessment in the Cereal and Derived Products Sector** 185
Pietro A. Renzulli, Jacopo Bacenetti, Graziella Benedetto, Alessandra Fusi, Giuseppe Ioppolo, Monia Niero, Maria Proto, Roberta Salomone, Daniela Sica and Stefania Supino

5 **Life Cycle Assessment in the Livestock and Derived Edible Products Sector** 251
Rocco Roma, Sara Corrado, Annalisa De Boni, Maria Bonaventura Forleo, Valentina Fantin, Michele Moretti, Nadia Palmieri, Andrea Vitali and Camillo De Camillis

6 Life Cycle Assessment in the Fruit Sector 333
Alessandro K. Cerutti, Gabriele L. Beccaro, Simona Bosco, Anna I. De Luca, Giacomo Falcone, Angela Fiore, Nathalie Iofrida, Agata Lo Giudice and Alfio Strano

Conclusions 389

List of Figures

Fig. 1.1 The China environmental labelling logo 7
Fig. 1.2 The Ukraine living planet logo 7
Fig. 1.3 The vitality leaf logo 8
Fig. 1.4 The ecomark label 9
Fig. 1.5 The new ecolabel logo 9
Fig. 1.6 The green yardstick logo 13
Fig. 1.7 EPD® in the world 14
Fig. 1.8 EPD Norge logo 15
Fig. 1.9 Earthsure® logo 15
Fig. 1.10 The EcoLeaf logo 16
Fig. 1.11 Reducing CO2 label (Carbon Reduction Label) 23
Fig. 1.12 Figure CO2 measured label (Carbon Reduction Label) 24
Fig. 1.13 Japanese carbon footprint label 26
Fig. 1.14 Thailand carbon footprint label 28
Fig. 1.15 Carbon emission certificate label 28
Fig. 1.16 Low carbon product certification label 28
Fig. 1.17 Taiwan Carbon Footprint label 28
Fig. 1.18 Casino carbon index label (*front* of the packaging) 38
Fig. 1.19 Casino Carbon Index label (*back* of the packaging) 38

Fig. 2.1 The olive oil supply chain and its main environmental consequences 70
Fig. 2.2 The goal and scope in the surveyed case studies 95
Fig. 2.3 The functional unit (FU) in the surveyed case studies 95
Fig. 2.4 The object of investigation in comparative case studies 96
Fig. 2.5 The system boundaries in the surveyed case studies 97

Fig. 3.1 Decline in vineyards of major European wine producers. (Adapted from: OIV 2013) 127
Fig. 3.2 System boundaries—overview of all methodologies examined 137
Fig. 3.3 Studies identified in the review of the LCA of the wine sector, published from 2001 to 2013 (last update on 31 July 2013) 144

Fig. 3.4 Percentage breakdown of the analysed studies according to the different Functional Units used . . . 146
Fig. 3.5 Percentage breakdown of the analysed studies according to the packaging inclusion in FU . . . 147
Fig. 3.6 Percentage breakdown of the analysed papers according to the System Boundary considered . . . 147
Fig. 3.7 Scores' attribution along the six key issues outlined in Table 3.6 (score 1–3; from 1 to 3 according to the increase of breadth and depth of information provided, see Table 3.6 for further details). The total number of studies for which the scoring was performed is 81 . . . 158
Fig. 3.8 Trend of scores for LCIA interpretation (score 1–3; from 1 to 3 according to the increase of breadth and depth of information provided, see Table 3.6 for further details) per number of studies reviewed. The total number of studies for which the scoring was performed is 81 . . . 162

Fig. 4.1 The supply chain scheme for cereals and cereal-based products . . . 189
Fig. 4.2 The main elements of sustainable cereals and cereal-based product supply chains . . . 189
Fig. 4.3 The world's food wastage (2009) . . . 192
Fig. 4.4 The cereal and cereal-based product supply chain: an environmental perspective . . . 193
Fig. 4.5 Percentages of reviewed studies regarding crops/derived products . . . 200
Fig. 4.6 Possible FUs related to cereal and cereal-derived product systems . . . 236

Fig. 6.1 Distribution of different work types across years . . . 345
Fig. 6.2 Main boundaries of a LCA study about fruit production and process involved in each sub-phase of the fruit life cycle: from cradle to grave . . . 354
Fig. 6.3 Time boundaries of the orchard life cycle (agricultural core phase). Real production line refers to an apple orchard in Cuneo province, northern Italy (Data from Cerutti et al. 2011b) . . . 356
Fig. 6.4 Breakdown of impact categories used in LCA fruit studies . . . 364
Fig. 6.5 Graphical representation of the orchard model with Gabi 4.0 . . . 369

List of Tables

Table 1.1 Current eco-labelling programmes available in the agri-food sector 6
Table 1.2 Ukraine living planet criteria for the agro-food sector. 8
Table 1.3 Current EPD systems available in the agri-food sector 12
Table 1.4 PCR (EPD®) for the "Food products" category 14
Table 1.5 PCR (EPD®) for the "Beverages" category 15
Table 1.6 Carbon footprint public initiatives 22
Table 1.7 PCR for the Japan carbon footprint 27
Table 1.8 Taiwan PCRs 30
Table 1.9 Water footprint of selected food products 32
Table 1.10 National Carbon Footprint studies in the dairy sector 37
Table 1.11 European LCA databases 43
Table 1.12 Asian LCI databases 43
Table 1.13 North America LCI databases 44
Table 1.14 Canada LCI databases 44
Table 1.15 Latin America LCI databases 44
Table 1.16 Oceania LCI databases 45
Table 1.17 Commercial LCA databases 45

Table 2.1 The olive oil market on the international scale (the average values of the 2007/2008–2012/2013 olive crop six-year period). . 59
Table 2.2 Articles reporting on the implementation of LCT tools in the olive industry 68
Table 2.3 Internal and external costs of the two systems (organic vs conventional) per functional unit (1 kg of extra virgin olive oil). 78
Table 2.4 Environmental performance referred to the olive oil EPDs. 89
Table 2.5 How to choose the functional unit when conducting an LCA of olive oil. 105

Table 3.1 Main constituents of wine. 126
Table 3.2 Illustration of the analysis results of the methodologies—Functional unit 133

Table 3.3 Analysis results of the methodologies—System boundary 134
Table 3.4 Illustration of the analysis results of the methodologies—Allocation, by-products, co-products and waste streams. 138
Table 3.5 Analysis results of the methodologies—use of resources and impact categories 142
Table 3.6 Key issues analysed in the present LCIA review (Sects. 3.3.6–3.3.9) with a description of the score properties . . . 151
Table 3.7 Summary of findings on wine environmental life cycle approaches 153

Table 4.1 Top ten world crop productions (2011). 188
Table 4.2 List of all the identified life cycle publications (prior to final screening and selection for review) concerning cereals or derived products 195

Table 5.1 Environmental assessment methods per product category and groups 256
Table 5.2 List of references included in the literature review and their main characteristics. 259
Table 5.3 List of references included in the literature review and their main characteristics. 271
Table 5.4 Articles reporting on the implementation of LCT tools in the sheep/goat sector 290
Table 5.5 Impact categories considered in the reviewed studies 297
Table 5.6 Summary of the reviewed studies on LCA pig production 302
Table 5.7 List of references included in the literature review and their main characteristics. 310

Table 6.1 European fruit production (in t of product and ha of cultivation) in 2011 compared with the world and Europe (data elaborated from FAOSTAT 2013) 336
Table 6.2 Parameters used to construct the synthesis matrix. 344
Table 6.3 List of references included in the literature review with main characteristics 346
Table 6.4 Percentage of reviewed studies which include downstream processes in main boundaries 356
Table 6.5 Percentage of reviewed studies about perennial fruit crops which include different phases of orchard life cycle within the time boundaries 357
Table 6.6 Overview of the reviewed studies which include some elements of carbon storage in the LCA analysis. 361
Table 6.7 FUs relevant to the audience and coherent with the typology of LCA study. 371
Table 6.8 LCA guidelines (system boundaries, functional unit, and reference period) for comparative assessments. 377

Contributors

Ioannis Arzoumanidis Department of Economic Studies, University "G. d'Annunzio", Pescara, Italy

Jacopo Bacenetti Department of Agricultural and Environmental Sciences, Production, Landscape, Agroenergy, University of Milan, Milan, Italy

Gabriele L. Beccaro Department of Agriculture, Forestry and Food Science, University of Turin, Grugliasco, Turin, Italy

Graziella Benedetto Department of Science for Nature and Environmental Resources, University of Sassari, Sassari, Italy

Annalisa De Boni Department of Agro-Environmental Science, University of Bari, Bari, Italy

Simona Bosco Scuola Superiore Sant'Anna, Institute of Life Sciences, via S. CeciliaPisa, Italy

Institute of Life Sciences, Scuola Superiore Sant'Anna, Pisa, Italy

Camillo De Camillis Food and Agriculture Organization of the United Nations (FAO), Agriculture and Consumer Protection Department Room C 535, Rome Italy

Giulio Mario Cappelletti University of Foggia, Foggia, Italy

Maurizio Cellura University of Palermo, Palermo, Italy

Alessandro K Cerutti Department of Agriculture, Forestry and Food Science, University of Turin, Grugliasco, Turin, Italy

Sara Corrado Institute of Environmental and Chemical Agriculture, Università Cattolica del Sacro Cuore, Piacenza, Italy

Anna I. De Luca Department of Agriculture, Mediterranean University of Reggio Calabria, Reggio Calabria, Italy

Giacomo Falcone Department of Agriculture, Mediterranean University of Reggio Calabria, Reggio Calabria, Italy

Valentina Fantin LCA and Ecodesign Laboratory-ENEA (Italian National Agency for New Technologies, Energy and Sustainable Economic Development), Bologna, Italy

Angela Fiore Department of Science, International Doctoral School in Bioecosystems and Biotechnology, University of Basilicata, Potenza, Italy

Maria Bonaventura Forleo University of Molise, Campobasso, Via F. De Sanctis, Italy

Alessandra Fusi Department of Agricultural and Environmental Sciences, Production, Landscape, Agroenergy, University of Milan, Milan, Italy

Nathalie Iofrida Department of Agriculture, Mediterranean University of Reggio Calabria, Reggio Calabria, Italy

Giuseppe Ioppolo SEAM Department, University of Messina, Messina, Italy

Agata Lo Giudice Ionian Department of Law, Economics and Environment, University of Bari, Taranto, Italy

Ornella Malandrino University of Salerno, Fisciano, SA, Italy

Paola Masotti Department of Economics and Management, University of Trento, Trento, Italy

Marina Mistretta University "Mediterranea", Reggio Calabria, Italy

Michele Moretti Department of Agro-Environmental Science, University of Bari, Bari, Italy

Elena Neri University of Siena, Siena, Italy

Giuseppe Martino Nicoletti University of Foggia, Foggia, Italy

Monia Niero QSA division, Department of Management Engineering, Technical University of Denmark, Kgs Lyngby, Denmark

Bruno Notarnicola Ionian Department of Law, Economics and Environment, University of Bari, Taranto, Italy

University of Bari, Taranto, Italy

Nadia Palmieri University of Molise, Campobasso, Via F. De Sanctis, Italy

Claudio Pattara University G. d'Annunzio, Pescara, Italy

Department of Economics, University "G. d'Annunzio", Pescara, Italy

Luigia Petti Department of Economic Studies, University "G. d'Annunzio", Pescara, Italy

Maria Proto Department of Management & Information Technology (DISTRAMIT), University of Salerno, Fisciano, Italy

Andrea Raggi Department of Economic Studies, University "G. d'Annunzio", Pescara, Italy

Pietro A. Renzulli Ionian Department of Law, Economics and Environment, University of Bari, Taranto, Italy

Rocco Roma Department of Agro-Environmental Science, University of Bari, Bari, Italy

Benedetto Rugani Centre de Recherche Public Henri Tudor, Centre de Ressources des Technologies pour l'Environnement (CRP Henri Tudor/CRTE)—6A, Esch-sur-Alzette, Luxembourg

Carlo Russo University of Foggia, Foggia, Italy

Giuseppe Saija University of Messina, Messina, Italy

Roberta Salomone University of Messina, Messina, Italy

SEAM Department, University of Messina, Messina, Italy

Daniela Sica Department of Management & Information Technology (DISTRA-MIT), University of Salerno, Fisciano, Italy

Alfio Strano Department of Agriculture, Mediterranean University of Reggio Calabria, Reggio Calabria, Italy

Stefania Supino Department of Management & Information Technology (DISTRA-MIT), University of Salerno, Fisciano, Italy

Giuseppe Tassielli Ionian Department of Law, Economics and Environment, University of Bari, Taranto, Italy

Manfredi Vale Aghetera Ambiente and Sviluppo, Venice, Italy

Andrea Vitali DAFNE Department, Università degli Studi della Tuscia, Viterbo, Italy

Chapter 1
Life Cycle Assessment in the agri-food sector: an overview of its key aspects, international initiatives, certification, labelling schemes and methodological issues

Bruno Notarnicola, Giuseppe Tassielli, Pietro A. Renzulli and Agata Lo Giudice

Abstract Sustainable development and, above all, sustainable production and consumption in the agri-food sector have been key issues since the 2000s, stimulating the creation of many international initiatives and strategies aimed at reducing environmental impacts deriving from food production and consumption and at finding more sustainable ways of production. This first chapter is designed to provide the reader with an as exhaustive as possible overview of the key concerns, applications, and methodological issues of agri-food life cycle assessment (LCA). On this scale the major international initiatives (with a special focus on two relevant and recent European ones), eco-labels and declarations, and footprints (at product level, based on an LCA approach) developed so far are reported. Some of the most important LCA initiatives developed by agricultural and livestock operators, the industry sector, logistics sector, trade, and the end of life of packaging and/or food waste operators are also described in the chapter. Considering that one of the key issues within the agri-food sector is the lack of reliable and up-to-date inventory data on food products and processes, the state of the art of the major existing international LCI databases is reported, and the national and international initiatives currently under development highlighted. Finally, the chapter takes into account dietary issues in the sense that in the context of food sustainability the importance of consumer behaviour and, in particular, dietary behaviour is becoming increasingly recognised, together with the product and its production chain.

B. Notarnicola (✉) · G. Tassielli · P. A. Renzulli · A. Lo Giudice
Ionian Department of Law, Economics and Environment, University of Bari,
Via Lago Maggiore angolo via Ancona, 74121 Taranto, Italy
e-mail: bruno.notarnicola@uniba.it

G. Tassielli
e-mail: giuseppe.tassielli@uniba.it

P. A. Renzulli
e-mail: pietro.renzulli@uniba.it

A. Lo Giudice
e-mail: agata.logiudice@uniba.it

B. Notarnicola et al. (eds.), *Life Cycle Assessment in the Agri-food Sector*,
DOI 10.1007/978-3-319-11940-3_1

Keywords Agri-food chains · Sustainable initiatives · Environmental labels · Databases · Methodological issues · Dietary issues

1.1 Introduction to Life Cycle Assessment (LCA) in the Agri-food Sector

During the last decades, scientific studies have shown that most food chains are not sustainable because of the environmental impacts occurring in different phases of their life cycle. In this context, in 2006 the European Science and Technology Observatory (ESTO) published a report on its project "Environmental Impact of Products" (EIPRO). One of the findings was that the contribution of food and drink products to the environmental impact of private consumption is between 22 and 34 %.

The Strategic Research Agenda 2006–2020 of the European Technology Platform Food for Life has defined sustainable food production as the most important challenge facing the European food industry. In general, sustainability tools and life cycle assessment (LCA) have been applied for more than 20 years to agricultural and food systems to identify methods of sustainable food production and consumption and as a means of supporting environmental decision-making via the identification of the environmental impacts throughout the systems' life cycles. One of the reasons for the growing consideration by the academic community of aspects regarding Food LCA is the fact that methodological issues (for example, the definition of the functional unit, difficulties in data collection, pesticides and their use, fertiliser dispersion models, impact categories of land use and water use) are different from the typical ones arising from LCAs of industrial products. Until now such topics have been tackled with many different approaches that do not represent standardised methods, hence much needs to be done to build a consistent, practical and life cycle science-based approach to product level sustainability and reporting of all food, beverage, and agricultural products.

It is in this context that this first chapter arises, being designed to provide the reader with a detailed overview of the key concerns, applications, and methodological issues of LCA with regard to the food sector. In the later chapters these aspects will be analysed in detail with specific regard to the chosen sectors (olive oil, wine, cereal and derived products, livestock and derived edible products, and fruit).

In particular, at the beginning of this chapter the major international initiatives, eco-labels and declarations, and footprints (at product level, based on an LCA approach) developed so far are reported. Among the international initiatives, a special focus is placed on two relevant and recent European ones which highlight governments' commitment toward issues of sustainable production and consumption and eco-labelling harmonisation. As far as the eco-labels/declarations and footprints are concerned, only the most important are reported.

In addition to governments, other actors in the supply chain play a fundamental role in the development and consolidation of the LCA methodology as an essential tool for the assessment of the environmental performance of food products. This aspect is highlighted in this chapter, which reports some of the most important LCA initiatives developed by agricultural and livestock operators, industry sector, logistic and trading sectors, and end of life of packaging and/or food waste operators.

As already mentioned, one of the key issues within the agri-food sector is the lack of reliable and up-to-date inventory data on food products and processes for developing not only accurate LCA studies but also for hotspot analysis, communication, and labelling. Consequently there is a growing need for comprehensive, clear, well-documented, and consistent data for increasing the accuracy and comparability of LCA studies. In this context, the state of the art of the major existing international LCI databases is reported, and the national and international initiatives currently under development are highlighted.

The final part of the chapter deals with dietary issues in the sense that in the context of food sustainability the importance of consumer behaviour and in particular dietary behaviour is increasingly being recognised, together with the product and its production chain. The dietary choices of the consumer and consumption style strongly affect results in terms of environmental sustainability.

1.2 International Initiatives, Labels, and Footprints in the Agri-food Sector Based on a Life Cycle Approach

1.2.1 Introduction

Agriculture and food production and consumption are arguably some of the most important drivers of environmental burdens, such as habitat change and loss of biodiversity, land use and soil degradation, climate change, water use and pollution, water scarcity, eutrophication of water bodies, and toxic emissions.

Nowadays, food production is becoming more and more globalised and industrialised, leading to its standardisation; agricultural practices, above all in the developed countries, have been intensified in order to increase the ratio yield/ha as much as possible. Furthermore, this globalisation has led to an increasing loss of local markets with a consequent increase in "food miles", i.e. the transport distances between farmers, industry, and consumers, with the consequences of social and environmental costs (Notarnicola et al. 2012a; Reisch et al. 2013).

Recent statistical studies have reported that the global population growth and the change in the dietary habits in emerging countries over the next 40 years will cause an increase in food (about 60 %), energy and water demands, the so-called energy-food-water nexus (Alexandratos and Bruinsma 2012). At the same time, the depletion of fossil hydrocarbons will increase the demand for biofuels and industrial materials, which may compete with food for biomass. All these changes will cause a destabilisation of the sustainable use of natural resources, possibly causing social and geopolitical tensions.

In this context, sustainable development and, above all, sustainable production and consumption in the agri-food sector have been key issues stimulating the creation of many international initiatives and strategies designed to reduce environmental impacts deriving from food production and consumption and to find more sustainable ways of production.

Since the 1980s, the European Union (EU) has been one of the main actors within the international context, showing high sensitivity to these issues, sustainable development being one of its key objectives in terms of continuously improving the quality of life and well-being of present and future generations. Starting from the year 2001, an interesting initiative on sustainable development was pursued by European governments to develop a strategy for strengthening and steering environmental politics towards a more ecological product market. It was followed, in 2003, by the so-called Strategy for Sustainable Consumption and Production (SCP), which aimed to reduce the environmental, social, and economic impact of products and services throughout their entire life cycle. The concept of SCP can be applied to all the existing products sold and bought on the market and hence also to food and drink products. These products in particular play a fundamental role in the everyday life of consumers, whose demand for high quality food has increased in the last few years. In a similar manner to other products, their production and consumption (from farm to fork and end of life) have environmental implications; nevertheless, because of specific aspects related to health, nutrition, well-being, cultural identity, and lifestyle, they need to be considered and treated differently from all other products.

During the same year the EU adopted a Communication on the Integrated Product Policy (IPP) (COM (2003) 302 final) with the aim of reducing the environmental impact of products and using, when possible, a market-driven approach combining competiveness with social concerns

In this context, the Directorate General Joint Research Centre/Institute for Prospective Technology Studies (DG JRC/IPTS) launched a project called Environmental Impacts of Products (EIPRO) in an attempt to identify those products with the most relevant environmental impacts throughout their life cycle, from cradle to grave, taking into account the full food production and distribution chain from farm to fork. The report, published in 2006, concluded that there are three areas which have the greatest impact: food and drink, private transport, and housing. Together they are responsible for 70–80 % of the environmental impact of consumption, and account for some 60 % of consumption expenditure. In particular, the food and drink sector accounts for 20–30 % of the environmental impact of private consumption (Tukker et al. 2006).

The Strategic Research Agenda (SRA) (2007–2020) of the European Technology Platform (ETP) Food for Life was published in 2007 and defined sustainable food production as the most important challenge that will be faced by the European food industry.[1]

[1] The newly revised Strategic Research and Innovation Agenda (SRIA) "2013–2020 and Beyond" now focusses specifically on innovation.

The consequence of the evolution of the IPP approach was the birth of a new European strategy in the SPC field that, during the 5 year period from 2007 to 2011, was considered a priority by the EU. In fact its action lines were implemented in environmental politics in order to prevent, reduce, and manage the impact of product life cycles. In this context, in 2008, the European Commission (EC) published the "Action Plan for Sustainable Consumption and Production and on the Sustainable Industrial Policy" (SCP/SIP) (COM(2008) 397 def) in order to define the interventions necessary for implementing the actual models developed for SPC: a dynamic framework was then proposed for improving the energy/environmental performance of products during their life cycle, increasing demand for better products and helping consumers to make decisions regarding such products (Lo Giudice and Clasadonte 2010).

It is important to underline that the SPC community, built on innovative instruments, should be able to boost the capabilities of both producers and consumers, in terms of making "sustainable" choices and influencing each other: these tools are based on a life cycle (systemic, cradle-to-grave) approach, using the LCA methodology.

1.2.2 Environmental Labels and Declarations

Today's consumer society has a strong impact on the environment, depending on the choices that consumers make to satisfy their needs. Choosing more sustainable products can certainly be decisive in terms of impact reduction, i.e. the selection of products that provide environmental, social, and economic benefits while protecting public health and environment over their whole life cycle, from the extraction of the raw materials until the final disposal. The consumer demand for environmentally friendly products is a powerful incentive for companies, who are thus stimulated to find new ways of producing more sustainable products, to intensify efforts at environmental management, and to improve product performance throughout the life cycle. It is important, therefore, to give consumers the right data for correct product choice, which means giving accessible, understandable, relevant, and credible information on the environmental quality and performance of the products.

Nowadays the increasing awareness of the effects of societies and lifestyle on the environment means consumers are inclined towards more sustainable behaviour. In this context, the information provided by the different certification/labelling systems, found on/with some kind of product/service/packaging, could be of crucial help. These systems are referred to as "eco-labelling" or "environmental labelling" and give information on the overall (the whole life cycle) environmental performance of the product/service/packaging or on one or more specific environmental aspects (for example, raw material origin and recyclability). In terms of sustainability, an eco-labelling system has a dual role in the market: first, by awarding seals of approval (in terms of environmental information—fewer impacts on the environment than functionally or competitively similar products) to products, it can influence the market's behaviour towards goods and services with lower environmental

Table 1.1 Current eco-labelling programmes available in the agri-food sector

Name	Website
China environmental labelling (CEL)	http://www.sepacec.com/cecen/
Living planet (Ukraine)	http://www.ecolabel.org.ua/
Vitality leaf (Russia)	http://www.ecounion.ru/en/site.php?&blockType=251
Ecomark (India)	http://www.cpcb.nic.in/Eco_Label.php

impact; second, by acknowledging those firms producing in a more sustainable way it ensures the environmental properties of their products and, in this way, they obtain added value compared with competitors (Udo de Haes et al. 2010).

The environmental assessment of product behaviour can be done through an independent quality assurance process (so-called certification) based on strict procedures and criteria.

According to the ISO Standard 14020:2002 (ISO 2002a), voluntary environmental labels/declarations aim at "*encouraging the supply and demand for those products and services able to cause low damage to the environment so that it will stimulate a continuous environmental improvement process managed by the market*".

Three types of labels/environmental declarations have been identified and regulated: Type I (ISO 14024) (ISO 2001), for example the EU Ecolabel, the most widespread and well-known Type I label; Type II (ISO 14021) (ISO 2002b), for example the "Mobius Cycle", related to the percentage of recycled material in a product; and Type III (ISO 14025) (ISO 2006a), for example the International EPD® system, the most widespread and well-known Type III declaration; there is also another category, not regulated by ISO standards, which has been defined as "environmental labels of Type IV", for example the trademarks Forest Stewardship Council (FSC), Dolphin Safe and Fair-trade Global.

1.2.2.1 Eco-labelling (Type I Labels)

In 2013, four Eco-labelling programmes were suitable for the agri-food sector, as reported in Table 1.1.

These Eco-labelling schemes are briefly discussed in the following paragraphs; in addition, a short description of the results of the feasibility study about the possible extension of the use of the "EU Ecolabel" to the agri-food sector is given.

China Environmental Labelling (CEL) (China) The "China Environmental Labelling" (CEL) programme (Fig. 1.1) is a public voluntary Chinese eco-label scheme established in 1993 by the State Environmental Protection Administration (SEPA, today the Ministry of Environmental Protection of China, MEP).

The programme aims at encouraging businesses to use resources and energy rationally to develop and produce environmentally friendly products, guide consumers to choose and identify sustainable green products, and provide a way for businesses and the public to participate consciously in environmental protection

Fig. 1.1 The China environmental labelling logo

Fig. 1.2 The Ukraine living planet logo

(IISD 1996). As far as the label types are concerned, two types, based on the criteria of ISO 14020 and ISO 14024, are available: Type I, for products within the scope of existing technical standards (issued by MEP); Type II for products not contained in the former: in this case, it is possible to generate a self-declaration that has to be verified by China Environmental United Certification Centre (CEC). The standards may be applied to many product categories, such as food, building materials, textiles, packaging supplies, etc. (International Trade Centre 2013). The only SEPA Technical Requirement Standard suitable for the agri-food sector is HJ/T 210–2005 (replacing HJBZ 13–1996), applicable to soft drinks (CEC n.d.).

Living Planet (Ukraine) This Eco-labelling programme was implemented in 2003 on the initiative of the all-Ukrainian non-governmental organisation Living Planet, with the assistance of the Committee of Verkhovna Rada (Parliament) of Ukraine on Ecological Policy and the Ministry of Environment and Natural Resources of Ukraine. The programme was developed with the aim of implementing an Eco-labelling programme of Type I (Fig. 1.2), according to the requirements of ISO 14024.

For the label development, the best practices of other Eco-labelling programmes, such as those of the EU, Germany, the USA, the Nordic countries and others, were taken into account (Berzina and Shevchenko 2011).

By the end of 2013, the following criteria were developed for the agri-food sector (Table 1.2).

Vitality Leaf (Russia) This Eco-labelling programme (Fig. 1.3) for products, work and services was developed, in 2001, by the non-commercial partnership Saint-Petersburg Ecological Union (SPbEU), a member of the Global Eco-labelling Network (GEN) since 2007. The system is based on the requirements of ISO 14024 and

Table 1.2 Ukraine living planet criteria for the agro-food sector

Standard number	Standard
OEM.08.002.03.010	Pasta
OEM.08.002.03.011	Vegetables and vegetable products
OEM.08.002.03.014	Cultivated mushrooms
OEM.08.002.03.015	Fruits and fruit products
OEM.08.002.03.016	Honey
OEM.08.002.03.018	Vegetable oils
OEM.08.002.03.023	Wine products
OEM.08.002.03.024	Vodka and alcoholic drinks
OEM.08.002.03.025	Bottled water
OEM.08.002.03.027	Bottled mineral water
OEM.08.002.03.031	Cereals
OEM.08.002.03.045	Instant cornflakes
OEM.08.002.03.046	Natural fermentation soft drinks
OEM.08.002.03.052	Food additives
OEM.08.002.03.054	Spreads and oily foods
OEM.08.002.03.053	Coffee and coffee drinks
OEM.08.002.03.056	Salt
OEM.08.002.35.069	Dairy and processed meat products

Fig. 1.3 The vitality leaf logo

it represents the only Russian eco-label recognised by the international community (Ecological Union 2013).

As far as the criteria for Eco-labelling are concerned, they include the following specific areas: level of environmental pollution; level of safety for human health; content of recyclable/recycled components; rational use of natural resources during the product's life cycle; use of renewable resources during the product's life cycle; waste management; and use of the best available technologies (NEASPEC 2012). In 2013, just three criteria existed for the agri-food sector: STO −56171713–1.01–2007 (Alcoholic beverages), STO VL 2.02.9730–11–1.0 (Vegetables), and STO VL 2.01.0131–10–1.0 (Drinking water).

Ecomark (India) To increase consumer awareness, in 1991, the Ministry of Environment & Forests (MoEF) launched, through the Central Pollution Control Board

Fig. 1.4 The ecomark label

Fig. 1.5 The new ecolabel logo

(CPCB), the Eco-labelling scheme Ecomark for easy identification of environmentally friendly products. The label (Fig. 1.4) is applicable to all goods which meet the specified environmental criteria and the quality requirements of Indian standards: the criteria follow a cradle-to-grave approach, i.e. from raw material extraction, to manufacturing, to disposal.

By the end of 2013, sixteen final criteria for product categories had been developed by the government of India: Soaps & Detergents, Paper, Food Items, Lubricating Oils, Packaging Materials/Packaging, Architectural Paints and Powder Coatings, Electrical/Electronic Goods, Food Additives, Wood Substitutes, Cosmetics, Aerosol Propellants, Plastic Products, Textiles, Fire Extinguishers, Finished Leather Goods, and Coir and Coir Products. Among them, just one is applicable to the agri-food sector: Food Items (Edible Oils, Tea, and Coffee).

A research report was published in 2007, highlighting that the scheme had not gained the expected appeal among consumers or industry; in fact only a few manufacturers of products like paper, pulp, leather, and wood particleboard had applied for and obtained the Ecomark licence, and they rarely used the symbol on their packaging as none of them had gained any benefit from it (Mehta 2007).

EU Ecolabel (Europe) The EU Ecolabel represents the best European recognition of products (and services) meeting specific environmental criteria and the highest environmental standards: these products are characterised by high performance and environmental quality, verified by a robust and independent certification process, and are recognisable by a specific logo represented by a flower (Fig. 1.5). Obtaining such a label can help a product to emerge and differentiate itself from its competitors on the market since the label certifies that it has a reduced environmental impact throughout its entire life cycle. As far as the eligibility criteria are concerned, the EC defines the groups of products/services that can be certified and, for each of them, the environmental criteria that must be met for the release of the label.

This voluntary scheme is an important component of the EU's Sustainable Consumption and Production Action Plan, and was launched in 1992 (with the adoption of the Council Regulation (EEC) n. 880/92) when the European Community decided to develop a Europe-wide voluntary environmental scheme that consumers could trust.

With its first review (Ecolabel II, Council Regulation (EC) n. 1980/2000) the application of this label was extended to services, and in 2010 the EC issued a new Regulation "Ecolabel III" (Council Regulation (EC) n. 60/2010) with the aim of: streamlining the developing path for eligibility criteria by focussing on the most significant environmental impacts throughout the product/service life cycle; ensuring that the top 10–20 % of environmental performers on the market could meet the criteria; reducing the label costs to encourage the interested stakeholders to undertake the certification path; widening the label application field by evaluating the possibility of including food (under conditions emerging from a feasibility study). Ecolabel III confirmed the application of environmental criteria to all consumer goods and services, with the exception of food, beverages, and medicines. It also foresaw the possibility of developing specific criteria for food and feed, depending on the results of a feasibility study to be conducted by the Commission by the end of December 2011. This study was conducted with the aim of evaluating:

- the feasibility of establishing reliable criteria applicable to the entire life cycle of food, feed, and drinks products, including the stages of cultivation;
- the impact and the added value of establishing these criteria and implementing this scheme in various sectors, and the possible impact on organically certified products (including the risk of consumer confusion);
- the possibility of restricting the label to organically certified products.

The feasibility study highlighted these main aspects.

1. The main environmental impacts (for example, biodiversity or soil fertility loss) linked to the food, feed and drink products life cycle are mostly owed to the primary production phase (or "extraction of the raw material"), even if dependent on the product category. Because of their nature, these impacts are not easily measurable and thus cannot be ranked in terms of environmental impact. The same can be said of ethical or social questions (for example, animal welfare, labour standards).
2. The environmental impact of food, feed and drink products in the "extraction of the raw materials" arises from the combination of the practice employed and the place where it takes place because of the use of physical elements such as land, water, etc. As a consequence of this, the environmental impacts for a particular product, on a specific site, using specific production technologies can vary significantly.
3. A deficiency within current labelling systems was highlighted by the study: existing labels focus only on the environmental impacts arising from the primary stage and not (or only to a limited extent) on the ones from the processing life cycle stage. This deficiency could represent a key point for the success of the

EU Ecolabel if focussed, thanks to its life cycle approach, on highly processed products considering the environmental impacts of processing, transport, and consumption while the environmental impacts of primary production could be dealt with via cooperation with the existing rigorous agri/fishery schemes.
4. Existing environmental labels are mainly based on input- or practice-based criteria with the disadvantage that they lead to a shift of environmental burdens when practices or ingredients are substituted, as well as hampering innovation. Conversely, output-based criteria can be more economically efficient, providing a transparent link with environmentally positive results. In this context, environmental footprint tools and multi-criteria methodologies are being developed in Europe as a basis for developing more output-based criteria for food, feed, and drink products.

On the basis of the feasibility study results, the following recommendations were made.

- It is necessary to develop a credible multi-criteria overall outcome-based assessment system for primary production, something which is missing at the moment.
- It is necessary to clarify the legality of using the current Ecolabel and the term "ECO" when referring to food, feed, and drink products.
- If the use of a label is extended to non-organic products, it is important to conduct an appropriate communication campaign in order to avoid consumer confusion (the use of a distinctive label could be a good solution).
- It is necessary to carry out an economic assessment regarding the full public and private costs of implementing the EU Ecolabel scheme.

The implementation of these recommendations could make possible the application of the EU Ecolabel to these product categories: yoghurt and cheese, bread, non-alcoholic beverages, and processed fish.

From October 2013, the EU Ecolabel certification has been applicable to 26 categories of products/services: however none of them belong to the food, feed, and drink sector (Oakdene Hollings Research and Consulting 2011).

1.2.2.2 Environmental Product Declaration (EPD)

An Environmental Product Declaration (EPD) belongs to the Type III programme and follows the requirements of ISO 14025:2006. It is a standardised tool, based on an LCA approach, which can be applied to any sector (including the agri-food one) and needs third party verification. An EPD summarises the whole sustainability history of a product in a single, written report: this report includes objectives, comparable and reliable information about the product's environmental performances during its life cycle, and its impacts on, for example, global warming, ozone depletion, water pollution, ozone creation, and greenhouse gas emissions. Depending on the need, it is also possible to include other impacts, such as human toxicity risk and corporate social responsibility (Meissner Schau and Magerholm Fet 2008).

Table 1.3 Current EPD systems available in the agri-food sector

Name	website
The international EPD® system	http://www.environdec.com/
EPD Norge	http://www.epd-norge.no/
Earthsure®	https://iere.org/programmes/earthsure/
Ecoleaf environmental label	http://www.ecoleaf-jemai.jp/eng/index.html
Sustainability measurement and reporting system (SMRS)	http://www.sustainabilityconsortium.org/smrs/

It is important to underline that, unlike Eco-labelling, the information carried by EPD is merely indicative, since neither the evaluation conditions nor the preference criteria nor the minimum performance levels which have to be respected are taken into account. With regard to the advantages deriving from the EPD implementation, whereas the adoption of an environmental management system (e.g. ISO 14001) allows the control and management of the environmental impacts connected with the activities and the processes developed for the production site, the EPD is seen as a means of communicating the product's environmental performances during its life cycle, with a cradle-to-grave approach. The EPD is, thus, an instrument which allows companies to give visibility to their own work, turning the environmental variables into competitive market factors.

The LCA study is the "heart" of an EPD and it must comply with certain product-specific calculations and requirements known as product category rules (PCRs). According to ISO 14025: 2006, a PCR is defined as a set of specific rules, requirements, and guidelines for developing environmental declarations for one or more products that can fulfil equivalent functions, determining what information should be gathered and how that information should be evaluated: it allows fair comparison between similar products (Lo Giudice and Clasadonte 2010; Del Borghi 2013).

The current EDP systems available in the agri-food sector are reported in Table 1.3.

What follows is a brief general introduction to the above-mentioned EPD systems. In the following chapters only the international EPD® system will be discussed with reference to the product systems described in this book.

International EPD® system (Sweden) The international EPD® system is the most widespread scheme. It was implemented in 1998 by the Swedish Environmental Management Council (SEMC) which, on behalf of the Global Type III Environmental Project Declaration Network (G.E.D. net), was appointed to oversee the harmonisation, at international level, of all the existing national programmes. This process led to the birth of the Guidelines MRS 1999:2 "Requirements for Environmental Product Declaration, EPD—an application of ISO/TR 14025 Type III Environmental Declarations", which were replaced, in 2008, by the new "General Programme Instructions for EPD".

The main differences between the two documents comprised: the organisational structure and the evaluation system; the logo (the new one is reported in Fig. 1.6);

Fig. 1.6 The green yardstick logo

the definition and identification of the product categories; the harmonisation and the consultation phase, at international level, regarding the PCR; the EPD contents; the possibility of obtaining an EPD with just one impact category; and the subdivision of the documentary checks into internal and external ones.

Furthermore, a specific declaration named "Climate Change" was introduced. This declaration is an extract of the climatic data of an EPD® and describes the emission of greenhouse gases (the carbon footprint) based on the following rules: all the greenhouse gases emissions (calculated for each step of the product life cycle) are included and expressed in CO_2 eq.; the information is separated according to the different stages of the life cycle; emissions are divided into fossil and biogenic ones; and the data related to other environmental effects are available in the EPD from which all data derive (Lo Giudice and Clasadonte 2010).

The changes aimed at making the product label consistent with the new ISO Standard 14025:2006 and also at encouraging global diffusion of EPD and harmonisation of the existing environmental labels/declarations.

As at February 2014, 352 EPD® (including the precertification) were registered in 19 different countries (Fig. 1.7):

Among them, 93 (Italy, Sweden, Denmark, and Greece) belong to the "Agriculture and fishery products", "Food products", and "Beverages" categories. Italy occupies first place with 65 certifications (53 "Food products", 10 "Beverages", and 2 "Agriculture and fishery products").

As far as the PCRs are concerned, for the "Food products" and "Beverages" categories the following have been developed (Table 1.4 and 1.5) (Environdec 2013):

EPD—Norge (Norway) The Norwegian EPD Foundation, a member of G.E.D.net, was established in 2002 by the Confederation of Norwegian Enterprises (NHO) and the Federation of Norwegian Building Industries (BNL) for the development of standardised and internationally valid EPDs for products and services (Fig. 1.8) (Magerholm Fet 2012).

By the end of 2013, EPDs had been developed for building material, furniture, electricity, chemicals, and packaging. Only one PCR was developed for the agri-food sector but it expired in 2009: NPCR 007 (Wild-caught Fish).

Earthsure® (USA) Earthsure® (Fig. 1.9) is the Ecolabel programme of the Institute for Environmental Research and Education (IERE) and the first EPD programme in North America, developed in compliance with ISO 14025:2006. It was developed in 2000 for food and agricultural product systems in order to assure businesses and consumers that the environmental benefits claimed by a manufacturer were genuine and substantial enough to warrant special recognition.

Subsequently, because of requests arising from other sectors, its application was expanded to cover the full range of products and services.

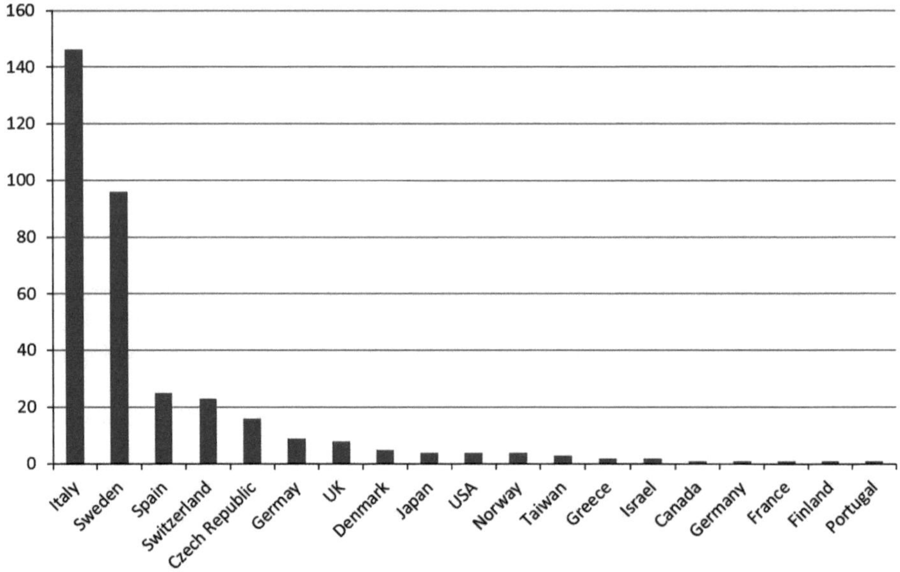

Fig. 1.7 EPD® in the world

Table 1.4 PCR (EPD®) for the "Food products" category

CPC[a]	Name
21	*Basic module: Meat, fish, fruit, vegetables, oils, and fats*
2111	Meat
2112	Poultry (update in progress)
2124	Fish, prepared or preserved; caviar and caviar substitutes (under development)
2131	Frozen vegetables, pulses, and potatoes
2132	Vegetable juices (under development)
2149	Other prepared and preserved fruit and nuts (under development)
21340	Table olives
21494	Jams, fruit jellies, and marmalades
21537	Virgin olive oils and their fractions (update in progress)
22	*Basic module: Dairy products and egg products*
221	Processed liquid milk and cream
2223	Yoghurt, butter, and cheese
23	*Basic module: Grain mill products, starches, and starch products; other food products*
231	Grain mill products
234	Bakery products
2371	Uncooked pasta, not stuffed or otherwise prepared
2372	Pasta, cooked, stuffed or otherwise prepared; couscous
23520	Refined sugar from sugar beet
23995	Sauces (update in progress)

[a] *CPC* Central Product Classification

Table 1.5 PCR (EPD®) for the "Beverages" category

CPC	Name
21	*24 Basic module: Beverages*
2143	Fruit juices
2431	Beer
24212	Wine of fresh grapes, except sparkling wine; grape must
24410	Bottled waters, not sweetened or flavoured (update in progress)

Fig. 1.8 EPD Norge logo

Fig. 1.9 Earthsure® logo

This programme verifies that a food product has undergone a life cycle assessment, taking into account several issues such as climate change, soil loss, and ecotoxicity caused by energy use in transport and manufacturing, and materials used in production and disposal (Earthsure 2012).

Two PCRs are available for the agri-food sector: Earthsure Meat—2006 (Meat) and Earthsure 50202201:2012 (Beer).

EcoLeaf Environmental Label (Japan) In 1998, the Japan Environmental Management Association for Industry (JEMAI, member of G.E.D. net), supported by the Japanese Ministry of Economy, Trade and Industry (METI), began developing a programme for environmental declarations with the aim of mitigating global warming. Between 1999 and 2000 its feasibility was studied in trial programmes. In 2002, the full implementation of the EcoLeaf programme began, in conformity with ISO 14025:2006 and focussing on the following goods: industrial goods, durable consumer goods, daily necessities, energy such as electricity, buildings, food, and services associated with these products.

Fig. 1.10 The EcoLeaf logo

The Guidelines for the Introduction of the EcoLeaf Environmental Label (First Edition 2002) represent the basis of this programme, setting out procedures for obtaining and verifying useful quantitative environmental data for manufacturers and distributors who register an EcoLeaf label (Fig. 1.10).

The EcoLeaf environmental label is based on the LCA methodology for quantitatively providing information about the environmental impact of goods throughout their whole life cycle, without making any judgment about whether the good meets any environmental quality standard. The label is composed of three sets of documents: (1) the product label, which shows a summary of the data, the Product Environmental Aspects Information Declaration (PEAID), and the background data; (2) the Product Environmental Information Data Sheet (PEIDS); and (3) the Product Data Sheet (PDS) (Kanzaki 2009).

Regarding the agri-food sector, the following PCR is currently available: 59 CP – Meat (http://www.ecoleaf-jemai.jp/eng/index.html).

Sustainability Measurement and Reporting System (SMRS) (USA) The Sustainable Consortium (TSC) is a global, academically led, multi-stakeholder organisation which carries out research and develops data, standards, systems, and tools for improving decision-making and driving sustainability in consumer goods. Its working sectors are: food, beverage, and agriculture; home and personal care; consumer electronics; toys, paper, and forestry products; and packaging. The TSC aims at reducing the distance between rigorous scientific analysis of the life cycle of consumer goods and the ability of the purchaser/consumer to learn and act on the environmental impact information provided by such analysis.

In this context, the TSC is developing, as a retailer initiative, the Sustainability Measurement and Reporting System (SMRS), which will represent a standardised framework (based on Type III declarations) for enabling the creation, analysis, and communication of comparable and standardised information about the life cycle of a product. Furthermore, it will be a common, global platform for companies to create, analyse, and communicate comparable and standardised information about the life cycle of a product, with the following advantages: reduction of the cost and time associated with the LCA and product declarations; common understanding across and between industry sectors of significant sustainability hotspots and improvement opportunities; sharing of sustainability data across the supply chain in a cost-effective and scalable manner (Mars et al. 2011).

This framework will lead to: actionable sustainability information (Level 1); and a large-scale system supporting standardisation and harmonisation of product LCAs over time (Level 2). In particular, the outcomes of the Level 1 SMRS are category sustainability profiles (CSPs), representing a summary of the best available, credible, and actionable knowledge regarding the sustainability aspects related to a product over its entire life span; apart from CSPs, Level 1 takes into account the key performance indicator (KPI), questions that retailers can use for assessing and tracking the performance of brand manufacturers on critical sustainability issues. It focussed on the environmental and social issues relevant to a single category or family of consumer goods. The KPI question sets are developed by TSC in collaboration with multiple stakeholder groups (i.e. companies, academics, civil-society organisations, and government agencies) and correspond directly to the issues that are highlighted in the corresponding CSPs. The CPSs, which apply to the product category level (e.g. laundry detergents, frozen beef, shoes), are not for product-level comparison but promote sharing of information and enable an informed conversation between merchant and retail buyer. The following CSPs and KPIs have been developed for the food, beverage, and agriculture (FBA) sector: beef, butter, cheese, coffee, cotton, cucumbers, leaf vegetables (lettuce), milk, potatoes, prepared salad, farmed salmon, sorghum, tea, tomatoes, and wine.

The outcomes of the Level 2 SMRS are product sustainability declarations (PSDs). These apply to the product level (e.g. JC's Frozen Beef Patties) and allow for a direct comparison of products against the product category baseline (including uncertainty). PSDs are built on a baseline LCA model plus PCRs and deliver transparent, science-based results (Redd 2011).

1.2.3 Footprint Indicators and Labelling

Sustainable development strategies have to face several environmental problems such as climate change, energy and water scarcity, biodiversity loss, deforestation, land and soil erosion, pollution, and desertification: these trends are expected to increase in the future if suitable countermeasures are not taken, contributing to food scarcity. In this context, specific indicators have been defined with the aim of assessing these impacts, trying to reduce them and, at the same time, maintaining economic and societal well-being.

Among these indicators, the "footprint" ones deserve to be highlighted. A "footprint" represents a quantitative measure of the natural resources consumed or of the pressure on the environment caused by human beings. In the last years, several footprints have been defined: environmental; social; economic; combined environmental, social, and/or economic; and composite. The most widespread ones in the food sector are the ecological, the carbon, and the water footprints (Čuček et al. 2012) and they will be discussed in detail in the next chapters with reference to the specific food product systems described in this book. What follows is a brief introduction to the current (end of 2013) international and national footprint standards.

These three indicators are complementary, since they measure completely different sustainability aspects and methodologically there are many similarities between them, but each has its own peculiarities related to the uniqueness of the substance considered. Furthermore, these footprints complement traditional analyses of human demand by taking into account both the producer's and the consumer's point of view. Recently, researchers proposed bringing together these three indicators (without creating a new one), with the aim of creating a robust and ready to use set of indicators useful for moving towards a multidisciplinary sustainability assessment: the so-called "Footprint Family". The "Footprint Family" is thus a set of indicators based on a consumption-based perspective through which it is possible to track human pressure on the surrounding environment (pressure seen as an appropriation of biological natural resources and CO_2 uptake, emission of greenhouse gases (GHGs), and consumption and pollution of global freshwater resources). Defined in this way, this indicator answers three specific questions (related to the monitoring of three key parts of the ecosystem—biosphere, atmosphere, and hydrosphere) and concurrently is a means of controlling the environmental pillars of sustainability; nevertheless, it does not represent a complete measure of sustainability because many environmental, economic, and social issues are not taken into account (Galli et al. 2012).

1.2.3.1 Ecological Footprint (EF)

The ecological footprint (EF) is a tool for assessing resources and emissions[2]. It was introduced by Wackernagel and Rees in 1996 and is the most widespread indicator for assessing the sustainability of humanity's demand on nature. In particular, it represents the human pressure on the Planet in terms of direct and indirect population demand for land and water, comparing resource depletion and waste absorption with how much is available on the Planet (biocapacity) on both local and global scales. It should be noted that EF focusses\on just one aspect of sustainability, human appropriation of the Planet's regenerative capacity, and it is expressed in units of world average bioproductive area, needed annually to provide (or regenerate) the resource flow, namely global hectares (gha) or hectares with global average productivity (Čuček et al. 2012).

International Initiatives

During the last decades, an increasing number of organisations, communities, and governments have adopted the EF as the key indicator of sustainable resource use.

[2] CO_2 is the only GHG taken into account.

Usually, such studies have been based on different approaches, leading to fragmentation and divergence. In order to overcome these issues, in 2006 the Global Footprint Network released the Ecological Footprint Standards, applicable to all footprint studies, including sub-national populations, products, and organisations. After 3 years a process of updating began, leading to the Ecological Standards 2009 which provided, for the first time, standards and guidelines for product and organisational footprint assessment. In 2012, a new update began.

The Ecological Footprint Standards ensure consistence of assessments because they are created according to community-proposed best practices; furthermore they guarantee that assessments are conducted, and communicated, in an accurate and transparent way (Global Footprint Network 2012).

1.2.3.2 Carbon Footprint (CF)

Vermeulen et al. (2012) report that in 2008 food systems (along the full supply chain: fertiliser manufacture, agriculture, processing, transport, retail, household food management, and waste disposal) accounted for 19–29% in terms of global anthropogenic GHG[3] emissions, emitting 9,800–16,900 Mt of carbon dioxide equivalent ($MtCO_2eq$).

Agriculture (characterised by a relevant regional variation) has the greatest impact, contributing between 7,300 and 12,700 of $MtCO_2eq$/year (including indirect emissions associated with land-cover change), equivalent to 80–86% of the total food systems' emissions and 14–24% of total global emissions. In particular: deforestation and land use change account for 2,200–6,600 $MtCO_2eq$/year, i.e. 30–50% of agricultural emissions and 4–14% of total global emissions; direct emissions, arising, for example, from activities like managing soils, crops, and livestock, contribute to the emission of 5,100–6,100 $MtCO_2eq$/year, i.e. 50–70% of agricultural emissions and 10–12% of total global emissions.

As a consequence of this, the whole food chain, excluding agriculture, accounts for 14–20% of food-related emissions and, at most, for 5% of global emissions.

During the last few years the carbon footprint (CF), that may be defined as an EPD focused on the climate change impact of a product, has become one of the key indicators of environmental sustainability aiming at identifying the main environmental hotspots and at stimulating emission reduction. In particular, CF is related to human pressure on the Planet in terms of the total amount of GHG emissions that are directly (on-site; internal) or indirectly (off-site; external; embodied; up/downstream) caused by an activity or have accumulated during the life cycle of a product. The CF is quantified by indicators such as global warming potential (GWP), which represents the amount of GHGs contributing to global warming and climate change, with a usual time horizon of 100 years. The CF results are expressed in $kgCO_2eq$,

[3] Carbon dioxide (CO_2), methane (CH_4), nitrous oxide (N_2O), hydrofluorocarbons (HFCs), perfluorocarbons (PFCs), and sulphur hexafluoride (SF_6): the six GHGs identified by the Kyoto Protocol.

obtained by multiplying the actual mass of a gas by its GWP factor, making the GW effects of the different GHGs comparable and additive (Čuček et al. 2012; Galli et al. 2012). It is important also to underline that CF is dependent on kWh in the sense that it is linked to the energy mix used of the country where the assessment is done. In this context, specific carbon labels can be adopted for communicating these results with the aim of helping all the interested stakeholders to make appropriate purchasing choices. There are, in fact, a growing interest and a demand for lower carbon products from the perspective of GHG emission reduction at both production and consumption level. This has led to the development and adoption by the different stakeholders of different analytical methods for calculating product CF. Currently (the end of 2013) all of them are voluntary footprint labels.

International Initiatives

At the moment just two standards have been developed by international stakeholders on the basis of international consultation processes: the "Greenhouse Gas (GHG) Protocol Product Life Cycle Accounting and Reporting Standard", launched by the World Resources Institute (WRI) and the World Business Council for Sustainable Development (WBCSD), and the new ISO 14067:2013 (ISO 2013).

Greenhouse Gas (GHG) Protocol The Greenhouse Gas (GHG) Protocol is a multi-stakeholder partnership of business, non-governmental organisations (NGOs), governments, and others created by the WRI and WBCSD. This partnership began in 1998 with the aim of developing internationally accepted GHG accounting and reporting standards and tools and promoting their adoption for developing a low-emission worldwide economy.

The GHG Protocol has implemented several different (but complementary) standards, protocols, and guidelines over the years. Among them, two standards (published in 2011) are based on an LCA approach: the Corporate Value Chain (Scope 3) and the Product Life Cycle Accounting and Reporting standards. The first, published as a supplement to the Corporate Standard (2004), represents a standardised methodology enabling companies to quantify and report their corporate value chain (Scope 3) GHG emissions. The second one is designed for companies and organisations of all sizes and in all economic sectors and is based on the framework and requirements of the ISO standards 14040 (2006b) and 14044 (2006c), the PAS 2050 and the ILCD Handbook.

The Product Life Cycle Accounting and Reporting standard covers the reporting of the six GHGs defined by the Kyoto Protocol. Furthermore, the standard provides guidance and requirements for all those companies and organisations interested in quantifying and publicly reporting an inventory of the GHG emissions associated with a certain product. A general framework is given so that the companies can make informed choices about reducing GHG emissions of the products they design, manufacture, sell, purchase, or use. It is thus possible to track a product's GHG inventory and emission reduction over time.

Apart from performance tracking, the standard is designed for product comparisons. In this case, additional specifications are needed to ensure its consistent application to a product or a product category: the specifications are provided within a "product rule", which is a document created by a group of stakeholders interested in a particular product/product category and in achieving consensus on the additional specifications for making a comparison or declaration regarding the product: a PCR (§ ISO 14025:2006) is an example of a "product rule". If there is no intention to perform product comparison but simply a desire to reach consensus on guidance for making a product GHG inventory in a certain sector, it is possible to refer to the so called "sector guidance" created by specific stakeholders and sector representatives (Chomkhamsri and Pelletier 2011). At the moment, the only PCR is the North American Product Category Rules (PCR) for ISO 14025 Type III Environmental Product Declarations (EPDs) and/or the GHG Protocol Conformant Product "Carbon Footprint" of Concrete.

Currently (at the end of 2013) there are no product rules or sectorial guidelines regarding the agri-food sector, but there are many ongoing initiatives that will eventually result in such rules over the next few years.

ISO 14067:2013 ISO/TS 14067:2013 "Greenhouse gases—carbon footprint of products- requirements and guidelines for quantification and communication", developed by the Technical Committee ISO/TC 207, was finally published in 2013. It aims at increasing transparency and quantifying and reporting CO_2 emissions over the entire life cycle of products and services. The standard focusses on GHGs, addressing just one impact category (climate change) and it ensures, for the first time, that CF data are comparable worldwide. Therefore the standard cannot be considered as an indicator of the overall environmental impact of products (Chomkhamsri and Pelletier 2011).

ISO 14067 was intended to be consistent with other standards such as ISO 14025, ISO 14044, and BSI PAS 2050, and specifies the principles, requirements and guidelines for the quantification and communication of the CF of a product (CFP); furthermore it specifies requirements and guidelines for the quantification and communication of a partial carbon footprint of a product (partial CFP). Finally, the standard provides for the development of CFP-PCR or the adoption of PCR developed in accordance with ISO 14025 and consistent with it (ISO 2013).

Public Initiatives

Several public initiatives have been developed over the years, with the joint participation of national governments, sometimes including international consultation and/or road testing (Table 1.6).

EUROPE PAS 2050 (United Kingdom)

The Carbon Trust was founded by the UK government in 2001 as a publicly funded independent company intended to help businesses transition to a low-carbon

Table 1.6 Carbon footprint public initiatives

Nation (Country)	Name of the initiative
UK (Europe)	PAS 2050
UK (Europe)	Carbon reduction label
France (Europe)	BPX 30-323-0
Japan (Asia)	JEMAI CFP project
Thailand (Asia)	TGO carbon footprint
Korea (Asia)	Korea PCF
Taiwan (Asia)	Taiwan product carbon footprint
China (Asia)	Low-carbon product certification
Quebec (North America)	Product carbon footprint pilot project
Chile (South America)	Under development
New Zealand (Oceania)	New Zealand GHG footprint strategy

economy. In response to the market need to identify the real drivers of emissions and reduction opportunities along product life cycles and to inform all the interested stakeholders about a product's carbon content, the Carbon Trust started an initiative with the aim of: developing a robust, consistent standard for assessing the GHG emissions during the life cycle of a product (PAS 2050); defining standardised framework requirements for communicating product carbon footprint (PCF) and reduction information (Code of Good Practice on Robust GHG Emissions and Reducing Claims); and developing a safe way for companies to share their PCF information publicly (Carbon Reduction Label).

In particular, the standard Publically Available Specification (PAS) 2050 was the first initiative ever developed for calculating GHG emissions during a product's life cycle. It was published in 2008 by the British Standards Institution (BSI) (and co-sponsored by the Carbon Trust and the UK Department for Environment). This standard has been adopted by many companies worldwide, and has had great influence on the development of other PCF methodologies.

PAS 2050 defines a framework for the assessment of GHG emissions of goods/services during their life cycle. The standard is based on existing life cycle assessment methods (ISO 14040 and 14044) and focusses on a single environmental issue: GHG emissions and their contribution to climate change. In particular, PAS 2050 can be applied to a wide range of goods and services and the guidelines provide principles for the preparation, and use, of "supplementary requirements" for specific industrial sectors or product categories to enable the consistent application of the standard within the particular sector or product category. Furthermore, it does not specify requirements for the disclosure or communication of the results of quantification of the life cycle GHG emissions of goods and services.

In 2011, with the aim of promoting harmonisation of standards, a review process began, resulting in a revised standard (PAS 2050:2011) largely consistent with the GHG protocol product standard and with other international footprint methods. Based on the consideration that PCF can be enhanced through additional

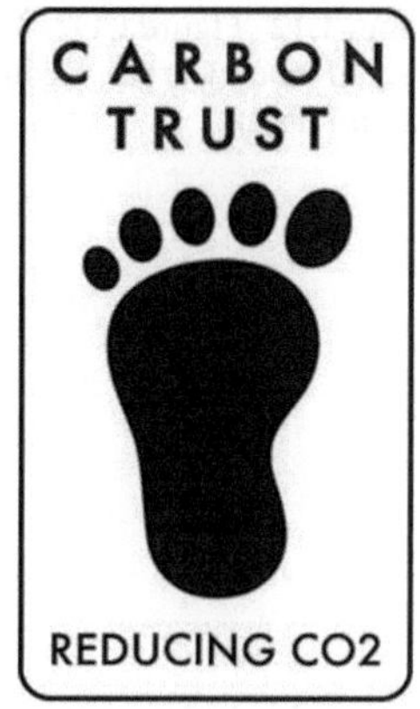

Fig. 1.11 Reducing CO2 label (Carbon Reduction Label)

category-specific rules, this new standard introduces a framework for the use of such "supplementary requirements" (BSI 2013).

As far as the agri-food sector is concerned, in 2012 the following supplementary requirements were published: PAS 2050-1 (Assessment of life cycle greenhouse gas emissions from horticultural products) and PAS 2050-2 (Assessment of life cycle greenhouse gas emissions—Supplementary requirements for the application of PAS 2050:2011 to seafood and other aquatic food).

In particular, PAS 2050-1:2012 was the first addition to PAS 2050:2011 that entailed assistance to the horticultural industry (including plants, production, flowers, fruits, nuts, vegetables, nursery trees, and orchards) in reducing GHG emissions, improving energy efficiency, and saving energy costs. After a couple of months, it was followed by the publication of PAS 2050-2:2012, which provides the global fish industry with a common approach for assessing GHG emissions associated with seafood and other aquatic food products (Blanke and Schaefer 2012).

Carbon Reduction Label (United Kingdom) In 2007 the Carbon Label Company, a subsidiary of the Carbon Trust, was launched "to help businesses measure, certify, reduce and communicate the life cycle GHG emissions of their products and services". In order to display the Carbon Reduction Label on the products, businesses must prove that the PCF has been measured by an internationally recognised method (the PAS 2050 standard). By the end of 2013, two labels were available:

- *Reducing CO_2 Label*: a simple and effective way for communicating that the PCF has been measured and certified; it also allows the company to communicate the carbon footprint measurement and its commitment to reducing it (Fig. 1.11);
- *CO_2 Measured Label*: used for improving a company's reputation by clearly communicating its accurate measurement of the PCF and disclosure of the results (Fig. 1.12).

The Carbon Reduction Label has been adopted around the world: for example, in many European countries, the USA, Canada, Australia, New Zealand and the Russian Federation (ITC 2012).

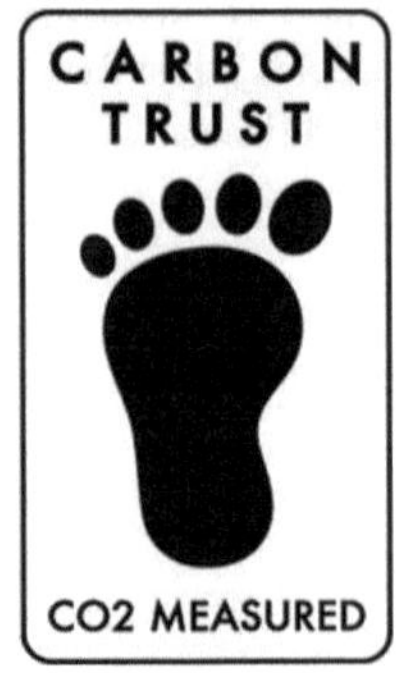

Fig. 1.12 Figure CO2 measured label (Carbon Reduction Label)

BPX 30-323-0 (France) The key aspects of the French labelling system, at a technical level, can be summarised as follows.

1. All the international approaches are currently (at the end of 2013) private, voluntary and with no legislative foundation. The French system is the only one with a legislative pillar for environmental labelling.
2. Most of the existing systems are focussed, above all, on the CF, at the risk of giving exclusivity to it using a mono-criterion label. The French system, conversely, is characterised by a multi-criteria environmental labelling approach: this means that a CF is required whatever the product category; other environmental indicators (at least one more, such as water footprint (consumption and quality) or resource depletion (biodiversity loss)) may be reported.
3. The French environmental labelling project recommends "a life cycle approach". This does not mean that ISO 14040 and 14044 must be followed slavishly but that an overall view of all impacts (of the same type) is sufficient.

Regarding the first point, two laws represent the legislative pillars of the French system.

1. Law no. 2009–967 regarding the implementation of the Grenelle Environmental Round Table (Grenelle 1). Article 54 of the law states: "…Consumers must have access to sincere, objective, and comprehensive environmental information on the overall characteristics of the product/packaging…".
2. Law no. 2010–788, regarding national undertaking to protect the environment (Grenelle 2). Article 228 of the law states: "…from 1 July 2011….a trial will be conducted for a minimum period of 1 year. The objective of this trial is to inform the consumers, gradually and by any suitable method, of the carbon footprint of products and their packaging, and the consumption of natural resources or impact on natural environments that are attributable to these products throughout their life cycle".

Apart from the legislative pillar, there is also a technical one: in particular, the Agence de l'Environnement et de la Maîtrise de l'Energie (ADEME) and the

Association française de normalization (Afnor) Platform have prepared a good practice reference and general document for elaborating Product or Sector Category Rules documents for environmental labelling. In this context, in 2009 the Platform published a document on general methodology named the BP X30-323 "General principles for environmental labelling of mass market products": it was inspired by PAS 2050 (but seen as a step forward in terms of its legislative basis and its multi-criteria approach) and had no major differences from the GHG Protocol.

This framework contains a detailed methodological appendix and it is accompanied by solid implementation tools (sector guidelines, public databases, etc.); furthermore, as already underlined, this methodology is aimed at multi-criteria labelling (not just a CF—but a Product Environmental Footprint) (Vergez 2012).

Agri-food sub-sectors are presently working on establishing PCRs specific to their product families: oil, wine and spirits, milk, mineral water, coffee, and pet food (Dron 2012).

Climate Declaration (Sweden) This declaration is an extract from the climatic data of an EPD® and describes the emission of GHGs (the carbon footprint) based on the following rules: all the greenhouse gas emissions (calculated for each step of the product life cycle) are included and expressed in CO_2eq.; the information is separated according to the different stages of the life cycle; emissions are divided into fossil and biogenic ones; and the data related to other environmental effects are available in the EPD from which all data derive (Lo Giudice and Clasadonte 2010).

ASIA JEMAI CFP Project (Japan)

In 2009, the Japanese government launched the trial of a Carbon Footprint System aimed at providing information on the life cycle emissions of GHGs associated with the production, processing, and use of consumer products.

In particular, a 3 year (2009–2011) pilot project (the "Carbon Footprint of Products Communication Programme" or "CFP Pilot Project"), as part of an "Action plan for achieving a low-carbon society", was launched. It was led by the Japan Environmental Management Association for Industry (JEMAI) of the Ministry of Industry and Innovation, in partnership with a distributor, Eon.

In the same year, the "Basic Guidelines of the Carbon Footprint of Products" (TS Q0010) and the "Guide for Establishing Product Category Rules" were established. The pilot project was completed in March 2012, with the development of a public database.

The Japanese labelling is a mono-criterion one, focussed on the carbon footprint. The official label (Fig. 1.13) indicates the impact in absolute values (grams of GHGs per gram of product) (JEMAI CFP Programme 2013).

By 2011, a total of 73 PCRs had been approved; of these, 23 belong to the agri-food sector based on the "Sector guidance for development of CFP-PCR: for perishables and for processed foods" (Table 1.7 (PCR Library 2013).

TGO Carbon Footprint (Thailand) Although Thailand does not have mandatory targets for the reduction of GHG emissions, national economic and social development clearly promotes low-carbon production and consumption; CF is also highlighted as a key strategy in the move towards a low-carbon society.

Fig. 1.13 Japanese carbon footprint label

In this context, in 2009, a pilot project for promoting the implementation of PCFs was launched by the Thailand Greenhouse Management Organisation (TGO, a public organisation) and the National Metal and Materials Technology Centre (MTEC). This project aimed at the implemention of PCFs and labelling and made Thailand, in 2010, the first country in south-east Asia (and from the developing world) to develop its own guidelines on CF, "The National Guideline Carbon Footprinting of Products". These guidelines are based on ISO 14040 and 14044, PAS 2050 and TS Q0010 (Mungkung and Gheewala 2012; Gheewala and Mungkung 2013).

As far as the PCRs are concerned, it is important to highlight that the practice of PCR development in Thailand takes place on two levels: PCR, company level; and PCR, national level. At the end of 2013, there were 12 PCRs at national level and 125 at company level. Regarding the agri-food sector, 4 are at national level (for jasmine rice, fruit and vegetables, chicken meat, livestock products) and almost 70 at company level (for the complete list, see http://thaicarbonlabel.tgo.or.th/carbon-footprint/index.php?page=6).

Furthermore, in 2009 the national CF labelling (Fig. 1.14) scheme was released. It takes into account the quantity of GHG emissions from each production unit throughout the whole life cycle of a product. The CF thus calculates the CO_2eq of the GHG emissions released by raw material acquisition, manufacture, use, waste management, and final disposal, including related transport in all stages (Gheewala 2012).

Korea PCF (KOREA) The Korea CF label was introduced in 2009, after a nine-month pilot programme, by the Environmental Industry and Technology Institute (KEITI). The labelling covers, inter alia, consumer goods, transport services, electronic appliances, and production goods. The assessment methodology is reported in the Guidelines for Carbon Footprint of Products and was published in 2009 by KEITI (Dawoon et al. 2013). The labelling programme considers only GHG emissions in terms of environmental issues and employs two levels of certification: Carbon Emission Certificate (level 1), related to the baseline emissions calculation for a product (Fig. 1.15) and Low Carbon Product Certification (level 2), indicating that the minimum GHG emission reduction defined by the government has been meet (Fig. 1.16) (Seo 2009).

Taiwan Product Carbon Footprint (Taiwan) In 2009, the Taiwan Environmental Protection Administration (EPA) began developing a Carbon Label System (Fig. 1.17).

Table 1.7 PCR for the Japan carbon footprint

PCR ID	PCR name
PA-CQ-01	Milk
PA-CP-01	Chicken
PA-CO-01	Seafood excluding aquaculture
PA-CN-01	Eggs
PA-CM-01	Processed seafood
PA-CH-01	Refined sugar
PA-CF-02	Pork
PA-BY-01	Raw milk (Intermediate Goods)
PA-BX-01	Soft drinks
PA-CG-02	Seasonings
PA-BW-02	Beer, happoshu, and beer-flavoured sparkling liquor
PA-BJ-03	Raw bananas
PA-BH-02	Instant noodles
PA-BF-04	Vegetables and fruits
PA-AM-02	Instant coffee
PA-AL-02	Chocolate (containing wafer)
PA-AJ-01	Rice biscuits
PA-AI-04	Hams and sausages
PA-AH-01	Cooked and sealed rice
PA-AG-01	Potato chips
PA-AB-02	Rapeseed oil
PA-AA-02	Rice
PA-BW-02	Mushrooms

One year later, the carbon labelling framework was established, together with several guidelines for assisting industries with CFP calculation, PCR drafting and carbon label application. The model was based on the PAS 2050 standard, the draft of the ISO 14067, and national conditions (Lin et al. 2013).

Regarding the agri-food sector, at the end of 2013 23 PCRs had been approved (Table 1.8).

Low-carbon Product Certification (China) The Carbon Trust's Carbon Reduction Label (CRL) has an indirect presence in China through its collaboration with multinational companies to stimulate low-carbon innovation and technology development. In particular, in the period 2009–2010 a pilot project for assessing the acceptance of the PCF and the feasibility of applying PAS 2050 was launched. Regarding the agri-food sector, in 2010 the product "sea scallops" was the first certified food product with a CF standard (according to ISO 14040).

Fig. 1.14 Thailand carbon footprint label

Fig. 1.15 Carbon emission certificate label

Fig. 1.16 Low carbon product certification label

Fig. 1.17 Taiwan Carbon Footprint label

At the same time, because Chinese companies working among the domestic markets are not so inclined to pay for the international CRL, in 2009 China's Ministry of Environmental Protection (MEP) signed a contract to cooperate with the German environmental bodies in certifying "low carbon-intensive products". This certification and labelling system is voluntary for Chinese manufacturers, and the targeted products are mostly daily necessities and are covered by China's environmental labelling procedure. At the moment the agri-food sector seems to be excluded from the application of this label (Brandi 2012).

North and Latin America Product Carbon Footprint pilot project (Quebec)

The Ministère des Finances et de l'Économie (MFEQ) was commissioned by the Québec government to implement a project for promoting the marketing of products whose carbon footprints had been measured and certified. In this context, in 2012 the MFEQ began a collaboration with the Interuniversity Research Centre for the Life Cycle of Products, Processes and Services (CIRAIG) to create a product CF pilot project with the aim of verifying the feasibility of large-scale CF labelling in Québec and guiding its eventual implementation.

The GHG Protocol Product Life cycle Standard was used as a basis for this study and the products submitted for testing were: aluminium ingots, cloud computing services, wood products, pulp and paper industry products, second-generation biofules, packaging products, and agri-food products. The pilot project has already ended and the main finding and recommendations have been submitted to the MFEQ but the results have not yet been disclosed to the public (CGF 2012).

Chile In 2009, a preliminary PCF initiative for food was launched by the Ministry of agriculture in Chile. In particular, the Research Institute for Agriculture (INIA) has begun developing a preliminary PCF methodology regarding the main agricultural export goods, such as wine and milk (PCF World Forum 2010).

Oceania New Zealand GHG Footprint Strategy (New Zealand)

The "New Zealand Greenhouse Gas Footprinting Strategy" for the Land-Based Primary Sectors was developed in 2007 as an initiative between the Ministry of Agriculture and Forestry (MAF) and the primary sector. The main goal of this strategy is not to label CFP, but to provide a standardised means of measuring and managing GHG emissions across the life cycle (including all transport—except the consumer's shopping trip—and emissions during the use phase) of a product. The aim is to help the different sectors to measure, manage, and mitigate GHG emissions across the supply chain.

The development of sector-specific approaches to GHG footprinting (according to PAS 2050 or other international standards) is a key aspect of this strategy. By the end of 2013, the following agri-food sectors were covered: dairy, fruit production (kiwifruit, and pipfruit), wine, arable crop production and meat (lamb). The following projects are under way: fruit production (summerfruit and berryfruit), vegetables (onions), meat (venison and beef/mutton), and beverages (zespri-water) (MPI 2013).

1.2.3.3 Water Footprints (WF)

Although the concept of CF is widespread among stakeholders, the same cannot be asserted for the water footprint (WF), albeit an even bigger challenge for consumer goods companies, in particular for all of those working in the food and drink sector.

The WF concept (closely linked to the virtual water concept) was introduced by Hoekstra in 2002 (Hoekstra 2003) with the aim of creating a consumption-based indicator of fresh water use. It was developed as an ecological footprint, accounting for the appropriation of natural capital in terms of direct and indirect water use deriving from the consumption/production of goods and services.

Table 1.8 Taiwan PCRs

No PCR	PCR name
10-008	Fruit juices V 1.0
10-014	Bottled water V 1.0
11-002	Packaged tea drinks V 1.0
11-003	Instant noodles (Frying process) V 1.0
11-004	Fresh milk V 1.0
11-005	Bread V 1.0
11-006	Store-prepared sweet potatoes V 1.0
11-007	Shell eggs V 1.0
11-008	Prepared eggs V 1.0
11-009	Pudding V 1.0
11-012	Packed grains and bean beverages V 1.0
11-015	Stuffed cakes and pastries V 1.0
12-001	Packed meals V 1.0
12-003	Rice sticks V 1.0
12-004	Meatballs V 1.0
12-006	Rice V 1.0
13-007	Fruit juices V 2.0
13-012	Bottled water V 2.0
Not available (n.a.)	Vegetables
N.a.	Uncooked noodles
N.a.	Fresh edible mushrooms
N.a.	Edible plant mill product
N.a	Jelly

A WF can be calculated for a process, a product, a consumer, a group of consumers (e.g. municipality, province, state or nation), or a producer (e.g. a public organisation, private enterprise).

Water use is measured in terms of water volume (in m^3) consumed (evaporated) and/or polluted per unit of time. The water footprint is a geographically explicit indicator that not only shows volumes of water use and pollution, but also the relative locations.

Regarding the food product and beverages sector, processed food and beverage production requires large amounts of water and what is consumed by humans every day makes up 50 % of the total WF, which includes the enormous volume of "virtual water" needed to produce the consumed food. Reducing WF is an environmental challenge that food and beverages companies should be prepared to meet if they want to maintain their competitive position and build a positive reputation among end-consumers.

International initiatives

ISO 14064 (DIS) (In development) ISO 14064 "Environmental management—Water footprint—Principles, requirements and guidelines" will be issued with the aim of providing principles, requirements, and guidelines related to water footprint assessment of products, processes, and organisations based on LCA. It will allow the evaluation of water footprints as stand-alone assessments or as part of more comprehensive environmental assessments. The result of a water footprint assessment will be a single value or a profile of indicator results (Sala et al. 2013).

The ISO draft has been registered, the ballot initiated, and the publication of the Standard is expected in 2014.

Water Use in LCA (WULCA) Water Use in LCA (WULCA) is a Life Cycle Initiative group project on the Assessment of Use and Depletion of Water Resources within LCA. It is an international project, launched in 2007, under the auspices of the UNEP/Society for Environmental Toxicology and Chemistry (SETAC) Life Cycle Initiative. It involves a great number of members from different countries in North America, Europe, and Australia as well as from different sectors (academia, industry, and consultant firms), with the aim of focussing on water use assessment and water footprinting via a life cycle perspective. In particular, the specific objectives of the project are to: develop a general assessment framework for water use including indicators that measure the environmental impacts on human health, ecosystems, and freshwater resources; establish adequate water inventory schemes and parameters; establish impact assessment methods for characterising water use and related environmental impacts; and develop recommended practice and guidance for developers and practitioners of the LCA methodology (Köhler and Aoustin 2008).

Water Footprint Network The Water Footprint Network is a global water footprint organisation focussing on the development of freely available standards and WF tools, with the aim of promoting sustainable, fair, and efficient fresh water use around the world. Huge emphasis is put on agriculture because it accounts for the vast majority of the WF. For this reason global and national scale footprints have been developed for many agricutural products (Table 1.9).

In 2009, the Water Footprint Network developed the first version of the Global WaterFootprint Standard. It was revised, in 2011, after intensive consultation with partners and researchers worldwide. The standard is contained in the Water Footprint Assessment Manual (Hoekstra et al. 2009).

1.2.4 EU Environmental Footprint (EF) Initiative

Considering the current situation of product environmental assessment, it is fair to say there has been a huge proliferation of methods and standards with the LCA methodology as a basis: ISO 14040, ILCD Handbook, BP X 30, PAS 2050,

Table 1.9 Water footprint of selected food products

Fruit and vegetables	Meat and animal origin food product	Processed products
Apple	Mutton	Sugar (from sugar beet)
Tomato	Pork	Sugar (from sugar cane)
Potato	Goat	Coffee
Peach or nectarine	Egg	Chocolate
Orange	Chicken	Rise
Olive	Cheese	Pizza margherita
Mango or guava	Butter	Pasta (dry)
Maize	Milk	Wine
Lettuce	Milk powder	Tea
Ground Nuts (in shell)	Beef	Bread (from wheat)
Dates		Beer (from barley)
Cucumber or pumpkin		
Cabbage		
Banana		

Ecological footprint, WBCSD/WRI etc. If, on the one hand, this proliferation has led to a progressive acceptance of the methodology, on the other, it has become clear that nowadays standardisation is not enough and that it is important to take a step forward to initiate a process of harmonisation among all the existing standards and methods. This is because there is confusion among the stakeholders regarding how to measure, make, and understand a claim about the environmental performance of a product. Furthermore, there is an evident lack of consistent and science-based multi-criteria environmental information covering the entire value chain.

All things considered, many initiatives with the aim of creating a "level playing field" for robust and scientifically valid applications have been implemented in the last few years.

In such context is the EU's Communication and the Commission Recommendation (2013/179/EU) "on the use of the Product Environmental Footprint (PEF) and Organisation Environmental Footprint (OEF) methods" proposed an action plan in which it: established the Product Environmental Footprint (PEF) and Organisation Environmental Footprint (OEF); recommended the use of these guidance documents to Member States, companies, private organisations, and the financial community; announced a 3 year testing period for developing product- and sector-specific rules through a multi-stakeholder process; established principles for communicating environmental performance, such as transparency, reliability, completeness, comparability, and clarity; aimed at supporting international efforts for more coordination in methodological development and data availability.

In particular, as far as the PEF is concerned, it entered a pilot phase (from 2013–16) launched through an open call for volunteers. The call is for all the stakeholders who wanted to propose a product category for which to develop specific PEFCRs: the only product category excluded (delayed until phase 2, in mid-2014) is the food and beverage one, because of the contemporary EU Food SCP Round Table initiative (the Envifood Protocol). At the beginning of October 2013 the EC announced the projects selected for the EF pilot phase. In particular, the following PEFCRs were proposed: batteries and accumulators, decorative paints, hot and cold water supply pipes, household detergents, IT equipment (servers, magnetic disk unit, switching equipment), leather, metal sheets, non-leather shoes, photovoltaic electricity generation, stationery, thermal insulation, t-shirts, uninterruptible power supply and intermediate paper products.

1.2.5 European Food Sustainable Consumption and Production RoundTable (EU Food SCP RT)

The ENVIFOOD Protocol, implemented by the European Food SPC RoundTable (EU Food SCP RT), is in line with the PEF Guide and represents a starting-point for developing PEFCRs among the food and beverage product category.

The EU Food SCP RT was launched in 2009 as an initiative co-chaired by the European Commission and food supply chain partners and supported by the UN Environment Programme (UNEP) and European Environment Agency (EEA).

The initiative is based on a harmonised life cycle approach (which requires, in equal measure, the involvement of all food value chain members) that allows an open and results-driven dialogue among all the stakeholders along the food chain. It was developed with the aim of supporting the EU policy objectives, above all the ones highlighted in the EU Action Plan on SCP and Sustainable Industrial Policy. In particular, the RT's main objectives are to: establish scientifically reliable and uniform environmental assessment methodologies for food and drink products, including product category specifications where relevant, considering their significant impact across their entire life cycle; identify suitable tools and guidance for voluntary environmental communication to consumers and other stakeholders; promote continuous environmental improvement measured along the entire food supply chain. One year after its launch and following a public consultation, the RT announced the adoption of 10 Guiding Principles on "*voluntary environmental assessment and communication of environmental information along the food chain, including to consumers*". The leading principle was: "*environmental information communicated along the food chain, including to consumers, shall be scientifically reliable and consistent, understandable and not misleading, so as to support informed choice*". The ten principles were formulated with the aim of promoting a coherent way of assessing and communicating, on a voluntary basis, the environmental performance of food and drink products, constituting the foundation from which to develop a harmonised framework methodology (ENVIFOOD Protocol) for the voluntary environmental assessment of food and drink products and for the communication of their environmental information along the food chain.

1.2.5.1 Environmental Assessment of Food and Drink (ENVIFOOD) Protocol

The ENVIFOOD Protocol is an initiative co-chaired by the European Commission and food supply chain partners (e.g. governments, business associations, NGOs).

It aims at establishing the food chain as a major contributor to sustainable consumption and production in Europe, representing the first developed (collectively agreed) sectorial and science-based methodology for assessing the environmental performance of food and drink products in Europe along their life cycle. In this context, this protocol could serve as a starting-point for developing:

- PEFCRs for food and food packaging by defining several product categories and related PEFCRs below the level of the Protocol;
- communication methods; product group/sub-group specific rules (PCRs);
- criteria, tools, datasets and assessments.

It is important to highlight that it is not a self-supporting guide and that, depending on the intended application, different additional requirements may apply (De Camillis et al. 2012).

The Protocol was developed by the Working Group 1 of the EU Food SCP RT following a stepwise procedure, in accordance with EU legislation and, as said before, it is based on the Guiding Principles of the Food SCP RT. It took into consideration:

- all the existing and upcoming international standards on LCA, environmental labels/declarations, and eco-design (ISO 14040-44; ISO/CD 14067; ISO 1402X, ISO/TR 14062);
- the International Reference Life Cycle Data System (ILCD) Handbook;
- the Commission's Product Environmental Footprint (PEF) Guide;
- other emerging methodologies/guidelines; critical review of environmental assessment case studies or data availability and requirements (Masoni et al. 2012).

After a consultation period, the Protocol was adopted in November 2012 (Envifood Protocol Draft Version 0.1) and the pilot project was launched in March 2013, 21 organisations agreeing to test the draft Protocol. Participants include a wide range of food and drink manufacturers, trade associations, and research institutes, and the project lasted for 6 months. Among the many products assessed with the ENVIFOOD Protocol were coffee, dairy, soy, chocolate, pet food, wine, and baby food products (EU Food SCP RT 2013). In the light of the results of this project, the Round Table agreed to adopt the final methodology and recommendations on voluntary communication of environmental information by the end of 2013.

1.3 LCA Initiatives Among the Actors of the Supply Chain

1.3.1 Agriculture and Livestock

The livestock sector is one of the subsectors with major impacts from an environmental perspective. Although on the one hand this sector has to face and respond to consumer demand for livestock products, on the other it has to improve its environmental performance and mitigate its impacts on climate.

In this context, several international initiatives are being developed. In particular, the Food and Agriculture Organization (FAO) of the United Nations has recently launched a partnership named Livestock Environmental Assessment and Performance (LEAP), which aims to improve environmental performances of the sector while taking into account social and economic issues as well. In particular, there is a desire to develop methodologies and sector-specific guidelines for the life cycle assessment of GHG emissions from livestock production; there is also a project for developing the GHG Life Cycle Inventory (LCI) database globally for major feed crop materials (FAO n.d.).

In 2013 the FAO published three reports on this theme: one developed by the Animal Production and Health Division (AGA) regarding "Greenhouse emissions from pig and chicken supply chains" (MacLeod et al. 2013); one regarding "Greenhouse emissions from ruminant supply chains" (Opio et al. 2013); and the last one an overall report providing a synopsis of GHG emissions for all livestock sectors (including dairy) and exploiting mitigation potential and options, named "Tackling climate change through livestock" (Gerber 2013). Finally, during the LEAP technical workshop held in September 2013 in Rome a draft of the FEED LCA Guide was presented: the guide aims at providing all the interested stakeholders with a robust and harmonised footprint methodology for feed and feed products (Fefac 2013).

At a European level, in 2010 the Joint Research Centre (JRC) published a final report regarding evaluation of the livestock sector's contribution to the 27-EU GHGs emission according to animal species, animal products, and livestock systems, following a food chain approach (Leip et al. 2010).

At national level, finally, in 2010 a report entitled "A Greenhouse Gas Footprint Study for Exported New Zealand lamb" was published. This study was commissioned by the New Zealand meat industry and its partners Landcorp Farming, Ballance Agri-Nutrients, and MAF. It aimed at identifying the most significant sources of GHG emissions in the lamb life cycle and those that could be addressed most easily in order to reduce emissions (Ledgard et al. 2010).

1.3.2 Industry Sector

As well known, the first application of the LCA methodology was carried out in 1969 in the context of a Resources Environmental Profile Analysis study commissioned by the Coca Cola Company regarding energy, material, and environmental issues linked to the life cycle of packaging.

Since then, LCA has gained growing consensus as a useful tool for environmental assessment of packaging. In this context it is important to highlight an interesting initiative developed in 2000 by the Sustainable Packaging Alliance (SPA), the Packaging Impact Quick Evaluation Tool (PIQET): the project aimed at providing the packaging supply chain with a quick and credible environmental assessment tool (LCA-based) which would assist them in making decisions on packaging development and innovation strategies. Nestlé is one of the most famous international companies to use this tool for ecodesign evaluation in the development phase of new food packaging (Notarnicola et al. 2012b). Other similar LCA-based analytical tools are: COMPASS (Comparative Packaging Assessment), Package Smart, PEAT (Packaging Environmental Assessment Tool), and InstantLCA Packaging (Dordan n.d.).

As far as the environmental impacts related to packaging are concerned, it is important to remember that they are minor compared with the overall impact of the packaged food product itself. In fact, apart from when the packaging manufacture accounts for a big percentage of the total (e.g. wine-making), the LCA of food products is not limited to analysis of packaging alone. In this context, and with reference to LCA-based tools useful for assessing the whole agri-food product system, it is important to underline the experience of Selerant, an ICT Italian company which, within its web-based product life cycle management (PLM) solution called DevEx, developed the Eco-Design Tool to help companies assess different packaged food product system designs from a life cycle perspective (Notarnicola et al. 2012b).

1.3.2.1 Dairy Industry

The dairy sector has been, and still is, extensively studied from an LCA perspective. Many studies regarding the GHG emissions from this sector have been developed (reviewed in chapter 6).

The International Dairy Federation (IDF) and the Food and Agriculture Organization of the United Nations (FAO), carried out a study, “Greenhouse Gas Emissions from the Dairy Sector—A Life cycle Assessment”, with the overall goal of providing estimates of GHG emissions from the dairy food chain (milk production and processing) in the main regions and farming systems of the world. Apart from this international study, others have been developed at national level (Table 1.10).

A study on the WF of cradle-to-farm gate milk production in the USA was commissioned by the Innovative Center for U.S. Dairy and completed by a consortium with expertise in dairy and water issues (Lessard et al. 2012).

1.3.2.2 Beverage Industry Environmental RoundTable (BIER)

The Beverage Industry Environmental RoundTable (BIER), a coalition of beverage industry companies and supporting partners working together on a variety of

Table 1.10 National Carbon Footprint studies in the dairy sector

Nation	Website
UK	http://www.dairyco.org.uk/non_umbraco/download.aspx?media=12018
Australia	http://www.dairyaustralia.com.au/Animals-feed-and-environment/Environment/Climate-redirect-page/MicroSite1/Home/Climate-and-greenhouse-basics/Greenhouse-gas-footprint/Dairy-footprint.aspx
Canada	http://www.agr.gc.ca/index_e.php
Chile	http://www.isr.qut.edu.au/downloads/chile_project2012_eng.pdf
USA	http://www.usdairy.com/Public%20Communication%20Tools/DairysEnvironmentalFootprint.pdf
South Africa	http://www.milksa.co.za/research/research-column/how-can-dairy-industry-limit-its-environmental-impact

environmental and stewardship initiatives, was launched in 2006. This coalition was created with the aim of defining a common framework for stewardship, driving continuous improvement practices and performance, and informing public policy in the areas of water, energy & climate, beverage container recycling, sustainable agriculture, and eco-system services. These goals have been achieved by developing leadership definitions of water stewardship in the beverage industry, best practice guidance tools on drought preparedness and management, facility water use, efficiency and conservation practices, benchmarking water use and efficiency, and mapping the state of the science of water footprinting practices. Recent agendas have focussed upon best practice sharing, benchmarking, developing sector guidance for GHGs and WF, and key stakeholder engagement.

In 2013 BIER published a protocol entitled "Beverage Industry Sector Guidance for Greenhouse Gas Emissions Reporting", aiming at estimating, tracking, and reporting GHG emissions within the beverage sector and at highlighting sector impacts on climate change and reduction priorities (BIER 2013). Furthermore, research studies regarding the CF of five beverage categories (beer, bottled water, carbonated soft drinks, spirits, and wine) were developed (BIER 2012a, b, c, d and e).

As far as the WF is concerned, in 2010 a specific working group was created with the aim of evaluating and addressing the increasing global efforts to develop WF methodologies, focusing on the beverage sector. As a result, in 2011 a report named "A Practical Perspective on Water Footprinting in the Beverage Sector" was published as a guide to the application of existing WF tools and development of new ones (BIER 2011).

1.3.3 Retailers, Distribution, Logistics, and Trading

As far as CF is concerned, several initiatives regarding the creation of specific standards by private business and retailers have been developed. Supermarket chains or retailers such as, for example, Casino and Leclerc (France), Migros (Switzerland), and Tesco (UK) have taken such action. In particular, the French retailer Casino has

Fig. 1.18 Casino carbon index label (*front* of the packaging)

Fig. 1.19 Casino Carbon Index label (*back* of the packaging)

launched a carbon labelling initiative for a selection of its private label products, using a colour code (Figs. 1.18 and 1.19).

The Casino Carbon Index was launched in 2008 by the supermarket chain Casino, in cooperation with ADEME and the private sector Bio Intelligence Service organisation, with the aim of evaluating the environmental impact of its own-brand products in France. This initiative covers 3000 products in the food, household products, toiletries, and perfume categories. The label informs the consumer about the GHG emissions involved in producing 100 g of the finished product. It takes into account the five main stages in the product life cycle (agricultural raw materials, packaging, processing, transport, and distribution) (Delahaye 2008).

In the same year Tesco launched the "Tesco carbon reduction label" (Tesco 2012), but it has been recently announced that the retailer is going to phase out the labels on its products because it is too time-consuming and expensive to justify. Concurrently the retailer claimed it would try find even better ways to communicate the carbon impact of products in a way that informs and empowers customers (The Grocers 2012). Usually, the pathway to CF assessment is not easily accessible nor in the public domain but the results are shown on both labels and websites, albeit customers sometimes find the CF labelling difficult to understand so alternative ways of communication are under consideration.

1.3.4 Consumers and Consumers' Organisations

As already underlined, consumers (as one of the final addressees of food products) play a fundamental role in orienting government and market decisions on the building of a more sustainable production and consumption society. Nevertheless, consumers still have limited knowledge about environmental sustainability issues and the most suitable tools to treat them, such as LCA-based ones. On the other hand, consumer associations' commitment is still very lukewarm in terms of supporting consumers and increasing their awareness about environmental issues.

All things considered, there is a need to create awareness and knowledge of these innovative tools to allow consumers to make informed purchasing decisions and to push the market and the decision-makers toward a greener society. In this context, consumers' associations can have a fundamental role in promoting the spread of knowledge and supplying evidence through testing products (Notarnicola 2011).

1.3.5 Food Waste and End of Life

Food waste derives from raw or cooked food materials, including food loss before, during, or after meal preparation in the household as well as food discarded in the process of manufacturing, distribution, retails and food service activities. It can, for example, constitute vegetable peelings, meat trimmings, and spoiled or excess ingredients or prepared food as well as bones, carcasses, and organs.

In 2010 a technical report entitled "Preparatory Study on Food Waste across EU 27" was published by the BIO Intelligence Service, a leading consultancy in France and Europe, commissioned by the European Commission (DG ENV) Directorate C—Industry. The study aimed at providing the EC with information about the situation of food waste in the EU in four sectors (manufacturing, wholesale/retail, food service, and household) to determine the scale of the problem and to identify appropriate measures to be taken. The report, excluding agricultural food waste, highlighted an annual food waste generation in the EU27 equal to about 89 Mt or 179 kg/capita.

Part of the study was dedicated to the assessment of environmental benefits from food waste reduction initiatives. In this context the environmental impact of the life cycle of the food waste was quantified (with a life cycle approach). An average was reported of at least 1.9t CO_2eq. /t of food waste produced in Europe during the whole life cycle of food waste, corresponding to an overall (at European level) environmental impact of at least 170 Mt of CO_2eq/year. The study results suggested the household sector had the greatest impact both per tonne of food waste (2.07 t CO_2eq/t) and at European level (78 Mt of CO_2eq/year).

The European Parliament passed a resolution on food waste avoidance in January 2012, asking the EC to take practical measures to halve food waste by 2025 (Bio Intelligence Service 2010).

1.4 Methodological Issues

The nature of food product systems is different from that of typical non-food based ones in that the system considered for the LCA usually also involves the agricultural and zootechnical phases. This inevitably implies that the biological processes representing the biosphere are considered in conjunction with those of the technical system of the technosphere. This in turn makes the system under analysis more complex and raises a number of specific methodological issues (Notarnicola et al. 2012b). Obviously this is also true for all the products reviewed in this book which, as detailed in the following chapters, in many cases represent a basis for the production of many staple foods in numerous countries.

Specifically, the above-mentioned complexity of the system boundaries of food products implies that the carbon cycle needs to be carefully considered in order to account not only for the fossil but also for the biogenic carbon flows. This is valid for both traditional LCAs and the more recent and en vogue carbon footprint. In fact, in theory, the biogenic carbon balance can be considered to be zero, but in reality agricultural practices such as composting and reduced tillage can have a carbon sequestration effect which can vary over the years and hence has to be accounted for and averaged over a period consistent with the characterisation factors for the global warming potential (GWP) considered in the specific study. Furthermore, such practices can also have a beneficial effect on biodiversity, which is seldom accounted for in LCAs since no standardised methodology is available.

Performing LCAs for the same product system with different functional units (FU), each based on a single different function of the product, has often been shown to generate completely different or even contrasting LCA results (Kim and Dale 2006). The selection of the FU is thus critical as it should best represent the scope of the assessment. Typically this selection is mainly guided by the aim of the investigation, the typology of impacts assessed, and the nature of processes analysed. Ideally, multiple FUs should be used for the same study for complete assessment of the product system from different perspectives (Seda et al. 2010). This would obviously improve the comparability of the LCA with other studies entailing a similar product system. Usually, FUs regarding food are based on mass, volume, cultivated area, energy or protein content, or economic value. These FUs however do not necessarily represent the quality of a product, which could play a crucial role in defining its main function (e.g. high-end food products such as certain types of wines or extra virgin olive oils).

The ever-growing use of pesticides and fertilisers in agriculture, adopted in order to reach higher production yields per unit of surface land, has made the impacts deriving from such additives considerable and at times (as illustrated in the following chapters) the main contributors to the overall impacts throughout the life cycle of the food system under analysis. However, data regarding their production are not always available. This leads either to the use of estimates for the assessment or to the complete exclusion of such a phase from the study. Furthermore, there is a large degree of uncertainty about the diffusion of fertilisers and in particular that of pesticides. The destination of a pesticide is not only dependent on the plant but also on site and time issues such as the type of soil, the weather conditions, the location of the water table, and farming practices. There are currently (2014) a number of pesticide diffusion modelling systems (e.g. Birkved and Hauschild 2006; Dijkman et al. 2012) that can be implemented but none has been universally accepted as a standard. The choice of diffusion model should be made carefully and be based on the available input data for the model in order to obtain the best possible results without the need to estimate and make too many assumptions about its implementation, hence improving the overall results of the LCA.

Global population growth is inevitably increasing the use of fresh water and land for activities that are related to agri-food systems. The assessment methodologies of the impacts of land and water use in LCAs are by no means standardised and are evolving constantly. Early attempts at assessment simply involved the quantification of the land or fresh water used by the product system. Recently more sophisticated land use approaches have been developed in terms of qualitative (e.g. biodiversity) and quantitative (e.g. quantity of organic matter) aspects. Similarly, water use assessment has evolved (Kounina et al. 2013) and now includes various methods (with relative characterisation factors) for evaluating loss of water quality and functionality including hybrid methods in conjunction with virtual water methodologies (Allan 1998). Water and land use assessment are also site-specific and hence there is a need for regionalised datasets. This problem has been partly addressed by the use of geographical information systems in conjunction with LCAs (e.g. Geyer et al. 2010), even though datasets of the required resolution are not always available. Such approaches should be implemented whenever possible to improve evaluation of the impacts of land and water use.

In LCAs the end-of-life phase is often omitted, thus excluding an important means of making more complete the evaluation of environmental impacts because of the omission of important results and failure to consider the possibilities of making a product system more recyclable or disposable.

Interpretation is seen as the LCA phase that particularly requires further methodological developments and practical guidelines (Zamagni et al. 2008). Indeed, neither ISO 14044 nor handbooks on the LCA methodology furnish specific indications on how to conduct this phase. Nonetheless, the above-mentioned ISO norm does distinguish some elements that should be considered during the interpretation phase, whenever possible, in order to improve the overall quality of an LCA: identification of the significant issues based on the results of the LCI and LCIA phases; evaluation that considers completeness, sensitivity, and consistency checks; conclusions, limitations, and recommendations.

Finally, relating to pesticides and fertilisers, data regarding their production are not always available. Similarly, data concerning the industrial phases of food production are not always accessible. Industrial companies are not always willing to reveal information about their industrial processes. Assumptions and estimates may therefore have to be made which do not reflect the reality of the product system under analysis.

In the following chapters the above-mentioned methodological issues will be discussed in relation to the LCA of the specific food products reviewed in this book.

1.5 LCI Databases for the Agri-Food Sector

One of the key issues in the agri-food sector is the lack of reliable and up-to-date inventory data on food products and processes for realising not only accurate LCA studies but also for hotspot analysis, communication, and labelling.

Several LCI databases have been developed but most of them are characterised by a lack of transparency, and they are often incomplete because they take into account only a few input-output flows; this lack of information concerning food-product impacts can lead to ambiguous interpretations and conclusions; furthermore, these databases are often outdated and not regionalised. Finally, the databases are often inconsistent with each other, because of their different approaches and assumptions.

Comprehensive, clear, well-documented, and consistent data are needed to increase the accuracy and comparability of LCAs in the food sector.

In the following paragraphs LCI databases applicable only to the agri-food sector are reviewed.

1.5.1 National Initiatives

1.5.1.1 Europe

The several European databases are described in Table 1.11.

1.5.1.2 Asia

Although at international level LCA has been recognised as a useful tool in the agri-food sector, this methodology is still not widespread in the Asian regions. As a result, in the last years the initiative on "LCA Agri-food ASIA" has been developed with the multi-national collaboration of government organisations, higher research and educational institutes, and private companies. In particular, the LCA Agri-food ASIA network has been launched with the aim, at regional level, of expanding the applications of LCA, developing collaborations, and creating the "Asian Food Database".

The Asian countries currently involved in the project are: Thailand, Japan, South Korea, Malaysia, Indonesia, and Vietnam.

Apart from the Asian Food Database, still under development, databases of individual Asian countries are being developed (Table 1.12).

1.5.1.3 North and Latin America

The following tables report the databases developed in North America (Table 1.13), Canada (Table 1.14), and Latin America (Table 1.15).

Table 1.11 European LCA databases

Database	Sectors of application	Website/Reference
LCADB. SUDOE (Southwest of Europe) database	Agriculture; construction; energy production; forest and forestry products; waste treatment; water; manufacture process; services; transport; use and consumption	Carles et al. 2013
LCAfood database (Denmark)	Crops and crop based products; milk and milk based products; vegetables; meat; fish; packaging	http://www.lcafood.dk
Agri-BALYSE database (France)	Plant production; animal production; tropical crops	Koch et al. 2011
Swiss Agricultural Life Cycle Assessment (SALCA) database (Switzerland)	SALCAcrop; SALCAfarm	http://www.agroscope.admin.ch
Ecoinvent v.3	Agriculture; energy supply; transport; biofuels and biomaterials; bulk and speciality chemicals; Construction materials; packaging materials; basic and precious metals; metal processing; ICT and electronics; waste treatment	http://www.ecoinvent.org/database/
CPM LCA Database (SPINE@CPM)	Agriculture; food and beverages; construction; energy production; forest and forestry products; waste treatment; water; manufacture process; services; transport; use and consumption; chemicals	http://www.cpm.chalmers.se/CPMDatabase/Start.asp

Table 1.12 Asian LCI databases

Database	Sectors of application	Website/Reference
Thai national LCI database (Thailand)	Energy, utilities, and transportation; Industrial materials; agriculture; commodity chemicals; recycling and waste management	http://www.thailcidatabase.net/
MY—LCID (Malaysia)	Energy; materials -including agricultural production means; systems; transport service	http://mylcid.sirim.my/sirimlca/
JALCA (Japan Agricultural Life Cycle Assessment) database	Agriculture products	Hayashi et al. 2012
MiLCA (Japan)	Agriculture and fisheries; mining; construction and civil construction and other non-manufacturing; food and beverages; textiles; chemicals; ceramics and building materials; metals; machinery, and other manufacturing; electricity gas; water; sanitation	http://www.milca-milca.net/english/index.php
Indian LCI database (India)	Not known	Wernet et al. 2011

Table 1.13 North America LCI databases

Database	Sectors of application	Website/Reference
U.S. LCI Database	Energy and fuels; transportation; water; transformation processes; infrastructure; metals; paper and paper products; Glass; plastics; chemicals and minerals; wood and wood products; agricultural and bio-based products; packaging; building products and assembles; textiles; end of life	http://www.nrel.gov/lci/
LCA digital commons	Agricultural products	http://www.lcacommons.gov/

Table 1.14 Canada LCI databases

Database	Sectors of application	Website/Reference
Canadian LCI database	Wood, pulp & paper; mines and metal; energy; waste management; non-metallic mineral materials; agri-food	http://www.ciraig.org/en/bd-icv_ca.php/

Table 1.15 Latin America LCI databases

Database	Sectors of application	Website/Reference
Mexicaniuh database (Mexico but also data collaboration from Cuba, Colombia, and Argentina)	Petroleum and petrochemical; electricity; minerals/metals; wood and construction materials; chemical; textiles and leather; agriculture; water; transport	Suppen and Felix 2013
Chilean Food & Agriculture LCA (Chile)	Fresh and processed fruits; aquaculture; meat; dairy; wine	Emhart et al. 2013
Chilean National LCI (Chile) database	Not known	Emhart et al. 2013
Brazilian LCI (Brazil) database	Management base data; construction; metals; energy; fuel; agriculture; plastics (Chemistry); electronics (End of Life)	http://www.cbcs.org.br/_5dotSystem/userFiles/comite-tematico/materiais/PBACV-CT2-GT2%20-%20ApresentacaoPBACV%20-%2021Jun2013.pdf

1.5.1.4 Africa

By the beginning of 2014, no African LCI databases had been developed. South Africa is one of the few countries showing growing interest in LCA applications and benefits.

Table 1.16 Oceania LCI databases

Database	Sectors of application	Website/Reference
AusLCI (Australian National Life cycle Inventory) database	Agriculture; transport; electricity; materials; bio- based materials; fuels	http://alcas.asn.au/AusLCI/
AusAgLCI (Australian Agriculture LCI) database	Sugar; grains; cotton; horticulture; red meat	Grant et al. 2012
NZ LCI Database (New Zealand)	Not known	Kellenberger 2007

Table 1.17 Commercial LCA databases

Database	Website/Reference
Ga.Bi 5	http://www.gabi-software.com/italy/databases/gabi-databases/
GEMIS v. 4.81	http://www.iinas.org/gemis-download-en.html
eVerdee	http://www.ecosmes.net/cm/navContents?l=EN&navID=info&subNavID=1&pagID=6
AMEE	https://my.amee.com/

1.5.1.5 Oceania

Two Australian LCI national databases have been launched. Regarding New Zealand, a feasibility study for the development of a NZ life cycle inventory database or a combined Australian/NZ database has been proposed (Table 1.16).

1.5.2 Other Databases

For an overall view of the current database reality, it is important to report the major existing commercial databases within which it is possible to find inventory data for the agri-food sector (Table 1.17).

Apart from these databases, in 2010 the UNEP/SETAC Life Cycle initiative launched a "LCI Database Registry" for connecting LCA providers and users looking for LCA data.

In 2012, Quantis, the Agroscope Reckenholz-Tänikon Research Station ART and some leading companies in the food sector launched the World Food LCA Database (WFLDB) project. The aim of the project is the development of an international database, as exhaustive and clear as possible and in line with current LCA standards (e.g. ISO 14040 and 14044) and other existing databases.

The WFLDB will include datasets about: agricultural raw materials, inputs (conventional and organic); infrastructures (agricultural buildings, equipment, and machinery); processes; processed food products; food storage, food transportation, and food packaging. The project is expected to be concluded in 2015.

The database will not be based on original data but on existing LCAs on food products (partners' LCA, ART and Quantis databases), literature review on LCA of food products, statistical databases of governments and international organisations (such as FAO), environmental reports from companies, technical reports on food and agriculture, partners' information on food processes, and collected primary data. The database will also include: information about agricultural production, manufacturing and end of life (where applicable); differences between production systems, regional differences (relevant aspects to be taken into account for reliable LCA assessments) and deforestation impacts (where relevant); and carbon land transformation, land use, and water consumption inventory flows.

The WFLDB has been designed to guarantee transparency, reliability, compliance with Ecoinvent's quality guidelines, and compatibility and coherency with existing software and databases (SimaPro, Gabi, etc.). Updating will be done once a year and will comply with quality requirements of major standards (ILCD Handbook, Sustainability Consortium, United Nations Environment Programme (UNEP), etc.) as well as with Ecoinvent (Lansche et al. 2013).

1.6 Dietary Issues

In the context of food sustainability, the importance of consumer behaviour and in particular dietary behaviour is increasingly recognised, together with the product and its production chain. The dietary choices of the consumer and the consumption style considerably affect results in terms of environmental sustainability. Therefore, many studies have aimed at investigating the fundamental aspects that link diets and the relative impact on the environment, using the LCA tool for this assessment. The factors of increasing importance in terms of eating habits are the amount of food and its origin.

Several studies compare different foods or compare various types of diets (Carlsson-Kanyama 1998; Jungbluth et al. 2000; Carlsson-Kanyama et al. 2003; Heller and Keoleian 2003; Davis and Sonesson 2008; Davis et al. 2010; Muñoz et al. 2010; Meier and Christen 2012; Notarnicola et al. 2014). A review by Heller et al. (2013) compared 32 LCA studies on diets. The results of all these studies pinpointed some key areas:

- in general, foods of animal origin show a worse environmental performance than those of plant origin
- consequently vegetarian diets have a better environmental profile than other diets
- there is a strong regional difference in food habits and production processes
- agriculture is often the most burdening phase in the life cycle of a food
- even the cooking phase is relevant in terms of environmental impacts.

In a study on the energy consumption associated with 150 foods, Carlsson-Kanyama et al. (2003) have shown that it can vary between 2 and 220 MJ per kg, and depends on many factors that include the origin (animal or vegetable), selected process or preparation, and transportation distance. The authors recommend the reduction of energy consumption and related emissions of greenhouse gases through diets that consume less meat and cheese and more seasonal and locally produced products. In Europe, diets are characterised by a high intake of proteins and unsaturated fats, mostly of animal origin. The main source of protein is meat; European diets involve a 40 kg intake of protein per year of which 62 % are of animal origin (De Boer et al. 2006; Davis et al. 2010). Therefore one of the key requirements is to focus on animal products. They have a significant impact on the environment as livestock production is a major source of greenhouse gases and nitrogen emissions.

However, there is strong variability in the data related to emissions from the production of the same food in different countries, as demonstrated by Notarnicola et al. (2014). In a project on environmental sustainability of food, the carbon footprint of the diet followed by the exploration team was calculated during a journey lasting 44 days from Shanghai to Milan, a trip of 13,000 km by electric motorcycle; the team crossed nine countries with completely different cultures and food traditions. A database containing the carbon footprint of 411 food products was set up; this indicator was calculated for the same 188 products in each country visited. For many products the value of the carbon footprint calculated varied widely from country to country because of different production techniques and different observed yields. For example, the impact of cow's milk from intensive farming in Europe compared with that of highly extensive farming in Mongolia showed a carbon footprint ranging from 1 kg CO_2eq/kg for European milk to 10 kg CO_2eq/kg for Mongolian. What emerges is that in colder countries high-caloric diets are based on the consumption of meat, milk, and dairy products with less eco-sustainability. In China, where the team consumed almost exclusively foods of plant origin, the daily carbon footprint result was much more positive and offset the Chinese electricity mix principally based on coal. This study shows that, in the assessment of food sustainability the electricity mix of the country plays a crucial role.

Consequently, one of the main dietary issues involves the replacement of animal foods with plant foods, but at the same time meat, fish, and dairy products are unique and specific sources of essential nutrients, whose replacement presents a variety of nutritional challenges (Millward and Garnett 2010). Furthermore, LCA results may significantly change when the quality of food is assessed by means of an appropriate functional unit, as shown by various studies.

Kägi et al. (2012) compare the environmental impact of some meals and their ingredients using different functional units: one meal (about 450 g); adjusted by the nutrient density score (NDS); adjusted by the nutrient-rich food index (NRF9.3). A comparison of the different meals per plate or meals adjusted by the NRF9.3 method shows that the most relevant impact comes from meat. With the NDS method, meat still shows a high impact but is not as dominant because of its high nutrient density.

Therefore, as claimed by Smedman et al. (2010), diet comparison cannot be based on daily intake or energy, protein, or fat content, but must consider much wider aspects such as the nutritional quality of a diet, measured for example by the NDS index.

Even Heller et al. (2013) describe a large number of functional units for the comparison of different foods and diets and conclude there is a necessity for a more sophisticated and comprehensive nutrition-based functional unit to link nutritional health and environmental objectives.

Some authors have identified the best ways to reduce the environmental impact of food, i.e. changing consumer behaviour. Jungbluth et al. (2012) report some of them such as buying food locally or seasonally, eating vegetarian food, buying organic food, or changing the diet.

In conclusion, it is necessary to affect the consumer's dietary behaviour to reduce the environmental impact of foods. Furthermore, actions are needed to increase efficiency in the production chain and reduce food waste. Moreover, dietary changes will reduce the environmental impact in the agri-food sector at the lowest economic cost (Meier and Christen 2012). However, the question that remains is: how easy is it to change the eating habits of a consumer? In fact, we eat not only to satisfy our hunger and foods are chosen not only for taste; many different factors come into play when food choices are made. Some people choose a specific diet to lose weight and others choose one to increase it; some people need an energy-intensive diet because of the nature of their job or because they practise specific sports; others choose a diet on the basis of affordability and others are influenced by advertising. Food consumption is thus a social and cultural factor in which traditions and ways of being play an important role. The challenge is therefore to stimulate consumers to choose foods that integrate their traditions with environmental and health aspects. At the same time, industry must produce foods more efficiently and with less environmental impact along the supply chain (Soussana 2012).

A proper assessment of the impacts, however, requires further research in the field of regionalisation of databases and life cycle inventories of agro-food products, with particular attention to agricultural practices and processing systems (Notarnicola et al. 2012b).

Conclusions

This first chapter was written as an introduction to the later ones, focussed on specific agri-food sectors, and as a reference basis for the readers of this book. Its aim is to provide an as exhaustive as possible overview of the key concerns, applications, and methodological issues of the LCA methodology applied to the agri-food sector.

The first part of the chapter was devoted to the description of the major international initiatives, eco-labels and declarations, and footprints developed at product level and based on an LCA approach. The significant hotspots which emerged in this context are:

- The relevant role and commitment of, above all, the European Commission and governments toward issues of sustainable production and consumption. The recent development of the "EU Environmental Footprint" (EF) initiative and the European Food Sustainable Consumption and Production Roundtable are proof of this.

- The huge number of eco-labelling and footprint systems developed may generate confusion and mistrust among consumers and all the involved stakeholders. It is evident that a step towards harmonisation among all the existing standards and methods is more than desirable. In this context, for example, the new EU Environmental Footprint system has the aim of "…establishing a common methodological approach …for assessing, displaying, and benchmarking the environmental performance of products/services/companies based on a comprehensive assessment of environmental impacts over the life-cycle".

The section devoted to the role played by the different actors of the supply chain in the development and consolidation of the LCA methodology as an essential tool for the assessment of the environmental performance of food products highlighted that:

- Many initiatives (at national and international level) have been, and are still being, developed by agricultural and livestock operators, the industry sector, logistic and trading, and the end of life of packaging and/or food waste operators. Within the industry sector, in particular, the dairy sector is one of the most proactive sectors not only at international but also at national level: many national LCA studies have been carried out over the years. The commitment of the logistic and trading sector is also considerable, and many initiatives have been undertaken by the major retail chains. Finally, there is a growing interest regarding the environmental impact arising from food waste, given that about one- third of the food produced in the world for human consumption every year (1.3 billion tonnes) is lost or wasted. In this regard, it is important to be aware that such waste is relevant and impactful not only from an economic but also from an environmental perspective. The last consideration concerns the consumer side: greater commitment by NGOs and other consumer associations is desirable in the sense of making LCA's applications, advantages, and opportunities more accessible and understandable for the consumer.

Impacts arising from the agri-food system tend to be different from the ones typically modelled in LCA (non-food based products). This necessitates a different methodological approach in the agri-food sector, related (for example) to the selection of the most suitable FU, system boundaries, allocation method, fertilisers, and pesticide dispersion models. Section 1.4 about the methodological issues focussed on the general issues arising from the development of an LCA study of a food product. Specific aspects related to the chosen sectors are analysed in the following chapters.

Accurate and consistent LCA studies require representative life cycle inventories acting as fundamental building blocks. Section 1.5 on LCI databases highlighted that:

- Albeit many LCI databases have been developed, most of them are characterised by a lack of transparency and are often incomplete because they take into account only a few input-output flows; this lack of information concerning the impact of food products can lead to ambiguous interpretations and conclusions; furthermore, these databases are often not up-to-date and not regionalised. Finally, the databases are frequently inconsistent with each other, because of different approaches and assumptions. There is a need for clear, well-documented, and consistent data to increase the accuracy and comparability of LCA in the food sector.

Finally, the main hotspots identified in the dietary issues section (1.6) are:

- the need to change the consumer's dietary behaviour to reduce the environmental impact of foods;
- the need for action to increase efficiency in the production chain and reduce food waste;
- the need for dietary changes to reduce environmental impacts of the agri-food sector at the lowest economic cost.

References

Alexandratos, N., & Bruinsma, J. (2012). World agriculture toward 2030/2050—the 2012 Revision. ESA working paper No. 12–03, Agricultural Development Economics Division, Agriculture Organization of the United Nations. http://www.fao.org/fileadmin/templates/esa/Global_persepctives/world_ag_2030_50_2012_rev.pdf. Accessed 12 Nov 2013.

Allan, J. (1998). Virtual water: A strategic resource global solutions to regional deficits. *Groundwater, 36,* 545–546.

Berzina, S., & Shevchenko, I. (2011, August). *Methodology of ecological criteria development for products under ecolabelling program of type I.* Paper presented at LCM 2011—Towards life cycle sustainability management, Berlin.

Beverage Industry Environmental RoundTable (BIER). (2011). A practical perspective on the water accounting in the beverage sector. http://media.wix.com/ugd/49d7a0_ad1fcedf02b04ea38015a976991c1137.pdf. Accessed 10 Jan 2014.

Beverage Industry Environmental RoundTable (BIER). (2012a). research on the carbon footprint of beer. http://media.wix.com/ugd/49d7a0_70726e8dc94c456caf8a10771fc31625.pdf. Accessed 10 Jan 2014.

Beverage Industry Environmental RoundTable (BIER). (2012b). research on the carbon footprint of bottled water. http://media.wix.com/ugd/49d7a0_824b8dcfeaa74427a56b57abb8e2417e.pdf. Accessed Jan 10 2014.

Beverage Industry Environmental RoundTable (BIER). (2012c). research on the carbon footprint of soft drinks. http://media.wix.com/ugd/49d7a0_7a5cfa72d8e74c04be5aeb81f38b136b.pdf. Accessed Jan 10 2014.

Beverage Industry Environmental RoundTable (BIER). (2012d). Research on the carbon footprint of spirits. http://media.wix.com/ugd/49d7a0_7643fd3fae5d4daf939cd5373389e4e0.pdf. Accessed 10 Jan 2014.

Beverage Industry Environmental RoundTable (BIER). (2012e). research on the carbon footprint of wine. http://media.wix.com/ugd/49d7a0_4d74ddfdfbd64d3a8c1b27c17f460e36.pdf. Accessed 10 Jan 2014.

Beverage Industry Environmental RoundTable. (BIER). (2013). Beverage industry sector guidance for greenhouse gas emissions reporting. http://media.wix.com/ugd/49d7a0_6339d006853c4d3bbdf6087b43d91580.pdf. Accessed 10 Jan 2014.

Bio Intelligence Service. (2010). Preparatory study on food waste across EU 27. European Commission (DG ENV). Directorate C-Industry. Executive summary. http://ec.europa.eu/environment/eussd/pdf/bio_foodwaste_sum.pdf. Accessed 10 Jan 2014.

Birkved, M., & Hauschild, M. Z. (2006). PestLCI—a model for estimating field emissions of pesticides in agricultural LCA. *Ecological Modelling, 198,* 433–451.

Blanke, M., & Schaefer, F. (2012, October). *Application of PAS 2050–1 supplementary requirements for horticultural products: Carbon footprint of pumpkin and asparagus*. Paper presented at the 8th international conference on LCA in the agri-food sector, Saint Malò.

Brandi, C. (2012). Low carbon standards and labels in rising powers. Working paper no. 5. http://www.risingpowers.net/sites/default/files/RisingPowers_WP5_Brandi.pdf. Accessed 12 Nov 2013.

British Standard Institute (BSI). (2013). PAS 2050—Specification for the assessment of the life cycle greenhouse gas emissions of goods and services. http://shop.bsigroup.com/Browse-By-Subject/Environmental-Management-and-Sustainability. Accessed 12 Nov 2013.

Carles, G., et al. (2013, August). *LCADB.SUDOE: Life cycle inventories database of the southwest of Europe*. Paper presented at the 6th international conference on life cycle management—LCM 2013, Gothenburg.

Carlsson-Kanyama, A. (1998). Climate change and dietary choices—how can emissions of greenhouse gases from food consumption be reduced? *Food Policy, 23*(3/4), 277–293.

Carlsson-Kanyama, A., Pipping Ekström, M., & Shanahan, H. (2003). Food and life cycle energy in-puts: Consequences of diet and ways to increase efficiency. *Ecological Economics, 44,* 293–307.

China Environmental United Certification Centre (CEC) (n.d.). China Environmental Labeling Program. http://www.city.sendai.jp/kankyou/kanri/icgps-e/pdf/special_1-4.pdf. Accessed 12 Nov 2013.

Chomkhamsri, K., & Pelletier, N. (2011). Analysis of existing environmental footprint methodologies for product and organizations: Recommendations, rationale, and alignment. European Commission—Joint research Centre—Institute for Environmental and Sustainability (EC-JRC-IES). http://ec.europa.eu/environment/eussd/pdf/Deliverable.pdf. Accessed 12 Nov 2013.

COM. (2003). 302 def. communication from the Commission to the Council and the European Parliament "Integrated Product Policy—Building on Environmental Life-Cycle Thinking".

COM. (2008). 397 def. communication from the Commission to the European Parliament, the Council, the European Economic and Social Commitee and the Committee of the Regions on the "Sustainable Consumption and Production and Sustainable Industrial Policy Action Plan".

Consumer Goods Forum (CGF). (2012). Initiatives in sustainability product measurement. http://www.theconsumergoodsforum.com/PDF/WorkingGroups-Accordion/Sustainability/11.CGF%20Sustainability%20Measurement%20-%20overview%20initiatives%20-%20June%202012.pdf. Accessed 12 Nov 2013.

Čuček, L., Klemes, J. J., & Kravanja, Z. (2012). A review of footprints analysis tools for monitoring impacts on sustainability. *Journal of Cleaner Production, 34,* 9–20.

Davis, J., & Sonesson, U. (2008). Life cycle assessment of integrated food chains—a Swedish case study of two chicken meals. *The International Journal of Life Cycle Assessment 13,* 574–584.

Davis, J., Sonesson, U., Baumgartner, D. U., & Nemecek, T. (2010). Environmental impact of four meals with different protein sources: Case studies in Spain and Sweden. *Food Research International, 43*(7), 1874–1884.

Dawoon, J., Beon-Young, P., Soohyun, C., & Touseef, A. (2013, August). CO_2 emission for the production of beef in South Korea. Paper presented at the 59th ICoMST international congress of meat science and technology, Izmir.

De Boer, J., Helms, M., & Aiking, H. (2006). Protein consumption and sustainability: Diet diversity in EU-15. *Ecological Economics, 59,* 267–274.

De Camillis, et al. (2012, October). *ENVIFOOD Protocol: Launch of the collectively-agreed sectorial methodology for assessing the environmental performance of food and drink products in Europe*. Paper presented at the 8th international conference on LCA in the agri-food sector, Saint Malò.

Del Borghi, A. (2013). LCA and communication: Environmental Product Declaration. *The International Journal of Life Cycle Assessment, 18,* 293–295.

Delahaye, M. G. (2008). Product and carbon footprint: The example of a major retail group. http://archives.dgcis.gouv.fr/2012/www.pme.gouv.fr/europe/pres2008/an/TR3/6_G_Delahaye_Anglais.pdf. Accessed 10 Jan 2014.

Dijkman, T.J., Birkved, M., & Hauschild, M. Z. (2012). PestLCI 2.0: A second generation model for estimating emissions of pesticides from arable land in LCA. *The International Journal of Life Cycle Assessment, 17,* 973–986.

Dordan (n.d.). How to assess sustainable packaging: An overview of the tools and resources available. http://leadwise.mediadroit.com/files/18826HowtoAssessSusPkgfordistribution.pdf. Accessed 12 Nov 2013.

Dron, D. (2012, October). *Sustainable food, a component of the green economy*. Paper presented at the 8th international conference on LCA in the agri-food sector, Saint Malò.

Earthsure®. (2012). Environmental product declarations general program instructions. http://iere.org/wp-content/uploads/Earthsure-General-Program-2012-Final.pdf. Accessed 12 Nov 2013

Ecological Union (2013). Life cycle ecolabelling program "Vitality leaf" for product, work and services. http://www.ecounion.ru/en/site.php?&blockType=251. Accessed 12 Nov 2013.

Emhart, C., Florenzano, A., Loyola, C., Bengtsson, J. (2013, August). Chilean food & agriculture lca database using Unep/Setac Shonan principles. Paper presented at the 6th international conference on life cycle management—LCM 2013, Gothenburg.

Environdec. (2013). http://www.environdec.com/en/PCR/. Accessed 12 Nov 2013.

European Food Sustainable Production and Consumption Round Table (EU Food SCP RT). (2013). Voluntary pilot testing of the ENVIFOOD protocol and communication tools. http://www.food-scp.eu/node/72. Accessed 12 Nov 2013.

FAO. (n.d.). http://www.fao.org/partnerships/leap/livestock-partnership/en/. Accessed 12 Nov 2013.

Fefac. (2013, July). Feed facts—International news. http://www.fefac.eu/files/47325.pdf.

Galli, A., Wiedmann, T., Ercin, E., Knoblauch, D., Ewing, B., & Giljum, S. (2012). Integrating ecological, carbon and water footprint into a "Footprint Family" of indicators: Definition and role in tracking human pressure on the planet. *Ecological indicators, 16,* 100–112.

Gerber, P. J., Steinfeld, H., Henderson, B., Mottet, A., Opio, C., Dijkman, J., Falcucci, A., & Tempio, G. (2013). Tackling climate change through livestock—A global assessment of emissions and mitigation opportunities. Rome: Food and Agriculture Organization of the United Nations (FAO).

Geyer, R., Stoms, D. M., Lindner, J. P., Davis, F. W., & Wittstock, B. (2010). Coupling GIS and LCA for biodiversity assessments of land use: Impact assessment. *The International Journal of Life Cycle Assessment, 14,* 692–703.

Gheewala, S. H. (2012, October). *Challenges for LCA in the agri-food sector, perspective from Thailand and Southeast Asia*. Paper presented at the 8th international conference on LCA in the agri-food sector, Saint Malò.

Gheewala, S. H., & Mungkung, R. (2013). Product carbon footprinting and labeling in Thailand: Experiences from an exporting nation. *Carbon Management, 4,* 547–554.

Global Footprint Network. (2012). Application standards. http://www.footprintnetwork.org/en/index.php/GFN/page/application_standards/. Accessed 12 Nov 2013.

Grant, T., Eady, S., & Winter, S. (2012, March). *Development of a Australian national inventory for agriculture*. Paper presented at the 2nd New Zealand life cycle assessment conference—LCA: A business compass for sustainable development, Auckland.

Hayashi, K., Shobatake, K., Makino, N., & Hukazono, S. (2012). *Developing of a life cycle inventory database for agricultural production systems in Japan: The JALCA database*. Paper presented at the 10th International Conference of Ecobalance, Yokohama.

Heller, M., & Keoleian, G. (2003). Assessing the sustainability of the US food system: A life cycle perspective. *Agricultural Systems, 76*(3), 1007–1041.

Heller, M. C., Keoleian, G. A., & Willett, W. C. (2013). Toward a life cycle-based, diet-level frame-work for food environmental impact and nutritional quality assessment: A critical review. *Environmental Science and Technology, 47*(22), 12632–12647.

Hoekstra, A. Y. (2003, December). Virtual water trade. Paper presented at the international expert meeting on virtual water trade, the Netherlands.

Hoekstra, A. Y., Chapagain, A. K., Aldaya, M. M., & Mekonnen, M. M. (2009). *Water footprint manual—State of the Art 2009*. Enschede: Water Footprint Network.
International Institute for Sustainable Development (IISD). (1996). Ecolabelling: Its implication for China. http://www.iisd.org/pdf/ecochina.pdf. Accessed 12 Nov 2013.
International Standard Organisation (ISO). (2001). *ISO 14024 Environmental labels and declarations type I environmental labelling. Principles and procedures*. Geneva: ISO.
International Standard Organisation (ISO). (2002a). *ISO 14020 Environmental labels and declarations—General principles*. Geneva: ISO.
International Standard Organisation (ISO). (2002b). *ISO 14021 Environmental labels and declarations—Self declared environmental claims*. Geneva: ISO.
International Standard Organisation (ISO). (2006a). *ISO 14025 Environmental labels and declarations—Type III environmental declarations*. Geneva: ISO.
Internationa Standard Organisation (ISO). (2006b). *ISO 14040. Environmental management - Life cycle assessment - Principles and framework*. Geneva: ISO.
International Standard Organisation (ISO). (2006c). *ISO 14044. Life cycle assessment—Requirements and guidelines*. Geneva: ISO.
International Standard Organisation (ISO). (2013). *ISO 14067 Greenhouses gases—carbon footprint of products- requirements and guidelines for quantification and communication*. Geneva: ISO.
International Trade Center (ICT). (2012). *Product carbon footprinting standards in the agri-food sector*. Technical paper. http://www.intracen.org/WorkArea/DownloadAsset.aspx?id=58175. Accessed 12 Nov 2013.
International Trade Centre (ICT) (2013). China environmental labelling—CEL. http://search.standardsmap.org/assets/media/ChinaEnvironmentalLabeling/English/AtAGlance_EN.pdf. Accessed 12 Nov 2013.
JEMAI CFP Program (2013). http://www.cfp-japan.jp/english/news/. Accessed 12 Nov 2013.
Jungbluth, N., Tietje, O., & Scholz, R. W. (2000). Food purchases: Impacts from the consumers' point of view investigated with a modular LCA. *The International Journal of Life Cycle Assessment, 5*(3), 134–142.
Jungbluth, N., Itten, R., & Schori, S. (2012). *Environmental impacts of food consumption and its reduction potentials*. Paper presented at the 8th international conference on LCA in the agri-food sector, St-Malò, France.
Kägi, T., Zschokke, M., & Dinkel, F. (2012). *Nutrient based functional unit for meals*. Paper presented at the 8th international conference on LCA in the agri-food sector, St-Malò, France.
Kanzaki, M. (2009). EcoLeaf label-LIfe-cycle analysis for ICT goods and services. OECD Conference on ICTs, the Environment and Climate Change. http://www.cdn.netamia.com/itst/ict2009/the_ecoleaf_label_life-cycle_analysis_for_ict_goods_and_services-190-1243422458.pdf. Accessed 12 Nov 2013.
Kellenberger, D. (2007, November). *Modelling principles for the collection of consistent and comprehensive LCI data*. Paper presented at the New Zealand sustainable building conference (SB07), Auckland, .
Kim, S., & Dale, B. (2006). Ethanol fuels: E10 or E85—life cycle perspectives. *The International Journal of Life Cycle Assessment, 11*(2), 117–121.
Koch, P., et al. (2011, August). *Methodological guidelines for LCA of French agricultural products*. Paper presented at LCM 2011—Towards life cycle sustainability management, Berlin.
Köhler, A., & Aoustin, E. (2008, May). Project group on assessment of use and depletion of water resources within LCA. UNEP/SETAC life cycle initiative SETAC Europe annual meeting, Warsaw Ecological Systems Design Group. http://www.wulca-waterlca.org/pdf/wulca.pdf. Accessed 12 Nov 2013.
Kounina, A., Margni, M., Bayart, J. B., Boulay, A. M., Berger, M., Bulle, C., Frischknecht, R., Koehler, A., MilÃ I Canals, L., Motoshita, M., Núñez, M., Peters, G., Pfister, S., Ridoutt, B., Van Zelm, R., Verones, F., & Humbert, S. (2013). Review of methods addressing freshwater use in life cycle inventory and impact assessment. *International Journal of Life Cycle Assessment, 18*(3), 707–721.

Lansche, J., et al. (2013, August). *The world food LCA database project: Towards more accurate food datasets*. Paper presented at the 6th international conference on life cycle management—LCM 2013, Gothenburg.

Ledgard, S. F., Lieffering, M., McDevitt, J., Boyes, M., & Kemp, R. (2010). A greenhouse gas footprint study for exported New Zealand lamb. http://www.fedfarm.org.nz/files/2010-Lamb-CarbonFootprint.pdf. Accessed 12 Nov 2013.

Leip, A., Weiss, F., Wassenaar, T., Perez, I., Loudjani, T., Tubiello, F., et al. (2010). Evaluation of the livestock sector's contribution to the EU greenhouse gas emission—final report. European Commission, Joint Research Centre.

Lessard, L., Vionnet, S., Dettling, J., Humbert, S., Henderson, A., Jolliet, O., & Wang, Y. (2012, October). *Water footprint of cradle to farm gate milk production in the U.S.* Paper presented at the 8th international conference on life cycle assessment in the agri-food sector, Saint Malo.

Lin, L. C., Pei, D. C., Chen, C. C., Yen, W. S., Lui, C. C., & Lee, W. S. (2013, August). *The industry views on the policy and development of carbon labeling in Taiwan*. Paper presented at the 6th international conference on life cycle management—LCM 2013, Gothenburg.

Lo Giudice, A., & Clasadonte, M. T. (2010). The EPD for the agro-food chain products. Calitatea-acces la success: *Facing the challenges of the future—excellence in business and commodity sciences, special issue, 11*, 472–480.

MacLeod, M., Gerber, P., Mottet, A., Tempio, G., Falcucci, A., Opio, C., Vellinga, T., Henderson, B., & Steinfeld, H. (2013). Greenhouse gas emissions from pig and chicken supply chains—A global life cycle assessment. Food and Agriculture Organization of the United Nations (FAO), Rome.

Magerholm Fet, A. (2012). Harmonization of documentation for EPD-generation—the Norwegian model. http://ww2.unime.it/emaf/images/eventi/Messina/ppt_convegno/6_Fet.pdf. Accessed 12 Nov 2013.

Mars, C., et al. (2011, May). *The Sustainability Consortium update: Type III product declaration development for laptops*. Paper presented at the 2011 IEEE International Symposium on Sustainable Systems and Technology (ISSST), Chicago.

Masoni, P., Fantin, V., & Zamagni, A. (2012). Matodi e certificazioni per misurare la sostenibilità. *Ecoscienza, 5*, 15–17.

Mehta, P. S. (2007). Why was India's Ecomark scheme unsuccessful?.CUTS Centre for International Trade, Economics & Environment. Jaipur Printers P. Ltd. http://www.cuts-citee.org/pdf/RREPORT07-01.pdf. Accessed 12 Nov 2013.

Meier, T., & Christen, O. (2012). Gender as a factor in an environmental assessment of the consumption of animal and plant based foods in Germany. *The International Journal of Life Cycle Assessment, 17*(5), 550–564.

Meissner Schau, E., & Magerholm Fet, A. (2008). LCA studies of Food Products as background for environmental products declarations. *The International Journal of Life Cycle Assessment, 13*, 255–264.

Millward, D. J., & Garnett. T. (2010). Plenary lecture 3: Food and the planet: Nutritional dilemmas of green-house gas emission reductions through reduced intakes of meat and dairy foods. *The Proceedings of Nutrition Society, 69*(1), 103–118.

Ministry for Primary Industry (MPI). (2013). Greenhouse gas (GHG) footprinting strategy. http://www.mpi.govt.nz/environment-natural-resources/climate-change/greenhouse-gas-footprinting-strategy.aspx. Accessed 12 Nov 2013.

Mungkung, R., & Gheewala, S. H. (2012, October). *Carbon footprinting and labeling of agri-food products: Practical issues for the development of Product Category Rules*. Paper presented at the 8th international conference on LCA in the agri-food sector, Saint Malò.)

Muñoz, I., MilÃ i Canals, L., & Fernández-Alba, A. R. (2010). Life cycle assessment of the average Spanish diet including human excretion. *The International Journal of Life Cycle Assessment, 15*, 794–805.

North-East Asian Subregional Programme for Environmental Cooperation (NEASPEC). (2012). http://www.neaspec.org/documents/eco_may_2012/Day1-Scheme_Products%20Category-Russia.pdf. Accessed Nov 12 2013.

Notarnicola, B., Hayashi, K., Curran, M. A., & Huisingh, D. (2012a). Progress working towards a more sustainable agri-food industry. *Journal of Cleaner Production, 28,* 1–8.
Notarnicola, B., Tassielli, G., & Renzulli, P. A. (2012b). Modeling the agri-food industry with life cycle assessment. In M. A. Curran (Ed.) *Life Cycle assessment handbook* (pp. 159–184). New York: Wiley.
Notarnicola, B., Tassielli, G., Renzulli, P. A., Lo Giudice, A., Colombo, N., Costantino, E. (2014). *Environmental sustainability of foods: The Meneghina Express project*. Paper presented at the XXVI National Congress of Commodity science, Pisa.
Oakdene Hollins Research and Consulting (2011). EU ecolabel for food and feed products—feasibility study (ENV. C.1/ETU/2010/0025). http://ec.europa.eu/environment/ecolabel/documents/Ecolabel_for_food_final_report.pdf. Accessed Nov 12, 2013.
Opio, C., Gerber, P., Mottet, A., Falcucci, A., Tempio, G., MacLeod, M., Vellinga, T., Henderson, B., & Steinfeld, H. (2013). Greenhouse gas emissions from ruminant supply chains—A global life cycle assessment. Food and agriculture organization of the United Nations (FAO), Rome.
PCR Library. (2013). Search for countries: Japan. http://pcr-library.edf.org.tw/product_country/japan.asp. Accessed Nov 12 2013.
PCF World Forum. (2010). International developments in product carbon footprinting and carbon labelling. PCF World Forum News 2/ March. http://www.pef-world-forum.org/wp-content/uploads/2010/03/pcf-world-forum-news2_march-2010.pdf. Accessed Nov 12 2013.
Redd, T. (2011). Eco-Babble: A science based, multi-stakeholder approach. The sustainability consortium. http://www.gmaonline.org/uploadFiles/184D0B0000002F.filename.REDD_EcoBabble.pdf. Accessed 12 Nov 2013.
Reisch, L., Eberle, U., & Lorek, S. (2013). Sustainable food consumption: An overview of contemporary issues and policies. *Sustainability: Science, Practice, & Policy, 9,* 7–25.
Sala, S., et al. (2013). Water footprint in the context of sustainability assessment. Report on the application of life cycle based indicators of water consumption in the context of integrated sustainability impact analysis. European Commission, Joint Research Centre, Institute for environment and sustainability. EUR 25781EN.
Seda, M., Antòn, A., & Munoz, P. (2010, September). Analysing the influence of the functional unit in agricultural LCA. Paper presented at the LCA Food 2010 Conference. Bari, Italy.
Seo, M. (2009). Korea carbon footprint labeling and its activities, KEITI. http://www.ftis.org.tw/active/download/1_6.pdf. Accessed 12 Nov 2013.
Smedman, A., Lindmark-Månsson, H., Drewnowski, A., & Edman, A. K. M. (2010). Nutrient density of beverages in relation to climate impact. Food & Nutrition Research 54.
Soussana, J. F. (2012, October). Research priorities for sustainable agri-food systems and LCA. Paper presented at the 8th international conference on LCA in the agri-food sector, St-Malò, France.
Suppen, N., & Felix, R. (2013, August). *The Mexican life cycle inventory database—Mexicaniuh.* Paper presented at the 6th international conference on life cycle management—LCM 2013, Gothenburg.
Tesco (2012, April). http://www.tescoplc.com/files/pdf/factsheets/footprint_gallery.pdf.
The Grocer. (2012). http://www.thegrocer.co.uk/companies/supermarkets/tesco/frustrated-tesco-ditches-eco-labels/225502.article?utm_medium=email&utm_source=groceronline%40wrbm.com&utm_campaign=1145974_Daily.
Tukker, A., Huppes, G., Guinée, J., Heijungs, R., de Koning, A., van Oers, L., et al. (2006). Environmental impacts of products (EIPRO). Analysis of the life cycle environmental impacts related to the total final consumption of the EU-25, Sevilla, Spain: European Commission, Joint Research Centre (JRC), Institute for Prospective Technological Studies (IPTS).
Udo de Haes, H. A., & de Snoo, G. R. (2012). Eco-labelling of agricultural food products. In U. Sonesson, J. Beril, & F. Ziegler (Eds.), *Environmental assessment and management in the food industry. Life cycle assessment and related approaches* (pp. 374–397). Cambridge: Woodhead Publishing.

Vergez, D. (2012). Display of the environmental footprint of products: French developments in the food sector. Department of the commissioner general fro sustainable development (CGDD). http://www.developpement-durable.gouv.fr/IMG/pdf/ED64EN.pdf. Accessed 12 Nov 2013.

Vermeulen, S. J., Campbell, B. M., & Ingram, J. S. I. (2012). Climate change and food systems. *Annual Review of Environment and Resources, 37,* 195–222.

Wackernagel, M., & Rees, W. E. (1996). *Our ecological footprint: Reducing human impact on the Earth.* Gabriola Island: New Society.

Wernet, G., Stucki, M., Shenoy, M., & Muthusezhiyan, N. (2011, August). *Establishing a data framework for life cycle management in India.* Paper presented at LCM 2011—Towards life cycle sustainability management, Berlin.

Zamagni A., Buttol P., Porta, P. L., Buonamici, R., Masoni, P., Guinée, J., Heijungs, R., Ekvall, T., Bersani, R., Bienkowska, A., Pretato, U. (2008). Critical review of the current research needs and limitations related to ISO-LCA practice. Deliverable D7 of work package 5 of the CALCAS project. ENEA. Rome.

Chapter 2
Life Cycle Assessment in the Olive Oil Sector

Roberta Salomone, Giulio Mario Cappelletti, Ornella Malandrino, Marina Mistretta, Elena Neri, Giuseppe Martino Nicoletti, Bruno Notarnicola, Claudio Pattara, Carlo Russo and Giuseppe Saija

Abstract The olive oil industry is a significant productive sector in the European Union and the related production process is characterised by a variety of different practices and techniques for the agricultural production of olives and for their processing into olive oil. Depending on these different procedures, olive oil production is associated with several adverse effects on the environment, both in the agricultural and in the olive oil production phase. As a consequence, tools such as LCA are becoming increasingly important for this type of industry. Following an overview of the characteristics of the olive oil supply chain and its main environmental problems, the authors of this chapter provide a description of the international state of the art of LCA implementation in this specific sector, as well as briefly describing other life cycle thinking methodologies and tools (such as simplified LCA, footprint labels and Environmental Product Declarations). Then, the methodological problems connected with the application of LCA in the olive oil production sector are analysed in depth, starting from a critical comparative analysis of the applicative LCA case studies in the olive oil production supply chain. Finally, guidelines for the application of LCA in the olive oil production sector are proposed.

Keywords Olive oil · Life cycle assessment · Life cycle costing · Environmental product declaration · Footprint labels

---

R. Salomone (✉) · G. Saija
University of Messina, Piazza S. Pugliatti 1, 98122 Messina, Italy
e-mail: roberta.salomone@unime.it

G. Saija
e-mail: giuseppe.saija@unime.it

G. M. Cappelletti · G. M. Nicoletti · C. Russo
University of Foggia, Largo Papa Giovanni Paolo II 1, 71121 Foggia, Italy
e-mail: giulio.cappelletti@unifg.it

G. M. Nicoletti
e-mail: giuseppe.nicoletti@unifg.it

C. Russo
e-mail: arloc81@gmail.com

B. Notarnicola et al. (eds.), *Life Cycle Assessment in the Agri-food Sector*,
DOI 10.1007/978-3-319-11940-3_2

2.1 Introduction

Olive oil production is an important agri-industrial sector (in terms of both production and consumption) in many Mediterranean regions (IOC 2013; Vossen 2007). Furthermore, the olive groves and olive production are increasing yearly (FAOSTAT 2013) and, recently, the importance of olive oil has also been growing in new producing countries located in America, Africa and Australia (IOC 2013). On a global scale, most olive cultivation areas[1] can be found in Mediterranean countries, such as Spain (2,503,675 ha), Italy (1,144,422 ha), Tunisia (1,779,947 ha), Greece (850,000 ha), etc. (FAOSTAT 2013). The leader of the international market is the EU, which produces over 70% of the world's olive oil. As concerns importing countries, the most important are the USA, Japan, etc. With regard to exports, the most relevant are the main EU countries, exporting over 440,000 t of olive oil, followed by Tunisia, Syria and others (Table 2.1).

Despite the economic importance of this food product in many countries, olive oil production is associated with several adverse effects on the environment that cause resource depletion, land degradation, air emissions and waste generation. The impacts may vary significantly as a result of the practices and techniques employed in olive cultivation and olive oil production (Salomone and Ioppolo 2012) and life cycle thinking approaches and assessment methods have increasingly been applied in order to gain a better understanding of their role from a life cycle perspective.

In the following sections, these different practices and techniques, along with the relative environmental consequences, are briefly described (Sect. 2.2). Then, a description of the international state of the art of life cycle thinking methodologies and tools, suitable for the environmental assessment of products and implemented in this specific sector, is presented (Sect. 2.3), with a specific focus on life cycle

[1] Data for the year 2011.

O. Malandrino
University of Salerno, Via Giovanni Paolo II, 132, Fisciano, 84084 SA, Italy
e-mail: ornellam@unisa.it

M. Mistretta
University "Mediterranea", Contrada Melissari—Feo di Vito, 89124 Reggio Calabria, Italy
e-mail: marina.mistretta@unirc.it

E. Neri
University of Siena, Via A. Moro 2, 53100 Siena, Italy
e-mail: elena.neri@unisi.it

B. Notarnicola
University of Bari, Via Lago Maggiore angolo via Ancona, 74100 Taranto, Italy
e-mail: bruno.notarnicola@uniba.it

C. Pattara
University G. d'Annunzio, Viale Pindaro 42, 65127 Pescara, Italy
e-mail: claudiopattara1@libero.it

Table 2.1 The olive oil market on the international scale (the average values of the 2007/2008–2012/2013 olive crop six-year period). (Source: Data (IOC 2013))

Country	Production (1000 t)	Imports (1000 t)	Exports (1000 t)	Consumption (1000 t)
EU[a]	2057.6	111.9	447.1	1819.3
Spain	1215.1	24.0	184.3	543.4
Italy	455.8	79.1	204.2	658.5
Greece	317.6	0.0	12.1	224.8
Portugal	58.4	1.4	41.0	80.9
France	5.3	5.4	1.6	108.7
Cyprus	4.9	0.0	0.0	5.5
Slovenia	0.5	0.1	0.1	1.9
Other EU countries	–	1.9	3.8	195.6
Tunisia	167.0	0.0	130.3	34.3
Syria	159.3	0.8	21.0	118.7
Turkey	149.2	0.0	22.9	124.0
Morocco	110.0	4.0	13.1	96.0
Algeria	47.4	0.2	0.0	47.0
Argentina	22.7	0.0	16.5	5.8
Jordan	20.8	3.6	1.5	20.7
Chile	15.4	0.8	6.2	10.7
Palestine	14.9	0.1	2.4	13.0
Lebanon	14.8	2.0	2.8	15.8
Libya	14.7	0.0	0.0	14.7
Australia	14.6	30.4	5.7	39.3
Israel	9.2	8.3	0.3	16.5
Albania	7.3	1.1	1.2	7.2
Egypt	5.8	1.9	1.4	6.5
Croatia	4.8	1.9	0.1	6.3
Iran	4.8	3.7	0.0	8.3
USA	4.3	270.2	3.7	271.3
Saudi Arabia	3.0	9.4	0.5	11.3
Montenegro	0.5	0.0	0.0	0.5
Other producing countries	15.0	3.0	5.5	13.1
Non-producing countries	–	250.8	–	250.8

[a] The import and export data of the EU countries are reported without intra-Community trade

assessment (LCA). The methodological problems connected with the application of LCA in the olive oil production sector are analysed in depth, starting from a critical comparative analysis of the applicative LCA case studies in the olive oil production supply chain (Sect. 2.4). Finally, guidelines for the implementation of LCA in this sector are proposed (Sect. 2.5), in order to deal with and manage best the methodological problems presented above.

2.2 The Olive Oil Supply Chain: Production Processes, Technologies, Product Characteristics and Main Environmental Problems

A supply chain is a network of organisations that are involved, through upstream and downstream linkages, in the different processes and activities that produce value in the form of products and services delivered to the ultimate consumer (Christopher 1992). According to this definition, the olive oil supply chain can be briefly described as follows (IOC 2013; Niaounakis and Halvadakis 2006; PROSODOL 2011), using the different life cycle phases of the olive oil product: cultivation, olive oil production, by-product management, product transportation and distribution, consumption and waste management.

The ***cultivation phase*** includes the cultivation of olives using different treatments, such as soil management, pruning, fertilisation, irrigation, pest treatment and harvesting. Each of these treatments[2] can be carried out in different ways depending on whether:

- the cultivation derives from centuries-old trees—traditional systems—or new intensive plants (in the latter option, the supply chain study must include plant breeding and tree planting);
- the irrigation system uses the dry farming or the drip irrigation method;
- the cultivation practices are conventional, organic or integrated, using different typologies of fertilisers and pest treatments;
- the soil management, pruning and harvesting are manual or mechanised.

Harvesting is a very important process, because changes in the acidity level of olives occur after harvesting and other changes occur depending on the harvest methods: hand harvesting is the best method, but very expensive, while mechanical harvesting, if properly conducted (avoiding the breaking of the fruit skin), can give good results. After harvesting, the olives are sent to olive oil mills and processed within 24 h, in order to avoid fermentation phenomena.

Because the cultivation of olives can be carried out by means of various treatments, the environmental impacts can be very different in the various olive farming areas. However, by simplifying, three types of plantation can be considered: low-input traditional plantations (randomly planted and/or terrace-planted ancient trees managed with few or no chemical inputs and high manual work input); intensified traditional plantations (they have the same characteristics as the first type together with an increase in the tree density and the weed control, soil management using artificial fertilisers and irrigation, the use of pesticides and mechanical harvesting); and intensive modern plantations (high small-tree density managed with extensive use of mechanised systems and irrigation). The low-input traditional plantations have the lowest environmental impact and, moreover, they play a role

[2] Some of these treatments and practices are managed similarly to other fruit cultivation (see Chap. 6).

in safeguarding the biodiversity and landscape value. Instead, the other two types of plantations can give rise to various environmental problems (i.e. soil erosion, run-off to water courses, degradation of habitats and landscapes and exploitation of scarce water resources) (Beaufoy 2000).

Much of the international olive production is transformed into olive oil. Different methods are used to extract oil from the olives and these processes create large volumes of liquid and solid waste. The waste stream is highly hazardous to the environment and presents a number of treatment challenges to olive oil producers.

The ***olive oil production phase*** includes two main phases: the preparation of a homogeneous paste and the oil extraction from the olives. First, the olives are classified and separated by quality; then, they are washed in order to remove the pesticides, dirt and impurities collected during harvesting (stems, leaves, twigs, etc.). A few olive oil mills do not wash olives, which are processed 'as they are' to overcome the problems connected with water consumption and the treatment of the polluted washing water. This is often motivated by the fact that extra moisture can involve problems (extractability and lower polyphenol content). However, these advantages should be cautiously compared with the disadvantages, since the pollution load of washing water demonstrates that olives need to be cleaned, otherwise pesticide and impurities remain on the olives and in the olive oil. After washing, crushing (tearing of the flesh cells to facilitate the release of the oil from the vacuoles) and malaxing (mixing the paste, allowing small oil droplets to combine into bigger ones) are essential steps. The next step consists of separating the oil from the rest of the olive components: oil is extracted using a press or a decanter, by pressing (the traditional or classical system) or by centrifugal separator (a continuous system), which can further use a three-phase or a two-phase decanter.

Traditional pressing (a discontinuous process) is still in use in some small mills that use a hydraulic press, but it is a relatively obsolete technology that has mainly been replaced with centrifugation systems, allowing lower manufacturing costs, better oil quality and shorter storage time of olives before processing. This process generates a solid fraction (olive husk or olive pomace) and an emulsion containing the olive oil, which is separated by decantation from the remaining wastewater.

Continuous centrifugation with a three-phase system, even though offering a higher production capacity with respect to traditional pressing, has some disadvantages, such as greater water and energy consumption (due to the addition of warm water to dilute the olive paste). This process uses a three-phase decanter that generates solid waste (olive husk or olive pomace), olive oil and wastewater.

Continuous centrifugation with a two-phase system allows the separation of oil from olive paste without the addition of water and this leads to the elimination of the problem of vegetable water. In fact, the two-phase system generates only olive oil and a semi-solid waste called olive wet husk or wet pomace (or two-phase olive mill waste).

Continuous centrifugation with a two-and-a-half-phase system (also called a modified system or water-saving system) exists; between a three-phase system and a two-phase one, it brings together the advantages of the two different systems (it requires the addition of a small amount of water and generates a solid fraction (olive wet husk or olive wet pomace) that includes part of the vegetation water and a smaller quantity of olive mill wastewater.

Another innovative technology is *oil extraction from de-stoned olives*. In the de-stoning process, the pits are removed before the kneading; some authors state that this process improves the quality of the extra virgin olive oil (better sensory qualities and shelf life) (Del Caro et al. 2006; Pattara et al. 2010). However, other authors (Di Giovacchino 2010) believe that this technology produces lower yields with a similar chemical sensory quality. Oil extraction from de-stoned olives can be made with both the three-phase and the two-phase system.

On average, the above-described techniques can produce around 200 kg of olive oil from 1 t of processed olives (Arvanitoyanni and Kassaveti 2008).

Therefore, as the average annual world production of olive oil in the 2007/2008–2012/2013 olive crop six-year period was equal to 2,862,800 t (IOC 2013), on the basis of the data available in the literature, it is possible to estimate that, on average in a year, the olive oil industry needs 572,560,000–1,674,738,000 kWh of energy and 1,431,400–16,045,994 m^3 of water, generating 5,725,600–8,588,400 t of solid waste and 8,588,400–17,176,800 t of wastewater (estimation from Arvanitoyanni and Kassaveti 2008).

The designation of *virgin olive oil* is solely recognised as the olive oil obtained from the fruit of the olive tree by mechanical or other physical means under conditions, particularly thermal conditions, that do not lead to alterations in the oil, which has not undergone any treatment other than washing, decantation, centrifugation and filtration, excluding oil obtained using solvents or re-esterification processes and any mixture with oils of other kinds (EC 1991, 2007, 2008, 2013a). In particular, in accordance with the standards of the International Olive Council (IOC n. d.) and the EC regulations, *virgin olive oils* are classified into:

- *extra virgin olive oil*, which is a higher quality olive oil with no more than 0.8 g per 100 g of free acidity (expressed as oleic acid) and a superior taste (fruitiness and no sensory defect). It must be produced entirely by mechanical means without the use of any solvents, and under temperatures that will not degrade the oil (lower than 30 °C);
- *virgin olive oil*, which has no more than 2 g per 100 g of free acidity and a good taste;
- *lampante olive oil*, which is virgin olive oil with free acidity, in terms of oleic acid, of more than 2 g per 100 g, and/or the other characteristics of which comply with those laid down for this category use.

Other classifications are related to the definition of *olive oil*, distinguishing:

- *refined olive oil*, obtained by the refining of virgin olive oil using methods that do not lead to alterations in the initial glyceridic structure; it has no more than 0.3 g per 100 g of free acidity;
- *olive oil*, which is a blend of refined oil and virgin oil (excluding the lampante virgin oil), fit for consumption as it is and having no more than 1 g per 100 g of free acidity;

- *olive pomace oil*, obtained by treating olive pomace with solvents or other physical treatments. This oil can be sold as *crude olive pomace oil*, which is intended for refining (then designated for human consumption) or for technical use, and *refined olive pomace oil*, which is obtained from crude olive pomace oil by refining methods, producing an oil with no more than 0.3 g per 100 g of free acidity.

In the olive oil production phase, the ***packaging process*** is also included, even though the olive oil is often sold unbottled (to final consumers or to national or multinational bottling companies) and only a few mills directly bottle olive oil with their own label. Olive oil is generally bottled in stainless steel containers or, better, in glass bottles (in order to preserve better the stability of virgin olive oil), although there are cases of the use of innovative packaging, e.g. bottles made of polyethylene terephthalate (PET), which are 100 % recyclable (Salomone et al. 2013a).

In the ***by-product management phase***, two methods are used to extract pomace oil. Olive pomace oil obtained from two-phase processing, with a moisture content close to 70 %, is physically extracted by centrifugation. The process also produces a residual water solution containing mineral salts, sugars and polyphenols (EC 2010). To extract pomace oil from the traditional and three-phase production methods, solvents are used. The olive pomace is mixed with the solvent hexane, which dissolves any residual oil. The exhausted pomace is then separated from the oil and hexane solution (called miscella) by filtration. Any hexane residues in the solid pomace are removed by means of a desolventiser, which evaporates the solvent (then captured for reuse). The oil and hexane solution is distilled, allowing the hexane to be recovered and reused, whilst the solvent-free oil undergoes further processing, such as refining. The solid waste from olive oil mills is also referred to as 'olive cake' and the liquid waste streams are termed olive mill wastewater. In recent years, the by-product management has been considered a strategic phase in the olive oil supply chain, because each of the different olive oil production methods creates different amounts and types of by-products, all of which are potentially hazardous to the environment. Therefore, the above-mentioned environmental problems have given rise to a series of studies for the development of methods for the treatment and valorisation of olive mill wastewater (Demerche et al. 2013; Kapellakis et al. 2008; Stamatelatou et al. 2012) and olive stones from de-pitted virgin olive oil (Pattara et al. 2010). In particular, the olive oil mill wastes have a great impact on land and water environments due to their high phytotoxicity (Roig et al. 2006) and their management is one of the main problems of the olive oil industry. Many options have been proposed for their treatment, disposal or valorisation (Niaounakis and Halvadakis 2006; Roig et al. 2006; Vlyssides et al. 2004):

- Olive mill wastewater (OMW), deriving from traditional pressing and from the three-phase system, is the main polluting mill waste. This is constituted by vegetable water from the olives and the water used in the oil extraction and its chemical composition is variable depending on the olive varieties, growing practices, harvesting period and oil extraction technology. In any case, it is highly polluting due to the presence of organic compounds (organic acids, lipids, alcohols and polyphenols), even though it also contains valuable substances such as nutrients (especially potassium). Untreated olive mill wastewater is a major ecological issue for olive oil producing countries due to its highly toxic organic loads. Olive mill wastewater can lead to serious environmental damage, ranging from colouring natural waters, altering soil quality, phytotoxicity and odour nuisance. Traditional olive oil processing methods are estimated to produce between 400 and 600 litres of *alpechin* (OMW—olive mill wastewater) for each ton of processed olives (Di Giovacchino 2010; EC 2010). The olive mill wastewater levels from three-phase processes are much higher, producing between 800 and 1000 L of OMW for each ton of processed olives. Virtually no wastewater is produced by the two-phase process, although its *wet pomace* waste streams tend to have high liquid contents that remain costly to treat. The olive mill wastewater is composed essentially of water (80–83 %), organic compounds (mainly phenols, polyphenols and tannins) that account for a further 15–18 % of the wastewater content and inorganic elements (such as potassium salts and phosphates) that make up the remaining 2 %. These proportions can vary depending on factors related to the climatic and soil conditions, farm management, harvesting methods and oil extraction processes. The presence of proteins, minerals and polysaccharides in OMW means that it has potential for use as a fertiliser and in irrigation. However, the reuse opportunities are restricted by the abundance of phenolic compounds, which are both antimicrobial and phytotoxic. These phenols are difficult to purify and do not respond well to conventional degradation using bacteria-based techniques. The olive oil mill polluting loads are therefore significant, revealing levels of both BOD_5 (biological oxygen demand in 5 days) and COD (chemical oxygen demand) between 20,000 and 35,000 mg per litre. This represents a notably large organic matter load compared with standard municipal wastewater, which exhibits levels between 400 and 800 mg per litre. Anaerobic digestion of *alpechin* results in only 80–90 % COD removal and this treatment remains insufficient to permit olive mill wastewater effluent to be discharged back into the environment. Discharging unsafe olive mill wastewater back into natural water systems can result in a rapid rise in the number of microorganisms. These microorganisms consume large amounts of dissolved oxygen in the water and so reduce the share available for other living organisms. This could quickly offset the equilibrium of an entire ecosystem. Further concerns are caused by the high concentrations of phosphorus in olive mill wastewater, since if released into water courses this can encourage and accelerate the growth of algae. The knock-on impacts include eutrophication, which can destroy the ecological balance in both ground and surface water systems. Phosphorous remains difficult to degrade and tends to be dispersed only in small amounts via deposits through food chains (plants–invertebrates–fish–birds, etc.). The presence of large quantities of phosphorous nutrients in olive mill wastewater provides a medium for pathogens to multiply and infect waters. This can have severe consequences for local aquatic life, as well as the humans and animals coming into contact with the water. Several other environmental problems can be caused by olive mill wastewater. These include lipids in the olive mill wastewater producing an impenetrable film on the surface of rivers, their banks and surrounding farmland.

At a glance, the most common treatment methods of OMW are:

a. evaporation in storage ponds in the open—this method produces sludge that may be disposed of in landfill sites or used as a fertiliser in agriculture (after a composting process with other agricultural by-products);
b. direct application to soil—this is a positive valorisation method of OMW considering its high nutrient content and its high antimicrobial capacity, but it also causes negative effects on soil associated with its high mineral salt content, low pH and presence of polyphenols. Land spreading of waste arising from olive processing is specifically regulated by law (e.g. in Italy by the Ministerial Decree—MIPAF 2005);
c. co-composting—this method refers to the co-composting of OMW with olive pomace or olive wet pomace; it allows the return of nutrients to cropland and avoids the negative effects previously cited when OMW is directly applied to soil (Cappelletti and Nicoletti 2006; Salomone and Ioppolo 2012);
d. the extraction of valuable organic compounds—the recovery of high-value compounds (phenolic compounds, squalene and tocopherols, triterpenes, pectins and oligosaccharides, mannitol, polymerin) or the utilisation of OMW as raw matter for new products is a particularly attractive way to reuse it, as the recovery process is of economic and practical interest (Fernández-Bolaños et al. 2006).

- Olive husk (OH), deriving from traditional pressing and from the three-phase system, is usually sent to oil factories (oil husk extraction mills) that, after a drying process, extract oil with specific solvents (traditionally hexane). This treatment process produces oil and a solid waste called *exhausted olive husk*, which is used as fuel since the dried OH presents high calorific power.
- Olive wet husk (OWH) derives from the two-phase system. In this case, olive vegetation waters are included in the OWH. Compared with the OH, the higher moisture level in the OWH creates more difficulties for its treatment in oil factories (mainly the higher energy demand for the drying process causing higher costs). For this reason, there are other methods for the treatment of the OWH and the most common are:

    1. Direct application to soil—due to its high potassium concentration and its low economic value, it can be directly applied to soil on land near the production site, but this practice could cause a negative effect on the soil even if it is less phytotoxic than wastewater (Cichelli and Cappelletti 2007);
    2. Composting (with or without the de-stoning process to obtain biomass for heat or electricity)—this method consists of the co-composting of OWH with other agricultural wastes (straw, leaves, etc.) or with manure used as a bulking agent. The compost obtained has a good degree of humification, no phytotoxic effect and a good amount of mineral nutrients (Cappelletti and Nicoletti 2006; Russo et al. 2008).

The ***packaging phase*** includes bottling olive oil in glass, tin or PET containers. As the average annual world consumption of olive oil in the 2007/2008–2012/2013 olive crop six-year period was equal to 2,862,800 t (IOC 2013), assuming that only containers capable of holding 1 kg of olive oil are used, the packages in circulation could amount to more than 2,860,000,000 per year.

The ***transportation and distribution phase*** includes the transport activities (related to raw materials, by-products and wastes) and the distribution of the product in local, regional, national or international markets. Transport activities can also occur elsewhere in the life cycle (other than those instances already mentioned), either between any two subsequent life cycle stages or within a given stage, depending on the site-specific means of processing and the level of supply chain integration.

The ***consumer phase***, in the case of olive oil, is certainly not significant from a life cycle perspective, considering that the product consumption does not need further preparation or treatments. Table 2.1 shows that the consumption of olive oil is quite widespread on the international scale in countries such as Italy, Spain, the USA, Greece, Turkey, Syria, etc.

Finally, the ***waste management phase*** (end of life) includes the treatment of bottles and packaging waste (cardboard boxes, etc.). This phase can also have great impacts on the environment depending on the method of waste management chosen (for example, reuse, recycling, landfilling, etc.).

The phases of the olive oil supply chain with the related main environmental consequences are synthetically represented in Fig. 2.1.

As far as the materials and energy balance related to the oil production are concerned, it is possible to highlight that the production (agricultural and industrial phases) of 1 kg of olive oil (double pressed) involves the consumption of 0.0264 kg of fertilisers (N_2, P_2O_5, K_2O), 0.019 kg of pesticides, 0.00855 kg of fuel, 0.243 kg of lube oil and 0.359 kWh of electrical energy (Nicoletti and Notarnicola 2000).

2.3 Life Cycle Thinking Approaches in the Olive Oil Production Sector: The State of the Art of the International Practices

As exhaustively reported in Chap. 1, the growing awareness of food sustainability is driving an increase in research activities in the agri-food sector and, among these studies, over the last 15 years or more, numerous life cycle thinking (LCT) approaches have been followed (mainly life cycle assessment studies), evaluating food products and processes in order to identify and pursue sustainable food production and consumption systems.

The specific sector of olive and the olive oil supply chain has been investigated by several LCT studies since 2000. A critical analysis and state of the art of LCA studies applied in the olive oil sector was, firstly, conducted in 2008 (Salomone 2008) and then updated in 2010 (Salomone et al. 2010a), but contained only a comparative analysis of Italian studies, with the aim of highlighting the features of and/or differences in the fundamental aspects of Italian LCA studies; the first review included 13 Italian LCA case studies, while the second one contained 23 case studies. On the contrary, the literature review presented in this paragraph is a wider and deeper analysis with respect to the previous ones, because it includes:

- international case studies, not only Italian ones;
- life cycle thinking tools, not only LCA ones;
- 'olive industry' case studies, not only olive oil ones.

In particular, this literature review includes LCA studies that directly or indirectly refer to the wider term 'olive industry', therefore including applications not only to olive oil production, but also to olives in general (for oil or table use), to olive oil mill waste treatment and valorisation, and to table olive and olive oil packaging. The review refers to book chapters and articles published in international and Italian scientific journals and conference proceedings from 2000 to 2013; grey literature or other published papers could be missing.

In Table 2.2, the identified articles are listed, specifying the LCT tool used for the analysis and the product being investigated: 42 used LCA, 7 applied both LCA and life cycle costing (LCC) or another kind of economic analysis, 2 implemented simplified LCA (S-LCA), 9 dealt with environmental footprints (the carbon, water or ecological footprint) or energy balance or analysis and carbon balance, 10 were EPDs (Environmental Product Declarations) or papers reporting on EPDs and 2 reported on the integrated use of LCA and multi-criteria analysis (MCA).

In the following, a state-of-the-art analysis of the literature on life cycle thinking studies implemented in olive and olive oil production is presented, discerning between scientific articles including only the LCA methodology and articles concerning other LCT tools (LCC, S-LCA, footprint labels, EPD, etc.).

2.3.1 *Life Cycle Assessment*

The literature review shows that the most-used LCA analysis applied in the olive oil sector presents a comparative nature. Indeed, the first LCA study applied in this sector dates back to 2000 (Nicoletti and Notarnicola 2000) and focuses on the comparison between irrigated and dry olive cultivation systems, together with different olive oil extraction techniques. The comparison allows the evaluation of six different systems, obtained from the combination of two agricultural practices (dry and wet systems) and three extraction processes (single pressure, double pressure and centrifugation). This analysis structure, differently combining various systems and methods, was lately applied in other papers (such as Busset et al. 2012; De Gennaro et al. 2005; Salomone and Ioppolo 2012; Salomone et al. 2010a), also adding alternative treatments of olive oil mill waste, thus offering an articulated comparative LCA of very different olive oil production scenarios. In particular, De Gennaro et al. (2005) analysed various processes of the olive oil production chain, combining different oil extraction methods of extra virgin olive oil and different disposal and/or

Table 2.2 Articles reporting on the implementation of LCT tools in the olive industry

Reference	LCA	Other tools	Product
Nicoletti and Notarnicola (2000)	✓		Olive oil
Raggi et al. (2000)		S-LCA	Olive husk
Nicoletti et al. (2001)	✓		Olive and sunflower seed oil
Mansueti and Raggi (2002)	✓		Olive husk
Salomone (2002)	✓		Olive oil
Abeliotis (2003)		S-LCA	Olive oil
Notarnicola et al. (2003)	✓	LCC	Olive oil
Notarnicola et al. (2004)	✓	LCC	Olive oil
Romani et al. (2004)	✓		Olive oil
Cecchini et al. (2005)	✓		Olive oil
De Gennaro et al. (2005)	✓		Olive oil
Olivieri et al. (2005a)	✓		Olive oil
Olivieri et al. (2005b)	✓		Olives
Nicoletti et al. (2007a)	✓		Table olives
Nicoletti et al. (2007b)	✓		Table olives packaging
Olivieri et al. (2007a)	✓		Olive oil
Olivieri et al. (2007b)	✓		Olive oil
Avraamides and Fatta (2008)	✓		Olive oil
Cappelletti et al. (2008)	✓		Table olives
Cini et al. (2008)	✓		Olive oil
Guzman and Alonso (2008)		EB	Olive oil
Olivieri et al. (2008)	✓		Olive oil
Russo et al. (2008)	✓		Olive husk
Salomone (2008)	✓	Review	Olive oil
Fiore et al. (2009)	✓		Olive oil
Russo et al. (2009)	✓		Olive oil
Salomone et al. (2009)	✓		Olive oil
Scotti et al. (2009)		EF	Olive oil
Cappelletti et al. (2010)	✓		Table olives
Cavallaro and Salomone (2010)	✓	MCA	Olive oil
Olivieri et al. (2010a)	✓		Olive oil mill wastewater
Olivieri et al. (2010b)	✓		Olive oil mill wastewater
Polo et al. (2010)	✓		Olive oil
Roselli et al. (2010)	✓	LCC	Olive oil
Russo et al. (2010)	✓		Table olives
Salomone et al. (2010a)	✓		Olive oil
Salomone et al. (2010b)	✓	Review	Olive oil
Cappelletti et al. (2011)	✓		Table olives
Christodoulopoulou et al. (2011)	✓		Olive oil
ECOIL (n.d.)	✓		Olive oil
Intini et al. (2011)	✓	CF	Olive oil mill waste
Nicoletti et al. (2011)	✓	EPD	Olive oil

Table 2.2 (continued)

Reference	LCA	Other tools	Product
Özilgena and Sorgüven (2011)		EA	Soybean, sunflower and olive oil
Recchia et al. (2011)	✓	MCA	Olive oil
Salmoral et al. (2011)		WF	Olive oil
Apolio (2012)		EPD	Olive oil
Assoproli (2012)		EPD	Olive oil
Busset et al. (2012)	✓		Olive oil
Cappelletti et al. (2012)	✓	EPD	Olive oil
Carvalho et al. (2012)	✓	LCC	Olive oil
De Cecco (2012)		EPD	Olive oil
De Gennaro et al. (2012)	✓	LCC	Olives
Farmers Groups (2012)		EPD	Olive oil
Intini et al. (2012)	✓		Olive husk
Lucchetti et al. (2012)		CF	Olive oil
Monini (2012a)		EPD	Olive oil
Monini (2012b)		EPD	Olive oil
Monini (2012c)		EPD	Olive oil
Monini (2012d)		EPD	Olive oil
Neri et al. (2012)	✓	EA	Olive oil
Russo et al. (2012)	✓		Table olives
Salomone and Ioppolo (2012)	✓		Olive oil
Testa et al. (2012)	✓		Olive oil
Chatzisymeon et al. (2013)	✓		Olive oil mill wastewater
El Hanandeh (2013)	✓		Olive oil mill waste
Iraldo et al. (2013)	✓		Olive oil
Kalogerakisa et al. (2013)	✓		Olive mill waste
Nardino et al. (2013)		CB	Olives
Notarnicola et al. (2013)	✓		Olive oil
Palese et al. (2013)		SM	Olives
Pergola et al. (2013)	✓	EA	Olives
Salomone et al. (2013a)	✓	S-LCA	Olive oil packaging

CB carbon balance, *CF* carbon footprint, *EA* energy analysis, *EB* energy balance, *EF* ecological footprint, *EPD* environmental product declaration, *LCC* life cycle costing, *MCA* multi-criteria analysis, *S-LCA* simplified LCA, *SM* sustainable model (economic and environmental analysis), *WF* water footprint

reuse treatments of pomace and other olive oil mill waste. The analysis indicates as the most eco-compatible production chain the one that uses continuous two-phase transformation and the pomace treatment for the production of fuel, while the least eco-compatible system is the system entailing three-phase continuous production, composting of the pomace and spreading on the ground of the oil mill waste water. In Salomone and Ioppolo (2012) and Salomone et al. (2010a), comparisons of eight different scenarios, including different combinations of cultivation practices,

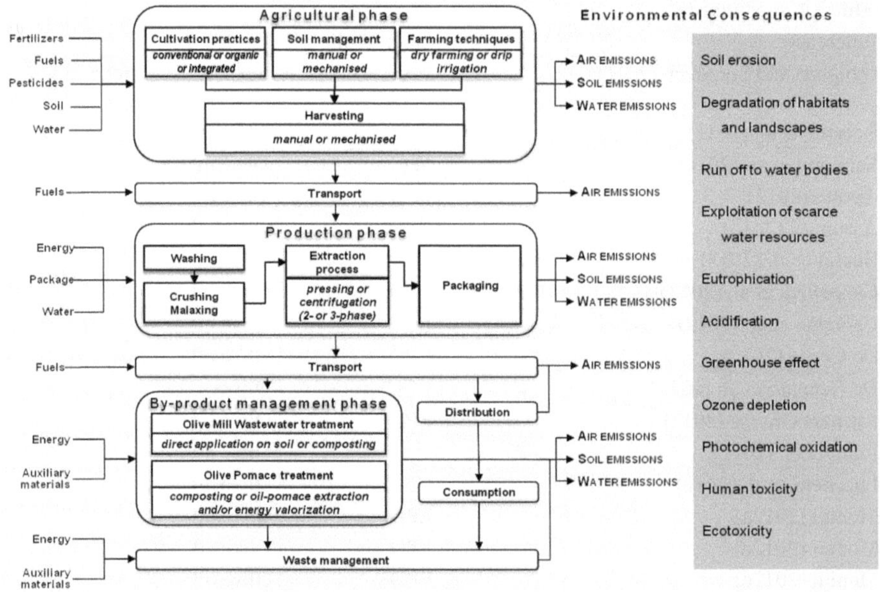

Fig. 2.1 The olive oil supply chain and its main environmental consequences

oil extraction methods and olive oil mill waste treatment, are presented. The analysis highlights a higher environmental load for conventional scenarios (except for impact categories associated with land use), an important environmental load associated with some sub-processes (such as fertilisation, the use of pesticides and the combustion of exhausted pomace), the higher environmental contribution of the sub-process of co-composting of olive wet pomace (OWP) with manure on fields rather than co-composting olive mill wastewater (OMW) and OWP with composter machines and a significant positive contribution (in terms of environmental credits for avoided production) associated with the use of by-products such as fuels or fertilisers. Busset et al. (2012) defined all the scenarios for olive oil production in France based on the different olive cultivation techniques, the different extraction processes and the different kinds of waste. Another paper, by Cavallaro and Salomone (2010), has a similar structure but provides new insights, because the LCA was implemented with MCA (see Sect. 2.3.6).

A different kind of comparative LCA study is dated 2001 (Nicoletti et al. 2001), presenting an evaluation of olive oil and sunflower seed oil. The results indicate that olive oil is more eco-compatible for all the categories except land use. The phase with the greatest impact is the agricultural phase for both systems (the main differences in this phase occur for the ODP category, caused by halon emission due to the production of pesticides from the sunflower cultivation). Concerning the industrial phase, higher impacts are connected with sunflower oil due to the VOC emissions occurring during sunflower oil chemical extraction. Another scientific article presents a comparison of olive oil with other kinds of vegetable oil (Özilgen and Sorgüven 2011), but it conducts an energy analysis rather than an LCA study (see Sect. 2.3.4).

Other comparative LCA studies focus on specific life cycle steps. For example, Russo et al. (2009) focused on the comparison of two processes of production of extra virgin olive oil using whole or de-stoned olives. The two processes present similar environmental performances, even though the process in which de-stoned olives are used has a lower impact: the real advantage of the de-pitting process is to obtain fragments of olive stone, which is an important by-product (much appreciated as fuel) both from an environmental and from an economic point of view. Similarly, Romani et al. (2004) focused on specific life cycle steps but reported on two diverse LCA applications: a comparison of organic and conventional virgin oil production and a comparison of two different applications for olive oil mill wastewater. The comparison of organic and conventional olive oil production is an aspect that had already been investigated by integrating the LCA and LCC methodologies (see Sect. 2.3.2), and other studies have been performed to compare these cultivation practices. Indeed, Cecchini et al. (2005) compared integrated, organic and conventional production of olive oil in Southern Italy and Neri et al. (2012) compared two organic and conventional farms in the central part of Italy, highlighting higher impacts for the agricultural phase in both case studies: in organic production, the impacts are related to a huge amount of fuel consumption (because of the use of old and low-efficiency machinery), whereas in conventional production, the main impacting input is the use of chemicals.

Olivieri et al. (2005a, b) applied LCA with a particular focus on the olive cultivation phase, for both conventional and organic farming, with the aim of quantifying numerically the environmental damage of the olive cultivation process and estimating the opportunities to reduce the impacts by comparison with organic olive cultivation (sensitivity analysis). Generally, these studies highlight higher impacts of conventional cultivation (except for the land use impact category). The other LCA presented by Romani et al. (2004), entailing the comparison of different wastewater treatments, falls into another widespread kind of LCA analysis that refers to waste treatment. Indeed, in 2002, the first LCA application not focusing on olive oil but on one of the main olive oil mill wastes (olive husk or olive pomace) was performed (Mansueti and Raggi 2002). In particular, this study reports the results of a comparative LCA between power generation from olive husk combustion and that from conventional technologies, highlighting that power generation from olive husk combustion (dark bars), for this specific case study, only deals with two kinds of impact categories: respiratory inorganic effects and acidification/eutrophication. As far as climate change is concerned, the olive husk combustion process has virtually no effect, since, as it is well known, CO_2 from biomass is not considered responsible for global warming.

After this paper, other LCA applications specifically focusing on olive oil mill waste were published and/or presented, such as the following studies:

- In Romani et al. (2004), a comparison of two different uses for olive mill wastewater (fertilisation–irrigation and optimised purified procedures able to recover higher quantities of polyphenols in view of possible future industrial application) is presented: a lower environmental load for the treatment allowing the recovery of polyphenols is highlighted;

- In Russo et al. (2008), an analysis of the environmental advantage deriving from the use of olive pits as fuel (by combustion in furnaces commonly fed with wood pellets or de-oiled pomace), comparing the environmental impact with that generated by the recovery of de-oiled pomace and the production of wood pellets, was performed. The results show that the recovery of olive pits offers environmental advantages with respect to other alternative fuels. This depends fundamentally on the higher net calorific value of the pit fuel and also on its simple recovery method (at the beginning of the process of olive oil extraction);
- In Olivieri et al. (2010a, b), an LCA study applied to a new integrated technology for olive oil mill wastewater (OMW) treatment and polyphenols recovery from a biphasic olive mill is presented. This method treats olive oil wastewater and, at the same time, produces novel products exploiting the antioxidant properties of polyphenols as a semi-manufactured good for 'novel food' (e.g. phytotherapy, cosmetics). The results of a sensitivity analysis show that the LCA of this process has less impact, with an overall percentage of 57 % with respect to the traditional process. The recovery of polyphenols from olive oil wastewater is important to add value to this waste as these substances can be an important source of new antioxidant products in 'novel food'. Moreover, the recovery of polyphenols helps to avoid phytotoxicity in soil;
- Intini et al. (2012) carried out an LCA in order to compare the environmental performance of using de-oiled pomace and waste wood as fuel. Only the global warming potential was calculated and compared with that of a plant for energy production that uses refuse-derived fuel (RDF) and that of one that uses coal. The LCA shows the important environmental advantages of biomass utilisation in terms of the reduction of greenhouse gas emissions;
- In Chatzisymeon et al. (2013), the LCA methodology was utilised to evaluate three different advanced oxidation processes for olive oil mill wastewater treatment (UV heterogeneous photocatalysis—UV/TiO_2; wet air oxidation—WAO; and electrochemical oxidation—EO). Both EO and WAO can be competitive processes in terms of COD, TPh and colour removal. EO was found to be a more environmentally friendly technique as it yields lower total environmental impacts, including CO_2 emissions to the atmosphere. The environmental impacts of all three treatments show that human health is primarily affected, followed by impacts on resource depletion. Overall, it was found that the environmental sustainability of these treatments is strongly related to their energy requirements and that their total environmental impacts decline according to the following order: UV/TiO_2 > WAO > EO;
- In El Hanandeh (2013), LCA was used to analyse the carbon emission reduction potential of utilising olive husk as a feedstock in a mobile pyrolysis unit. Four scenarios, based on different combinations of pyrolysis technologies (slow versus fast) and end-use of products (land application versus energy utilisation), were compared and the results show that all the scenarios result in significant greenhouse gas emission savings;

- In Kalogerakisa et al. (2013), an LCA of the extraction of compounds, such as hydroxytyrosol and tyrosol as well as total phenols (TPh), from real olive oil mill wastewater (OMW) was performed, in order to provide the best available and most sustainable extraction technique using ethyl acetate, chloroform/isopropyl alcohol and diethyl ether. The use of ethyl acetate yields low environmental impacts and high antioxidant recovery performance and, therefore, it is assumed to be the best option, from both an environmental and a technical point of view, while the chloroform/isopropyl alcohol mixture was found to impose detrimental effects on the ecosystem, human health and fossil resources.

Another kind of LCA study in this sector relates to the analysis of the main life cycle phases. Indeed, in 2002, a paper presenting a cradle-to-gate analysis was presented (Salomone 2002), including the cultivation of olives, olive oil production, olive husk treatment and transport between these treatment phases. This was the first study to include pomace treatment in the LCA analysis of the product 'olive oil'. The motivation was to avoid allocation, as suggested by ISO 14044 (at the time of the research ISO 14041), expanding the system in order to include the treatment of this by-product of olive oil production as well. After this, other papers studied the olive oil production chain, including the reuse of by-products and waste, such as the above-mentioned comparative LCAs of different olive oil production scenarios (Salomone and Ioppolo 2012; Salomone et al. 2010a), but in these cases the motivation was mainly connected with a vision of integrated environmental management of the whole olive oil production chain (thus including by-product treatment and valorisation). Furthermore, in Cini et al. (2008), LCA was used to evaluate the environmental impact of olive oil production considering different possibilities for the by-product reuses, but (similarly to De Gennaro et al. 2005) the paper does not include the cultivation step, just taking into account the extraction process of olive oil following different methods: the extraction process with oil production and pomace treated as waste; the extraction process with oil production and pomace used as a fertiliser; the extraction process with oil and pomace stone production; the extraction process with oil and pomace stone production; and the use of pomace residue as a fertiliser.

In 2007, table olive production also started to be investigated, mainly because of the growing interest in this specific sector caused by the increase in their cultivation and processing activities, as well as the relevant amount of wastes generated by the connected processing industries. Different papers have analysed this particular kind of production and its various aspects in depth: green olive cultivation and olive processing using the Spanish-style method (Cappelletti et al. 2010; Nicoletti et al. 2007a); black olive cultivation and olive processing using the Californian-style method (Cappelletti et al. 2008); a comparison of three different methods used for processing ripe table olives—two different methods of the Californian-style and the Spanish-style method (Russo et al. 2010, 2012); a comparison of the different packaging systems (Nicoletti et al. 2007b); and a study (Cappelletti et al. 2011) focusing on the production processes, the characteristics of wastewater and the pollution prevention technologies (in this case, the LCA results underline that eutrophication is a very important impact for the table olive processing industries, and it derives from the pollution of the wastewater).

Some LCA applications in the olive oil sector relate to the analysis of olive oil production in specific geographic areas, such as:

- Avraamides and Fatta (2008)—LCA was used to evaluate the consumption of raw materials and emissions of pollutants from olive oil production in Cyprus (Greece) and to identify the processes causing the most significant environmental burdens. The interpretation results were organised in an interesting classification of the individual processes in priority categories according to their potential optimisation: fertilisation and oil extraction processes should be considered as priority 1 processes, irrigation and pruning are classified in priority 2, pest control and soil management in priority 3 and tree planting, collection and transportation of olives to the processing unit (as their contribution to all the environmental flows considered was less than 0.5 %) in priority 4;
- Fiore et al. (2009)—in this paper, the results of an LCA application to the Sicilian (Italy) olive oil production, obtained from olives cultivated by an intensive managing system, are described. The study highlights the environmental burden deriving from the agricultural phase as well as the packaging phase, which involves an environmental impact due to the glass bottle production;
- Christodoulopoulou et al. (2011)—a comprehensive LCA was carried out on olive oil of extra virgin quality, produced from 487 olive groves by 3 groups of 68 olive growers in southern Greece. The first goal of the study was to assess the environmental performance of olive oil in order to use it for an Environmental Product Declaration (EPD) according to PCR 21537 of Environdec. The second goal was to use the LCA as a starting point for the continuous improvement procedure with regard to the environment, by identifying the areas with the most significant impacts and by taking measures for their control;
- Busset et al. (2012)—an LCA study of the French olive oil production sector is presented: it was elaborated partly in order to reduce the carbon footprint and to optimise the waste management of the olive oil sector in the SUDOE area (Spain, Portugal and France). The first results permitted the definition of all the scenarios for olive oil production in France based on the different olive production techniques (with or without irrigation, mechanical or not, organic or not), the different extraction processes (pressing, centrifugation in two phases or centrifugation in three phases) and the different waste management schemes (incineration or spreading). The expected result was a comparison of all the scenarios in order to identify the parameters that influence the environmental consequences of olive oil production;
- Salomone and Ioppolo (2012)—the LCA methodology was applied to investigate the olive oil sector and identify useful information for taking strategic decisions aimed at the improvement and optimisation of a local olive oil chain in the province of Messina (Italy), directly involving a sample of companies of the local association of oil producers;

- Notarnicola et al. (2013) analysed the cultivation phase of olives for the production of olive oil performed on 63 farms in the northern area of the city of Bari in Puglia (Italy), with the aim of assessing the variability of the LCA results. This is one of the few papers to analyse with the same inventory more than 60 data sets. The results indicate great variability within the management methods of the olive orchards, with agronomical practices differing from producer to producer (even from the same area). This is reflected in the high degree of variability of the inventory and impact assessment results.

Other interesting applications of the LCA methodology within the field of olive oil production include its use for supporting the definition of environmental management strategies and the integration of tools. In this category of studies, three cases could be included:

- The first one is the case reported in different papers discussing integrated environment quality–HACCP systems aimed to realise useful guidelines for the acquisition of a territory product mark (Olivieri et al. 2007a, b, 2008). LCA was used to characterise environmental critical states in the cultivation and production of virgin oil; the most important problems identified are the use of fertilisers, the use of pesticides for olive fly capture and land use in conventional olive cultivation;
- The second one is the case of a study specifically focused on the design of a model of a Product-Oriented Environmental Management System (POEMS) for agri-food companies (Salomone et al. 2013a), which includes the use of the LCA methodology for the product orientation of integrated management systems; one of the case studies reported is the comparison of two different packaging systems of extra virgin olive oil: glass vs PET bottle. The overall comparison highlights higher scores for the glass bottle system compared with the PET bottle system, except for the fossil depletion category, in which the higher score is linked to the PET bottle system, caused by PET production;
- The third one is an LCA applied to the production of extra virgin olive oil in the Val di Cornia, Tuscany, Italy (Testa et al. 2012; Iraldo et al. 2013). The LCA study is intended to support the experimental implementation of a system of environmental qualification of a product, managed locally, which combines the features of type I and type III eco-labels. The agricultural phase is the most impactful of all the categories, in particular due to acidification, eutrophication and water consumption. The major impacts result from the production of pesticides. However, the use of pruning residues as a fertiliser and for domestic heating brings significant benefits for certain impact categories. In the extraction phase, olive mill waste water recovery as a fertiliser leads to a reduction in water consumption, eutrophication and global warming.

Another application of LCA presenting new insights is provided by Salomone et al. (2009), in which a comparison between a conventional extra virgin olive oil and a high-quality extra virgin olive oil with the characteristic of excellence is presented. The new element consists of an attempt to integrate the environmental impacts and the quality characteristics of the product into the LCA methodology by inserting an impact category called 'cardiovascular risk' defined on the basis of the contribution that the phenols, contained in olive oil, make to increasing HDL cholesterol (which helps to reduce cardiovascular risks); the aim of the study was to match the potential environmental impacts of the entire product life cycle with strategies for quality exploitation of the same and to assess the potential ways of integrating environmental aspects and quality improvements into the strategic decision making of firms.

Finally, the LCA methodology was part of different research projects in the olive oil sector, such as the ECOIL project in Greece (ECOIL n.d.), the OiLCA project in the SUDOE area (Spain, Portugal and France) (Busset et al. 2012), the EMAF project in Italy (Salomone et al. 2013b) and the Life+ECCELSA project in Val di Cornia, a rural area in the south of Tuscany, Italy (Iraldo et al. 2013).

2.3.2 Life Cycle Costing

Economic tools, also in the agro-food sector, can be combined with LCA in several ways (though not completely integrated) as a separate complementary analysis, within a toolbox or as a way of expanding it.

Generally speaking, these tools can play two main roles in life cycle management (LCM): on the one hand, they can provide ways of accounting for costs within the same boundaries and with reference to the same functional unit (FU) as in LCA (microeconomic-oriented accounting tools); on the other hand, macroeconomic-oriented accounting tools, such as input–output tables, either in monetary or in physical terms (in the latter case leading to material flows analysis—MFA), aim to study the way in which materials and substances flow through the economy.

As far as the accounting for costs at the microeconomic level is concerned, although life cycle costing (LCC) is not as standardised, as LCA is, there is a significant body of literature that addresses its conceptual framework and methodology. Thus, applications to food products, being applications of more generalised concepts, might seem not to pose major methodological problems: there is, in fact, evidence that LCC is also being used as a decision support tool within the LCA of food products. However, the literature provides few applications of LCC to food products and, more generally, to non-durable products: in this sector, applications of traditional LCC make sense only if an investment in a brand new food production plant is being evaluated. Furthermore, the approaches adopted when LCC is used within environmental management may vary significantly: cost elements, especially subsidies and external costs, are expected to affect the ranking of alternative options heavily, unless one specific option is found to be both environmentally sustainable and cost effective compared with the others.

On the contrary, examples of expansion of LCA by means of combined environmental–economic analyses include applications of input–output analysis along with MFA and LCA. In this case, as stated before, macroeconomic-oriented accounting tools, such as input–output tables, are used. Either they can be used in hybrid LCA of food products to extend the system boundaries to include all the complex transactions that characterise the entire economic system (such an approach has been used even at the institutional level to support integrated product policies) or they can be used to reveal the importance of understanding the physical structure underlying any food production system. The combination of macroeconomic analysis and LCA may prove to be particularly useful since, compared with detailed life cycle inventories, many models of entire economies employ a much smaller number of categories for representing production and consumption activities (Settanni et al. 2010).

As regards the application of LCC, or another kind of economic analysis, in the olive oil sector, the literature review highlighted seven studies (from 2003 to 2013): in particular, five of them are about the integrated application of LCA and LCC (Carvalho et al. 2012; De Gennaro et al. 2012; Notarnicola et al. 2003, 2004; Roselli et al. 2010), one is about the application of a sustainable model (economic and environmental analysis) (Palese et al. 2013) and the last one is about an energy, economic and environmental analysis (Pergola et al. 2013). Of the seven papers, just four were reviewed in depth because three of them (Notarnicola et al. 2003; Pergola et al. 2013; Rosselli et al. 2010) are parts of other papers (respectively: Notarnicola et al. 2004; Palese et al. 2013; De Gennaro et al. 2012).

As with the LCA studies (see Sect. 2.3.1), most of the LCC studies are of a comparative nature (organic vs conventional extra virgin oil; different olive-growing systems; alternative agronomical techniques vs conventional ones). Regarding the geographical boundaries of the examined papers, four focus on Italian case studies (De Gennaro et al. 2012; Notarnicola et al. 2004; Palese et al. 2013; Pergola et al. 2013) and one on a European case study (Carvalho et al. 2012). Furthermore, just two papers focus on the food product 'olive oil', respectively extra virgin olive oil (Notarnicola et al. 2004) and olive oil (Carvalho et al. 2012), while the others are about olive-growing models or agronomical techniques.

The paper by Carvalho (2012) was developed within the OiLCA international project with the aim of improving the competitiveness of the olive SUDOE space (Spain, Portugal and the south of France) and reducing the environmental impact of olive oil production through the application of the principles of eco-efficiency. This paper does not develop a comparative study, aiming to identify opportunities for waste management among the olive oil production using cutting edge technology that takes into account economic aspects, encouraging the modernisation of the sector and contributing to improving the quality of the final product. The management of these residues represents a big challenge because of their predominance and unavoidable production; it is thus important to take into account the available or emerging technologies, which may result in both economic and environmental benefits. The study was conducted by coupling the LCA and LCC methodologies, with 1 L of olive oil as the FU and the following phases as system boundaries: cultivation, oil production and packaging. Accordingly, it was possible to identify improvement solutions with their associated investment and production costs, providing business people with useful tools for making decisions based on economic (and environmental) criteria. These solutions have not yet been disclosed to the public.

Regarding the comparative studies, the one by Notarnicola et al. (2004) aimed to compare the production systems of organic and conventional extra virgin olive

Table 2.3 Internal and external costs of the two systems (organic vs conventional) per functional unit (1 kg of extra virgin olive oil). (Source: Notarnicola et al. 2004)

Agricultural phase	Organic	Conventional
Pesticides	0.171	0.117
Fertilisers	0.268	0.181
Lube oil	0.023	0.011
Electrical energy	0.143	0.085
Water	0.077	0.046
Diesel	0.084	0.048
Labour	4.344	2.864
Organic certification cost	0.064	–
Total (1)	5.174	3.352
Transports	0.0784	0.039
Industrial phase		
Electrical energy	0.014	0.024
Labour	0.089	0.045
Water	0.002	0.022
Packaging	0.298	0.298
Waste authority	0.015	0.015
Organic certification costs	0.009	–
HACCP certification costs	0.0009	0.0009
Total (2)	0.428	0.405
Total (1+2)	5.680	3.796
External costs of energy	0.664	0.533
External costs of fertilisers and pesticides	0.439	9.870

oil in order to assess their environmental and cost profiles and to verify whether the two dimensions (environmental performance and costs) move in the same direction. For the cost assessment, in particular, the LCC methodology was applied with the same FU and system boundaries as the LCA study: 1 kg of extra virgin olive oil and all the direct (agriculture practices, harvesting, transport and oil extraction) and indirect (production and transport of the pesticides, fuels, etc.) activities. The transportation of chemicals (from the factories to the agricultural fields), of materials and of the workers involved in the harvesting and pruning operations (from town to orchard) and of olives (from the orchard to the oil mill) were also included in the system boundaries. All the related internal and external costs of the two systems are reported in the study (see Table 2.3), showing (for example) that the damage caused by conventional agriculture due to the use of fertilisers and pestidicides (in terms of reclamation and decontamination) costs more than 22 times that of organic agriculture or that the organic system is characterised by higher production costs due to the lower organic yields (this higher cost is, then, reflected in a higher market price).

Regarding the obtained outcomes, in the LCA–LCC comparison between conventional and organic extra virgin oil, if the external costs are not taken into account, the organic olive oil has a higher cost profile; on the contrary, if these costs are added to the conventional (internal) company costs and to the less tangible, hidden and indirect company costs, the organic olive oil has a lower total cost in comparison with the conventional one. All that considered, it is important to account for external costs, as the European Commission is already doing in several projects, for example the ExternE project (ExternE 2013). As far as the LCA results are concerned, the study demonstrated that the organic olive system is more eco-compatible than the conventional one by a factor of five due to the great difference in the TETP and FAETP categories.

Another comparative study is the one by De Gennaro et al. (2012), about the integrated assessment (environmental and economic) of two innovative olive-growing systems, 'high density' (HDO, over 200 tree/ha) and 'super high density' (SHDO, over 1,500 trees/ha), during their life cycle. The system boundaries included the phases of planting, cultivation, growing production, full production and plant removal and disposal, with an FU of 1 t of olives. The production of fertilisers and pesticides was also included, while transformation, distribution and consumption were excluded because they are the same for the two systems. The economic assessment was performed as requested by the LCC methods using, as criteria, the net present value (NPV) and the internal rate of return (IRR). This analysis shows that the HDO could be considered more convenient than the SHDO (the most innovative system): in fact, despite the lower operating costs of the latter, due to the complete mechanisation of pruning and harvesting operations, these costs are counterbalanced by the higher initial investment costs that the company has to face (which result as three times those of the HDO system). Furthermore, the HDO model achieves better performance (in terms of NPV and IRR) than the SDHO model: this result is mainly driven by the lower plantation costs, longer production cycle, higher productivity of olives and greater efficiency in the use of inputs that characterise the HDO model. Furthermore, the full production phase represents the major impact for both systems (more than 75 % of the whole impact in all the impact categories in HDO, between 50 and 75 % in SHDO). Regarding the environmental assessment, this analysis also shows a better performance of the HDO system for all the impact categories (Global Warming Potential GWP, Ozone Depletion Potential ODP, Acidification Potential AP, Photochemical Ozone Creation Potential POCP, Human Toxicity Potential HTP, Freshwater Aquatic Ecotoxicity Potential FAETP, Marine Aquatic Ecotoxicity Potential MAETP, Terrestrial Ecotoxicity Potential TETP, Nutrification Potential NP, Abiotic Depletion Potential ADP), with a percentage ranging from 21 to 37 %. The superior performance of the HDO system is mainly linked to the lower use of energy but also to lower chemical inputs and higher olive yields. As far as the energy use is concerned, the full production phase is characterised by the highest energy consumption, with 87.4 % (HDO) and 75.1 % (SHDO). Finally, the study highlights that the results remain the same even if a sensitivity analysis (modifying the olive yields of the two systems) is carried out.

Finally, the paper written by Palese et al. (2013) focuses on the proposal for a sustainable system (SS) for the management of olive orchards (156 plant ha^{-1} with a distance of about 8 m × 8 m) located in semi-arid marginal areas. This new model presents two key aspects: the reuse of urban wastewater distributed by drip irrigation and the use of soil management techniques based on the recycling of the polygenic carbon sources internal to the olive orchard. Economic (and also environmental) analysis was performed to evaluate the sustainability of the proposed method when compared with the conventional management system (CS). In particular, the economic results were expressed at constant values by the formula:

$$\text{Gross Profit (GP)} = \text{Total Output (TO)} - \text{Production Costs (PC)}$$

with:

TO representing the income from sales of oil and table olives
PC showing the sum of fixed and variable costs, gross of taxes and overheads.

Data were evaluated for a period of 8 years, showing that the annual TO (€ ha^{-1} $year^{-1}$), calculated at constant values, was strongly affected by the extent of the crop load measured in the examined period. In particular, the TO of the SS was shown to be constantly positive and greater (about three times, mostly due to the higher quality of the olive production–table olives) than the CS value. Regarding the PC, the SS showed higher values than the CS. Both systems presented a positive value of the GP/ha, but the SS was four times more profitable than the CS. Finally, the SS produced quite a regular income over the considered period thanks to the annual yield, while the CS guaranteed a GP in alternative years. The environmental assessment was focused, above all, on the CO_2 stocks in plants and soil as well as the anthropogenic and natural CO_2 emissions. It demonstrates that from this point of view the SS system is the most sustainable as well. By comparing the mean annual fluxes of CO_2 (net primary productivity—NPP—total emissions), the SS system shows positive data with an important gain of CO_2 sequestered from the atmosphere (15.45 t/ha^{-1}/$year^{-1}$), while the CS has total emissions that are higher than the NPP; the SS shows an annual gain of 3.85 CO_2 t/ha^{-1} in the first 0–0.6 m soil layer; on the contrary, the CS shows an important mean annual loss equal to 5.10 CO_2 t/ha^{-1}. Finally, the SS is able to fix a higher amount of CO_2 than CS (more than double). All that considered, the SS appears sustainable not only from the economic but also from the environmental and social points of view.

2.3.3 Simplified Life Cycle Assessment (S-LCA)

The practical use of environmental LCA methods and software tools in industry has revealed the need for simplifications of many applications. Hence, streamlined LCA methods have been derived from experience with the complex full methods (Hauschild et al. 2005). Simplified LCA (S-LCA), also known as streamlined LCA, emerged as an efficient tool for evaluating the environmental attributes of a product's, process's or service's life cycle (Hayashi et al. 2006). The aim of S-LCA is to provide, essentially, results that are the same as or similar to a detailed one, i.e. covering the whole life cycle using qualitative and/or quantitative generic data, followed by a simplified assessment, thus significantly reducing the expenses and expended time. It has to include all the relevant aspects, but good explanations can, to some extent, replace resource-demanding data collection and treatment (Schmidt and Frydendal 2003). The assessment should focus on the most important environmental aspects and/or potential environmental impacts and/or stages of the life cycle and/or phases of the LCA and undertake a thorough assessment of the reliability of the results. S-LCA studies can be conducted to make a quick assessment of a product: the challenge is to adapt the LCA methodology and simplify its use, but to a more advanced LCA stage than for a screening LCA. S-LCA has to be interpreted as an 'adapted' LCA, depending on the effort that the LCA practitioner wants to put in for every life cycle stage. The minimum requirements can be summarised as follows:

- the goal and scope;
- the life cycle stages included, as well as a clear definition of the system boundaries;
- the input materials/items included and excluded, with justification, as well as processes for energy, water, etc.;
- an overview of the calculation rules and comments on the degree of approximation/uncertainties;
- the impact categories considered (with justification);
- the limitations;
- the life cycle impact results and interpretation;
- a statement regarding consistency;
- the results.

The data used in a simplified study should, as far as possible, provide the existing time and budget constraints related to the country where the products are produced or being used. However, as this is not always possible, it is also acceptable to use assumptions, for example using data that represent a country with a similar electric energy grid mix and manufacturing technology. The data should represent the technology used as closely as possible.

In the olive oil production sector, LCA studies are, generally, aimed at identifying the environmental burdens associated with the processes involved and at proposing actions for further environmental improvements. Nevertheless, such goals are often complex tasks, mainly due to the lack of reliable input data related to the whole life cycle of the assessed system, thus affecting the accuracy and the significance of the study. An S-LCA procedure can make possible studies based on information that is already available, e.g. at the early conceptual design stage or when the input data do not allow the assessment of sources of environmental burdens.

The scientific literature in the sector includes a few studies that specifically apply a simplified procedure. Among them, Abeliotis (2003) focused on the analysis of a three-phase olive oil mill. It is not a comparative analysis but it aimed to assess the greatest environmental burdens of the production system examined. In each production stage, the input and output streams of mass and energy were identified (inventory phase) and the environmental impacts associated with the process were grouped together into a number of environmental impact categories (global warming potential, acidification, eutrophication and photo-oxidant formation, etc.). The boundaries of the system start with the fertilisation of the olive trees and end with the extraction of olive oil. Region-specific and agricultural phase LCI data were not available. For some processes, such as fertiliser and pesticide application, although site-specific data were desirable, estimates of emission factors and estimation techniques from the literature were used. The data for the mass and energy balance at the extraction stage were derived from the examined production process, but no experimental data were available with regard to the organic load of the effluent olive mill wastes from the treatment step, the N_2O emissions and the energy embedded in fertilisers. Thus, these data were deduced from the literature sources and adapted to the analysed process.

This study shows that the most significant impact arising from the assessed process is the GWP, attributed to the electricity required for the olive oil extraction process as well as the energy used for the fertiliser production. However, two relevant impacts are not taken into account (land use and human toxicity), due to the lack of specific data about several sources of environmental burdens, such as the use of pesticides and the presence of phenols in the effluent olive mill wastes. Furthermore, no data about the treatment of the olive mill wastes are available.

Another example of an S-LCA study is presented in Raggi et al. (2000), in which the production and use of olive husk bricks, as a fuel for residential heating, were screened and a preliminary comparison of such a technology with natural gas combustion was carried out. The system boundaries were defined to cover all the steps from olive husk handling and pressing to its combustion in households, including the production of packaging and ancillary materials. The environmental burdens related to the oil extraction from olive cake were allocated in total to the extracted oil. With regard to the data quality, primary data were collected on-site directly from the economic factors involved in the product life cycle, while the literature and international databases were used for secondary data. The study presents a partial life cycle impact assessment, since only the GWP and AP were investigated. The results highlight that the most significant contribution to the GWP arises from transport, followed by the energy requirement in the husk-processing activities. No contribution of the CO_2 from the combustion of olive husk was considered, assuming it to be 'virtually' equal to the CO_2 absorbed by the plants during their vegetative cycle. With regard to the AP, the most significant contribution derives from the combustion of the biomass, due to the sulphur content in olive husks and the NO_x released from the boilers.

The olive husk as a fuel in residential heating was compared with the performance of natural gas technology, with regard to the GWP and AP, in order to assess the environmental benefits and drawbacks associated with the biomass use, but the related primary energy saving was not considered. The assessed husk-based heating system contributes much less to the GWP than the use of fossil fuels, unless husk is transported over longer distances. However, the authors do not provide any specification about such distance. This study evidences the need for higher quality data in order to avoid estimations, since many of them are missing or inaccurate, such as the emission factors of husk combustion.

In the two above-cited studies, the S-LCA procedure is applied as a preliminary tool to assess different products of the olive oil chain. The former is aimed at evaluating the environmental burdens of olive oil produced in a three-phase mill, identifying and quantifying material and energy consumption and releases into the environment at the mill stage; the latter shows a preliminary LCA study of olive husk used as biomass in residential heating, comparing it with a fossil fuel, i.e. natural gas. Both the studies highlight the critical issues of the assessed production processes, such as the contribution to the GWP impact category, even though a more accurate analysis would also require the assessment of other impacts, such as the life cycle energy requirement in terms of primary energy, which is strictly connected to the GWP.

2.3.4 Footprint Labels (Carbon Footprint, Water Footprint, Ecological Footprint)

The term 'footprint' has become a popular means of indicating a quantitative measure of human beings' appropriation of natural resources (Hoekstra and Chapagain 2008). All three indicators, the carbon footprint, water footprint and ecological footprint, are aimed at evaluating environmental impacts in terms of the appropriation of natural resources needed to sustain the supply chain of a generic product. Specifically, the three indicators highlight the effect of resource consumption on different environmental compartments: air (in terms of greenhouse gas emissions), water (in terms of the volume of water consumed and/or polluted) and land (in terms of land use) (Neri et al. 2010). The joint use of more than one indicator should provide a full sustainability diagnosis (Bastianoni et al. 2013).

In particular, the *carbon footprint* (CF) methodology is commonly defined as the quantification of greenhouse gas emissions associated with the life cycle of a good or service. Referring to the life cycle, the carbon footprint derives from the LCA methodology, but focuses exclusively on issues related to the phenomenon of global warming (Weidema et al. 2008). The PAS 2050 (BSI) was one of the first standards introduced in this context to standardise a similar methodology in 2013. Later on, the ISO published the international standard rules related to this method in May 2013—ISO/TS 14067:2013 (ISO 2013a). The unquestioned acceptance of the carbon footprint by retailers and the media has been possible thanks to its ease of comprehension and immediacy (even for non-experts) and the explicit reference to the problem of global warming. Its diffusion has been achieved thanks to the interest arising from different sectors, including the agro-industrial one, which immediately saw the carbon footprint as a tool for product/image/marketing improvement and strategic communication when it comes to the consumer.

Within the olive oil sector, the IOC (International Olive Council) is taking steps to draft guidelines for the correct and uniform application of the new ISO 14067. The carbon footprint-related scientific literature includes a small number of studies specifically related to it. Among them, only two (Lucchetti et al. 2012; Polo et al. 2010) focus on the analysis of the carbon footprint of 1 kg of olive oil (albeit only for the bottling stage): Nardino et al. (2013) carried out an empirical and tool-related assessment of the ability to fix the atmospheric carbon from the olive grove, while in Intini et al. (2011), a comparative evaluation of the use of de-oiled pomace, fossil fuels and wood biomass (in the operation of a power plant for the production of electricity and heat) was carried out. Polo et al. (2010) applied the carbon footprint methodology to five agro-industrial products, including two types of olive oil (1 L in glass bottles and 5 L in PET ones). The analysis shows that the CFPs are of 1.1 and 5.5 kg of CO_2 (respectively for bottles of 1 and 5 L). Furthermore, Özilgen and Sorgüven (2011) carried out an evaluation of three different methods (energy, exergy and carbon dioxide emissions) for three different oils (soybean, sunflower and olive) using 1000 kg of raw material product as a functional unit (soybean, sunflower and olive). In this study, the agricultural phase is responsible for most of the carbon dioxide emissions due to the excessive use of fertilisers (Ozilgen and Sorgüven 2011). The total CO_2 emissions for producing oil from 1 ton of olives is 323.1 kg CO_2, of which 164.9 kg is linked to the agricultural phase, 123.3 kg to the oil production phase, 31.9 kg to the packaging phase and 3.0 kg to the transportation phase.

Three of the five works analysed (Intini et al. 2011; Lucchetti et al. 2012; Nardino et al. 2013) are representative of the Italian scenario, demonstrating the attention given, at a scientific level, to the agro-industrial production in Italy. On the other hand, one (Özilgen and Sorgüven 2011) was developed in Turkey, while the last, though not precisely defined, is believed to have been carried out in Spain.

No article takes into account the olive oil product from cradle to grave. Specifically, Lucchetti et al. (2012), during their analysis of the bottling process, do not use calculation software but use emission factors directly (published by government agencies and electricity producers). Furthermore, not all the GHGs provided by the IPCC are highlighted; only CO_2 and CH_4 are considered in the study. Intini et al. (2011) carried out an assessment of the benefits arising from the possible use of de-oiled pomace, for energy, taking into account both the current technologies that are already widespread and the nationwide availability of this product. The analysis shows the possible avoided emissions of GHGs if all the de-oiled pomace is destined not for residential users (as happens today) but for electricity and heat production plants. The analysis undertaken by Polo et al. (2010), although very interesting for the results achieved, does not show how the data collection was conducted and which software or database was used for the calculation of the carbon footprint. The study by Nardino et al. (2013), while making explicit reference to the carbon budget within an olive grove, does not use the specific methodology of the carbon footprint to assess the total mass of CO_2 stored. Indeed, some methods were proposed by Nardino et al. (2013) based on the study of gas exchange between the atmosphere and tree cultivation and compared (to assess their significance) with empirical methodologies. From this work, it is apparent that olive groves are useful for carbon storage and for biomass production destined for energy purposes (values between 10 and 15 t (C) ha^{-1} $year^{-1}$). In the study by Özilgen and Sorgüven (2011), the source of the emission coefficients and whether the study included all GHGs or just carbon dioxide were not clear. In general, referring to the carbon footprint methodology, it can be said that it is not, at least in the olive oil sector, a frequently applied tool, due to both the small number of papers in the literature and the lack of comprehensive studies related to our subject of interest. The reasons for this refer to the recent standardisation of the method (ISO/TS 14067:2013 was published only in May 2013), to the scientific limitations of the tool, even though it allows strong communication, and to the fact that the olive oil sector, over the past 10 years, has invested more in improving the quality of the product (acidity, content of antioxidants, etc.) and in certification of origin (protected designation of origin PDO and protected geographical indication PGI), leaving out the communication of connected environmental aspects.

The *water footprint* (WF) is also to be noted, being a water use indicator that considers both the direct and the indirect content related to a process or good and referred to as the volume of fresh water used per unit of the product. It is divided into three components (Hoekstra and Chapagain 2008): the blue WF (blue water, surface or underground), the green WF (rainwater that is stored temporarily in the soil or vegetation) and the grey WF (the volume of fresh water required to assimilate the load of pollutants).

For the olive oil sector, the literature review highlighted just one paper; only the contribution of Salmoral et al. (2011) was assessed in analysing the WF of olives and olive oil produced in Spain. The analysis was conducted over several years (1997–2008) and on data aggregated at the provincial and national levels. It was found that the average value of the WF at the national level is: 8250 to 3470 $L\,L^{-1}$ for the green WF (without irrigation), 2770 to 4640 $L\,L^{-1}$ for the green WF (with irrigation), 1410 to 2760 $L\,L^{-1}$ for the blue WF (with irrigation) and 710 to 1510 $L\,L^{-1}$ for the grey WF. Since the relevant literature on this subject was found to be limited, no comparative evaluation can be undertaken with other producing nations (e.g. Italy).

The third indicator belonging to the footprint family (Galli et al. 2012) is the *ecological footprint* (hereafter EF). It is evaluated by considering all the direct and indirect inputs that are associated with the analysed system during its entire life cycle (Bastianoni et al. 2013). Each of these inputs is converted in terms of the global hectares (gha) needed to support its production. In particular, the EF of a final, or intermediate, product is defined as the total amount of resources and waste assimilation capacity required in each of the phases necessary to produce, use and/or dispose of that product (Global Footprint Network 2009). If the EF is considered as a stand-alone indicator within LCA, it is defined as the sum of time-integrated direct land occupation and indirect land occupation, related to nuclear energy use and to CO_2 emissions from fossil energy use and cement burning (Huijbregts et al. 2008). Therefore, the EF provides a more differentiated and complete picture of the environmental impact due to the combination of fossil CO_2 emissions, nuclear energy use and direct land occupation in one common metric, 'global hectares' (Huijbregts et al. 2008). One important difference from the original EF approach (Wackernagel et al. 2005) is that the Huijbregts approach considers product-specific yield factors applied to forestry, pasture and crops to obtain the direct land occupation instead of the global average yields (Ecoinvent Centre 2004).

Despite its diffusion and popularity, product EF applications are still scarce and, especially regarding the olive oil sector, there is no adequate background of case studies to highlight the appropriation of natural capital, the efficiency of natural resource use and the environmental pressure related to this sector. Indeed, up to now, studies focusing on the EF of olive oil production processes and phases by phase assessments have still not been published. The olive oil product is always grouped into the category 'oils and fats' related to per capita consumption in territorial footprint assessments, without any clear reference to each individual component. The only available data, obtained by using the original EF approach (Wackernagel et al. 2005), highlights the requirement for 905 g m^2 per capita for the annual consumption of 12 kg olive oil (75.4 g m^2 per capita for 1 kg olive oil consumed), of which 89.3 % is due to the cropland area type and the remaining 10.7 % is due to the CO_2 area type (Scotti et al. 2009). This study refers to the municipality of a northern area in Italy; therefore, it is a very specific and local outcome.

Deeper studies on EF application to the olive oil sector are desirable to monitor the combined impact of anthropogenic pressures that are more typically evaluated independently and could thus be used to understand, from multiple perspectives, the environmental consequences of human activities. In this sense, it would be interesting to know how big the EF related to the agricultural practices, oil mills and waste management could be, highlighting the phase that requires more biologically productive area in terms of the earth's regenerative capacity. From the comparison between EF and biocapacity (i.e. the ecological balance), related to olive oil production, it would be possible to assess the size of the deficit. It is likely that the reuse of part of the wastes as fertilisers may reduce the overshoot and decrease the farm dependence on additional external goods. The main strength of the EF methodology is its ability to explain, in simple terms, the concept of ecological limits, thus helping to safeguard the long-term capacity of the biosphere to support mankind and understand how resource issues are linked with economic and social issues (Bastianoni et al. 2013). In this sense, the EF could be an effective and immediate tool to communicate how much the agricultural and transformation practices in the olive oil sector exceed the ecological limits and how to manage and use the available resources in a sustainable way.

The olive oil sector is also assessed using methodologies other than footprint labels. For example, *emergy, energy and exergy evaluations* can provide a set of information on the human 'processes' 'un'-sustainability from other viewpoints (e.g. the eco-centric viewpoint), which LCA does not take into account (e.g. human labour). In particular, emergy (Odum 1996) provides an estimate of the environmental work required to generate goods and services from a 'donor perspective' (Ridolfi and Bastianoni 2008). Applications of these three methods to the olive oil production chain are scarce. Recently, Neri et al. (2012) compared organic and conventional production in Italy using emergy evaluation. This study highlights that both systems present higher values related to the agricultural phase, even though the organic farm shows a higher environmental performance for all the phases. The conventional system uses 4 % renewable resources, while the organic system uses 12 %. Human labour represents 4.33 % and 25.10 % of the total emergy flow for conventional and organic systems, respectively. This study is the only one to show the importance of human labour, which is a fundamental topic in the olive oil sector, along with the agricultural, transformation and packaging phases.

Agriculture is also the most energy- and exergy-intensive process, with diesel being the dominant energy and exergy source (Özilgen and Sorgüven 2011). In this study, the use of waste vegetable oils converted into biofuel, as an alternative to diesel in heating oil burners, is proposed as an improvement.

A comparison between organic and conventional systems is also provided by an energy use assessment in Spain (Guzman and Alonso 2008). This case shows the lower energy efficiency of irrigated land as opposed to dry land (i.e. non-irrigated) regardless of their style of management and, on the other hand, the greater non-renewable energy efficiency of organic olive growing in comparison with conventional production. The use of '*alperujo*' (olive wet husk) compost and temporary plant covers and the reduction of machinery use to when it is strictly necessary are proposed as possible improvements.

These studies highlight the importance of resource valorisation and the renewability of different forms of production management.

2.3.5 Product Category Rules (PCRs) and Environmental Product Declarations (EPDs)

An Environmental Product Declaration (EPD) is a verified document containing the quantification of the environmental performance of a product or service according to the appropriate categories of parameters calculated using the LCA methodology (ISO 2006a). This methodology allows the EPD to provide objective information by which all the aspects that lead to continuous improvement of environmental conditions related to the production of a product or service can be identified. The EPD communicates the environmental performance of products and services with key characteristics and guidelines that result in a number of advantages for organisations that use the EPD and for those using EPD information (Environdec 2014). The requirements for EPDs of a certain product category are defined in Product Category Rules (PCRs). PCRs are sets of rules, requirements and guidelines for developing an EPD for one or more product categories that can fulfill equivalent functions. PCRs ensure that similar procedures are used when creating EPDs, allowing the comparison between different EPDs.

As far as food products are concerned, numerous PCRs have been developed, including the one for the product category 'virgin olive oil and its fractions' made according to the definition provided on the International Olive Council website (Environdec 2014) and according to Regs. EC 1019/2002, EC 796/2002 and subsequent amendments. On the contrary, 'lampante' virgin olive oil and olive pomace olive oil are excluded. This PCR expired on 31st December 2013; the updated document has been published in April 2014. On the basis of what is reported in the reference PCR, when developing the EPD, the functional unit of 1 L of virgin olive oil must be declared as a unit of the product including the packaging; information on the end-of-life phase of the packaging is also necessary.

The system boundaries included in the PCR provide general upstream, main and downstream processes. In particular, the 'upstream processes' must include the flow of raw materials and energy necessary for the production of virgin olive oil. In the 'processing' of raw materials, the extraction of virgin olive oil from the olive fruits, waste management, storage of olive oil and primary packaging (including transportation) must be included in the 'main process'. Finally, the downstream processes must include transportation from the production site/retailer to the final storage, waste management/recycling, the use of the product by the customer or consumer and recycling or waste management of packaging/materials after use.

In the EPD, the environmental performances associated with each of the three phases of the life cycle are reported separately. In addition, all the data reported in the EPD are subjected to independent verification of the declaration and data, according to the ISO standard 14025:2006 (ISO 2006a). Furthermore, the declaration has to be updated every year and reviewed every three years.

After the issuing of the PCR for olive oil by the International EDP® System, the interest among olive oil industries in the EPD increased. At the 30th September 2013, 8 EPDs were registered, 7 of which refer to Italian olive oil industries or associations. In particular, the first experience involved 68 Greek olive growers from the Peloponnese and Crete, organised by 3 farmers' organisations: Nileas, Pezea Union and Mirabello Union. This experience was soon followed, in the Italian context, by the EPDs achieved by the firm APOLIO (Cappelletti et al. 2012) and afterwards by the association ASSOPROLI Bari and by the firms De Cecco and Monini; the latter certified 4 different types of extra virgin olive oil: 'Granfruttato', 'Classico', 'Poggiolo' and 'Delicato'.

Through a deep analysis of the data referring to the environmental performances reported in the eight EPDs, some differences can be highlighted. These are due not only to the variety of the systems analysed (olive grove management, olive oil extraction system, packaging and transportation), but also to the different assumptions made when the system boundaries were defined. In relation to this issue, indeed, even though there are some differences in the inventory data, all the EPDs include the agricultural phase (upstream phases). Regarding the downstream phases, not all the EPDs consider the use phase and the end of life of the packaging material.

Since references to specific indications are lacking in the PCRs, in some cases only the transportation from the olive oil mill to the retailer are included. In other cases, the use phase and the end of life of the packaging material are also included.

These different assumptions contribute to increasing the variability of the total results (as highlighted in Table 2.4). Starting from the upstream phase, it should be pointed out that a comparison among the different types of olive cultivation cannot be made due to the lack of detailed information. Sure enough, the EPDs give information about the olive grove management system, but there are no quantitative and qualitative data as far as the agricultural practices are concerned: these details could be very useful, especially regarding the business relations with large-scale retailers.

Table 2.4 Environmental performance referred to the olive oil EPDs. (Source: www.environdec.com)

Impact catgory	Unit	F.G.N. P.U. and M.U. 2012.	Apolio 2012	Assoproli 2012	De Cecco 2012	Monini Gran-fruttato 2012c	Monini Classico 2012a	Monini Poggiolo 2012d	Monini Delicato 2012b	Min	Max	Average	St. Dev.
Non ren. material	kg	0.4	0.2	0.3	1	0.6	0.7	0.7	0.7	0.2	1	0.6	0.3
Non ren. energy	MJ	18.6	8.1	40.6	17.3	59	47.5	45.7	53.3	8.1	59	36.3	18.9
Renew-able material	kg	0.1	9.8	0.1	0.0	3,789.3	11,953.1	12,525.8	10,106	0.0	12,525.8	4,798	57,582
Renew-able energy	MJ	1.5	0.3	3.1	2.3	1	1.3	1.3	1.2	0.3	3.1	1.5	0.8
Use of water	m^3	0.3	0	0.5	0.5	3.8	12	12.5	10.1	0	12.5	5	5.6
Use of electricity	MJ	8.3	0.1	4.3	3.1	1.3	1.9	1.9	1.8	0.1	8.3	2.8	2.5
ADP	kg SbEq.	0	0	0	0	26.2	15.4	19.8	23.4	0	26.2	10.6	11.7
GWP	kg CO_2Eq.	2.5	1.3	2.8	2.5	4	3.2	3	3.6	1.3	4	2.9	0.8
ODP	kg CFC −11Eq.	0	0	0	0	0.4	0.8	0.8	0.7	0	0.8	0.4	0.4
AP	kg SO_2Eq.	0	0	0	0	28.3	17	15.8	20.6	0	28.3	10.2	11.5
POCP	kg C_2H_4Eq.	0	0	0	0	6.7	2.9	2.5	4	0	6.7	2	2. 5
EP	kg PO_4Eq.	0	0	0	0	6.6	15.6	16.5	13.7	0	16.4	6.5	7.6

Table 2.4 (continued)

Impact catgory	Unit	F.G.N. P.U. and M.U. 2012.	Apolio 2012	Assoproli 2012	De Cecco 2012	Monini Gran-fruttato 2012c	Monini Classico 2012a	Monini Poggiolo 2012d	Monini Delicato 2012b	Min	Max	Average	St. Dev.
FAETP	kg 1,4-DBEq.	0.8	0	0.7	0.1	106	397.9	421.54	334.3	0	421.5	157.7	192.8
MAETP	kg 1,4-DBEq.	1,692	563.6	1,129.6	221.4	2554,953	888,356.9	938,502.6	751,467.7	221.4	938,502.6	354,610.9	429,897.6
TETP	kg 1,4-DBEq.	0	0	0	0	2.5	7.2	7.5	6.2	0	7.5	2.9	3.5
HTP	kg 1,4-DBEq.	5.402	0.175	2.733	0.339	2,156.038	1,489.442	1,417.620	1,683.583	0.175	2,156.038	844.417	926.341
Land use	$m^2 \times yr$	9.723	23.064	23.391	2.055	0.058	0.060	0.061	0.060	0.058	23.391	7.309	10.354

As regards the analysis of the core phase, the information about the olive oil extraction processes is not always complete. The processes are often described in a generic way and unclear aspects are presented in the related environmental performance.

By analysing the packaging phase, the choice of the container is a further aspect that influences the variability of the results. Indeed, although the functional unit is always one litre of extra virgin olive oil, the glass container used is sized 0.5 L in some cases (ASSOPROLI Bari), in other cases 0.75 L (the group Nileas, Pezea Union and Mirabello Union) and in others again 1 L (APOLIO, De Cecco, Monini). In all the EPDs, the high environmental impact related to the production of the glass container (bottle) is highlighted. This entails, by considering the same functional unit, the biggest container being advantaged (fewer kilograms of glass per litre of extra virgin olive oil) (Cappelletti et al. 2007).

In the downstream phases, the environmental performance, in some cases, is exclusively related to the transportation from the olive oil mill to the retailers (the group Nileas, Pezea Union and Mirabello Union), while, in others, the transportation to the consumer and the packaging disposal are also considered (ASSOPROLI Bari, APOLIO, De Cecco, Monini). However, the environmental impacts related to the phases mentioned above have very little influence on the total impacts declared. As far as the disposal of packaging is concerned, it must be considered, furthermore, that the environmental performance is influenced by the assumptions made for the packaging phase (the type of container) and by the behaviour of the consumers. Therefore, to calculate the environmental impact estimates, data deriving from the literature are principally used.

Definitively, the comparison of the eight EPDs shows that the environmental performance is declared by following the scheme defined by the PCRs and the GPIs (General Programme Instructions). In most of the cases, the evaluation methods are clearly described as well as the impact categories (as defined by the PCRs).

By comparing the aggregated data referring to the environmental performance among the EPDs registered for the olive oil, significant variability in the results is highlighted. In all the cases, the results underline the high environmental impact of the agricultural phases: for almost all the impact categories analysed, indeed, over 50 % of the total impact derives from the olive cultivation phase. This is an aspect that is frequently observed when an LCA study is carried out in the olive oil sector (Salomone et al. 2010b); this also represents a typical hot spot of the agro-food sector, towards which further efforts should be oriented in order to reduce the environmental impact (Salomone et al. 2013a) and decide on the best practices to be applied to the whole sector.

2.3.6 Other Tools

The assessment of the eco-profile of a food product system is a complex task due to its huge overlap with other product systems and due to uncertainty, which often affects the results of the analysis (Avraamides and Fatta 2008).

Olive oil is one of the most representative products of the food sector in the Mediterranean area, and related environmental LCA studies show significant environmental impacts associated with the resource consumption and waste releases from the relative agricultural stage and production processes (Ardente et al. 2010; Cellura et al. 2012). Due to the complexity and heterogeneity of the agricultural processes, such as the crop variety, and the different levels of mechanisation in the field, suitable methodologies to quantify the environmental sustainability of the olive oil chain are needed. In such a context, decision-making support tools, in particular multi-criteria analysis (MCA), could aid LCA experts in selecting the option, among several, that attains the best environmental performance, according to a set of criteria defined by the decision maker (Beccali et al. 2002a, b).

Within the specific literature, there are some studies on the integration of LCA and multi-criteria analysis (MCA) methods as effective tools to analyse the olive oil production chain (Beccali et al. 2003). Among these, Recchia et al. (2011) assessed different scenarios concerning the agricultural phase, the olive transport from the grove to the mill and the extraction phase. The application of MCA identified five optimal scenarios, according to the evaluation criteria defined by the following rules: (1) five environmental criteria, preferring scenarios characterised by a low level of field mechanisation, a short transport distance and highly efficient extraction plants exploiting reused energy from field and plant wastes (pruning and pomace stone); and (2) three economic criteria, taking into account harvesting and pruning costs and olive productivity. The weighed ranking derived by the MCA shows that the highest score is assigned to the high-intensity scenario characterised by a high score for the economic criteria, due to the mechanised field management and a significant olive yield, and a medium score for the environmental criteria, due to the reuse of pomace and pruning residues. Within the ranked scenarios, the one characterised by economic drawbacks, due to a low level of mechanisation and low olive productivity, shows the lowest environmental impact, due to the presence of traditional groves containing an olive mill. Then, LCA was applied to the above five scenarios in order to identify the one with the lowest environmental impacts, in terms of global warming potential (GWP) and global energy requirement (GER). The results of the LCA endorse the results of the MCA: the traditional grove scenario involves the lowest GWP and GER. This outcome is essentially due to the absence of organic fertilisation and irrigation plants and to the reuse of prunings as biofuel for the mill's energy requirement. Conversely, the worst eco-profile was found in the high-intensity scenario, in which by-products (pomace and vegetation water) are treated as waste.

In other studies, multi-criteria analyses have been conducted for the interpretation of LCA results. Among these, in Cavallaro and Salomone (2010), the joint use of LCA and a multi-criteria algorithm was developed and applied to the olive oil chain. The tool derives from PROMETHEE (Preference Ranking Organization Method on Enrichment Evaluation) (Brans and Vincke 1985), using the outranking approach based on a pair-wise comparison of alternatives for each criterion. Such a tool was applied to eight scenarios of conventional and organic olive oil production and assessed following a life cycle approach. The results show that the preferable scenario is conventional tree cultivation, oil extraction with a three-phase system and co-composting of olive husk and olive mill wastewater (the obtained compost is considered as avoided production of fertiliser and stones as avoided production of fuel). On the contrary, the worst environmental performances are related to two scenarios: one with organic olive tree cultivation and the other involving considerable use of pesticides and chemical fertilisers.

In conclusion, although there are few studies in the literature, the integration between LCA and MCA has proven to be particularly useful in gaining a better understanding of complex comparisons among different scenarios of olive oil production, which are generally characterised by many differences in single processes (e.g. pest treatment, cultivation management, olive oil extraction technologies, etc.).

2.4 Methodological Problems Connected with the Application of Life Cycle Assessment in the Olive Oil Production Sector: Critical Analysis of the International Experiences

In order to highlight the main methodological problems that emerge when LCA is applied to the production of olive oil, a previous analysis conducted only on Italian case studies (Salomone et al. 2010b) was widened and deepened, performing a critical analysis of the international experiences of LCA in this specific sector. The critical analysis followed three basic steps of investigation:

1. Mapping of the international LCA studies on olive oil—on the base of the state-of-the-art analysis presented in Sect. 2.3, it emerged that, as of the 30th September 2013, 72 studies have been published on olive oil, olives in general (for oil or table use), olive oil mill waste treatment and valorisation and table olive and olive oil packaging (see Table 2.1). With the aim of clearly identifying the specific applicative and methodological problems encountered when LCA is applied in the olive oil sector, the critical analysis presented hereafter focuses only on the applicative case studies that used the LCA methodology connected directly or indirectly with the olive oil production supply chain, so that papers reporting literature reviews, methodological discussions, application in the table olive sector and the application of LCT tools other than LCA were excluded (resulting in the inclusion of 50 scientific articles in the following analysis).
2. Data collection concerning the applicative and methodological aspects related to the identified case studies—after the mapping, all the data relevant to the comparative analysis were collected for each study by using a dual input channel information flow:
   - a checklist, following the ISO 14044:2006 requirement structure (ISO 2006c), for the collection of the most important information contained in the published study;

- a questionnaire, aimed to highlight the main issues not directly deductible from the paper; pursuing this goal, the questionnaire was directly completed by the authors of each study and it was therefore used to gather the information not contained in the published work, but essential for the correct understanding of the most important issues concerning the applicative and methodological aspects encountered by applying LCA to this specific sector of analysis (it is necessary to clarify, however, that 24 % of the questionnaires were not considered because the authors did not reply to the request for collaboration with this research);

3. Implementation of the comparative critical analysis—the collected data were then organised into a database in order to simplify the comparative and critical analysis of the international experiences gathered and to highlight the common features and/or differences connected to the investigation of the fundamental aspects of LCA studies.

The 50 analysed case studies show very heterogeneous characteristics in size, content and depth of analysis; they report the results, more or less exhaustive, of applicative case studies carried out on the cultivation of olives, olive oil extraction, olive oil packaging and/or treatment of waste in the olive oil industry. As far as the form of publication and the methodology used are concerned, these studies show, however, more homogeneous features. In fact, the papers were mostly published in conference proceedings (42 %) and in scientific journals (30 %), while 12 % are Environmental Product Declarations (EPDs) and the remainder (about 16 %) consists of other types of documentation, such as book chapters or reports. As explained in Sect. 2.3, grey literature could be missing.

The LCA methodology was used as a single tool in 66 % of the papers (including two cases of simplified LCA) or in conjunction with other assessment methods—such as life cycle costing or another kind of economic analysis (10 %) and carbon footprint and emergy analysis (4 %)—or communication tools (indeed, papers containing EPD descriptions or EPDs represent 20 % of the gathered documents).

Focusing on the ISO 14044's specific requirements, LCA case studies present various characteristics that are briefly described in the following sub-paragraphs with the aim of highlighting how the main applicative and methodological aspects were dealt with in the international case studies.

2.4.1 The Goal and Scope

The goal and scope of an LCA shall be clearly defined and consistent with the intended application–ISO 14044:2006, 4.2.1 (ISO 2006c), because the choice of the functional unit, the identification of the system boundaries, the time horizon of the study and, in more general terms, the depth and direction of the whole study will depend on its delineation. As shown in Fig. 2.2, most of the papers surveyed have as their scope the evaluation of the potential environmental impacts (60 %), the identification of the environmental burdens (58 %), the identification of hot spots (35 %)

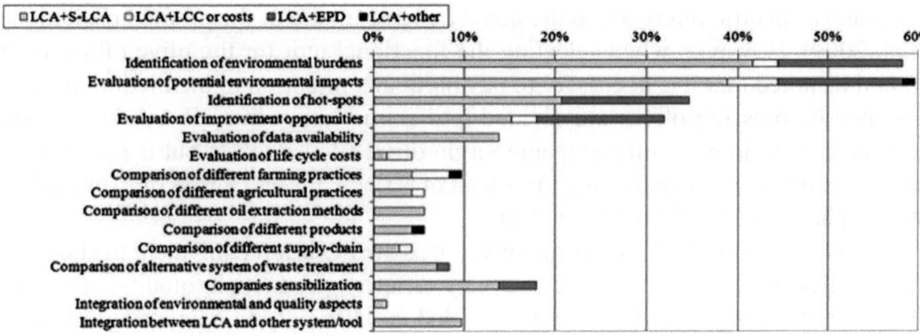

Fig. 2.2 The goal and scope in the surveyed case studies

and the evaluation of improvement opportunities (32 %) (each study may have more than one goal). Furthermore, the various kinds of comparative evaluation (totalling about 39 %) and the company sensitisation (18 %) are among the main goal and scope of the surveyed case studies. While all the studies unambiguously state the reason for carrying out the study, none clearly define the intended audience, except EPDs, which obviously are disclosed to the public.

2.4.2 The Functional Unit

Figure 2.3 shows the functional unit (FU) adopted in the case studies surveyed. The FU should be consistent with the goal and scope of the study—ISO 14044:2006, 4.2.3.2 (ISO 2006c). In most of the papers, the FU is a certain amount of olive oil (1 kg, 1 L or 0.75 L) with different dictions (olive oil, virgin olive oil, extra virgin olive oil or simply oil), but it does not seem that the diction has a specific load in the goal and scope of the analysis, except in a few cases, for example the one

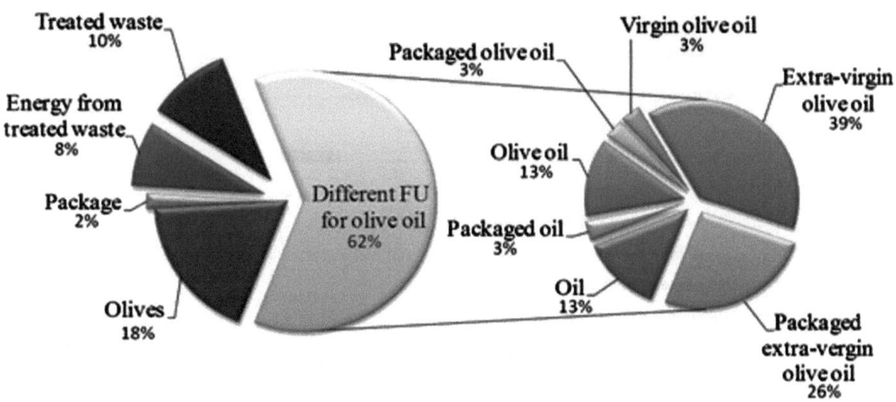

Fig. 2.3 The functional unit (FU) in the surveyed case studies

including a specific reference to the quality characteristic of the product (Salomone et al. 2009). However, when selecting the functional unit for the olive oil chain, it should be noted that it is necessary to pay particular attention to the diction: the oils obtained by pressing olives are divided into extra virgin olive oil, virgin olive oil and current virgin olive oil (lampante virgin olive oil also exists but is not a food), while the diction olive oil is used in a blend of refined oils and virgin oils (excluding the lampante virgin oil) (see Sect. 2.2).

Therefore, choosing 1 L of virgin olive oil as the FU is not equivalent to choosing 1 L of olive oil, because they are two very different products in qualitative terms. However, the analysis of the studies revealed the difficulty in comparing oils with completely different organoleptic characteristics and yields (which also depend on cultivars, harvesting and oil extraction). The investigation performed involving the authors of the case studies allowed us to highlight that 24 % of the responding authors declared that they had encountered difficulties in choosing the FU, mainly linked to the comparison of completely different olive oils. Indeed, 50 % of the analysed papers report comparative studies mainly focusing on the comparison of cultivation practices and of the different olive oil extraction methods (see Fig. 2.4). Exploring the answers of the authors participating in the investigation, further information can be outlined; for example, it can be observed that 36 % of the authors of comparative studies declared themselves to have faced problems in the definition of the goal and scope requirement, while only 16 % of the authors of non-comparative studies encountered problems in this phase of the LCA study. Examining the comparative studies in more detail, the main problems in goal and scope definition were mainly linked to the choice of a proper FU (78 %): the chosen solution was often simplification and the functional unit selected was a certain amount of generic oil or olive oil in order to include olive oils with different organoleptic properties.

Another difficulty when choosing the functional unit was the identification of a common element when considering the whole production chain, including olive oil waste treatment. In this case, a certain amount of olive oil or of olives was chosen as a functional unit. Olives as the FU were generally selected when the analysis was limited to the cultivation phase or when the whole production chain, including olive

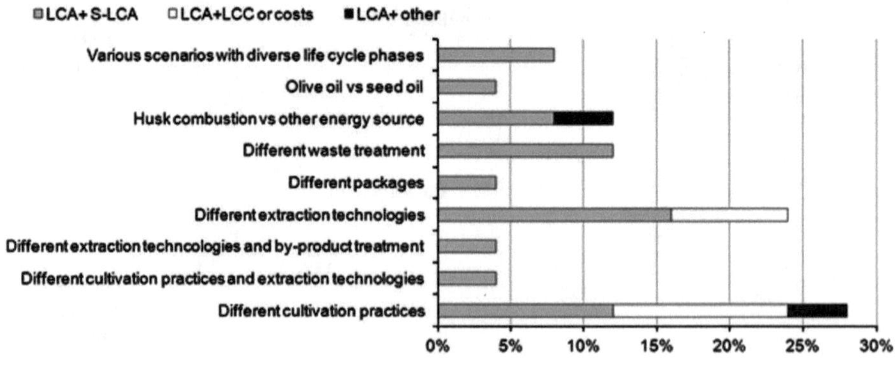

Fig. 2.4 The object of investigation in comparative case studies

oil waste treatment, was included. The choice of the functional unit, however, was strongly related to the purpose of the study and to the system boundaries.

2.4.3 The System Boundaries

When choosing the system boundaries, the surveyed studies adopted different methods; thus, general conclusions cannot be drawn from the results reported in the various scientific articles, but common issues can be identified. Indeed, the main problems encountered by the authors, concerning the definition of the system boundaries, were determined by the lack of significant data about some processes of the chain (e.g. the combustion of olive husk and pits, characteristics of the quality of husk compost and different types of husk, waste/by-product processing, end-life of the olive groves), which caused these processes to be excluded from the system boundaries. In other cases, doubts regarding the attribution of some treatment processes of olive oil waste were detected, such as the processes in the oil husk industry. These problems were solved using several methods: exclusion from the system, inclusion in the system and appropriate allocation among the various products of the oil husk industry and/or appropriate choice of the functional unit (e.g. the quantity of olives processed).

Despite these differences, however, it was possible to verify, as shown in Fig. 2.5, the chain phases that have received the most attention: cultivation, olive oil production, transport linked to these processes and olive oil mill by-product/waste treatment (including both the treatment in the olive oil husk mill and the other types of treatment of olive oil mill by-products/waste).

The deletion of life cycle stages, processes, inputs or outputs is only permitted if it does not significantly change the overall conclusion of the study; any related decision must be clearly stated and the reasons and implications for their omission must be explained—ISO 14044:2006, 4.2.3.3 (ISO 2006c). Of the analysed studies, 74 % specified the exclusion of some processes from the system boundaries, mainly

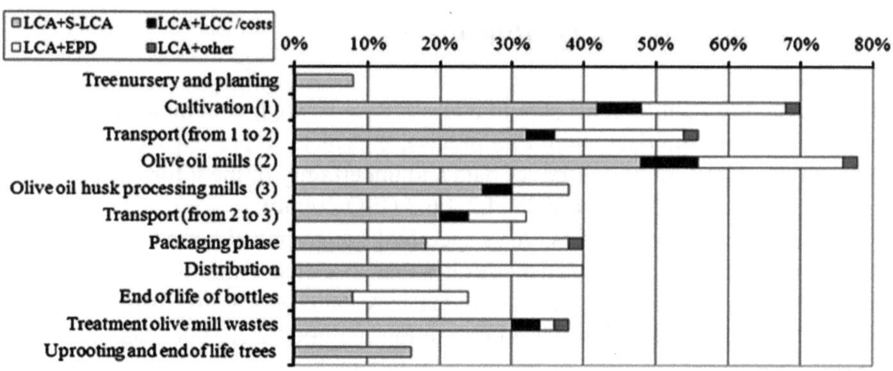

Fig. 2.5 The system boundaries in the surveyed case studies

because, being comparative studies, the processes were common to the systems analysed (24 %), while in the other cases, the reasons were mainly linked to missing data and/or incomplete information (24 %), but rarely were the implications clearly stated.

Even if the system boundaries and exclusions were not clearly detailed in all the studies, the analysis revealed that 70 % of the studies included in the analysis the cultivation phase, which was organic cultivation in only one case, 11 % of the cases integrated cultivation and 31 % included conventional cultivation, while 29 % of the studies included a comparison of two or three farming systems (conventional, integrated, organic); on the contrary, in the other studies including the cultivation phase, the farming practice typology was not specified. In 54 % of the case studies, the cultivation systems were also differentiated according to the agronomic technique (dry 37 %, irrigated 37 % or both 15 %). Furthermore, only 11 % of the studies that included the agricultural phase in the system boundaries also accounted for olive grove planting, while 40 % explicitly stated that this phase was excluded, mainly due to missing data (43 %), the consideration of the cultivation of olive trees more than 25 years old (36 %) or the comparative nature of the studies (14 %).

Concerning the olive oil extraction phase, the analysis revealed that 46 % of the studies that included this phase analysed the three-phase continuous system (including three cases of the de-stoning process), 8 % the two-phase continuous system, 8 % the discontinuous system and 5 % continuous centrifugation with a two-and-a-half-phase system (also called the modified system or water-saving system); 23 % investigated a comparison of different olive oil extraction methods, while the remainder did not specify the technology used (therefore failing to comply with the data quality requirements—see Sect. 2.4.4).

Focusing on the 76 % of case studies including olive oil mill by-product/waste treatment, 50 % included the treatment of olive oil husk in olive oil husk mills, while the remainder referred to other treatments of olive oil wet husk and olive oil wastewater. In particular, only 8 % of the studies focused on this phase of the life cycle, while in the other cases two or more life cycle phases were considered together with the waste/by-product treatment.

2.4.4 Availability and Quality of Data

The data quality requirements should address time-related, geographical and technology coverage; the data should be precise, complete, representative, consistent and reproducible; and the sources of data and uncertainty of the information should be clearly stated—ISO 14044:2006, 4.2.3.6.2 (ISO 2006c).

Of the analysed case studies, 76 % specified geographical boundaries, whereas 54 % specified temporal ones (all the studies that specified temporal boundaries also specified geographical ones). Technological coverage was almost always specifically stated when different olive oil extraction methods were considered (as previously observed, only 10 % of the case studies including the olive oil extraction phase did not specify the method).

A total of 94 % of the analysed papers used primary data collected from various companies of the olive oil sector, 86 % used an LCA database and 21 % used data available in the literature. The most commonly used databases were Ecoinvent (54 %), the SimaPro Database (30 %), Buwal 250 (21 %), ETH-ESU 96 and IVAM LCA 3 (both cited in 16 % of the studies) and the PE International Database (9 %). In 40 % of the studies, the data quality was verified with various methods of analysis, 80 % using sensitivity analysis.

Concerning the data availability, the inventory phase of the agro-industrial sector still suffers from a lack of data availability and data uncertainty (especially for certain types of materials, such as herbicides and pesticides), as well as problems related to emissions estimates of nitrogen and phosphate compounds and the dispersion of pesticides, the use of agricultural machinery and the CO_2 emission balance.

The comparative analysis conducted on the studies of LCA, considering only the applied studies including the agricultural phase, confirmed these critical issues:

- 53 % of the authors who responded to the investigation lamented the lack of data in databases about the production of pesticides; 13 % of these excluded the process from the system boundaries and 75 % of the studies used data in the database for similar compounds and weighted the results based on the active ingredient;
- 40 % of the authors commented on the lack of data on fertiliser production in the databases and their solution was always to use data from the databases modified according to the content of N, P and K;
- 7 % of the authors lamented the lack of data concerning the production of herbicides in the databases and the lack of data regarding the emissions from herbicides; their solution was always to exclude them from the system boundaries;
- 53 % of the authors of these studies referred to the lack of data regarding emissions due to pesticide use and the difficulty in calculating the pesticide dispersion in soil, air and water; the solution was to use models to estimate emissions in 25 % of the studies (such as the successive enhancement of the model; Birkved and Hauschild 2006; Dijkman et al. 2012; Hauschild 2000), in 50 % of the cases the emissions were estimated using literature data or were considered to be similar to other compounds and in 13 % of the cases they were excluded;
- 43 % of the authors lamented the lack of data regarding emissions from fertiliser use and the difficulty in calculating the dispersion in soil, air and water; the solution in 31 % of the cases was to use estimation models, such as the Brentrup model (Brentrup et al. 2000) for nitrogen compounds and data from the literature regarding the behaviour of phosphorus and potassium fertilisers; in 62 % of the cases the substances contained in the fertiliser were calculated using estimations from the literature (e.g. using the ratio between the real weight and the molecular weight and then estimating the emissions to the air, water and soil);
- 37 % of the authors had problems calculating the emissions from the use of agricultural machinery based on the type of work, due to insufficient data or uncertain data sources; the solution was mainly (82 %) to consider the emissions to be derived from fuel consumption.

Other issues encountered in these studies are connected to the balance of CO_2 emissions and the lack of characterisation methods. The balance of CO_2 emissions was difficult to determine for 33 % of the responding authors due to a lack of specific data, and the solution was to use generic data collected from the database, if available, estimation from the literature or exclusion.

2.4.5 Allocation Methods

Whenever possible, allocation should be avoided by dividing the unit process to be allocated into two or more sub-processes or expanding the product system. When allocation cannot be avoided, the allocation procedures should be clearly stated and explained and, whenever several alternatives seem applicable, a sensitivity analysis should be conducted—ISO 14044:2006, 4.3.4 (ISO 2006c). Among the applied studies, 36 % used some form of allocation: of these analyses, 56 % used allocation methods for olive oil and for olive oil husk; 17 % for olive oil, husk and olive stones; 11 % for the various products of the oil husk industry; and 6 % for sunflower oil and meal. Some (11 %) studies also applied allocation to husk and wastewater or to different products resulting from wastewater treatment. In the studies including allocation, this was calculated in 33 % of the cases based on the price, in 11 % of cases by mass and in 28 % of cases by price and mass; the remainder did not mention the allocation method.

Allocation, especially in systems in which the various waste treatment technologies are included, was considered a problem by 37 % of the authors who responded to the investigation. These authors cited different motivations mainly connected to how to allocate the environmental load of the olive oil extraction process (among olive oil and the other by-products), but also difficulties in the representation of the reuse of pruning residues as natural fertilisers. The most commonly cited solutions were the expansion of the system boundaries in order to include the process and calculate the advantage obtained from the avoided product or the application of allocation using different methods coherently with the scope of the analysis.

2.4.6 Life Cycle Impact Assessment (LCIA)

The LCIA should be carefully planned to achieve the goal and scope of the study. The mandatory elements of the LCIA include the selection of impact categories, classification and characterisation, while the optional elements are normalisation, grouping, weighting and data quality analysis—ISO 14044:2006, 4.4 (ISO 2006c).

Regarding the impact assessment, only 12 % of the studies reported all the phases of LCIA. Classification and characterisation results were described in 90 % of the cases, normalisation in 36 % of the cases, grouping evaluation in 16 % of the cases and weighting evaluation in 20 % of the cases.

The identification of the selected impact categories and related assessment methods was particularly complex because 14 % of the papers lacked sufficient elements to be able to detect the full data. Focusing only on the papers in which the information was specified, the most frequently used evaluation method was the CML in its various versions (28 %), followed by Eco-Indicator 99 (26 %), EPS 2000 (16 %), ReCiPe in its various versions (14 %), IPPC 2007 (12 %), Impact 2002 (9 %) and EDIP 96 (7 %). Sometimes, the CML was applied with modifications and/or additions, such as updates of the characterisation factors (IPCC for GWP) or the addition of the land use, the energy content or weight factors that take economic aspects into account. On the contrary, the changes to the method Eco-Indicator 99 (particularly the E/E) mainly described the costs and benefits of olive oil for human health. The most commonly used impact categories were global warming (92 %), acidification (82 %), ozone layer depletion (78 %), photochemical oxidation (74 %) and human toxicity (60 %).

2.4.7 *Interpretation and Tools Supporting the Interpretation Analysis*

The life cycle interpretation phase comprises several elements, such as the identification of the significant issues, an evaluation that considers completeness, sensitivity and consistency checks and conclusions, limitations and recommendations ISO 14044:2006, 4.5 (ISO 2006c).

All the reviewed studies report information on the interpretation phase, though with different levels of depth. In all of these, it was possible to identify the significant issues, but papers reporting conclusions, recommendations and limitations are scarce. Moreover, the reported elements are too fragmented and poorly defined to allow us to achieve important comparative results: different choices of functional units and system boundaries did not enable unequivocal conclusions to be reached. However, it can certainly be outlined that 51 % of the studies that accounted for both the agricultural and the other stages of the life cycle (with or without the intermediate stage of transport) identified the agricultural phase as the most polluting. In the agricultural phase, the agronomic practices with the greatest environmental impact were the spreading and use of fertilisers and the spraying and use of pesticides. The most important impact categories were eutrophication, acidification and ecotoxicity (in its various forms) and the most polluting substances were fertilisers, pesticides and energy consumption.

Only 16 % of the analyses used sensitivity analysis for the evaluation of interpretation results.

2.4.8 *Critical Review*

A critical review (CR) by experts is a process that seeks to ensure that the LCA study is aligned with the requirements of ISO 14044:2006, is scientifically and technically valid, is consistent with the goal and scope of the study and is transparent and consistent (ISO 2006c). Except for EPDs, none of the other examined studies present elements suggesting that a critical review was carried out by external independent experts. Even though a CR undoubtedly improves the credibility of a study, it is still rarely practised, maybe due to the additional costs incurred, and only organisations working with environmental labelling or product declarations push themselves to demonstrate the quality of their LCA results with a CR. The International Organisation for Standardisation recently published (May 2014) a technical specification based on the critical review process in order to specify better the requirements contained in the ISO 14044 (ISO 2014a).

2.5 The Implementation of the Life Cycle Assessment Methodology in the Olive Oil Production Sector: Lessons Learned

The state of the art and literature review of the international experiences of the LCT approaches applied in the olive industry (presented in Sect. 2.3) and the critical comparative analysis of the applicative LCA case studies in the olive oil production supply chain (presented in Sect. 2.4) allowed a better understanding of the specific methodological and applicative issues that a practitioner might encounter when applying the LCA methodology in the sector of olive oil production, and many points for reflection and improvement emerged.

When performing an LCA study, the first preliminary suggestion is to gain a clear and deep knowledge both of the supply chain to be studied and of the full LCT methodological panorama currently available.

General methodological guidelines already exist, such as:

- the ISO standards on the LCA methodology, in particular ISO 14040 (ISO 2006b), ISO 14044 (ISO 2006c) and the related technical reports and technical specifications;
- the ISO standards on environmental labels and declarations, in particular ISO 14020 (ISO 2000), ISO 14021 (ISO 1999), ISO 14024 (ISO 1999) and ISO 14025 (ISO 2006a);
- the ILCD (International Reference Life Cycle Data System) Handbook (EC, 2012);
- the ISO technical specification on the carbon footprint of products ISO/TS 14067: 2013 (ISO 2013a) and the forthcoming standard on water footprint (ISO 2014b);
- the Ecological Footprint Standard (Ecological Footprint Standard, 2009).
- Furthermore, some guidelines specifically focus on food products, such as:
- the Envifood Protocol—Food and Drink Environmental Assessment Protocol (European Food Sustainable Consumption & Production Round Table, 2013);
- Product Category Rules (PCR) and Product Environmental Footprint Category Rules for food and drink products (PEFCRs).

All the above guidelines highlight the importance of taking into account the life cycle approach, including all the stages from raw material acquisition through processing, distribution, use, end-of-life processes and all the relevant related environmental impacts.

This chapter aims to deepen the research and further to suggest best practices as actions that could be easily implemented by stakeholders, when developing LCAs in the olive oil production sector. In the following, the lessons learned from the literature review and the critical comparative analysis are briefly presented in order to summarise not only the issues emerging from the current practice, but also the needs for further research work aiming to improve the LCA implementation in this specific agri-food sector; we suggest that practitioners carrying out LCA studies on olive oil should follow the subsequent suggestions at the level of both methodological issues and hot spots.

2.5.1 The Goal and Scope

Goal and scope definition is the first step of an LCA analysis and should set the overall context of the study, defining its aims, methods of impact assessment and intended application. Furthermore, the scope should include the definition of the functional unit and the system boundaries, referring them to the aim of the study. The goal and scope of an LCA implemented in the olive oil sector (as in any other sector) should be clearly defined and unambiguously state the reason for carrying out the study. This task seems particularly simple but, considering that the goal and scope delineation will affect the choice of the functional unit, the identification of the system boundaries, the time horizon of the study and, in more general terms, the depth and direction of the whole study, caution should be applied when defining them; in particular, some elements that deserve to be highlighted are:

- when presenting the scope, the reasons for such a choice should also be explained (e.g. if the scope is the identification of hot spots, the purpose of their identification and their use should also be clarified);
- the intended audience should be defined, in order to understand clearly whom the results target and the kind of use the audience may make of these results.

2.5.2 The Functional Unit

Choosing the functional unit (FU) is one of the very first critical tasks encountered when carrying out an LCA study and the keystone of the whole project. The choice of the FU may vary according to the aim of the LCA study and may be determined in different terms, such as functionality, nutritional value, portion size or other criteria. A functional unit is defined by the ISO 14044 norm as the 'quantified performance of a product system for use as a reference unit'. In addition, the ISO 14040 norm indicates that: 'The functional unit defines the quantification of the identified functions (performance characteristics) of the product. The primary purpose of a functional unit is to provide a reference to which the inputs and outputs are related. This reference is necessary to ensure comparability of LCA results. Comparability of LCA results is particularly critical when different systems are being assessed, to ensure that such comparisons are made on a common basis.'

As highlighted in Sect. 2.4.2, when selecting the FU for the olive oil chain, particular attention should be paid to the diction of olive oil, which may indicate very different products in qualitative terms. In this sector, although the general LCA guides allow a certain amount of flexibility, with regard to the olive oil production processes the European Food Sustainable Consumption & Production Round Table (European Food Sustainable Consumption & Production Round Table, 2013) suggests that weight or volume are the most suitable; however, due to the extremely wide variability in the quality of the oils (the price of an extra virgin olive oil rises from a few euros per litre to a few tens of euros per litre), it is very important also to include the product quality in the functional unit. But how can the quality of olive oil be defined? What defines the quality of olive oil? Certainly, the quality of an olive oil depends on characteristics such as acidity, flavour and the E vitamin and tocopherol content. Hence, how can the most appropriate FU be identified? Indeed, different authors of LCAs in this specific productive sector have encountered difficulties in choosing the proper FU, mainly when performing comparisons of completely different olive oils or when considering the whole production chain, including olive oil waste treatment.

Keeping in mind that the choice of the FU is strongly related to the purpose of the study and to the system boundaries, though, some guidelines could be suggested, as summarised in the following sub-paragraphs and in Table 2.5.

When the LCA aims to analyse the whole olive oil chain, a certain amount of olive oil can be used (e.g. 1 L or 1 kg), paying particular attention to the diction of the different types of olive oil (extra virgin, virgin, etc.), and the packaging should be included, especially if the LCA results should be declared in an EPD (as indicated in the PCR 'virgin olive oil and its fractions' of the International EPD System®).

When the LCA focuses on one or two specific phases of the life cycle of olive oil production, the FU should be chosen in order to provide better the reference to which the input and output data of these phases will be normalised (e.g. a certain surface of the olive grove—1 hectare—for the cultivation phase or a certain amount of waste—1 kg of wastewater—for waste treatment processes).

When considering the whole production chain, including olive oil waste treatment, the difficulty lies in choosing an FU that represents a common element; in this case, a certain amount of olives might be the most suitable choice.

In comparative analysis between different oils (e.g. olive oil and seed oil), quality indicators could be used in quantitative ways: for instance, due to the much stronger taste of the extra virgin olive oil than the seed oil (e.g. sunflower), one can state that the FU could be the quantity of oil needed to mix a portion of salad: in this case, experimentally one can identify the two quantities that carry out the same function, which will be, for example, one unit of extra virgin olive oil versus four units of sunflower oil.

Table 2.5 How to choose the functional unit when conducting an LCA of olive oil

Requirement	Possible choices	Recommended when
Functional unit	Hectare	The system boundaries include only the cultivation process
	Olives	The system boundaries include all the phases from cultivation to waste treatment
	Oil	In a comparative study of olive oil and other seed oil
	Olive oil	In a comparative study of olive oils with very different organoleptic characteristics
	Extra virgin olive oil Virgin olive oil	In a single product study or in a comparative study of olive oils with very similar organoleptic characteristics
	Antioxidants (polyphenols and tocopherols)	If the nutritional characteristics of the product are of primary importance for the description of the system
	Olive mill waste	The system boundaries include only waste treatment processes

In comparative analysis among extra virgin olive oils (the best quality of olive oils), indicators of the olive oil quality should be taken into consideration, as the prices or, if available, the score that the olive oil has received in the panel test (EC 1991).

When the nutritional characteristics of the product are at the core of the goal and scope of the study, the quantity of antioxidants (polyphenols and tocopherols) present per litre/kg of extra virgin olive oil could be considered. A functional unit of this kind allows researchers to consider not only the yields per hectare (which greatly affect the environmental impact attributable to the FU as oil, olive oil and extra virgin olive oil), but also the quality of the product, which is sometimes overlooked in industrial production. Another suggestion to follow is to use a set of different functional units (quantity or volume, price, panel test score, etc.) and to assess the variability of the results on the basis of the use of the different FUs in the sensitivity analysis. Therefore, the choice of an appropriate FU for an LCA study in the olive oil sector seems to be an issue requiring particular attention; in Table 2.5, some suggestions are highlighted.

2.5.3 *The System Boundaries*

The choice of the processes that should be included in or excluded from the study depends on the defined goal and scope, according to the availability and quality of data related to the analysed processes. As a consequence, no specific guidelines can be drawn for this topic. In any case, it should be noted that, for EPD communication purposes, the system boundaries are clearly indicated in the PCR ‘virgin olive oil and its fractions' of the International EPD System®, which specifies the requirements for the definition of system boundaries (divided into upstream, core and downstream processes), geographical and time boundaries, boundaries to nature and boundaries to other product life cycles. In general, the system boundaries should, as far as possible, include all the relevant life cycle stages and processes; they should be defined following general supply chain logic, including all the stages: agricultural, industrial, by-product management, transportation/distribution and consumer shopping, food preparation and cooking, consumption and waste management. Human digestion and excretion should be included in the system boundaries, even if they remain the least-studied life cycle stages of all food products. Concerning the carbon balance, one should try to avoid it equalling zero, but focus on the real verification of the carbon balance, which can be modified depending on which effect overrides the other (sequestration or emission). Of course, the effect of sequestration prevails in the majority of studies that follow this approach and therefore the total carbon balance is negative (thus good for the environment).

The literature review and the critical comparative analysis presented in the previous paragraphs highlight that, regarding the definition of the system boundaries, the main problems encountered by the authors of the surveyed studies were determined by the lack of significant data on some specific processes of the chain, which caused these processes to be excluded from the system boundaries, which in turn caused the need to redefine and recalibrate the goal and scope of the study (according to the iterative nature of LCA methodology). This means that one of the most significant issues on which further research work should be focused is the availability of LCI data, especially for some kinds of processes for which there is still a lack of complete and reliable data, as considered more extensively in Sect. 2.5.4.

2.5.4 Quality of Data

Data availability and data quality constitute one of the main problems of LCAs applied in the agri-food industry; with particular reference to the olive oil sector, the literature review and critical comparative analysis reported above revealed that there is still a lack of complete and reliable data for many kind of processes located differently in the various life cycle phases.

As in other agri-food production (Notarnicola et al. 2012), in olive oil production, most of the problems also concern the agricultural step specifically. Hence, this phase is often partially assessed due to different reasons, almost always linked to the unavailability of data, such as:

- The production of some specific kinds of fertilisers, herbicides and pesticides—this problem is usually tackled by excluding the production of these inputs or including the production of a generic fertiliser/herbicide/pesticide (present in the available databases), by entering the quantitative data on the effective consumption of the input weighted according to the active ingredient of the fertiliser/herbicide/pesticide in the database;

- The dispersion of compounds into the environment (air, water and soil) deriving from the use of fertilisers, herbicides and pesticides—this problem is usually approached by excluding these emissions or estimating them using a specific model of dispersion, such as the Brentrup model for fertiliser dispersion, and the successive enhancements of the PestLCI model (Birkved and Hauschild 2006; Dijkman et al. 2012; Hauschild 2000) for pesticide dispersion. In general, the direct emissions from chemicals should be stressed more in environmental assessments of the cultivation phase, but a few times they have been included in the calculation. It is also important to underline that a more complete database on chemicals should lead to a more 'realistic' evaluation of the potential impacts;
- The balance of CO_2 emissions—this calculation is generally omitted (thus implicitly considering the carbon balance as net zero); more recently, a number of studies have begun to include the carbon balance in the boundaries, but, due to a lack of specific data and characterisation methods, generic data collected from commercial and free databases or estimations from the literature were used (Carvalho et al. 2012; Iraldo et al. 2013; Nardino et al. 2013; Palese et al. 2013; Sofo et al. 2005); it should also be highlighted that the CO_2 absorbed by the plants during their vegetative cycle (the age of plants plays an important rule) should be taken into account;
- The emissions from the use of agricultural machinery—these emissions may also significantly change based on the type of machinery, the type of work and the type of ground, but due to insufficient or uncertain data only the emissions deriving from fuel consumption are generally included;
- The use of pruning residues as natural fertilisers—frequently the destination of pruning residues should also be included, because they are often used as fertiliser or for domestic heating, bringing significant benefits for certain impact categories;
- Plant breeding and tree planting—few studies include the establishment of olive groves, generally because they consider new cultivation with young trees; furthermore, the PCR 'virgin olive oil and its fractions' of the International EPD System® specifies the inclusion of this process only 'if the olive grove life time is expected to be less than 25 years', but, even if the PCR does not mention it, in this case the end-life of the olive groves should also be considered;
- Double counting and incorrect attributions—when collecting primary data in an agricultural firm that produces mainly olives/olive oil (with or without a private mill), generally data related to different processes (mechanical processing of the soil, phytosanitary treatments, canopy management, fertilisation, etc.) will be available and will often be detailed and precise; however, if the company has many cultivars, special care should be paid to reporting all the data obtained for the FU choice, thus avoiding double counting and incorrect attributions.

Concerning the other life cycle phases, problems of data quality and availability may occur: in the olive oil extraction phase (e.g. because in the commonly available data sources many of the industrial processes involved are lacking, so that emissions are only related to energy consumption) and above all in the waste treatment phase (which is still lacking relevant data for many processes that characterise the olive oil production supply chain). Examples of processes characterised by a general lack of data are: the combustion of olive pomace and pits; the quality characteristics of pomace compost and the different types of pomace; emissions from composting activities; emissions from the combustion of exhausted pomace; emissions from the spreading of OMW on soil, etc. In particular, OMW is a significant potential pollutant (high phytotoxicity, see Roig et al. 2006), but also contains valuable substances such as nutrients that could be reused in cropland and avoid the negative effects; the OMW should be considered as a new raw material necessary to make a new product and it should be valorised in LCA studies.

In the case of a cooperative oil mill, the choice of the FU becomes critical and it must made especially in relation to the availability of data. The collection of primary data related to the agricultural phase is the bottleneck of the whole study, because the correct assignment of each datum to the functional unit must be undertaken with extreme caution. The variety of cultivars and the variety of all the management operations of the olive grove are the variables that need to be taken into account when choosing the functional unit. If there are doubts about the availability of correct data related to a single cultivar, the FU also has to be defined on the basis of this variable. The same consideration must be made as regards the phase of oil extraction. The variability of the oil and water content, the kneading time frames and other factors that affect the extraction process should be considered for the choice of the FU. The possibility of measuring the energy and heat consumption of the extraction system and of linking these data to the FU in a precise manner should be taken into account in any revision of the FU.

While issues related to the agricultural phase are often common to other food products, and therefore have probably already been discussed by the scientific community, the issues related to the waste treatment of this sector (primarily for pomace and wastewater) are more specific to this area and inevitably more attention is necessary for this aspect.

Generally, little attention is paid to the transport phase. The EU produces over 70 % of the world's olive oil, and the most important countries that import the product are the USA, Brazil and Japan. In this respect, the following question arises: is it best to produce the most environmentally effective olive oils with low impact and transport them for thousands of kilometres or is it best to produce olive oils with conventional impacts and consume them locally? Therefore, the transport phase of the packaged final product, to the market or to the consumer, should be more frequently included in the assessment.

Another aspect that is generally not considered in LCA studies for reasons of lacking data is human labour, which in the olive oil production sector is a fundamental input. It plays a primary role, especially concerning traditional and organic systems, in the soil and the management, pruning and harvesting phases. In this direction, it could be important to integrate other methodologies (e.g. emergy evaluation) with the LCA in order to obtain a more complete and coherent view of the unsustainability of systems. For example, the combination of macroeconomic analysis and LCA may prove to be particularly useful since, compared with detailed life cycle inventories, many models of entire economies employ a much smaller number of categories to represent production and consumption activities (Settanni et al. 2010).

The joint use of more than one indicator should provide a full sustainability diagnosis (Bastianoni et al. 2013); therefore, it is very important to highlight outcomes obtained through other methodologies, different from LCA, as well. For example, the EF could be an effective and immediate tool to communicate how much agricultural and transformation practices in the olive oil sector exceed the ecological limits and how to manage and use the available resources in a sustainable way.

In general terms, the use of literature data can be suggested for the background system and plant/field-specific data for the foreground. High-quality data are the basis of any high-quality product environmental assessment. According to ISO 14044, the dimensions of data quality are: time-related coverage, geographical coverage, technology coverage, precision, completeness, consistency, reproducibility, source of data and uncertainty of the information. Preference should be given to primary and secondary data that are compliant with the ILCD Data Network entry-level requirements (EC 2012). Secondary data should be country-specific. To assess the data quality, the PEF data quality indicator (EC 2013b) should be used. The data and calculations need to be transparent, enabling external peer reviews to be undertaken. Estimations are very frequently not accurate; therefore, if possible, they should be avoided, even if this could cause the exclusion of phases from the system boundaries. In addition, assumptions made due to a lack of data should be clearly declared because they often cause high variability and incomparability among different case studies.

Moreover, the results should be presented in as disaggregated a form as possible to facilitate comparisons and to understand better which inputs/processes are included (e.g. packaging materials, with or without transport, and so on). However, starting from the assumption that missing data should not be ignored (unless they are within the defined cut-off criteria), when data gaps are filled with similar or estimated data (using data for analogous processes or materials or using estimation and/or characterisation methods, etc.), data quality checks should be made in order to increase the value of the LCA findings for decision making or comparative assertions.

Finally, by considering the site-specific characteristics of agricultural activities—in contrast to the site-independent nature of the LCA methodology (Notarnicola et al. 2012; Salomone and Ioppolo 2012)—and the variability of data in this specific sector (stressed by Notarnicola et al. 2013), a consistency check of the data quality should be carried out in any case.

2.5.5 Allocation Methods

Following the ISO requirements—ISO 14044:2006, 4.3.4 (ISO 2006c), whenever possible, allocation should be avoided by dividing the unit process to be allocated into two or more sub-processes or expanding the product system. In this case, in olive oil production, the most common solutions cited by authors are the expansion of system boundaries in order to include the process connected to by-product treatment and calculate the advantage obtained from the avoided product or the application of allocation using different methods coherently with the scope of the analysis.

When allocation cannot be avoided, the inputs and outputs of the system should be partitioned between its different products or functions in a way that reflects the physical relationships between them; i.e. they should reflect the way in which the inputs and outputs are changed by quantitative changes in the products or functions delivered by the system. Whenever it is unclear whether allocation based on the underlying of physical relationships is appropriate, economic allocation should be performed as a sensitivity analysis.

However, the allocation procedures in the olive oil sector should take into account the fact that the systems of this sector are characterised by one main product (olive oil) of generally high quality and thus high value as well as a large quantity of low-value by-products (pomace, OMW) that can be used as fuel and/or for composting purposes, which means that allocation using only the mass quantities or only the economic value could be misleading. Indeed, the allocation procedure should take into account both the mass and the economic value of the by-products (a weighting between the mass and the economic value is needed in order to balance the quantities of by-products obtained with their low economic value).

2.5.6 Life Cycle Impact Assessment (LCIA)

The impact assessment should be carefully planned to achieve the goal and scope of the study, by choosing the impact categories coherently and carrying out classification, characterisation, and, if necessary, grouping and weighting. As far as the choice of the mid-point and end-point impact categories are concerned, there are many evaluation methods that allow the highlighting of the environmental performance in the olive oil chain.

Furthermore, the way in which the results are shown can underline particular aspects of the environmental assessment. Percentage results, for example, could be shown in order to highlight the contribution of the sub-phases to the total environmental impact or, alternatively, after the grouping and weighting phase, the contribution of each impact category. The absolute values are useful in order to quantify the results for each impact category in a simply and understandable way.

In an LCA study of the olive oil chain, the consumption of water, energy and other resources should be indicated, and the following emissions should be considered: greenhouse gases; ozone-depleting gases; acidification gases; gases that contribute to the creation of ground-level ozone; the emission of substances to water contributing to oxygen depletion; and emission linked to human and eco toxicity. Other impact categories that should be evaluated due to their importance in the olive oil sector are land use and water used.

Beyond the impact indicators, inventory data can provide information about the assessed product's environmental performance. The use of energy, divided by the energy source, can be established as an indicator if considered significant. Water use should be assessed as part of the resource depletion category and, given its importance for the olive oil sector, particularly in the agricultural step, the water use indicator should be reported separately from other resource use indicators.

Together with data availability and data quality, life cycle impact assessment (LCIA) is the other issue in which the major LCA methodological problems occur. The main reasons are linked to the fact that standardised and universally accepted impact assessment methodologies for some impact categories are still lacking, or at least require further refinements and improvements to measure the environmental problems they are intended to represent consistently. This is, for example, the case of land use, for which it is actually not possible to perform a complete assessment of all the connected impacts (essentially due to the lack of data); land use is currently assessed using a few key impacts and for a complete assessment further research is necessary to deal with the unresolved problems.

In addition, water use impact assessment, of more recent interest in LCA with respect to land use, needs improvements in environmental assessment schemes. Water use has been increasingly considered important since climate change and different assessment methods have begun to be developed, but improved inventory data and agreement on which LCIA methods should be used for the assessment of relevant aspects are necessary.

In general, it can be observed that the problems of LCIA for olive oil production coincide with those of the wider agri-food sector and therefore the same considerations expressed in Chapter 1 and in the other chapters on the further agri-food chain analysed in this book are of interest for the olive oil production sector.

2.5.7 Interpretation

By following the ISO standards, the interpretation phase should identify the significant issues and evaluate the strength and consistency of the results.

In the olive oil sector, considering the unresolved problems previously mentioned, partly specific to this production and partly in common with the general agri-food sector, in order to obtain a reliable and consistent interpretation of the LCA results, sensitivity checks on uncertain data and on 'sensitive' methodological choices should be performed.

For the olive oil sector, uncertain data and 'sensitive' methodological choices could be:

- the choice of the functional unit;
- the production of some specific kinds of fertilisers, herbicides and pesticides;
- the dispersion of compounds into the environment (air, water and soil) deriving from the use of fertilisers, herbicides and pesticides;
- the balance of CO_2 emissions;
- the emissions from the use of agricultural machinery;
- data concerning many waste/by-product treatments;

- allocation methods;
- some impact methods (such as land use and water use);
- and all the other data of an uncertain source or of an estimated nature.

2.5.8 Critical Review

In order to assess the scientific and technical validity of the study and improve its credibility, a critical review could be carried out by an external independent expert.

The analysis performed put in evidence the issue that the CR of experts is still rarely practised (maybe due to the additional costs incurred), and only organisations working with environmental labelling or product declarations push themselves to demonstrate the quality of their LCA results with a CR. For these reasons, a critical review by independent experts should be practised for each LCA study, on one hand to reduce the variability and subjectivity and on the other hand to increase the credibility (e.g. ISO/TS 14071:2014).

The role of the expert review is also essential for reducing errors and uncertainty in the LCA data, so that new solutions to encourage greater use of external reviews should be found: the recent ISO 14071 helps to find solutions in this direction.

Conclusions

The critical comparative analysis allows some general hot spots of the olive and olive oil supply chain to be highlighted:

- When comparing different kinds of vegetable oil, olive oil resulted as more eco-compatible than sunflower seed oil for all the categories except for land use, and for both systems the phase with the greatest impact is the agricultural one (Nicoletti et al. 2001);
- When performing a cradle-to-gate or a cradle-to-grave LCA analysis, the agricultural phase results as the one with the greatest impact in almost all the impact categories (Avraamides and Fatta 2008; Christodoulopoulou et al. 2011; Iraldo et al. 2013; Salomone 2002; Testa et al. 2012);
- When focusing on the cultivation phase, the environmental impacts are mainly due to the use of fertilisers that cause eutrophication and acidification (Nicoletti and Notarnicola 2000; Salomone 2002), as well as the use of pesticides and land use in conventional olive cultivation (Olivieri et al. 2005b, 2007a). Considering different practices, it can be observed that the irrigation system is more eco-compatible than the dry system thanks to its higher olive productivity (Nicoletti and Notarnicola 2000) and conventional scenarios highlight higher environmental loads than organic ones (except for the impact categories associated with land use) (Olivieri et al. 2005a, b; Salomone and Ioppolo 2012; Salomone et al. 2010a);

- When focusing on the olive oil extraction phase (even if the agricultural stage is more significant than the processing one) the processing stage is of primary importance when it comes to groundwater contamination, mainly due to the particular management practice of effluent disposal to evaporation ponds (Avraamides and Fatta 2008). Considering the different olive oil extraction methods and their by-product treatments, the double-pressure system resulted as more effective than single pressure and centrifugation (Nicoletti and Notarnicola 2000), and even with a wider scenario analysis the most eco-compatible production chain is the one that uses continuous two-phase transformation (De Gennaro et al. 2005);
- When focusing on olive mill by-product treatment, significant positive contributions are obtained in terms of environmental credits for avoided production, associated with the use of by-products as fuels or fertilisers, and different examples were analysed in the studies, e.g. olive mill waste water recovery as fertiliser (Testa et al. 2012), energetic exploitation of pomace stone (Cini et al. 2008) and the recovery of olive pits used as fuel (Russo et al. 2008), the co-composting of OWP with manure on fields or co-composting of OMW and OWP with composter machines (Salomone and Ioppolo 2012), etc.

The analysis also revealed interesting points for reflection. The processes identified as those with a greater environmental impact are also those with the least data, such as the production and use of pesticides, herbicides and fertilisers; therefore, uncertainties and variability remain in the data. Thus, how can a more efficient and environmentally friendly local olive oil production chain be designed and how can LCA be used as a chain-focused management tool?

In order to develop LCA as a useful predictive tool for restructuring supply chains with the aim of improving their environmental performance, the lessons learned allow us to highlight that in this sector research is needed to increase the credibility of the existing LCA data and the priority is the improvement and expansion of databases for these substances; however, the models that estimate their dispersion in water, air and soil must also be simplified. Despite these limitations, this study can help us to understand better how useful the LCA methodology can be in the decision-making process connected to the definition of an environmental chain strategy and it certainly stresses the main gaps in the current knowledge concerning where future research and developments should be concentrated. However, the olive oil chain should not be interpreted as simple olive processing and olive oil production, followed by the problem of disposal and waste management. The whole olive oil chain must include the systems, treatment plants and waste recovery to obtain biomass for energy use, to produce compost and other substances that are useful to the cosmetic and pharmaceutical industries. Thus, this sector is multi-product and each option must be properly assessed considering the whole chain from both environmental and economic points of view, and LCA should be used as a starting point for the continuous improvement procedure with regard to the environment, identifying the inputs, processes or phases with the most significant potential impacts and considering measures for their control.

References

Abeliotis, K. (2003, October). A simplified life cycle assessment of olive oil production in a three-phase mill in Evia, Greece. (Paper presented at the International Symposium on the "Olive Tree and the Environment", Chania, Greece).

Apolio (2012). Dichiarazione ambientale di prodotto olio extra vergine di oliva denocciolato. Accessed 12 Dec 2013. http://gryphon.environdec.com/data/files/6/8574/epd346it.pdf.

Ardente, F., Cellura, M., Mistretta, M., & Longo, S. (2010, Sept). Life cycle assessment (LCA) of protected crops: An Italian case study. (Paper presented at the 7th International Conference on Life Cycle Assessment in the Agri-Food Sector, Bari, Italy).

Arvanitoyannis, I. S., & Kassaveti, A. (2008). Olive oil waste management: treatment methods and potential uses of treated waste. In I. S. Arvanitoyannis (Ed.), *Waste management for the food industries* (pp. 453–568). New York: Elsevier.

Assoproli (2012). Environmental product declaration for oasis extra virgin olive oil packaged in glass bottles of 0.5 l. Accessed 12 Dec 2013. http://gryphon.environdec.com/data/files/6/8781/epd367_Assoproli_oasis_extra_virgin_olive_oil.pdf.

Avraamides, M., & Fatta, D. (2008). Resource consumption and emissions from olive oil production: a life cycle inventory case study in Cyprus. *Journal of Cleaner Production, 16*, 809–821.

Bastianoni, S., Niccolucci, V., Neri, E., Cranston, G., Galli, A., & Wackernagel, M. (2013). Sustainable development: Ecological footprint in accounting. In S.E. Jørgensen (Ed.), *Encyclopedia environmental management* (pp. 2467–2481). New York: Taylor and Francis.

Beaufoy, G. (2000). *The environmental impact of olive oil production in the European Union: Practical options for improving the environmental impact.* Brussels: European Commission, Environment Directorate-General.

Beccali, M., Cellura, M., & Mistretta, M. (2002a, January). A decision support system software based on multi-criteria analysis for the selection of urban sustainability scenarios. (Paper presented at the International Conference "Rio 02 World Climate & Energy Event", Rio de Janeiro, Brasil).

Beccali, M., Cellura, M., & Mistretta, M. (2002b, May). An user-friendly software for sustainable management of urban areas. (Paper presented at the 9th International Conference on the modelling, monitoring and management of environmental problems "Envirosoft 2002". Environmental studies, Bergen, Norway).

Beccali, M., Cellura, M., & Mistretta, M. (2003). Decision making in energy planning. Application of the Electre Method at regional level for the diffusion of renewable energy technology. *Renewable Energy, 28,* 2063–2087.

Birkved, M., & Hauschild, M. Z. (2006). PestLCI – a model for estimating field emissions of pesticides in agricultural LCA. *Ecological Modelling, 198,* 433–451.

Brans, J. P., & Vincke, Ph. (1985). PROMETHEE A new family of outranking methods in MCDM. *Management Science, 31,* 647–656.

Brentrup, F., Küsters, J., Lammel, J., & Kuhlmann, H. (2000). Methods to estimate on-field nitrogen emissions from crop production as an input to LCA studies in the agricultural sector. *The International Journal of Life Cycle Assessment, 5,* 349–357.

Busset, G., Belaud, J. P., Clarens, F., Espi, J., Montréjaud-Vignoles, M., & Sablayrolles, C. (2012, Oct). Environmental evaluation of French olive oil production. (Paper presented at the 8th International conference on LCA in the agri-food sector, Saint-Malo, France).

Cappelletti, G. M., & Nicoletti, G. M. (2006, Sept). Compost from oil mill by-products. (Paper presented at the XV IGWT symposium, "Global Safety of Commodity and Environment, Quality of Life", Kiev, Ukraine).

Cappelletti, G. M., Nicoletti, G. M., & Russo, C. (2007, Sept). Analisi comparata dei vari sistemi di packaging nel settore delle olive da mensa, mediante l'utilizzo dell'LCA. (Paper presented at the XXIII National Conference of Commodity Science, Terracina, Rome, Italy).

Cappelletti, G. M., Nicoletti, G. M., & Russo, C. (2008, Nov). LCA of californian black-ripe table olives. (Paper presented at the 6th International Conference on Life Cycle Assessment in the agri-food sector, Zurich, Switzerland).

Cappelletti, G. M., Nicoletti, G., & Russo C. (2010). Life cycle assessment (LCA) of Spanish-style green table olives. *Italian Journal Food Science, 22,* 3–14.

Cappelletti, G. M., Nicoletti, G. M., & Russo, C. (2011). Wastewater from table olive industry. In F.S. Garcia Einschlag (Ed.), *Waste water, evaluation and management* (pp. 351–376). Rijeka: Intech.

Cappelletti, G. M., Nicoletti, G. M., Leone, A., Lopriore, G., & Russo, C. (2012, Oct). EPD of extra-virgin olive oil: An Italian experience. (Paper presented at the 8th International Conference on LCA in the agri-food sector, Saint-Malo, France).

Carvalho, J., Ribeiro, A., Araújo, J., Batista, M., Castro, F., & Vilarinho, C. (2012, Sept). OILCA – enhancing the competitiveness and reducing the carbon footprint of the olive oil sector through waste management optimization and the establishment of an ecological label. (Paper presented at the 4th International Conference on Engineering for Waste and Biomass Valorisation, Porto, Portugal).

Cavallaro, F., & Salomone, R. (2010, Sept). Interpretation of life cycle assessment results using a multicriteria tool: Application to olive oil chain. (Paper presented at the VII International Conference on Life Cycle Assessment in the Agri-Food Sector, Bari, Italy).

Cecchini, F., Franchini, F., Neri, P., & Guerra, E. (2005). Valutazione dell'Impatto ambientale dovuto alla produzione biologica, integrata e convenzionale di 1 l di olio d'oliva di aziende del Sud d'Italia. Technical report PROT-P135–091, Rome: ENEA.

Cellura, M., Longo, S., & Mistretta, M. (2012). Life Cycle Assessment (LCA) of protected crops: an Italian case study. *Journal of Cleaner Production, 28,* 56–62.

Chatzisymeon, E., Foteinis, S., Mantzavinos, D., & Tsoutsos, T. (2013). Life cycle assessment of advanced oxidation processes for olive mill wastewater treatment. *Journal of Cleaner Production, 54,* 229–234.

Christodoulopoulou, G. M. L., Manolaraki, G. G. C., Malliaraki, S., Aggelaki, K., & Zontanou, E. (2011). Life Cycle Assessment of Extra Virgin Olive Oil produced by three groups of farmers in south Greece. RodaxAgro. http://www.rodaxagro.gr/main/downloads/en/lca_oil_in_greece_summary.pdf. Accessed 18 Dec 2013.

Christopher, M. L. (1992). *Logistics and supply chain management.* London: Pitman Publishing.

Cichelli, A., & Cappelletti, G. M. (2007, May). Valorization of olive residues by spreading on agricultural land: technical assets. (Paper presented at the VII International Conference "Economie Europeană: prezent si perspective", Suceava, Romania).

Cini, E., Recchia, L., Daou, M., & Boncinelli, P. (2008, Sept). Human health benefits and energy saving in olive oil mills. (Paper presented at the International Conference Innovation Technology to Empower Safety, Health and Welfare in Agriculture and Agro-food Systems, Ragusa, Italy).

De Cecco (2012). Environmental product declaration De Cecco Extra Virgin Olive Oil. http://gryphon.environdec.com/data/files/6/9226/epd410_Dececco_olive_oil.pdf. Accessed 12 Dec 2013.

De Gennaro, B., Notarnicola, B., & Tassielli, G. (2005). Ricadute ambientali nella filiera dell'olio di oliva. *Estimo & Territorio, 12,* 12–19.

De Gennaro, B., Notarnicola, B., Roselli, L., & Tassielli G. (2012). Innovative olive-growing models: an environmental and economic assessment. *Journal of Cleaner Production, 28,* 70–80.

Del Caro, A., Vacca, V., Poiana, M., Fenu, P., & Piga, A. (2006). Influence of technology, storage and exposure on components of extra virgin olive oil (Bosana cv) from whole and de-stoned fruits. *Food chemistry, 98,* 311–316.

Demerche, S., Nadour, M., Larroche, C., Moulti-Mati, F., Michaud, P. (2013). Olive mill wastes: Biochemical characterizations and valorization. *Process Biochemistry, 48,* 1532–1552.

Di Giovacchino, L. (2010). *Tecnologie di lavorazione delle olive in frantoio. Rese di estrazione e qualitÃ dell'olio.* Milan: Tecniche Nuove.

Dijkman, T. J., Birkved, M., & Hauschild, M. Z. (2012). PestLCI 2.0: A second generation model for estimating emissions of pesticides from arable land in LCA. *The International Journal of Life Cycle Assessment, 17,* 973–986.

EC (1991). Commission Regulation (EEC) No 2568/ 91 of 11 July 1991 on the characteristics of olive oil and olive-residue oil and on the relevant methods of analysis. *Official Journal of the European Union, L, 248,* 1–83.

EC (2007). Commission Regulation (EC) No 702/2007 of 21 June 2007 amending Commission Regulation (EEC) No 2568/91 on the characteristics of olive oil and olive-residue oil and on the relevant methods of analysis. *Official Journal of the European Union, L,* 161, 11–27.
EC (2008). Commission Regulation (EC) No 640/2008 of 4 July 2008 amending Commission Regulation (EEC) No 2568/91 on the characteristics of olive oil and olive-residue oil and on the relevant methods of analysis. *Official Journal of the European Union, L,*178, 11–16.
EC (2010). *Life Focus/ LIFE among the olives: Good practice in improving environmental performance in the olive oil sector*. Luxembourg: Office for Official Publications of the European Union.
EC (2012). *The International Reference Life Cycle Data System (ILCD) Handbook: Compliance rules and entry-level requirements*. Luxembourg: European Commission, Joint Research Centre, Publications office of the European Union.
EC (2013a). Commission Implementing Regulation (EU) No 299/2013 of 26 March 2013 amending Regulation (EEC) No 2568/91 on the characteristics of olive oil and olive-residue oil and on the relevant methods of analysis. *Official Journal of the European Union, L,* 90, 52–70.
EC (2013b). Product Environmental Footprint (PEF) Guide, Annex II to the Recommendations of the Commission on the use of common methods to measure and communicate the life cycle environmental performance of products and organizations.
ECOIL (n.d.). LCA of Olive Oil using SimaPro 6. ECOIL PROJECT—University of Cyprus. http://www.ecoil.tuc.gr/SimaPro%206.pdf. Accessed 8 Aug 2013.
Ecoinvent Centre (2004). Ecoinvent data v1.2. Final reports Ecoinvent 2000, vol. 1–15. Swiss Centre for Life Cycle Inventories. Dübendorf: Ecoinvent Centre.
Ecological Footprint Standard (2009). Global Footprint Network. http://www.footprintnetwork.org/images/uploads/Ecological_Footprint_Standards_2009.pdf. Accessed 10 Jan 2014.
El Hanandeh, A. (2013). Carbon abatement via treating the solid waste from the Australian olive industry in mobile pyrolysis units: LCA with uncertainty analysis. *Waste Management and Research, 31,* 341–352.
Environdec (2014). http://www.environdec.com/. Accessed 14 Feb 2014.
European Food Sustainable Consumption & Production Round Table (2013). ENVIFOOD Protocol Environmental Assessment of Food and Drink Protocol. Version 1.0, November. http://www.food-scp.eu/files/ENVIFOOD_Protocol_Vers_1.0.pdf. Accessed 10 Jan 2014
ExternE (2013). External costs of energy. http://www.externe.info/externe_d7/. Accessed 1 March 2014
FAOSTAT (2013). Olives – production – crops. FAOSTAT – FAO's corporate database, http://faostat3.fao.org/. Accessed 24 Dec 2013
Farmers Groups of Nileas, Peza Union and Mirabello Union (2012). EPD for one pack of 0.75 Litre of Extra Virgin Olive Oil. http://gryphon.environdec.com/data/files/6/8059/epd274.pdf. Accessed 12 Dec 2013
Fernández-Bolaños, J., Rodríguez, G., Rodríguez, R., Guillén, R., & Jiménez, A. (2006). Potential use of olive by-product. Production of interesting organic compounds from olive oil waste. *Grasas y Aceites, 57,* 95–106.
Fiore, M., Breedveld, L., Arrivas Bajardi, C., Giaimo, L., & Notaro A. (2009). Certificazione ambientale di prodotti agroalimentari. LCA dell'olio d'oliva. *ARS,* 122, 12–20.
Galli, A., Wiedmann, T., Ercin, E., Knoblauch, D., Ewing, B., & Giljum, S. (2012). Integrating ecological, carbon and water footprint into a "Footprint Family" of indicators: Definition and role in tracking human pressure on the planet. *Ecological Indicators, 16,* 100–112.
Global Footprint Network (2009). Ecological Footprint Standards 2009. http://www.footprintstandards.org. Accessed 17 July 2011
Guzman, G. I., & Alonso, A. M. (2008). A comparison of energy use in conventional and organic olive oil production in Spain. *Agricultural Systems, 98,* 167–176.
Hauschild, M., 2000. Estimating pesticide emissions for LCA of agricultural products. In: B. Weidama & M. J. G. Meeusen (Eds.), *Agricultural Data for Life Cycle Assessments*, *2* (pp. 64–79). The Hague: LEI.

Hauschild, M., Jeswiet, J., & Alting, L. (2005). From life cycle assessment to sustainable production: Status and perspectives. *CIRP Annals–Manufacturing Technology, 54,* 1–21.

Hayashi, K., Gaillard, G., & Nemecek, T. (2006). Life cycle assessment of agricultural production systems: Current issues and future perspectives. *Good agricultural practice (GAP) in Asia and Oceania* (pp. 98–110). Taipei: Food and Fertilizer Technology Center .

Hoekstra, A.Y., & Chapagain, A.K. (2008). *Globalization of Water: Sharing the planet's freshwater resources*. Oxford: Blackwell Publishing Ltd.

Huijbregts, M. A. J., Hellweg, S., Frischknecht, R., Hungerbühler, K., & Hendriks, A. J. (2008). Ecological footprint accounting in the life cycle assessment of products. *Ecological Economics, 64,* 798–807.

Intini, F., Kühtz, S., & Rospi, G. (2011). Energy recovery of the solid waste of the olive oil industries–LCA analysis and carbon footprint assessment. *Journal of Sustainable Energy & Environment, 2,* 157–166.

Intini, F., Kühtz, S., & Rospi G. (2012). Life cycle assessment (LCA) of an energy recovery plant in the olive oil industries. *International Journal of Energy and Environment, 3,* 541–552.

IOC – International Olive Council (2013). World Olive Oil Figures. http://www.internationaloliveoil.org/estaticos/view/131-world-olive-oil-figures. Accessed 24 Dec 2013.

IOC – International Olive Council (n.d.). Designations and definitions of olive oils. http://www.internationaloliveoil.org/web/aa-ingles/oliveWorld/aceite.html. Accessed 24 Dec 2013.

Iraldo, F., Testa, F., & Bartolozzi, I. (2013). An application of Life Cycle Assessment (LCA) as a green marketing tool for agricultural products: The case of extra-virgin olive oil in Val di Cornia, Italy. *Journal of Environmental Planning and Management*. (in press). doi:10.1080/09640568.2012.735991.

Iraldo, F., Testa, F., & Bartolozzi, I. (2014). An application of life cycle assessment (LCA) as a green marketing tool for agricultural products: the case of extra-virgin olive oil in Val di Cornia, Italy. *Journal of Environmental Planning and Management, 57*(1), 78–103.

ISO (1999a). *ISO 14021 environmental labels and declarations—self-declared environmental claims (Type II environmental labelling)*. Geneva: International Organisation for Standardisation.

ISO (1999a). *ISO 14024 Environmental labels and declarations—Type I environmental labelling—Principles and procedures*. Geneva: International Organisation for Standardisation.

ISO (2000). *ISO 14020 Environmental labels and declarations—General principles*. Geneva: International Organisation for Standardisation.

ISO (2006a). *ISO 14025 Environmental labels and declarations – Type III environmental declarations—Principles and procedures*. Geneva: International Organisation for Standardisation.

ISO (2006b). *ISO 14040 Environmental management—Life cycle assessment—Principles and framework*. Geneva: International Organisation for Standardisation.

ISO (2006c). *ISO 14044 Environmental management—Life cycle assessment—requirements and guidelines*. Geneva: International Organisation for Standardisation.

ISO (2013a). *ISO/TS 14067 Greenhouse gases—Carbon footprint of products—Requirements and guidelines for quantification and communication*. Geneva: International Organisation for Standardisation.

ISO (2014a). *ISO/TS 14071 Environmental management—Life cycle assessment—Critical review processes and reviewer competencies: Additional requirements and guidelines to ISO 14044:2006*. Geneva: International Organisation for Standardisation.

ISO (2014b). *ISO/DIS 14046.2 Environmental management—Water footprint—Principles, requirements and guidelines*. Geneva: International Organisation for Standardisation.

Kalogerakisa, N., Politia, M., Foteinisa, S., Chatzisymeona, E., & Mantzavinos, D. (2013). Recovery of antioxidants from olive mill wastewaters: A viable solution that promotes their overall sustainable management. *Journal on Environmental Management, 128,* 749–758.

Kapellakis, I. E., Tsagarakis, K. P., & Crowther J. C. (2008). Olive oil history, production and by-product management. *Reviews in Environmental Science and Bio/Technology, 7,* 1–26.

Lucchetti, M. C., Romano, I., & Arcese, G. (2012, June). Carbon Footprint: un'analisi empirica per la produzione di olio. (Paper presented at the VI Convegno della Rete Italiana LCA, Bari, Italy).

Mansueti, A., & Raggi, A. (2002). *A comparative LCA of Olive Husk Combustion for Power generation: An Italian case study*. Paper presented at the Fifth International Conference on EcoBalance: Practical tools and thoughtful principles for sustainability Tsukuba, Japan. (November).

MIPAF (2005). Decree 6 July 2005 – Criteri e norme tecniche generali per la disciplina regionale dell'utilizzazione agronomica delle acque di vegetazione e degli scarichi dei frantoi oleari, di cui all'articolo 38 del decreto legislativo 11 maggio 1999, n. 152. *Official Journal of the Italian Republic, 166,* 74–79.

Monini (2012a). EPD Extra virgin Olive Oil "Classico". http://gryphon.environdec.com/data/files/6/9629/epd384_monini_classico.pdf. Accessed 12 Dec 2013.

Monini (2012b). EPD Extra virgin Olive Oil "Delicato". http://gryphon.environdec.com/data/files/6/9631/epd386_monini_delicato.pdf. Accessed 12 Dec 2013.

Monini (2012c). EPD Extra virgin Olive Oil "GranFruttato". http://gryphon.environdec.com/data/files/6/9628/epd383_monini_granfruttato.pdf. Accessed 12 Dec 2013.

Monini (2012d). EPD Extra virgin Olive Oil "Poggiolo". http://www.environdec.com/en/EPD-Search/?&query=olive%20oil. Accessed 12 Dec 2013.

Nardino, M., Pernice, F., Rossi, F., Georgiadis, T., Facini, O., Motisi, A., & Drago, A. (2013). Annual and monthly carbon balance in an intensively managed Mediterranean olive orchard. *Photosynthetica, 51,* 63–74.

Neri, E., Niccolucci, V., Rugani, B., & Bastianoni, S. (2010). *The CEW environmental label for typical products: The case of Cinta Senese pig*. Paper presented at the Footprint Forum 2010-Meet the Winners of the 21st Century-The State of the Art in Ecological Footprint Theory and Applications, Colle Val d'Elsa, Siena, Italia .(June).

Neri, E., Rossetti, F., Rugani, B., Niccolucci, V., Pulselli, F. M., Bastianoni, S., & Marchettini, N. (2012). Life Cycle Assessment and eMergy application to the comparison between organic and conventional productions. Paper presented at the 8th International Conference on LCA in the Agri-Food Sector, Saint- Malo, France. (October).

Niaounakis, M., & Halvadakis, C. P. (2006). *Olive Processing Waste Management: Literature Review and Patent Survey*, (2nd ed., Vol. 5). Oxford: Elsevier.

Nicoletti, G. M., & Notarnicola, B. (2000). L'impatto ambientale della filiera dell'olio di oliva: pratiche agricole e tecniche di estrazione. Paper presented at the XIX Congresso Nazionale di Merceologia, "La sfida per il terzo millennio: tecnologia, innovazione, qualitÃ e ambiente", Sassari, Italy. (September).

Nicoletti, G. M., Notarnicola, B., & Tassielli, G. (2001). *Comparative LCA of virgin olive oil vs. seed oils*. Paper presented at the International Conference LCA in Foods, Goteborg, Sweden. (April).

Nicoletti, G. M., Cappelletti, G. M., & Russo C. (2007a). LCA of "Spanish-style" green table olives. Paper presented at the 5th International Conference LCA in Foods 2007, Gothenburg, Sweden.(April).

Nicoletti, G. M., Cappelletti, G. M., & Russo C. (2007b). Analisi comparata dei vari sistemi di packaging nel settore delle olive da mensa, mediante l'utilizzo dell'LCA. Paper presented at the XXIII Congresso Nazionale di Merceologia, Terracina, Italy. (September).

Nicoletti, G. M., Cappelletti, G. M., Leone, A., Lopriore, G., & Russo, C. (2011). *Criteri per l'applicazione della Dichiarazione Ambientale di Prodotto agli oli extravergini d'oliva*. Roma: Aracne Editrice.

Notarnicola, B., Tassielli, G., & Nicoletti, G. M. (2003). *LCC and LCA of extra-virgin olive oil: organic vs conventional*. Paper presented at the 4th International Conference on Life Cycle Assessment in the Agri-food sector, Bygholm, Denmark. (October).

Notarnicola, B., Tassielli, G., & Nicoletti G. M. (2004). Environmental and economical analysis of the organic and conventional extra-virgin olive oil. *New medit, 3,* 28–34.

Notarnicola, B., Tassielli, G., & Renzulli P. A. (2012). Modeling the agri-food industry with life cycle assessment. In M. A. Curran (Ed.), *Life Cycle Assessment Handbook: A Guide for Environmentally Sustainable Products* (pp. 159–184). Hoboken: Wiley.

Notarnicola, B., Tassielli, G., & Renzulli P. (2013). *La variabilitÃ dei dati nella LCA della produzione olivicola*. Paper presented at the VII Convegno della Rete Italiana LCA, Milano, Italy. (June).

Odum, H. T. (1996). *Environmental Accounting: Emergy and Environmental Decision Making*. New York: John Wiley and Sons.

Olivieri, G., Falconi, F., Pergreffi, R., Neri, P., & Romani A. (2005a). *Life Cycle Assessment for environmental integrated system in the olive oil tuscan company*. Paper presented at the European Meeting of the International Society for Industrial Ecology, Bologna, Italy. (January).

Olivieri, G., Romani, A., & Neri, P. (2005b). Analisi del ciclo di vita (LCA) della coltivazione convenzionale e biologica di olive toscane. Technical Report PROT-P135–070. Rome: ENEA.

Olivieri, G., Neri, P., Bandini, F., & Romani, A. (2007a). *Life cycle assessment of Tuscany olive oil production*. Paper presented at the AGS Annual Meeting, Barcelona, Spain. (March).

Olivieri, G., Neri, P., Bandini, F., & Romani, A. (2007b). Analisi del ciclo di vita (LCA) della coltivazione e della trasformazione delle olive toscane per un sistema integrato HACCP–ISO 9000–ISO 14000. *Ricerche e Innovazioni nell'industria alimentare, 7,* 605–609.

Olivieri, G., Bandini, F., Borsacchi, L., Tacconi, D., & Romani A. (2008). Life Cycle Assessment applicato alla produzione di olio d'oliva in toscana: buone pratiche per un'efficace gestione integrata igienico-ambientale. In S. Poretta (Ed.), *Ricerche e innovazioni nell'industria alimentare. Atti dell'8° Congresso Italiano di scienza e tecnologia degli alimenti* (pp. 701–705). Pinerolo: Chiriotti Editore.

Olivieri, G., Romani, A., & Falconi, F. (2010a). *The preliminary LCA of membrane processes for olive oil wastewater polyphenols recovery*. Paper presented at the VII International Conference on Life Cycle Assessment in the Agri-Food Sector, Bari, Italy. (September).

Olivieri, G., Romani, A., & Neri, P. (2010b). Life Cycle Assessment of polyphenol extraction wastewater: Critical review and update. *Journal of Commodity Science Technology Quality, 49,* 95–103.

Özilgen, M., & Sorgüven, E. (2011). Energy and exergy utilization, and carbon dioxide emission in vegetable oil production. *Energy, 36,* 5954–5967.

Palese, A. M., Pergola, M., Favia, M., Xiloyannis, C., & Celano, G. (2013). A sustainable model for the management of olive orchards located in semi-arid marginal areas: Some remarks and indications for policy makers. *Environmental Science & Policy, 27,* 81–90.

Pattara, C., Cappelletti, G. M., & Cichelli, A. (2010). Recovery and use of olive stones: Commodity, environmental and economic assessment. *Renewable and Sustainable Energy Reviews, 14,* 1484–1489.

Pergola, M., Favia, M., Palese, A. M., Perretti, B., Xiloyannis, C., & Celano, G. (2013). Alternative management for olive orchards grown in semi-arid environments: An energy, economic and environmental analysis. *Scientia Horticulturae, 162,* 380–386.

Polo, J. A., Salido, J. A., & Mourelle, A. (2010). *Calculation and verification of carbon footprint in agricultural products*. Paper presented at the VII International Conference on Life Cycle Assessment in the Agri-Food Sector, Bari. (September)

PROSODOL (2011). Production Units and Factories. http://www.prosodol.gr/?q†= node/450. Accessed 9 Oct 2013.

Raggi, A., Cichelli, A., Pagliuca, G., & Costantino, A. (2000). *A screening LCA of olive husk combustion for residential heating*. Paper presented at the 1st World Conference on Biomass for Energy and Industry, Sevilla. (June)

Recchia, L., Boncinelli, P., Cini, E., Vieri, M., Garbati Pegna, F., & Sarri, D. (2011). Olive oil production chain. In L. Recchia, P. Boncinelli, E. Cini, M. Vieri, F. Garbati Pegna, & D. Sarri (Eds.), *Multicriteria analysis and LCA techniques: With applications to Agro-Engineering* (pp. 99–128). London: Springer.

Ridolfi, R., & Bastianoni, S. (2008). Emergy. In S. E. Jørgensen & B. D. Fath (Eds.), *Encyclopedia of ecology: Ecological indicators* (pp. 1218–1228). Amsterdam: Elsevier.

Roig, A., Cayuela, M. L., & Sánchez-Monedero, M. A. (2006). An overview on olive mill wastes and their valorization methods. *Waste Management, 26,* 960–969.

Romani, A., Olivieri, G., Neri, P., Tacconi, D., Pinelli, P., Lapucci, C., & Ieri, F. (2004) Paper presented at the XXI Congresso Nazionale di Merceologia, "Risorse naturali e sviluppo economico-sociale. Contributo delle scienze merceologiche", Foggia. (September)

Roselli, L., De Gennaro, B., & Cimino, O. (2010). *Innovative olive-growing models: an environmental and economic assessment*. Paper presented at the 119th EAAE Seminar, Capri. (June)

Russo, C., Cappelletti, G.M., & Nicoletti, G.M. (2008). *LCA of Energy Recovery of the Solid Waste of the Olive Oil Industries*. Paper presented at the 6th International Conference on Life Cycle Assessment in the Agri-Food Sector – Towards a sustainable management of the Food chain, Zurich. (November)

Russo, C., Cappelletti, G. M., & Nicoletti, G. M. (2009). *Confronto, mediante metodologia LCA, fra la produzione di olio extravergine di oliva da olive intere e denocciolate*. Paper presented at the XXIV Congresso Nazionale delle Scienze Merceologiche, Torino. (June)

Russo, C., Cappelletti, G. M., & Nicoletti, G. (2010). Life cycle assessment (LCA) used to compare two different methods of ripe table olive processing. *Grasas Y Aceites, 61,* 136–142.

Russo, C., Cappelletti, G. M., & Sabatini, N. (2012). *Assessment by the use of LCA methodology of the process of CO2 neutralization for Spanish Style Green table olives*. Paper presented at the 18th IGWT Symposium "Technology and innovation for a sustainable future: a commodity science perspective", Roma. (September)

Salmoral, G., Aldaya, M. M., Chico, D., Garrido, A., & Llamas, M. R. (2011). The water footprint of olives and olive oil in Spain. Spanish *Journal of Agricultural Research, 9,* 1089–1104.

Salomone, R. (2002). *Applicazione della metodologia Life Cycle Assessment alla produzione di olio di oliva: alcuni risultati preliminari*. Paper presented at the Euroconference on University and Enterprise, Rome. (September)

Salomone, R. (2008). Applicazione della Life Cycle Assessment in aziende di produzione di olio di oliva: analisi comparativa degli studi italiani. In L. Morselli (Eds.), *Ecomondo 2008: Ambiente & Territorio:* Vol. 2, (pp. 454–459). Santarcangelo di Romagna: Maggioli Editore.

Salomone, R., & Ioppolo, G. (2012). Environmental impacts of olive oil production: A life cycle assessment case study in the province of messina. *Journal of Cleaner Production, 28,* 88–100.

Salomone, R., Ioppolo, G., & Saija, G. (2009). *Integrazione di aspetti ambientali e strategie di qualitÃ di prodotto in un'ottica di ciclo di vita: life cycle assessment di oli extra-vergini di oliva di alta qualità*. Paper presented at the XXIV Congresso Nazionale delle Scienze Merceologiche, Torino. (June)

Salomone, R., Ioppolo, G., & Battaglia, G. (2010a). *Environmental impacts of olive oil production: an LCA case study in the province of messina*. Paper presented at the VII International Conference on Life Cycle Assessment in the Agri-Food Sector, Bari. (September)

Salomone, R., Cappelletti, G. M., Ioppolo, G., Mistretta, M., Nicoletti, G., Notarnicola, B., Olivieri, G., Pattara, C., Russo, C., & Scimia, E. (2010b). *Italian experiences in life cycle assessment of olive oil: a survey and critical review*. Paper presented at the VII International Conference on Life Cycle Assessment in the Agri-Food Sector, Bari. (September)

Salomone, R., Ioppolo, G., & Saija, G. (2013a). The implementation of product-oriented environmental management systems in agri-food SMEs. In R. Salomone, M. T. Clasadonte, M. Proto, & A. Raggi (Eds.), *Product-Oriented Environmental Management Systems (Poems): Improving Sustainability and Competitiveness in the Agri-Food Chain with Innovative Environmental Management Tools*, (pp. 303–330). Dordrecht: Springer.

Salomone, R., Clasadonte, M.T., Proto, M., & Raggi. A. (Eds.) (2013b). *Product-Oriented Environmental Management Systems (POEMS)—Improving Sustainability and Competitiveness in the Agri-Food Chain with Innovative Environmental Management Tools*. Dordrecht: Springer.

Schmidt, A., & Frydendal, J. (2003). Methods for calculating the environmental benefits of 'green' products. In: C. Erdmenger (Ed.), *Buying into the environment: experiences, opportunities and potential for eco-procurement*, (pp. 136–165). UK: Greenleaf Publishing.

Scotti, M., Bondovalli, C., & Bodini A. (2009). Ecological Footprint as a tool for local sustainability: The municipality of Piacenza as a case study. *Journal of Environmental Impact Assessment Review, 29,* 39–50.

Settanni, E., Tassielli, G., Notarnicola, B. (2010). Combining LCA of food products with economic tools. In U. Sonesson, J. Berlin, & F. Ziegler (Eds.), *Environmental assessment and management in the food industry: Life Cycle Assessment and related approaches*, (pp. 207–218). Cambridge: Woodhead publishing.

Sofo, A., Dichio, B., Xiloyannis, C., & Masia, A. (2005). Antioxidant defences in olive trees during drought stress: changes in activity of some antioxidant enzymes. *Functional Plant Biology, 32(1),* 45–53

Stamatelatou, K., Blika, P. S., Kopsahelis, A., & Lyberatos, G. (2012). Integrated management methods for the treatment and/or valorization of olive mill wastes In A. McElhatton & P. J. do Amaral Sobral (Eds.), *Novel technologies in food science: Their impact on products, consumer trends and the environment–series: Integrating food science and engineering knowledge into the food chain:* Vol. 7 (pp. 65–118). New York: Springer.

Testa, F., Bartolozzi, I., & Iraldo, F. (2012). *Un'applicazione dell'Analisi del Ciclo di Vita come strumento di green marketing: il caso dell'olio extravergine di oliva della Val di Cornia.* Paper presented at the VI Convegno della Rete Italiana LCA, Bari. (June)

Vlyssides, A. G., Loizides, M., & Karlis, P. K. (2004). Integrated strategic approach for reusing olive oil extraction by-products. *Journal of Cleaner Production, 12,* 603–611.

Vossen, P. (2007). Olive oil: History, production, and characteristics of the world's classic oils. *Hortscience, 42,* 1093–1100.

Wackernagel, M., Monfreda, C., Moran, D., Wermer, P., Goldfinger, S., Deumling, D., & Murray, M. (2005). National Footprint and Biocapacity Accounts 2005: the Underlying Calculation Method. www.footprintnetwork.org/download.php?id=5. Accessed 15 Sept 2013

Weidema, B. P., Thrane, M., Christensen, P., Schmidt, J., & Løkke, S. (2008). Carbon Footprint. A Catalyst for Life Cycle Assessment?. *Journal of Industrial Ecology, 12,* 3–6.

Chapter 3
Life Cycle Assessment in the Wine Sector

Luigia Petti, Ioannis Arzoumanidis, Graziella Benedetto, Simona Bosco, Maurizio Cellura, Camillo De Camillis, Valentina Fantin, Paola Masotti, Claudio Pattara, Andrea Raggi, Benedetto Rugani, Giuseppe Tassielli and Manfredi Vale

Abstract Currently, stakeholders' increasing attention to quality is driving the wine sector to rethink and change its own production processes. Amongst product quality dimensions, the environment is gaining ever-growing attention at various levels of policy-making and business. Given its soundness, the use of Life Cycle Assessment (LCA) has become widespread in many application contexts. Apart from applications for communication purposes, LCA has also been used in the wine sector to highlight environmental hot spots in supply chains, to compare farming practices and to detect improvement options, *inter alia*. Case studies whose focus is the wine industry abound in high quality publications.

This Chapter has a two-fold focus: firstly, an analysis of the methodologies and standards of the Life Cycle Thinking concept, related to wine, and secondly, a

L. Petti (✉) · I. Arzoumanidis · C. Pattara · A. Raggi
Department of Economic Studies, University "G. d'Annunzio",
Viale Pindaro 42, 65127 Pescara, Italy
e-mail: l.petti@unich.it

I. Arzoumanidis
e-mail: i.arzoumanidis@unich.it

C. Pattara
claudiopattara1@gmail.com

A. Raggi
a.raggi@unich.it

G. Benedetto
Department of Science for Nature and Environmental Resources, University of Sassari
Via Enrico De Nicola 1, 07100 Sassari, Italy
e-mail: gbenedet@uniss.it

S. Bosco
Scuola Superiore Sant'Anna, Institute of Life Sciences, via S. Cecilia, 3 56127 Pisa, Italy
e-mail: s.bosco@sssup.it

M. Cellura
University of Palermo, Viale delle Scienze, 90128 Palermo, Italy
e-mail: maurizio.cellura@unipa.it

B. Notarnicola et al. (eds.), *Life Cycle Assessment in the Agri-food Sector*,
DOI 10.1007/978-3-319-11940-3_3

critical analysis of wine LCA case studies in order to compile a list of scientifically-based environmental hot-spots and improvements.

The chapter also expands the knowledge on LCA's application to the wine industry by discussing how best to contribute to:

- the identification of the critical environmental issues of the wine supply-chain and the essential elements that an LCA case study in the sector should consider;
- the identification of an optimal set of indicators and methodologies for the evaluation of the environmental impacts of wine;
- the comparability of results;
- the improvement of the environmental research quality in this sector.

Keywords Life cycle assessment · Wine · Life cycle-based tools · Life cycle-based methodologies · Case-studies

C. De Camillis
Food and Agriculture Organization of the United Nations (FAO), Agriculture and Consumer Protection Department, Room C 535, Viale delle Terme di Caracalla, 00153 Rome, Italy
e-mail: Camillo.DeCamillis@fao.org

European Commission (EC), Joint Research Centre, Institute for Environment and Sustainability, Sustainability Assessment Unit, Via Enrico Fermi 2749, 21027 Ispra, Italy

V. Fantin
LCA and Ecodesign Laboratory, ENEA (Italian National Agency for New Technologies, Energy and Sustainable Economic Development), Via Martiri di Monte Sole 4, 40129, Bologna, Italy
e-mail: valentina.fantin@enea.it

P. Masotti
Department of Economics and Management, University of Trento, via Inama 5, 38122 Trento, Italy
e-mail: paola.masotti@unitn.it

B. Rugani
Centre de Recherche Public Henri Tudor, Centre de Ressources des Technologies pour l'Environnement (CRP Henri Tudor/CRTE)—6A, avenue des Hauts-Fourneaux, L-4362, Esch-sur-Alzette, Luxembourg
e-mail: benedetto.rugani@tudor.lu

G. Tassielli
Ionian Department of Law, Economics and Environment, University of Bari, Via Lago Maggiore angolo via Ancona, 74121 Taranto, Italy
e-mail: giuseppe.tassielli@uniba.it

M. Vale
Aghetera Ambiente and Sviluppo, San Polo 3082/c, 30125 Venice, Italy
e-mail: vale@aghetera.eu

3.1 Introduction

3.1.1 Background

3.1.1.1 Nutritional, Cultural and Functional Aspects

The origins and history of the beverage obtained by the fermentation of grapes are strongly linked to those of European people and their civilisations. Indeed, even before the Bible, the civilisations of the Middle East knew the beverage and they considered it as a gift from the gods to those who had founded their society (Austin 1985). Even in Ancient Greece, the religious mystical concept of wine (as a means of communication with the gods) appears early on in the conceptions of Homer and Plato, who see it as a pleasure to be enjoyed slowly (Austin 1985). Plato himself repeatedly stresses the importance of social drinking during feasts; wine is also an essential element in the original Socratic method of seeking the truth in a group (Austin 1985). It is worth mentioning that no one ever saw Socrates drunk, although he could outdrink anyone.

From the Middle Ages until about 1600, the consumption of wine declined in favour of beer (Babor 1986); this was because of the costs of production (beverages obtained from cereal crops were cheaper).

The development of viticulture and the availability of wine at affordable prices were welcomed by the Mediterranean populations, who had never given up completely on the most beloved and traditional of beverages. In fact, the conviction that "good wine makes good blood" became proverbial, "blood" having a double meaning: physical, as it nourishes the organs, and mental, influencing behavioural attitude and disposition (mood).

In fact, wine is often considered not just as food in the popular tradition but as a medicine. Although not specific to certain diseases, it is nonetheless applicable to the replenishment and renewal of one's strength (a "tonic" or "restorative", ignored by mainstream medicine) and in general to the recovery and maintenance of well-being. The link between wine and health, and the positive effect of the regular consumption of wine (albeit in moderate amounts) received sensational affirmation between 1980 and 1990 from the medical profession in the form of the "French Paradox" (Leger et al. 1979). Statistical and epidemiological investigations documented a reduced incidence of cardiovascular diseases and relative complications in some regions of southern France, in spite of the high consumption of atherogenic fats (Leger et al. 1979). This was in sharp contrast to the high impact of these diseases in other European and North American populations, who consumed equivalent amounts of the same fats. The difference was attributed to the habitual consumption of wine by the French as a protective factor against atherosclerosis and the atherosclerotic cardiovascular damages related to it. Numerous researchers (Rimm et al. 1996; Kauhanen et al. 1999; Criqui and Ringel 1994; Artaud-Wild et al. 1993; Nigdikar et al. 1998) believe there is a negative correlation between moderate consumption of red wine (one to two glasses per day) and coronary heart disease; however, the question is not entirely clear.

Table 3.1 Main constituents of wine. (Source: Cozzani 2005)

Constituents	Quantity (g/l)
Water	750–900
Ethyl alcohol	70–130
Methyl alcohol	0.02–0.2
Higher alcohols	0.1–0.5
Glycerol	4–15
Sugars	Traces in dry wines
(Glucose and fructose)	Varying amounts in sweet wines
Tartaric acid	2–5
Malic acid	0–7
Citric acid	0.1–0.5
Succinic acid	0.5–1.5
Lactic acid	1–5
Acetic acid	0.2–0.9
Phenolic compounds (tannins, etc.)	0.2–3
Nitrogen compounds	0.05–0.9
Minerals (such as ash)	2–3

Brief Review of the Main Constituents The chemical components of wine number several hundred, but common chemical analyses determine only the main constituents, which are useful for characterising the product from a commodity perspective and verifying its compliance with legal regulations, quality classifications and related disciplines. Table 3.1, which summarises the main chemical components of wine, shows that wine does not provide considerable amounts of any of the major nutrients: neither protides (proteins, peptides, amino-acids), nor lipids (fats, oils), nor glycides (simple or polymeric sugars), with the exception of sweet wines. The vitamin-related content of wine is almost negligible. The only components which can be regarded as important for both nutrition and health are alcohols and in particular ethanol (or alcohol/spirit or ethyl alcohol by definition), which is present in wine at average concentrations of 10% (weight/volume).

3.1.1.2 The Wine Supply Chain

World wine production in 2012 was lower compared with the previous year, declining by about 10% and reaching 252 Mhl of wine produced (OIV 2013). Consumption now appears to be stable at about 243 Mhl, despite a significant decrease in taxes in some countries (OIV 2013). Nevertheless, in recent years, the wine industry has been affected by continuous changes in terms of technology, product quality and consumer requirements. In this dynamic environment, European countries certainly remained the principal actors, accounting for 64% of world production (OIV 2013).

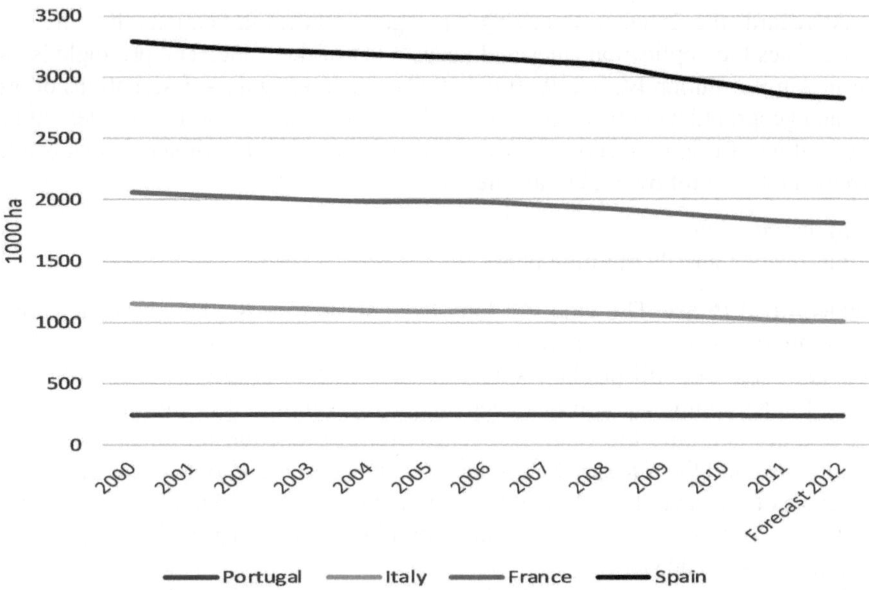

Fig. 3.1 Decline in vineyards of major European wine producers. (Adapted from: OIV 2013)

On the other hand, with regard to international dynamics, France is the clear leader in terms of value of goods exported, and Italy predominates in terms of quantities and volumes exported, followed by Spain.

The world trade has grown dramatically, reflecting a consumption of wine that is no longer merely local. It brings together the "old producers" with the new world of wine, which includes the United States, Australia, Argentina, Chile, New Zealand and South Africa, thus creating a new competitive structure.

As regards the area covered by vineyards worldwide, estimates prepared by the OIV (International Organisation for Vine and Wine) show a slowing-down in the sector (OIV 2013). That report shows a decrease for the year 2012 in territories occupied by vineyards (see Fig. 3.1). Vineyards covered an area of about 7528 Mha worldwide in 2012, including those not yet producing or harvested. Although there is a slight decline in the years 2011–2012, this is still lower than in previous years.

As regards the regulation about the production and sale of wine, the OIV establishes the general principles to which every state should refer, but national regulation may vary from country to country. With regard to the terminology adopted, every state that produces grapes or wine generally incorporates the definitions set by the OIV. Specifically, wine is defined (OIV 1992) as "the beverage resulting from full or partial alcoholic fermentation of fresh grapes, whether crushed or not, or of grape must. Its actual alcoholic strength may not be less than 8.5 % vol. However, considering the conditions of the climate, the terroir and the grape variety, quality factors of special or particular traditions of some vineyards, the minimum total alcoholic strength may be reduced to 7 % vol. according to the specific regulations of the region concerned" (OIV 1992).

As regards the designation of origin and geographical indications, it is the OIV that defines the application rules and keeps a list of the same. The principle is enshrined in Resolution ECO 2–92 (OIV 1992) which recognises designations of origin and geographical indications. The product "wine" is the result, regardless of the designation of origin or geographical indication, of a number of stages that can be grouped into the following broad categories:

- agricultural phase
- vinification and distribution phase

Agricultural Phase The vine is a long-lived shrub (50–70 years in some cases). There are cases of vineyards (e.g., in Maribor, Slovenia) that were planted about 400 years ago but still produce very low quantities of grapes. When it comes to LCA, the agricultural phase has been frequently simplified in the literature, taking into account only the year(s) of actual grape production for quantifying input requirement and the release of emissions. Yet it is important to consider the overall life cycle of a vineyard, including its planting, the first unproductive years, the productive years and then senescence and disruption, as in the case of every perennial crop (Marenghi 2005).

Every new vineyard planting is characterised by the work needed for the preparation of the soil. The first stage concerns the physico-chemical analysis of the soil, which determines all the main indicators of the soil (texture, organic matter, pH, nutrient deficiencies). A preliminary analysis of soil characteristics, along with the knowledge of the climatic condition of a territory, allows technicians to choose the cultivars best suited to the area and decide which precautions should be taken during the first planting. These preliminary steps must also consider the plantation density (number of plants per hectare), which ranges from 1500 to 10,000 vines per hectare. The planting density will affect the intensity of treatments for pest management and the harvest costs. The planting needs deep tillage in order to allow the root system to grow unimpeded. After that, vine support is performed. The vine, being a climbing plant, requires a supporting infrastructure; these may be structures with stakes of wood, concrete or metal. Afterwards, vine cuttings can be planted. The vine needs two or three years to start producing grapes. During this period, the vineyard management is fully operative; fertilisation, pesticide treatments and soil management, with the exclusion of harvest, are needed. The vineyard grape yield grows for the first six to eight years and then stabilises. Vineyard management in the productive years is strongly dependent on the microclimate of the area, the characteristics of the soil, the field slope and grape quality. Pruning is usually carried out both in wintertime and in spring. Weed management can be performed by mechanical weeding or by chemical weeding; pest control is crucial in vineyard management to prevent attacks by pathogens and consequent reduction in grape quantity and quality. When the grapes reach optimal maturity (in terms of sugar level, level of acidity, and colour) they are collected through manual or mechanised processes (mechanical harvester), and then they are conveyed into trailers and transported to the winery for vinification. Grapes cannot be stored because of decay-related problems, so the process of vinification must be initiated immediately (Reynier 2011).

Vinification and Distribution Phases Once the grapes arrive at the winemaking facilities, all the quality parameters of the product are controlled and the phase of vinification can start. Each bunch of grape is deprived of the stem (in order to avoid problems of fermentation and tannic flavours) (Ribéreau-Gayon and Peynaud 1979) and pressed to promote the fermentation of the entire mass. The must is pumped into fermenters (fermentation tanks), where yeast is added and fermentation occurs whereby sugars are converted into alcohol and carbon dioxide. During this phase, an exothermic reaction takes place causing the temperature to increase, usually between 26 and 30 °C. Control of the reaction temperature affects the quality of wine to a significant degree (especially white wines); this operation thus entails the highest energy consumption in the vinification process. However, the fermentation process can differ depending on the type of wine. For example, in the case of red wine the must is fermented with the skins, whereas for white wine skins are removed.

When the entire sugar component has been transformed, the wine is separated from the skins. This process can be performed by different techniques (draining, pressing, etc.) and allows wine to be obtained, which is then transferred for ageing. These techniques include:

- the formation of homogeneous masses required for large volumes of wine in order to ensure a uniform quality standard;
- the ageing in casks or barrels, intended primarily for small quantities of valuable wine (because of the high cost).

When winemakers deem that the product is suitable for the market, the wine can either be sold in bulk, i.e. without any type of packaging, or be bottled and packaged before distribution to retailers or end consumer.

3.1.1.3 Main Environmental Problems

The wine industry is a productive activity and, as such, cannot be considered environmentally impact-free. For example, the phase of agriculture in the wine life cycle can generate a remarkable impact on climate change (Arzoumanidis et al. 2013a; Pattara et al. 2012a; Petti et al. 2010a), which is caused by the use of fossil fuels for cultural practices, pesticides and herbicides used for crop protection and fertilisers applied to maintain high yields. Nonetheless, the industrial phase also imposes environmental loads that cannot be ignored in the framework of an overall assessment of the life cycle of wine. What follows is a summary (by no means exhaustive) of the main environmental issues related to the life cycle stages of wine. The impact categories and related indicators enumerated below are analysed in Sect. 3.3:

- Land use and land use change. These land-based indicators can be effective for the impact assessment of the vineyard plantation, as the land was previously used for other crops or forest and may be used for purposes other than wine production after the dismantling of the vineyard.

- Climate change. It is well known that climate change is related to the emission of greenhouse gases (IPCC 1997) generated by the use of machinery in the agricultural phase and during the industrial production of electricity consumed in the winemaking process.
- Ozone depletion. The reduction of the ozone layer is caused mainly by chlorine and bromine, which are contained in many substances and compounds. Amongst these, refrigerant gases can be identified, which were used until the 1980s for temperature control in the winemaking process (industrial refrigerators). Currently, CFCs and HCFCs are banned by the EU (Reg. CEE 3952/92). However, it is still possible to find them as refrigerants in old structures.
- Photochemical ozone formation.
- Resource depletion. Water consumption in the wine production process is related to the agricultural phase (use of water for plant protection treatments, irrigation) and in the industrial phase (washing of fermentation and storage tanks); other renewable resources such as wood and cork, and non-renewable ones, such as fossil fuels and minerals, are also directly and indirectly consumed in the wine life cycle.
- Eutrophication. The fertilisers used in the field are not completely absorbed by the roots of the plants, and as a result of atmospheric precipitations they leach into surface- and groundwater. This is one of the most significant impact categories in wine production.
- Acidification. This impact category refers to all the factors that contribute to the reduction of the pH of the soil or water. Acidification may be caused by the emission into the atmosphere or the release into the soil of precursor compounds (e.g. NO_x, SO_x, NH_3).

3.2 Life Cycle Assessment Methods for Measuring and/or Communicating the Environmental Performance of Wine and Wineries

As mentioned in Sect. 3.1, the environmental relevance of the wine sector has been growing over the last decades, rendering it an important contributor to a series of environmental impacts (see e.g., Arzoumanidis et al. 2013b).

The environmental performance of wine has been thoroughly examined in an array of LCA case studies (see Sect. 3.3). In this Section, methodologies that are based on the life cycle thinking concept and that are related to the wine sector are characterised. These methodologies can be divided into two categories: (1) those which are product-related and (2) those which are organisation-related.

The life cycle methodologies at the product level that were identified and that will be analysed in detail are (last update in July 2013): (1) product category rules (PCRs) issued by the International Environmental Product Declaration (EPD) system; (2) the Beverage Industry Sector Guidance for GHG Reporting; (3) the Sustainability Consortium methodology; (4) Sustainability Assessment Methodology for Wine (Italian Ministry for the Environment, Land and Sea); (5) the OIV (Organisation Internationale de la Vigne et du Vin) GHG Accounting Protocol for the Vine and Wine Sector.

On the other hand, methodologies at the organisation level comprise: (1) the OIV GHG Accounting Protocol for the Vine and Wine Sector; (2) the Joint Research Centre's (JRC) low carbon farming practices methodological guidelines; (3) the Beverage Industry Sector Guidance for GHG Emissions Reporting by the Beverage Industry Environmental Roundtable.

The methodologies were thoroughly characterised and analysed in order to provide an overview of the methodological specifications addressed both at product and at organisation level. This detailed analysis may facilitate the harmonisation of the assessment rules and act as a stepping-stone towards consolidation of environmental assessment methods. This would be useful also for delivering some insightful information regarding the "lessons learnt", as discussed in Sect. 3.4.

As a first step, the methodologies were characterised using the template developed and collectively agreed at the world level in the framework of the PCR Development Initiative (PCR Development Initiative 2013). The analysis thus included aspects as follows:

- General information: name of the methodology, date of expiration, product category, standards conformance, etc.
- Goal and scope: functional unit, system boundaries, data quality requirements, etc.
- LCI: primary and secondary data collection requirements, requirements regarding allocation, etc.
- LCIA: impact indicators, justification for their selection.

As well as what is in the PCR template, methodologies were also screened to identify what are considered as co-products, by-products and waste streams.

To this end, the identified methodologies were examined and separate characterisation sheets were produced for each one of them.

It must be noted, however, that the Italian Sustainability Assessment Methodology for Wine (Sustainability in the Italian Viticulture 2014) and the methodology developed by the Sustainability Consortium (TSC 2014) were excluded from this study, because they were not publicly available at the time of the review.

Finally, a brief description of simplification in LCA and simplified LCA tools is outlined in Subsection 3.2.3.

3.2.1 Brief Description of the Methodologies and Standards

In this section, the various methodologies identified are briefly presented for organisation and product level.

The International EPD® System Two methodological guidelines were identified as relevant for this review: (1) PCR of wine of fresh grapes, except sparkling wine;

grape must (EPD 2013, 2010:02) and (2) PCR of packaged sparkling red, white and rosé wines (in any kind of container and closure system) (EPD 2006, 2006:03). These methodologies, which are both at the product level, were issued by the International EPD® System (ENVIRONDEC 2014). The International EPD® system, which is based on international standards such as ISO 9001, ISO 14001, ISO 14040, ISO 14044, ISO 14025, ISO 21930, is one of the organisations supporting the development, release and update of PCRs. These PCRs provide, amongst other things, product-specific rules ranging from goal and scope definition to minimum data quality requirements for LCA studies instrumental to EPDs®, business-to-business shaped environmental communication systems according to ISO 14025 (EPD 2006; EPD 2013). In this context, supporting LCA studies are conducted with the attributional data modelling approach (De Camillis et al. 2013); see also Sect. 3.3.

Beverage Industry Sector Guidance for Greenhouse Gas Emissions Reporting The Beverage Industry Environmental Roundtable issued the second version of the Guidance for Greenhouse Gas Emissions Reporting in 2010 (BIER 2010), both at an organisation and at a product level. The overall aim of this roundtable, which was founded in 2006, is to identify ways to reduce water use, energy consumption and GHG emissions across the value chains of associated organisations and across the life cycles of products of the beverage sector (BIER 2010, p. ii). The specific objective of the guidelines under study is to estimate, track and report GHG emissions within the beverage industry.

The Sustainability Consortium The Sustainability Consortium (TSC) is an organisation that aims at developing methodologies, tools, and strategies to drive a new generation of products and supply networks that address sustainability-related issues about particular product categories (TSC 2014). Wine-specific and fruit-specific (thus including grapes) guidelines at the product level are under development by the TSC. These will also be used to derive key performance indicators to be used by retailers to classify wineries.

The OIV GHG Gas Accounting Protocol for the Vine and Wine Sector The International Organisation of Vine and Wine is an intergovernmental body, the aim of which is, amongst others, to contribute to harmonising existing technical documents and practices as well as to exploring the chance to proactively develop new technical specifications from the very beginning (OIV 2014). Being a sector-specific organisation, the OIV acknowledges the necessity of harmonising the international existing GHG accounting standards (for instance, the International Wine Carbon Protocol, the ISO 14040, 14044 and 14064 standards and others) in the vine and wine sector. For this reason, OIV issued the GHG Accounting Protocol for the Vine and Wine Sector in 2011 (OIV 2011), focusing on both the organisation and the product level.

The JRC Low Carbon Farming Practices Methodological Guidelines The Institute for Environment and Sustainability of the Joint Research Centre (JRC) of the European Commission along with Solagro, a non-profit organisation based in France, issued a set of guidelines for enhancing low carbon farming practices in 2013

(Bochu et al. 2013). The GHG emission measurement tool, called the "Carbon Calculator", calculates emissions at farm scale and delivers results at the organisation level for a reporting period of one year. The guidelines underpinning the Carbon Calculator are based on ISO 14044 and European reference methods (i.e. the Organisation Environmental Footprint Guide and Envifood Protocol). The JRC supported the development of this tool in response to the European Parliament's request for a project on the certification of low carbon farming practices in the EU.

The Sustainability Assessment Methodology for Wine The Italian Ministry for the Environment, Land and Sea launched a project for the evaluation and labelling of the sustainability performance of wine in July 2011. The project aims, amongst others, at issuing guidelines for the sector, building on existing methodologies, such as the OIV and the EU indications (Sustainability in the Italian Viticulture 2014). At present, these sector-specific guidelines are under development and specific matrices for e.g. water and carbon footprinting accounting are recommended for use. Particular emphasis in this project is given to the assessment of the impact on landscape.

3.2.2 Key Issues

The key issues resulting from the analysis for the aforementioned (see Subsection 3.2.1) methodological issues are reported. It must be noted that the results presented are not an exhaustive representation of what can be found in the methodologies, and they are only related to the objectives of this review. These comprise the following aspects: functional unit; system boundary; allocation and by-product, co-product and waste streams; use of resources and impact categories. The following Tables (3.2 and 3.3) include direct citations to the methodological documents examined.

As far as the functional unit selection is concerned (see Table 3.2), most of the methods refer to volume (1 L of wine), which appears to be confirmed also by the selection of weight, as it can be easily transformed into volume by using the density of the product under study.

Table 3.2 Illustration of the analysis results of the methodologies—Functional unit

Methodology	Functional unit
BIER—Beverage Industry Sector Guidance for Greenhouse Gas Emissions Reporting (BIER 2010)	Different for different types of beverages
OIV—Greenhouse Gas Accounting Protocol for the Vine and Wine Sector (OIV 2011)	1 kg of grapes or 0.75 L of wine
JRC—low carbon farming practices (Bochu et al. 2013, p. 19)	"Area or weight"
EPD®—PCR—Wine of fresh grapes, except sparkling wine; grape must (EPD 2013, p. 6)	"1 L of wine including packaging"
EPD®—PCR—Packaged sparkling red, white and rosé wines (EPD 2006, p. 3)	"1 L of wine"

Table 3.3 Analysis results of the methodologies—System boundary

Methodology/reference	System Boundary
BIER 2010, pp. 9–13, 18	*Enterprise inventory approach*
	" Use the operational control approach as defined by The GHG Protocol to define Scope 1 and 2 emissions
	Emissions from non-beverage operations such as entertainment, media, or food businesses are not addressed within this Guidance
	Report GHG emissions from operationally controlled sources as Scope 1 emissions
	Beverage industry GHG emissions sources included under Scope 2 (indirect emissions) generally fall into one of the following two categories
	(a) Emissions from directly purchased utilities &
	(b) Emissions from indirectly purchased utilities
	Scope 3 emissions include any emissions in the company's value chain not accounted for under Scopes 1 and 2. The distinction between scopes is unique to each beverage company depending on its operational boundaries. Reporting of Scope 3 emissions is currently voluntary"
	Product CF approach
	"[…] Boundaries are not drawn within the value chain to assign emissions to scopes. Instead, all emissions within the value chain boundary of a specific product are accounted for and parceled to a functional unit, which could be a specific container, serving size, or case of product
	The areas of the value chain are the same as those described for enterprise reporting, and include the GHG emissions associated with raw material inputs, transportation streams, manufacturing, and disposal/recycling of beverage materials"
OIV 2011, pp. 7–9	*Enterprise Protocol* (*EP*)
	Primary boundaries
	"All emissions classified as scope 1 (direct GHG emissions) or scope 2 (purchased power utility), are included."
	Secondary boundaries
	"[…] All the activities which are not under the control of the company but on which the company depends for its normal activity are included in the secondary boundaries. Examples of such emissions are: infrastructures, purchased consumables, waste."
	"[…] The vitivinicultural companies are only responsible for the emissions that are included into the primary boundaries
	The emissions classified into the secondary boundaries will be calculated in the case that the companies evaluate the global GHG emissions, related to their activities"
	Product Protocol (*PP*)
	"The boundaries […] are based in the life cycle of the product (business-to-consumer or 'cradle to grave'):"
	Grape production

Table 3.3 (continued)

Methodology/reference	System Boundary
	Wine processing
	Distribution and retail
	End-life-phase (covering disposal and recycling)
	"All emissions directly linked with the production process or life cycle of the vitivinicultural product should be included
	Examples […]: fuel and energy used (even from not owned machinery) in vineyard operations (ex. harvesting, vineyard treatments, etc.); fuel and energy used (even from not owned machinery) in winemaking and processing (ex. bottling…); fuel and energy used in the product transport; input production; waste disposal"
	"Emissions related to business travels are not included in the PP as they are not directly linked with the wine or grape life cycle
	Even if inside the wine life cycle boundaries, the consumption phase is not considered in the PP due to its negligible impact"
Bochu et al. 2013, pp. 13–15	"The Carbon Calculator assessment has to be carried out at farm level over a reporting period of one year
	Organisational boundaries
	[It] focuses on the main farming systems of the EU −27
	The farm is a physical land area with crops, livestock, buildings, machinery and inputs
	"Control" approach (100 %): the farm is owned by the farmer (financial) or the owner Controls the farmer
	Data for activities are available (the "farmer" knows them)
	In most of the cases: inputs purchased are used on the farm
	The Carbon Calculator is not designed for the following specific farms or on-farm activities
	Processing and distribution of agricultural products; agritourism, offices, sale of heat; specific agricultural products with specific inputs and emission factors (EF); rice cultivation and other waterlogged farming systems; forest activity (Carbon Calculator is only restricted to trees and hedges along crops or grassland plots); fishery, and the lists of EF are not complete (for lack of specific research), especially for: organic fertilisers for conventional or organic farming if not produced on farm; organic fertilisers for greenhouse nutritive solutions; specific inputs such as plastic pots, plants (vegetables, horticulture…) or seeds; specific machineries or buildings
	Environmental footprint boundaries
	The Carbon Calculator takes direct and indirect activities and associated GHG impacts into account. It uses a "cradle to farm-gate" approach including
	Direct emissions on the site/farm: emissions for energy used, CH_4 and N_2O (livestock, soils), C storage variations (soil, land use changes, farmland features like trees and hedges) and HFC emissions
	Indirect emissions (downstream emissions, not on the site) from

Table 3.3 (continued)

Methodology/reference	System Boundary
	Agricultural inputs; end-of-life of plastics and organic matter output as waste; NH_3 volatilisation, leaching and run-off (N_2O) The Carbon Calculator does not include emissions out of farm-gate and up to trailers and consumers: distribution, storage by industries, transportation of farm products, and processing out of the farm"
EPD 2013, pp. 7–8	"*Up-stream processes*
	The upstream processes include the following inflow of raw materials and energywares needed for the production of 1 L of wine of fresh grapes (except sparkling wine) or grape must The production of the grapes in agriculture and at the farm or at the well from the cradle Generation of energy wares used in agriculture, at the farm, and in production Production of other ingredients used in wine of fresh grapes (except sparkling wine) or grape must, detergents for cleaning, etc. Production of primary, secondary and third tier packaging materials Use of fertilisers" *Core processes*
	"The core processes include the production and the packaging of the final wine of fresh grapes (except sparkling wine) or grape must. The core processes include external transport of raw materials and energy wares to final production and internal transportation at the production site" *Downstream processes*
	" Transport from final production to an average distribution platform recycling or handling of packaging materials after use In the EPD, the environmental performance associated with each of the three life-cycle stages above are reported separately"
EPD 2006, p. 4	"*Production phase*
	Field activities (setting up/managing vineyards, irrigation, fertilisation, harvesting crops, transport to pressing facilities) Pressing Vinification (may occur in several phases in different locations) Bottling and packaging (may occur in several phases in different locations)" *Use phase*
	"Distribution of the product (transport to dealers) Use of the product and disposal of packing materials"

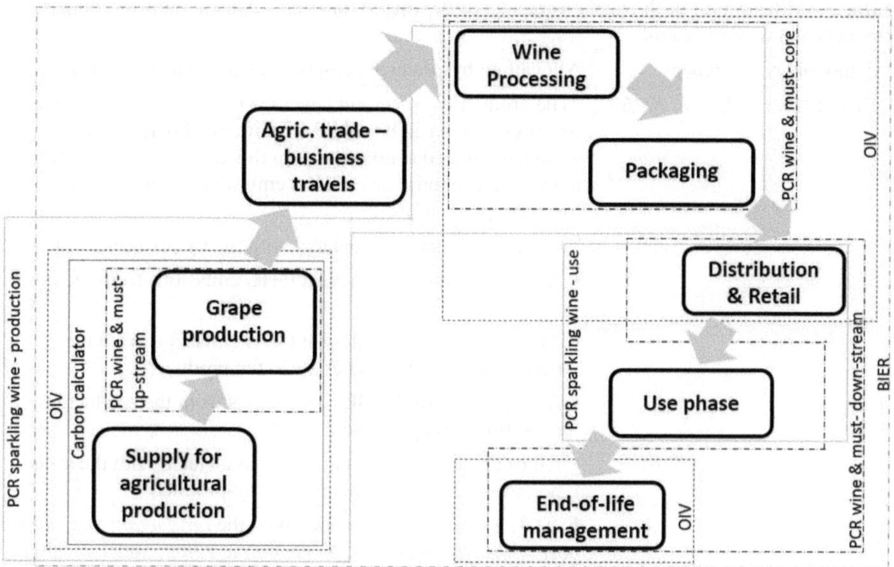

Fig. 3.2 System boundaries—overview of all methodologies examined

As regards the system boundary, this was separately defined for organisation- and product level, where applicable (see Table 3.3). The low carbon farming practices methodology (Bochu et al. 2013) obviously focused on the farm level.

The organisation-related methodologies include the Beverage Industry Guidance (BIER 2010), which uses the same rules for Scope 1, 2 and 3 as the GHG Protocol (see Table 3.3). Similarly, the OIV Guidance distinguishes primary (Scopes 1 and 2) and secondary boundaries (Scope 3), clarifying that vitivinicultural companies are only responsible for the emissions that are included in the primary boundaries (OIV 2011).

Regarding the methodologies at the product level, the PCRs issued by the International EPD System focus on dividing the life cycle phases into upstream, core and downstream ones for non-sparkling wine and grape must and into production and use phase for sparkling wine (see Table 3.3). The Beverage Industry Guidance (BIER 2010) sets the boundaries for the product CF not drawn within the value chain but all emissions within the value chain boundary of a specific product are accounted for and parcelled out to the functional unit. Finally, the OIV Guidance (OIV 2011) includes all life cycle phases, such as grape production, wine processing, distribution and retail, and end-life phase (disposal and recycling), but nevertheless excludes the consumption phase (see Table 3.3) and emissions related to business travel. For an overview of the life cycle stages covered by all the methods characterised in this chapter, please refer to Fig. 3.2.

Table 3.4 illustrates in detail the different approaches adopted by the different methodologies/guidelines with regard to allocation and by-products, co-products and waste streams. In most cases, and where mentioned, allocation is normally

Table 3.4 Illustration of the analysis results of the methodologies—Allocation, by-products, co-products and waste streams

Methodology/reference	Allocation, by-products, co-products and waste streams
BIER 2010, pp. 23–24, 59–60	"The production of certain beverage types may generate **by-product(s)** that can be sold for commercial purposes (such as an animal feed supplement). In this case, a portion of the relevant greenhouse gas (GHG) emissions should be allocated to the by-product itself The GHG emissions associated with the by-product include • An allocation of the relevant GHG emissions from the raw materials • An allocation of the relevant GHG emissions from the transport of the raw materials to the producer • An allocation of the GHG emissions from the production operations (Scope 1 and 2); and • All of the downstream emissions associated with the transportation, storage and sale of the by-product For GHG emissions associated with the *production and transport of the raw materials*, an ***economic value*** model should be used for allocating the relevant GHG emissions between the primary product and the by-product 1. Select the base unit for the raw material (e.g., bushels or tons) 2. Calculate the production yield for both the primary product and by-product (e.g., gallons of product per bushel of raw material) 3. Using the value of the product and by-product, calculate the total revenue per unit of raw material; and 4. Calculate the percentage of revenue contributed by the by-product and use this as the allocation percentage for GHG emissions from raw material production and transportation" "While the GHG emissions of the by-product are not allocated to the life cycle GHG emission of the primary product, beverage producers should calculate the by-product life cycle emissions in order to understand which emissions should be allocated to their products" The **waste** transport "…must be considered at each point up to and including the ultimate disposal location. GHG emissions associated with the incineration or landfilling of wastes are also included in the product carbon footprint The beverage production process also generates a number of **by-products**, which are often beneficially reused, such as bagasse, pumice, spent grains, spilled product, and wastewater Need to account for "waste products" that become **co-products** by virtue of them having a beneficial use (such as composting or feed material) up to the point of product differentiation. […] *Any emissions associated with transporting or further processing of that co-product are allocated to the co-product and not the original product from which it was derived*

Table 3.4 (continued)

Methodology/reference	Allocation, by-products, co-products and waste streams
	Evaluate **wastewater streams** coming from a beverage production facility or other locations in the life cycle to identify the energy demand associated with wastewater treatment. In some cases, wastewater treatment will be performed at a company-controlled facility, and the purchased energy used in wastewater treatment is considered a Scope 2 emission. However, when wastewater is sent off site to a third-party treatment site, such as publicly owned treatment works, include the energy use associated with transportation and treatment in Scope 3 emissions In the case of materials which are **recycled for reuse** in another product's life cycle (such as PET, which may be used in future PET bottles or for another use), use an **allocation** method ***based on the market recycling rate***. Depending on local market conditions, this approach affords the environmental benefits of recycling either to the recyclers or to the beverage producer" For the case of wine, no co-products are mentioned other than wine/grape
OIV 2011, pp. 25–26	*Waste disposal*
	"GHG emissions from **aerobic waste treatment, both solid and liquid**, (arising from the biogenic carbon fraction of the waste) are considered part of the *short term carbon cycle and are excluded from the PP* [*and EP*]. The emissions arising from the vine biogenic carbon fraction are included as part of the vine carbon cycle." **Energy consumed in the disposal** is *included* in the PP (and for the EP, if outside the company boundaries), is included in the secondary boundaries *Direct reuse*:
	Emissions related to the **reuse of wine byproducts or waste** are *included in the EP* if inside the boundaries of the company are *excluded from the PP* and should be integrated in the life cycle of the new product in which they are integrated as an input "In the vine and wine industry, examples of reuse included in the PP and EP (when inside the company boundaries) are: pruned canes ground for soil amendment; preparation and burning of wood residues or grape marc for energy purposes; compost preparation; distillation of wine or grape marc." *Recycling*
	Emissions related to the **recycling of wine by-products or waste** are *included in the PP and in the primary boundaries of the EP*, when the company is responsible for the recycling process

Table 3.4 (continued)

Methodology/reference	Allocation, by-products, co-products and waste streams
	PP: "A special case in the vine and wine industry is the recycling of the glass bottles: In order to avoid double accounting, and taking into account that glass from bottles can be recycled infini, the recycling GHG emissions *are already included in the glass production emissions figures*. Note: if this rule is not applied, the cullet production emissions would be assigned *twice*: first as glass recycling (of the previous bottle) and second as raw material use for the production of the successive bottle"
	EP: "If the company is responsible of the recycling of glass bottles, the recycling emissions should be carefully studied due to its importance when applying the EP. Taking into account that glass from bottles can be recycled infini, and in order to simplify the calculation, the recycling GHG emissions used could be the upstream ones (recycling figures of the bottle before the company use it)"
Bochu et al. 2013, pp. 19, 106	*Multiple outputs*
	"The Carbon Calculator systematically uses the **protein or energy allocation** key to distribute GHG emissions between: Milk and meat from dairy animals (cow, sheep, goat); Eggs and poultry meat for laying hens
	As processing is outside the boundaries of the Carbon Calculator, there is *no possibility to allocate GHG emissions between co-products resulting from processing*."
	"Distribution of GHG emissions **between products and co-products** throughout the supply chain are determined according to the three main rules:
	Type 1: direct assignment during the data input. For example, the GHG emissions (manufacturing) of mineral fertilisers applied on a crop will be directly attributed to this product (depending on the end-use of the crop)
	Type 2: automatic allocation. For example, on a specialised dairy farm (products = milk and meat from dairy animals) an automatic allocation rules 85–15 % base on protein content for enteric fermentation will be implemented
	Type 3: assignment made by the user himself. For example, in case of propane gas used on a farm, the user will distribute the percentage/quantity of use of this input between different available products"
	The user cannot select these co-products, as they are automatically created
	For the case of wine, no co-products are mentioned other than wine/grape
EPD 2013, pp. 9, 11	"Allocation between different products and co-products **shall be based on product mass**."
	"The potential environmental impacts and benefits of **recycling** of primary packaging shall be illustrated in the EPD

Table 3.4 (continued)

Methodology/reference	Allocation, by-products, co-products and waste streams
	Impacts could be calculated taking into account a typical scenario of the area in which wine is mainly distributed" For the case of wine, no co-products are mentioned other than wine/grape
EPD 2006, p. 6	"For each type of product belonging to the product category (packaged sparkling red, white and rosé wines) it is necessary to prepare specific Environmental Declarations. In case two types of product happen to be produced at the same site, the data regarding the specific production activity must be *allocated proportionately according to the following formula*: **(Total production of the type of product/total output of the site) * 100 = Percentage of allocation** In vinification phase the word 'production' means: amount of product obtained from grapes pressing plus addition of musts or wines coming from other plants if any, plus starting goods on hand minus final goods on hand" Here, dregs and pomace are mentioned as by-products

performed by mass. Protein or energy allocation is also mentioned in the low carbon farming practices methodology (Bochu et al. 2013).

As far as the use of resources and the selection of impact categories are concerned, please refer to Table 3.5. For GHG-related methodologies, the impact category taken into consideration is obviously global warming. The two PCRs, nonetheless, apart from the greenhouse effect, cover a broader range of environmental impact categories such as acidification, stratospheric ozone depletion, formation of oxidising photochemicals, eutrophication, etc. In addition, these PCRs tackle a series of resources, such as renewable and non-renewable resources, water use, electricity consumption, etc.

3.2.3 *Simplified LCA Tools*

The widespread use of LCA amongst public and private economic sectors has rendered it a powerful tool for the assessment of the environmental performance of products. For example, the application of LCA has become necessary for many firms, in particular Small and Medium-sized Enterprises (SMEs), which in most cases are related to wine production facilities (Arzoumanidis et al. 2013c). These firms often have to cope with lack of time, knowledge and resources, thus finding a full LCA application a difficult task (Arzoumanidis et al. 2014a). Therefore, simplification in LCA may often occur within the phase of LCI, LCIA or both (Arzoumanidis et al. 2013c), limiting the inclusion of processes or environmental impact categories to be considered.

Table 3.5 Analysis results of the methodologies—use of resources and impact categories

Methodology/reference	Use of resources and impact categories
BIER 2010	Greenhouse effect (GWP) in t CO_2 equiv
OIV 2011	Greenhouse effect (GWP)
Bochu et al. 2013	Greenhouse effect (GWP) in t CO_2 equiv
EPD 2013, pp. 11–12	"*Use of resources*
	The consumption of natural resources and resources shall be reported in the EPD
	Input parameters, extracted resources
	Non-renewable resources/Renewable resources
	Material resources
	Energy resources (used for energy conversion purposes)
	Water use
	Electricity consumption (electricity consumption during manufacturing and use of goods, or during service provision)"
	Potential environmental impacts
	"Emissions of greenhouse gases (expressed in global warming potential, GWP, in 100-year perspective)
	Emission of ozone-depleting gases (expressed as the sum of ozone-depleting potential in CFC 11-equivalents, 20 years)
	Emission of acidification gases (expressed as the sum of acidification potential expresses in SO_2-Eq.)
	Emissions of gases that contribute to the creation of ground level ozone (expressed as the sum of ozone-creating potential, ethene-equivalents)
	Emission of substances to water contributing to oxygen depletion (expressed as PO_4-Eq.)
	Other indicators
	"The following indicators shall be reported in the EPD
	Material subject for recycling
	Hazardous waste, kg (as defined by regional directives)
	Other waste, kg
	Toxic emissions
	Land use, m^2a for land occupation"
EPD 2006, pp. 7–8	"Use of renewable resources
	Without energy content
	With energy content
	Use of non-renewable resources
	Without energy content
	With energy content
	Consumption of electrical energy
	Categories of emissions
	Gas with greenhouse effect (GWP) kg CO2 equiv. (100 years)
	Acidification (AP) kmol H +

Table 3.5 (continued)

Methodology/reference	Use of resources and impact categories
	Reduction of stratospheric ozone (ODP) kg CFC-11 equiv. (20 years)
	Formation of oxidising photochemicals (POCP) kg ethane equiv
	Eutrophication (NP) kg O_2
	The above categories comply with enclosure A of MSR 1999:2
	Wastes
	Hazardous wastes, kg
	Non-hazardous wastes, kg"

The application of simplified LCA tools may require, in general, limited time and resources (Arzoumanidis et al. 2013c). Simplified tools appear to have clear and easy to understand calculation and visualisation methods and can be considered as suitable for an effective communication of the environmental performance of products and services (Arzoumanidis et al. 2013c). Simplified LCA tools normally offer characteristics such as user friendliness along with the life cycle thinking orientation, as well. Several opportunities were identified that could render such tools more easily adoptable: a proactive approach as regards the strategic management of the environmental variable, sensitivity of management to environmental issues and an interest in eco-labelling initiatives on the side of the market (Salomone et al. 2012).

In contrast, simplified LCA tools are characterised by their difficulty in incorporating the methodological differences across firms and sectors. Furthermore, several weaknesses can be identified for such tools. For example, reduced scope and increased subjectivity can be considered as weaknesses of simplified LCA tools (Arzoumanidis et al. 2013c). In addition, external threats are mostly connected to a general lack of environmental awareness by the firms combined with a central focus on short-term problems, mainly due to market pressure (Arzoumanidis et al. 2013c). Besides that, the fact that environmental management tools are not normally perceived as an opportunity for SMEs has also to be taken into consideration (Masoni et al. 2001). Technical staff's lack of willingness and/or time were two other critical issues identified. Finally, it must be noted that environmental issues are often perceived as limitations and a source of additional and often unknown (or hidden) costs (Masoni et al. 2004).

When choosing the most suitable simplified LCA tool, the objectives of a study, and more importantly the characteristics of the product under study, are issues to be considered (Arzoumanidis et al. 2014a). The modelling of one tool can be for instance more suitable for creating a phase of the life cycle. Finally, the degree to which the incorporated database can contain most of the processes that are needed for the study can play a quite important role in the selection of the most suitable tool (Arzoumanidis et al. 2014a). As regards wine, for instance, the existence of specific processes related to the agricultural and/or vinification phase may play an important role in the selection or not of a certain tool.

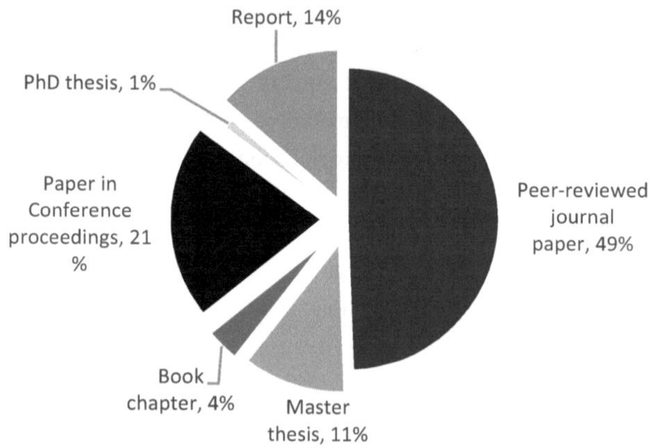

Fig. 3.3 Studies identified in the review of the LCA of the wine sector, published from 2001 to 2013 (last update on 31 July 2013)

3.3 Critical Analysis of Life Cycle Assessment: Case Studies in the Wine Sector

This chapter reports the results of a comprehensive critical analysis in the domain of the LCA of wine. An extensive search was conducted to select studies from the international literature that could encompass all the issues related to the LCA of wine. Following a screening process, 81 papers published between 2001 and 2013 (last update at July 2013) were finally selected and analysed, including papers available in peer-reviewed journals and conference proceedings, official reports, such as analyses commissioned by private or public institutions, and thesis reports (grey-literature). Figure 3.3 shows the percentage breakdown of the 81 investigated studies according to the typology.

The complete list of the reviewed studies, including the summary of findings on wine Environmental Life Cycle Approaches, is available in Table 3.7.

In order to outline the main peculiarities of an LCA study in the wine sector, examine which improvements could be made and suggest a number of lessons learnt, nine aspects were identified and analysed: (1) *Goals* (Sect. 3.3.1); (2) *Functional Unit* (FU) (Sect. 3.3.2); (3) *System boundary* (Sect. 3.3.3); (4) *Data issues* (Sect. 3.3.4); (5) *Handling multi-functional processes* (Sect. 3.3.5); (6) *Life Cycle Impact Assessment* (*LCIA*): *impact categories, assessment methods and indicators* (Sect. 3.3.6); (7) *Interpretation* (Sect. 3.3.7); (8) *Critical analysis* (Sect. 3.3.8); (9) *Comparative analysis* (Sect. 3.3.9).

The investigation performed on the first five aspects applied a mere conceptual approach (no quantitative results were generated for those issues discussed from Sect. 3.3.1 to 3.3.5). In contrast, the analysis carried out from Sect. 3.3.6 to 3.3.9 had a more quantitative nature.

It should be noted that for the critical analysis related to the aspects dealt with in Sect. 3.3.1–3.3.5, only 59 papers were considered among those included in Table 3.7. The excluded papers comprise papers regarding only one or few subsystems of the wine value chain, such as: packaging (Bengoa et al. 2009; Cleary 2013;

González-García 2011a, b; Latunussa 2011; Patingre et al. 2010; Woodward 2010); packaging and transportation (Cholette and Venkat 2009; WRAP 2007); closure systems (Gabarell et al. n.d.; Kounina et al. 2012; Rives et al. 2011, 2012, 2013); fertilisers (Ruggieri et al. 2009); waste management (Dillon 2011).

3.3.1 Goals

In almost all cases, the studies had the general aim of assessing the environmental impacts of wine. After the various papers related to wine were analysed, it was evident that the objective of the study in most cases was to estimate the environmental impacts in order to identify the hot spots in the life cycle of wine and to assess the effect of potential improvement options/possibilities.

Many studies dealt with comparative assessments; for instance, comparisons concerned white and red wines (Notarnicola et al. 2003), high quality and average quality wines (Notarnicola et al. 2003), wines from different regions (Vàzquez-Rowe et al. 2013).

A few studies compared different farming strategies: conventional and organic (Barberini et al. 2004; Cecchini et al. 2005b; Kavargiris et al. 2009; Niccolucci et al. 2008); industrial, organic and biodynamic (Eveleth 2013); organic and semi-industrial (Pizzigallo et al. 2008); biodynamic, conventional and an intermediate biodynamic conventional wine-growing plantation (Villanueva et al. 2013). However, for example, the goal of Notarnicola et al. (2003) was not a direct comparison of different wines, such as red and white wine or high quality and medium quality wine, as these are not "perfect substitutes", but different types of wines. Therefore, the differences in each of the environmental profiles wine had to be examined.

Other comparisons were made between different wineries to assess performances. For example, in Pattara et al. (2012a), one of the goals of the study was to assess which of two wineries had the highest performance when it came to CO_2-eq emissions associated with the production of Montepulciano d'Abruzzo DOC.

Moreover, the suitability of simplified LCA tools, such as VerdEE (Morgante et al. 2004; Petti et al. 2006), for the evaluation of the environmental performance of wine was also assessed, supporting the use of simplified tools as previously highlighted in Sect. 3.2.4.

Environmental assessments were also made by calculating the Carbon Footprint (BIER 2012; Bosco et al. 2013; Fearne et al. 2009; Pattara et al. 2012b; Vázquez-Rowe et al. 2013) and the Water Footprint (e.g. Ene et al. 2013; Pina et al. 2011) of wine, or by comparing the Ecological Footprint of different wines (e.g. Niccolucci et al. 2008).

On the other hand, Colman and Päster (2009) did not explicitly use the Carbon Footprint method, but developed a similar model for quantifying greenhouse gas emissions from the production and distribution of a bottle of wine to determine the phase with the greatest impact in terms of global warming.

Another goal, detected in Cecchini et al. 2005a, was the evaluation, through the application of different characterisation methods such as Eco-indicator 99,

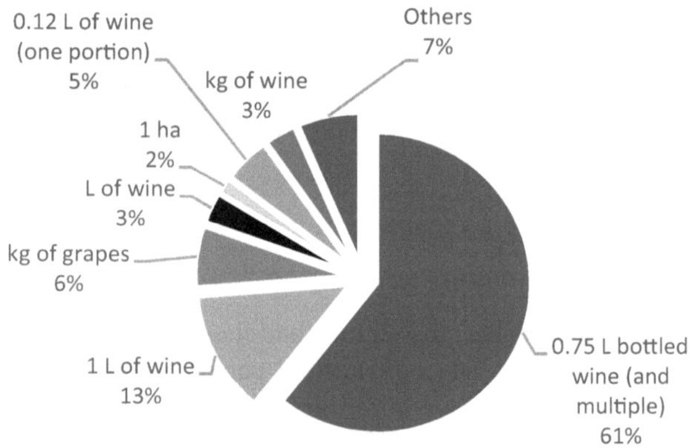

Fig. 3.4 Percentage breakdown of the analysed studies according to the different Functional Units used

EPS2000 and EDIP96 applied with SimaPro 5.0® software for the LCIA phase, of the impacts on the environment and on human health caused by the processes involved in wine bottle production.

3.3.2 Functional Unit

Amongst the 61 studies analysed, the Functional Unit (FU) was typically determined in terms of product mass or cultivated area units. In particular, as pointed out also in previous reviews (Benedetto 2013; Petti et al. 2010b; Rugani et al. 2013), 61 % of the studies (Fig. 3.4) define the FU as a 750 ml bottle of wine (or multiples thereof). Most of the authors consider this amount of wine as an FU because that is the most usual way of delivering the finished product to the market.

On the other hand, other authors (e.g. Arcese et al. 2012) considered 1 L of wine as an FU (13 % of the studies analysed), due to the fact that they aimed to avoid accounting for possible differences in packaging strategies within the same company, and thus to focus only on the quantity of the final product purchased by the customer (see also Fig. 3.7).

However, many authors, who generally focused on the agricultural phase referred to other FUs, such as 100 L of wine (Pattara et al. 2012a), kg of grapes (6 % of the studies took various amounts of grapes expressed in kg as an FU) or kg of wine (used in 3 % of the analysed cases).

Three case studies (5 % of the total) considered 0.12 L of wine (one portion) as an FU. This unusual choice probably reflects one restaurant serving corresponding to 1.3 standard portion (12 g of pure alcohol × 1.3 = 15.6 g). This corresponds to 12 cl of wine with an alcohol content of 13 % (Mattila et al. 2012b).

In Mann et al. (2010) the FU used was the amount of impact due to the 2009 Cru vineyard's production; in Niccolucci et al. 2008 two types of FUs are defined: a unit mass for the bottle and a unit area for vineyards; while in Notarnicola et al. (2007, 2008) the FU is related to enrichment of must by one alcoholic degree. These cases are categorised as "Others" for the particular nature of the FU.

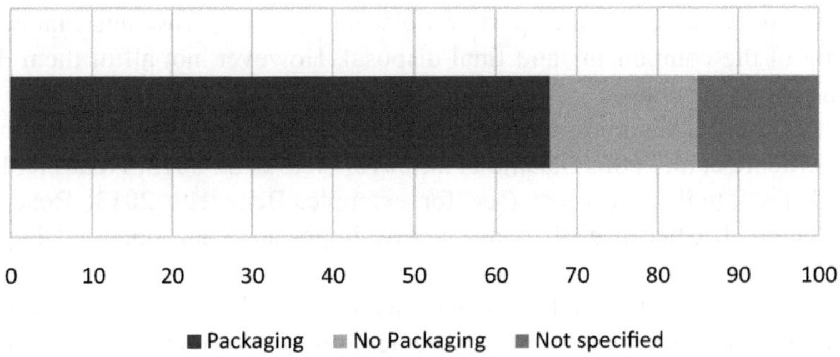

Fig. 3.5 Percentage breakdown of the analysed studies according to the packaging inclusion in FU

Fig. 3.6 Percentage breakdown of the analysed papers according to the System Boundary considered

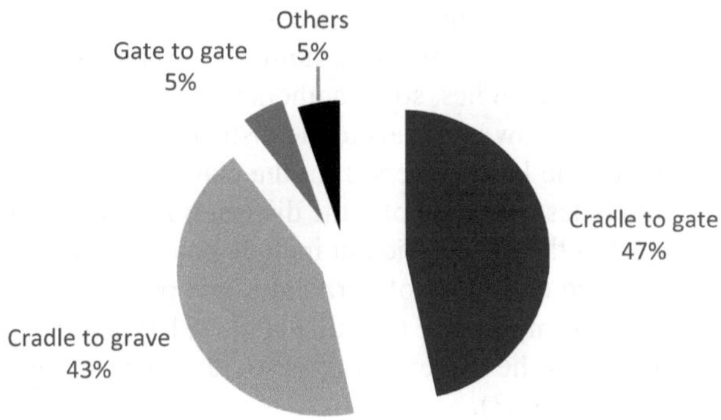

Only one study (Kavargiris et al. 2009) used 1 ha as an FU to compare organic and conventional vineyards.

Sixty seven percent of the studies analysed considered some form of packaging in the FU. Figure 3.5 shows that packaging was not considered in 18 % of cases, and 15 % of the studies did not explicitly indicate whether the packaging was included or not in the FU.

3.3.3 System Boundary

According to ISO 14044 (ISO 2006, p. 8): "The system boundary determines which unit processes shall be included within the LCA". As pointed out by Rugani et al. (2013), the variability of impacts across different case studies of wine may be strongly influenced by the system boundary identification. As a likely consequence, the choice of the relevant and irrelevant processes to be included or not in the system boundary could represent a problem in the definition of environmental performance of wine (Notarnicola et al. 2003).

Forty three percent of the studies (Fig. 3.6) claimed to consider the complete life cycle from "cradle to grave", including the extraction and the processing of raw materials, production, transport and distribution, use, reuse and maintenance, recycling of the components and final disposal. However, not all of them defined the same life cycle phases.

Most of the LCA studies analysed did not consider the vineyard-planting subphase (Ardente et al. 2006; Gazulla et al. 2010; Neto et al. 2013; Notarnicola et al. 2003). A few studies, however (see for example, Benedetto 2013; Bosco et al. 2011), included it because of its agronomic importance and potential impact on GHG emissions.

Furthermore, the consumption phase was not considered in most of the papers because of the lack of relevant data or the negligible environmental impacts of this phase, e.g. transport from the point of sale to the place of consumption or refrigeration, if any (Gazulla et al. 2010; Neto et al. 2013).

Forty seven percent of the analysed papers assessed the life cycle of wine from "cradle-to-gate", which included the impacts deriving from the phases of grape cultivation, wine production and storage, bottling and packaging activities. Within this percentage of studies, some authors referred to the LCA system boundary from the extraction of raw materials to the distribution phase, with a "cradle-to-market" perspective. The latter differs from the classical "cradle-to-gate" perspective in that it also includes the phase of wine distribution (Arzoumanidis et al. 2013b).

Most of the studies did not include phases such as wastewater treatments or production and emissions of herbicides and pesticides because of the lack of relevant data and/or importance (Gazulla et al. 2010); another reason could be the difficulty of modelling the dispersion of pesticides and nutrients in the environment (Notarnicola et al. 2003).

A few authors (5 %) focused their study on a single life cycle stage, with the aim to address specific research or policy questions. This is the case, for example, with the wine enrichment phase (Notarnicola et al. 2007) and the agricultural phase (Villanueva et al. 2013).

Lastly, other studies (5 %) considered a system boundary "from gate-to-gate", they focused their attention on specific phases. For example, Reich-Weiser et al. (2010) considered only scenarios of transport to New York after wine had been packaged for sale in Napa, California, and Bordeaux.

3.3.4 Data Issues

The LCA studies on wine highlighted the importance of obtaining significant onsite data for the processes included in the system (as reported in Petti et al. 2010b).

In practice, a great number of brands in Europe base their grape production phase on a broad number of vine-growers who sell their grapes to the wineries every year. This situation renders environmental evaluations on viticulture complicated, since multiple data for multiple facilities have to be handled. The use of average values for this type of multiple dataset usually entails large standard deviations that may impede adequate interpretation of the results (Reap et al. 2008; Rugani et al. 2013; Weidema and Wesnæs 1996). In other words, the use of average inventory data for analysing a multiple set of vine-growing plantations is likely to be subject to significant data variability, distorting the individual performance of each of the assessed vineyards (Vázquez-Rowe et al. 2012a).

In the studies analysed many of the data collected for the processes of grape-growing, winemaking and bottling are from primary sources. These sources include data collected (often by means of questionnaires) from vineyard and winery staff and power company sources (e.g. Benedetto 2013; Bosco et al. 2013; Gazulla et al. 2010; Herath et al. 2013a; Kavargiris et al. 2009).

Data are often derived from secondary sources, as in the studies on the distribution phase of the final product to the consumer (e.g. Barry 2011), for the fuel and electricity supply chain, the manufacturing and transport of agrichemicals, winemaking additives and glass bottles (LCI databases such as BUWAL®, ecoinvent®, SimaPro®, GaBi professional®, IDEMAT®, amongst amongst others).

3.3.5 Handling Multi-functional Processes

In the specific case of wine, by-products such as grape residues and fermentation sediments are obviously impossible to produce separately; therefore, it would make no sense to divide the winemaking process into two or more independent sub-processes.

Most of the studies did not refer to the allocation of by-products and co-products. Others did not consider the allocation process because marc and lees obtained from the vinification process were excluded from analysis (and in one case they were returned to the soil; Rallo 2011) or because some allocation procedures were automatically included within the LCA software calculations (Arcese et al. 2012).

Vázquez-Rowe et al. (2012b) did not perform any allocation because, although wine was not the only product derived from winery transformation, grapes were the only product acquired from the cultivation phase, since all by-products were obtained once the grapes were delivered at the winery. Conversely, a series of residues were generated during the wine production stage. These products were incorporated into the vineyard as fertiliser. Therefore, no allocation was considered in this stage since the only marketable product was wine.

The problem of how to allocate the different co-products of winemaking (skins, pips and stalks, etc.) is tackled in the literature by allocation of the environmental burden by mass (Bosco et al. 2011), economic value (Cecchini et al. 2005a, b) or a combination of both (Nicoletti et al. 2001).

As regards allocation by mass, the co-products leaving the systems, such as stalks, lees and marc, were considered as solid waste for which there was no disposal treatment, since they became raw materials for other productions, respectively compost for stalks and tartaric acid for marc (Notarnicola et al. 2003).

Arguably, Gazulla et al. (2010) chose economic allocation because it reflected the actual thrust behind the entire wine industry in a better way than mass- or energy-based allocations could do; since the main product was obviously wine itself, and not any of its by-products.

3.3.6 Life Cycle Impact Assessment (LCIA): Impact Categories, Assessment Methods and Indicators

The general aim of Sect. 3.3.6 was to analyse and discuss how and to what extent LCIA is applied to the wine sector in a wide range of the reviewed literature. To this end, six key issues were selected as intrinsically related to the LCIA sphere, and then analysed along with the 81 studies considered in Table 3.7. These six key issues were: (1) LCIA method(s); (2) LCIA phase(s); (3) LCIA results; (4) LCIA result quality; (5) Interpretation phase; (6) Indicator(s)/method(s) other than LCIA. As illustrated in Table 3.6, each key issue was assigned a score from one to three. This operation was performed to compare every aspect of the LCIA sphere across all the selected studies, normalising any qualitative (e.g. quality of results, interpretation profiles, etc.) or quantitative (e.g. number of impact categories considered, LCIA results, etc.) information on a common semi-quantitative metric. The rationale behind the '1–3' scoring range in Table 3.6 was the same for all six issues and reflected the breadth and depth of information provided by each analysed study with regard to the key issue considered. For example, concerning key issue 1, score 1 was attributed to the studies that did not apply any LCIA method, score 2 to the studies that included only one single-score method (e.g. carbon footprint), and score 3 to the studies that carried out an LCIA with a multi-score perspective (e.g. application of two or more LCIA indicators). A complete description of the properties and assumptions behind this scoring approach is reported in Table 3.6. The key issues numbered 1–4 are discussed in Sect. 3.3.6, key issue 5 is discussed in Sect. 3.3.7 and key issue 6 in Sect. 3.3.9 (Table 3.6).

Table 3.6 shows the main topics investigated in this section: the type of LCIA method adopted and the phases reached in the analysis, the number and type of the indicators used to outline a general profile of the impact assessment associated with wine production and the quality of the impact results obtained.

The results of the review, carried out following the methodology described above, are synthetically presented in Fig. 3.7.

From the review of the 81 papers on the LCA method adopted (see key issue 1 in Table 3.6), it emerged that 20% of the papers did not carry out the LCIA phase. In many cases, the study provided an overview of key drivers for wineries to move towards sustainability practices and outlined actual environmental practices: e.g. Dodds et al. (2013) for the New Zealand wine industry and Ardente et al. (2006), who presented a preliminary analysis of an environmental management scheme (EMS) and environmental product labelling potential in the winery sector (POEMS) with a Sicilian wine production case study. The reason why LCIA is not explicitly included in the scope of the LCA-based study may be because of the need to consider criticalities and environmental aspects that are not usually dealt with by typical LCIA approaches. For example, some papers focussed on a detailed inventory (Pizzigallo et al. 2008; Reich-Weiser et al. 2010; Notarnicola et al. 2007) or presented a comparison of published studies on the LCA of wine on the basis of their methodologies and results, as in Woodward (2010), with regard to packaging

Table 3.6 Key issues analysed in the present LCIA review (Sects. 3.3.6–3.3.9) with a description of the score properties

Key issue	Score range	Scoring description
1. LCIA method(s)	1–3 From 1 to 3 according to the increase of breadth and depth of information provided	1: *LCIA method (s) not applied* 2: *Single-score method* (only one impact issue evaluated) 3: *Multi-score method* (more than one impact criteria evaluated)
2. LCIA phase(s)		1: *LCIA not performed* or *inventory results used otherwise* 2: *Only characterisation is performed* 3: *Characterisation+ normalisation+ weighting*
3. LCIA results		1: *Lower granularity* = only qualitative analysis or quantitative but with scarce resolution (low detail of information or only relative contribution results) 2: *Medium granularity* = quantitative results with good resolution (absolute values provided by process phases) 3: *Higher granularity* = quantitative results with wider resolution and transparency (absolute values by detailed/site-specific inventory process)
4. LCIA result quality		1: *Lower quality* = incomplete system boundary + not sufficiently representative LCI data + uncertainty neither considered nor evaluated 2: *Medium quality* = more complete system boundary + sufficiently representative LCI data + uncertainty or variability considered but not necessarily evaluated 3: *Higher quality* = almost complete system boundary + sufficiently representative LCI data + uncertainty and/or variability evaluated
5. Interpretation phase		1: Reporting of this basic information *Identification of the significant issues based on the LCI and/or LCIA results* 2: Reporting of this additional information *Conclusions, limitations, and recommendations* 3: Reporting of further evaluations about *Completeness, sensitivity and consistency checks*

Table 3.6 (continued)

Key issue	Score range	Scoring description
		Appropriateness of the definitions of the system functions, the functional unit and system boundary and/or identification of the limitations identified due to data quality assessment and sensitivity analysis
6. Indicator(s)/method(s) other than LCIA		1: *Conventional LCA* = only LCIA method(s) applied
		2: *Only use environmental assessment metric(s) other than those typically included in LCIA methods applied*
		3: *Comparative/combination purpose* = application of LCIA + other (complementary) environmental assessment metric(s)

options. Finally, other studies concerned the application of GHG emissions' inventory to the wine supply chain (Arzoumanidis et al. 2013b; Kavargiris et al. 2009; WRAP 2007). Interestingly, Kavargiris et al. (2009) performed a comparison between conventional and organic white wines in Greece based on energy balance and carbon-related emissions, and Reich-Weiser et al. (2010) analysed the GHGs impact analysis of shipping and distribution systems. On the same subject, Arzoumanidis et al. (2014b) explored biogenic accounting emissions in the case of wine, which is not yet included and defined in the international standards (BSI 2011; ISO 2013). However, when authors do not include LCIA, they tend to address aspects typically outwith the environmental sphere, as shown by the preliminary evaluation on social LCA in the wine sector performed by Sanchez Ramirez (2011), who identified 26 indicators according to the UNEP/SETAC LCI framework (UNEP/SETAC 2009). Soosay et al. (2012) investigated the worth of sustainable value chain analysis (SVCA) as a tool for promoting better alignment between the allocation of resources in the supply chain industry and consumer preferences in a specific target market.

Twenty-eight per cent of the selected studies performed the LCIA but from a single-score perspective, thus addressing only one aspect of the cause-effect chain (EC 2010). More specifically, the majority of those studies (18 out of 23) analysed the global warming potential (typically referred to as "Carbon Footprint–CF"), whereas the water footprint (WF) was considered in just three cases (Ene et al. 2013; Herath et al. 2013a, b) and land use in just one case (Mattila et al. 2012b). Moreover, many international organisations for wine production, such as the International Wine Carbon Calculator (IWCC) and the OIV, are working to standardise the CF estimation protocols and guidelines currently under development (Pittock et al. 2003; Hayes and Battaglene 2006; Webb et al. 2007; see also Sect. 3.2.1). This is because their focus is explicitly on the continuous improvement of the wine life cycle and new technology options might also offer opportunities to mitigate impacts

Table 3.7 Summary of findings on wine environmental life cycle approaches

Author	Year	Type of publication	Wine geographic location	Type of wine	Functional Unit (FU)	System boundaries[a]
Amienyo	2012	PhD thesis	UK	Red wine	1000 L of beverage, 0.75 L red wine glass bottle	C-G
Aranda et al.	2005	Peer-reviewed journal	Spain	n.s.	0.75 L bottled wine	C-G
Arcese et al.	2012	Peer-reviewed journal	Italy	n.s.	1 L of wine	C-R
Ardente et al.	2006	Peer-reviewed journal	Italy	Red wine	0.75 L bottled wine	C-R
Arzoumanidis et al.	2013b	Book chapter	Italy	Red wine	0.75 L bottled wine	C-M
Barberini et al.	2004	Master thesis	Italy	Red wine	0.75 L bottled wine	C-G
Barry	2011	Master thesis	New Zealand	White wine	0.75 L bottled wine	C-G
Benedetto	2013	Peer-reviewed journal	Italy	White wine	0.75 L wine	C-W
Bengoa et al.	2009	Conference proceedings	Canada	n.s.	To hermetically hold 0.75 L of table wine for two years	C-G
BIER	2012	Report	Europe and North America	n.s.	0.75 L glass bottle, six-pack	C-Co
Bosco et al.	2011	Peer-reviewed journal	Italy	Red wine	0.75 L bottled wine	C-G
Bosco et al.	2013	Peer-reviewed journal	Italy	Red wine	0.75 L bottle of wine	C-G
Burja and Burja	2012	Peer-reviewed journal	Romania	White wine	1 kg of grape	n.s.
Camilleri	2009	Conference proceedings	n.a.	n.s.	n.a.	n.a.
Carballo Penela et al.	2009	Peer-reviewed journal	Spain	White wine	0.75 L bottled wine	C-G
Carta	2009	Master thesis	Italy	Red wine, mixed red/white	0.75 L bottled wine	C-G
Catania and La Mantia	2006	Conference proceedings	Italy	n.s.	0.75 L bottled wine	C-G
Cecchini et al.	2005a	Master thesis	Italy	Red wine	0.75 L bottled wine	C-G
Cecchini et al.	2005b	Master thesis	Italy	Red wine	0.75 L bottled wine	C-R

Table 3.7 (continued)

Author	Year	Type of publication	Wine geographic location	Type of wine	Functional Unit (FU)	System boundaries[a]
Cecchini et al.	2006	Master thesis	Cile	Red wine	0.75 L bottled wine	C-R
Cholette and Venkat	2009	Peer-reviewed journal	USA	n.s.	6-bottles box transported	W-R
CIV	2008	Report	Italy	Sparkling red wine	1 L bottled wine	C-G
Cleary	2013	Peer-reviewed journal	n.s.	n.s.	The packaging required for 1 L of young, non-sparkling wine and that for 750 ml of spirits	C-G
Colman and Päster	2009	peer-reviewed journal	Australia, France	Red and white	0.75 L bottled wine	C-G
Comandaru et al.	2012	Peer-reviewed journal	Romania	White wine	0.75 L bottled wine	C-G
Del Principe	2013	Report	n.s.	n.s.	1 L of wine including packaging	C-G
Dillon	2011	Master thesis	South Africa	n.s.	1000 L of wine	n.a.
Ene et al.	2013	Peer-reviewed journal	Romania	n.s.	0.75 L bottled wine	C-G
Eveleth	2013	Report	n.s.	n.s.	1 L of wine	C-R
Fearne et al.	2009	Conference proceedings	Australia	n.s.	n.s.	C-G
Gabarell et al.	n.d.	Conference proceedings	n.s.	n.s.	One million of natural cork stoppers	P-G
Gazulla et al.	2010	Peer-reviewed journal	Spain	Red wine	0.75 L bottled wine	C-G
Gonzalez et al.	2006	Report	France, Sweden	Red wine	1 L of wine	C-G
Gonzalez-Garcia	2011a	Peer-reviewed journal	n.s.	n.s.	1 kg of wood based products (among which wine boxes)	n.a.
Gonzalez-Garcia	2011b	Peer-reviewed journal	n.s.	n.s.	One wood box (1.35 kg) for the storage of three standard wine bottles	P-R

Table 3.7 (continued)

Author	Year	Type of publication	Wine geographic location	Type of wine	Functional Unit (FU)	System boundaries[a]
Greendelta	2011	Software guideline report	France	n.s.	1 kg of rapes	n.a.
Greenhaigh et al.	2011	Report	New Zealand	White wine	0.75 L bottled wine	C-G
Herath et al.	2013a	Peer-reviewed journal	New Zealand	Super-premium wines	0.75 L bottled wine	C-W
Herath et al.	2013b	Peer-reviewed journal	New Zealand	Super-premium wines	0.75 L bottled wine	C-W
Jimenez et al.	2013	Peer-reviewed journal	Spain	Red wine	1000 kg grape	C-G
Kavargiris et al.	2009	Peer-reviewed journal	Greece	White wine	1 ha	C-W
Kounina et al.	2012	Peer-reviewed journal	n.s.	n.s. (average from literature)	0.75 L bottled wine	C-G
Leonardi et al.	2006	Report	Italy	Red, white and rosè wines	1 L of wine	C-G
Latunussa	2011	Master thesis	n.s.	n.s.	Packaging and distribution of 1000 L of wine	C-G
Mann et al.	2010	Conference proceedings	USA	Aperitif wine	Total amount of bottles of Cru production for the year 2009	C-W
Mattila et al.	2011	Peer-reviewed journal	Spain	Red wine	One portion of wine equal to 0.12 L	C-Co
Mattila et al.	2012a	Peer-reviewed journal	Spain	Red wine	One portion of wine equal to 0.12 L	C-R
Mattila et al.	2012b	Peer-reviewed journal	Spain	Red wine	One portion of wine equal to 0.12 L	C-R
Morgante et al.	2004	Conference proceedings	Abruzzo region	Red wine	6 bottles of 0.75 L with primary and secondary packaging	C-G
Neto et al.	2013	Peer-reviewed journal	Portugal	White wine	0.75 L bottled wine	C-W
Niccolucci et al.	2008	Peer-reviewed journal	Italy	Red wine	Unit mass of bottle & unit area of vineyard	C-R

Table 3.7 (continued)

Author	Year	Type of publication	Wine geographic location	Type of wine	Functional Unit (FU)	System boundaries[a]
Nicoletti et al.	2001	Conference proceedings	Italy	Red wine	1 L of wine (not bottled)	C-W
Notarnicola et al.	2003	Chapter in edited volume	Italy	Red wine and white wine	0.75 L bottled wine	C-W
Notarnicola et al.	2007	Chapter in edited volume	Italy	n.s.	The enrichment of 1000 L of must by 1 alcoholic degree	V
Notarnicola et al.	2008	Conference proceedings	Italy	n.s.	1000 L of must for which the aimed enrichment is 1 alcoholometric degree from 10° to 11°	C-G
Notarnicola et al.	2010	Conference proceedings	Italy	red wine	133 bottles of 0.75 L	C-R
Patingre et al.	2010	Report	Different countries	n.s.	Packaging and distribution of 1000 L of wine	P-G
Pattara et al.	2012a	Conference proceedings	Italy	Red wine	100 L red wine without packaging	C-W
Pattara et al.	2012b	Peer-reviewed journal	Italy	Red wine	0.75 L bottled wine	C-G
Petti et al.	2006	Conference proceedings	Italy	red wine	0.75 L bottled wine	C-R
Pina et al.	2011	Poster	Portugal	White wine	0.75 L bottled wine	C-B
Pizzigallo et al.	2008	peer-reviewed journal	Italy	red wine	1000 kg of wine	C-G
Point et al.	2012	Peer-reviewed journal	Canada	Red and white	0.75 L bottled wine	C-G
Rallo	2011	Master thesis	Italy	Passito	0.75 L of Passito di Pantelleria	C-G
Reich-Weiser et al.	2010	Peer-reviewed journal	France	n.s.	0.75 L bottled wine	W-R
Rives et al.	2011	Peer-reviewed journal	Spain	All wine tipes	One million standard natural cork stoppers	P-G
Rives et al.	2012	Peer-reviewed journal	Spain	All wine tipes	One million champagne cork stoppers	P-G

Table 3.7 (continued)

Author	Year	Type of publication	Wine geographic location	Type of wine	Functional Unit (FU)	System boundaries[a]
Rives et al.	2013	Peer-reviewed journal	Spain	All wine tipes	One million cork stoppers and 1 t cork	P-G
Rugani et al.	2009	Conference proceedings	Italy	Red wine	100 kg bulk wine	C-W
Ruggieri et al.	2009	Peer-reviewed journal	Spain	n.s.	1 kg N to vineyard lands	
Sanchez Ramirez	2011	ppt	Italy	Red wine	0.75 L bottled wine	C-G
Schlich	2010	Conference proceedings	Germany, Europe, South-Africa	n.s.	0.75 L bottled wine	C-G
Soosay et al.	2012	Peer-reviewed journal	Australia	Oxford landing wine	n.s.	C-G
Vázquez-Rowe et al.	2012a	Peer-reviewed journal	Spain	White wine	0.75 L bottled wine	C-B
Vázquez-Rowe et al.	2012b	Peer-reviewed journal	Spain	White wine	1.1 kg harvested grape (=750 ml Rías Baixas wine)	C-F
Vázquez-Rowe et al.	2013	Peer-reviewed journal	Luxembourg	1 Red wine, 1 Sparkling and 1 White wine	0.75 L bottled wine	F-W
Venkat	2012	Peer-reviewed journal	USA	Red and white wine	1 kg of grape	C-F
Villanueva et al.	2013	Peer-reviewed journal	Spain	Ribeiro wine	1.1 kg of grape	C-F
Woodward	2010	Report	South Africa	All wine tipes	Many studies, different FUs	P-G
WRAP	2007	Report	Australia, France, UK	n.s.	0.75 L wine	C-G
Zabalza et al.	2003	Conference proceedings	Spain	n.s.	100 L wine	C-R

[a] *C* cradle, *G* grave, *F* farm gate, *W* winery gate, *B* bottling, *R* retailer, *Co* consumer, *P* packaging production, *n.a.* not available, *n.s.* not specified, *n.d.* no date

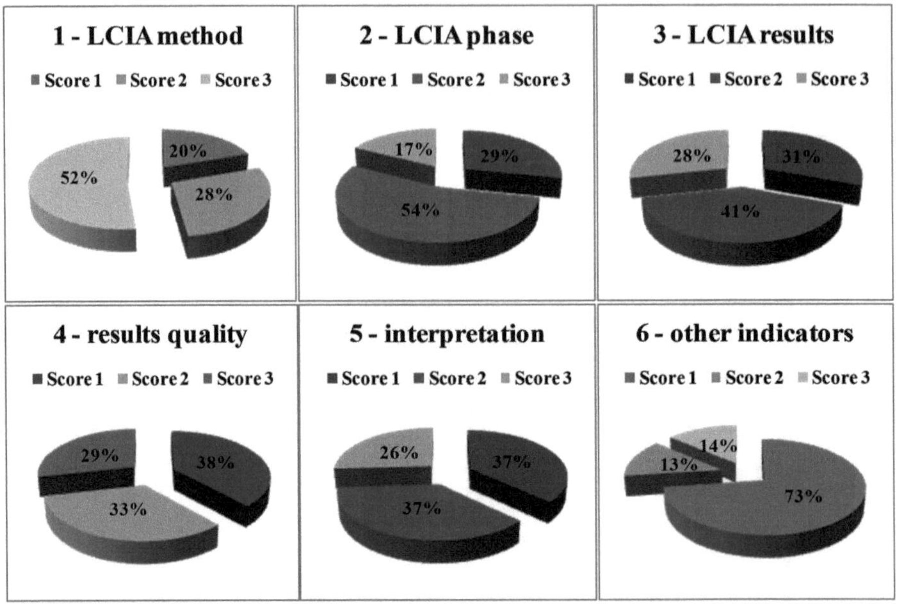

Fig. 3.7 Scores' attribution along the six key issues outlined in Table 3.6 (score 1–3; from 1 to 3 according to the increase of breadth and depth of information provided, see Table 3.6 for further details). The total number of studies for which the scoring was performed is 81

of priority relevance for human well-being, such as "climate", "water" and available "land". As a result, IWCC (Wine Institute 2010) was tested by some scholars for comparing cooperative wineries in central Italy (Pattara et al. 2012a) and for a first application of the OIV-GHGAP system to the wine life cycle (Pattara et al. 2012b). In both cases, a detailed investigation of the advantages and limitations of the protocol and the carbon calculator software was carried out in comparison with conventional LCA-based GWP assessments. The British standard PAS 2050 or the GHG protocol were followed in three other cases (BIER 2010; Cholette and Venkat 2009; Soja et al. 2010).

With regard to the land use issue, this was chosen as a unique indicator in a comparative case study of beer vs. wine production by Mattila et al. (2012b). This is one of the first studies that compares land use and land use change impact indicators in detail, an important issue for all agricultural-based production systems, including wine.

A multi-score perspective in the LCIA of wine was adopted in 52 % of the total of reviewed papers. More than half (22 out of 42) of these studies applied the CML method (Guinée et al. 2002), which proved to be effective in identifying impacts related to a large spectrum of environmental effects at different scales. Ten studies applied the Eco-indicator99 method, three of them comparing it with EPS 2000 and EDIP 96 methods (Cecchini et al. 2006; Cecchini et al. 2005a, b). Two studies applied the IMPACT+2002 method (Cleary 2013; Bengoa et al. 2009) and one the more recent Recipe approach (Arzoumanidis et al. 2013b). The CML method was probably applied because of its maturity (it was developed at the beginning of 2000) and because it embraces up to 10 different impact categories including potentials of global warming, acidification, eutrophication and human toxicity. However, it includes neither land use nor water consumption-related indicators. Since no robust multi-score LCIA method exists, which is capable of including a complete range of impacts of the cause-effect chain, it is reasonable to apply single-score and multi-score methods in combination, strengthening the evaluation of any possible issues and impacts arising from the production of wine.

As regards the LCIA phase in which results were elaborated and presented (see key issue 2 in Fig. 3.2), in 28 % of the cases the inventory results were used otherwise (not for LCIA purposes), and 54 % of the studies presented an LCIA composed of the mandatory phases of classification and characterisation of impacts, with the application of mid-point indicators (from the CML method, as previously observed). Only the remaining 17 % added the normalisation and weighting steps to the characterisation and, in this case, the most frequently used endpoint methods were Eco-indicator99 (Cecchini 2005a, b; Jiménez et al. 2013; Mann et al. 2010; Comandaru et al. 2012; Cecchini et al. 2006; Aranda et al. 2005; Gonzalez et al. 2006; Della Giovampaola and Neri 2004; Catania and La Mantia 2006), IMPACT+02 (Cleary 2013; Bengoa et al. 2009) and Recipe (Arzoumanidis et al. 2013b). The reason why so few studies considered the normalisation and weighting steps is unknown, but one could hypothesise that the authors calculated normalised and weighted scores simply because of the choice of impact methods, which typically provide endpoint targets.

As regards the phases with the greatest impact on the wine production chain, 40 % of the reviewed studies explicitly identified the phase underlying the highest impacts of the wine life cycle, i.e. they performed a contribution analysis. Among these, 34 % demonstrated that impacts are typically generated by packaging production (31 %) during the winery and bottling phase, followed by the agricultural phase (19 %) and transport for distribution to consumers (13 %). This review also shows that several studies did not directly aim at assessing the potential impacts associated with the functional unit of wine (e.g. 0.75 L bottle) but rather products normally included in the wine life cycle, such as natural cork stoppers (Rives et al. 2011, 2012, 2013) or wood boxes for wine bottle storage (González-García et al. 2011a, b).

The majority of the studies presented a good level of granularity in terms of absolute values and a detailed description of the life cycle phases and indicators (41 % with assigned score 2). Nevertheless, 31 % of the studies presented only qualitative analysis, or quantitative but with scarce resolution (little detailed information or only relative contribution results). This is the typical problem of insufficient data being available for a complete LCA study (Ardente et al. 2006; Notarnicola et al. 2007). In certain cases, authors only aimed at informing local stakeholders or companies about the environmental performance of their wine supply chain, and it is usually better in such cases to communicate via ranked scores or percentage numbers (Comandaru et al. 2012; Dodds et al. 2013) Only 28 % of the studies had a high level of granularity in terms of the results, showing quantitative results with wide resolution and transparency, and presenting absolute values through a detailed/site-specific inventory process.

The results' quality is strictly linked to the methodology used for data collection, the depth of the analysis and the representativeness of the elements (site-specific information, local technological or ecological parameters, etc.) included in the evaluated system. The significance of the results could be improved by an attempt to assess the level of uncertainty of the model and the parameters and by testing the robustness and variability of results with a sensitivity analysis. With regard to the outcomes of the scoring attribution for key issue 4, the majority of the studies presented a low level in terms of result quality (38 %), including an incomplete system boundary description and insufficiently representative LCI data. Furthermore, neither an uncertainty nor a sensitivity analysis was carried out, probably because the scope of the analysis was narrowed to assessment of wine life cycle performance only at a preliminary stage (Aranda et al. 2005; Pizzigallo et al. 2008, Carballo Penela et al. 2009; Ruggieri et al. 2009;; Point et al. 2012; Herath et al. 2013) In most cases, the authors used both primary and secondary data from the literature or only a small part of the data for LCI collected from a real case study. For example, in the case of Woodward (2010) the objective was to study and summarise the packaging options available to the wine industry, including the positives and negatives of traditional glass and alternative media. This was based on a review of available research, literature and reports and on the opinions of local and international industry stakeholders.

Finally, 33 % of the studies showed a good level of quality in the results presentation, with more complete system boundary assumptions and representative LCI data. The uncertainty or variability of the model dataset was taken into account but not necessarily evaluated. In effect, only a few quantitative evaluations exist: for example, grapes' yield variability over time (Barry 2011; Bosco et al. 2011) or the amount of fertilisers and pesticides used (Neto et al. 2013).

With regard to the characterisation of uncertainty and variability, only a few authors have reported quantitative assessments. In Kounina et al. (2012), data were mostly collected from the literature, and an uncertainty analysis (Monte Carlo simulation) was performed to assess the variability of uncertain parameters linked to the wine test changes regarding the use of stoppers (impact of wine, cork stoppers, screw caps and replacement rates). Similarly, Cleary (2013) investigated the importance of uncertainty associated with key data input for wine packaging and considered alternatives to conventional single-use glass bottles. The results of this study show that data uncertainty was relatively low.

A high level of result quality was obtained by 29 % of the studies, where an almost complete system boundary (from cradle-to-grave, including details of specific internal processes) was performed, with sufficiently representative LCI data and uncertainty and/or variability evaluations. For example, Mattila et al. (2012b) analysed the impact of uncertainties in LCI and LCIA of wine and beer; for wine detailed uncertainty information was reported mainly for N2O emissions from soil. Moreover, in the Beverage Roundtable (BIER 2010) data uncertainty was assessed by applying the methodology and guidance provided by the Greenhouse Gas Protocol document, namely quantitative inventory uncertainty (WRI and WBCSD 2011). Finally, Bosco et al. (2013) performed a sensitivity analysis to evaluate the robustness of the LCA model and to identify the key parameters and main factors related to the soil organic matter (SOM), whose role is essential in the overall CF (see Sect. 3.3.8). The sensitivity analysis highlighted that the glass bottle was the most important parameter influencing the final results, in agreement with the literature on wine chain LCAs, and for the vineyard phase, the main influential parameters were related to grape yield and the amount of organic matter inputs (cover crops, residues) and, secondly, the mineralisation and humification coefficients.

3.3.7 Interpretation

In the present review, the selected 81 studies were scored from one to three and discussed according to the approach illustrated in Table 3.6 with regard to the "interpretation phase" issue.

In the context of wine LCA, the interpretation phase is typically carried out at different scales of depth and breadth, as the scope of each study and the elaborated LCA results largely differ from one to another. Therefore, we have analysed interpretation as one independent key issue of the LCA methodology, awarding scores from one to three to each of the 81 selected studies (see Table 3.6). In accordance with the principles of ISO 14044, we assigned score 1 to those works that identified only the significant issues based on the LCI and/or LCIA phases, as this is the basic procedure performed by any author. Then, we attributed score 2 if the study reported additional information about the limitations of the analysis and the appropriateness of the definitions and assumptions, allowing for specific recommendations on how to improve the LCA and/or the impact profile in the conclusions section. Finally, the score of three was assigned when the study also included uncertainty-related issues in LCI and/or LCIA data or carried out an evaluation of the completeness and sensitivity of flows and processes, so improving the consistency of the results obtained.

As shown in Fig. 3.7 (key issue 5), studies are similar in terms of scores 1 and 2 (37 % of the studies), suggesting that most authors tended not to advance their interpretation phase from the mere identification of the significant issues and the reporting of limitations and recommendations. In effect, only about a third of the reviewed papers (26 %) present an evaluation of the reliability and robustness of the LCA profile. These latter are mainly investigated by means of a sensitivity analysis on the most relevant issues determined at the beginning of the interpretation (those flows or processes which are more significant in terms of inventory and/or impact on results), such as activities related to packaging (e.g. Barry 2011; Bosco et al. 2011; Catania and La Mantia 2006; González-García et al. 2011b; Cecchini et al. 2006) or transportation (e.g. Amienyo 2012; BIER 2012; Cholette and Venkat 2009; Cleary 2013). Some authors also highlighted the need to improve current LCI practices and explore additional features associated with LCIA and its interpretation by implementing accounting strategies for carbon sequestration and biogenic

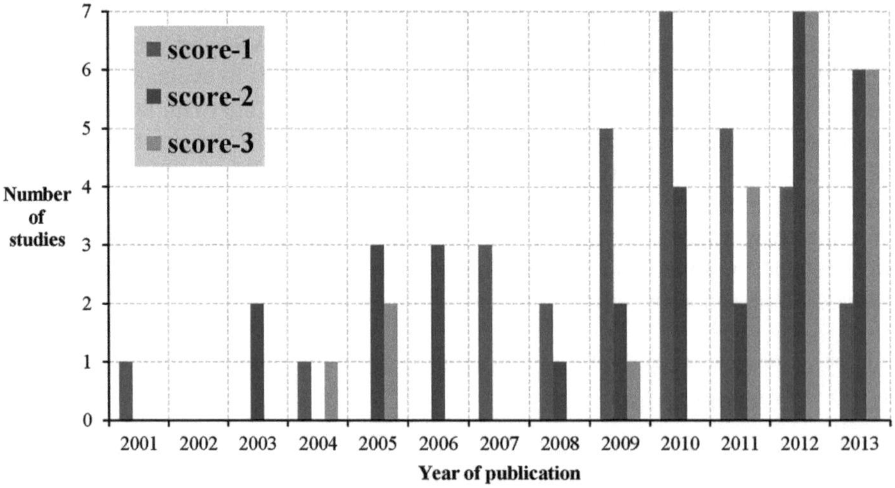

Fig. 3.8 Trend of scores for LCIA interpretation (score 1–3; from 1 to 3 according to the increase of breadth and depth of information provided, see Table 3.6 for further details) per number of studies reviewed. The total number of studies for which the scoring was performed is 81

carbon emissions (Arzoumanidis et al. 2014b; Colman and Päster 2009; Soosay et al. 2012), and changes in SOM (Bosco et al. 2013).

Studies that dealt with uncertainty evaluations or carried out sensitivity analyses and cross-check validations are certainly worthy of further consideration, as they are at the cutting edge in terms of the wine LCA state of the art. They also enable us to understand the extent to which the complexity of the wine LCA interpretation has progressed over time. In this context, the number of studies with a score of three has grown much more lately (2011 to 2013) than the number of studies with a score of one, which have decreased (see Fig. 3.8). This reflects the increased attention of researchers and companies in the investigation of the relevant LCA issues, and the role of the interpretation phase is becoming more and more important for the consistent reporting of information related, in particular, to: (1) the analysis of life cycle hotspots; (2) the determination of weak elements of the methodological approach; and (3) the uncertainty and variability of elementary flows and impact scores.

With regard to item (1), scholars who performed scenario analyses show that strategies of improvement in the use of fertilisers, the lowering of glass bottle weight and the management of transportation of the bottled wine, which are usually the most sensitive parameters of the wine life cycle profile, can provide great benefits in terms of impact reduction (Cleary 2013; Jiménez et al. 2013; Rugani et al. 2013). In contrast, studies which aimed at developing new methodological approaches or calculation tools (referring to item (2)), rather than pure case study of LCA-related analyses, proved that synergies among the use of different approaches exist and can reveal hidden environmental consequences. This is the case, for instance, with the implementation of WF characterisations in LCIA (Herath et al. 2013a, b) and the use of eco-design tools in order to improve the elements raised in the interpretation (González-García et al. 2011b). Finally, it is worth noting that, when authors undertook quantitative uncertainty analyses (e.g. Monte Carlo simulations), the reliability and specificity of the interpretation phase strongly increased (Bosco et al. 2011, 2013; Cleary 2013; Kounina et al. 2012; Mattila et al. 2012b), allowing the scope of the assessment to expand and enrich the geographical and temporal variability of the studies (see, for example, the large spread of results provided in Vázquez-Rowe et al. 2012, 2013). Of course, the opportunity to perform such a robustness evaluation typically depends on the representativeness of the dataset's attributes and on the implementation of site-specific technological and local climate parameters in the LCI stages, whose usage, quality and completeness remain generally quite low in the reviewed studies (cf. Sect. 3.3.6).

Therefore, the most interesting challenge in terms of enhancing the *interpretation* of the wine LCA is to make sensitivity and uncertainty analyses more routinised and operational, and this can be established as long as the case studies and LCI profiles of wine production are readily accessible in the literature.

3.3.8 Critical Analysis

All the seven aspects introduced so far contain an extensive number of methodological elements that comprehensively frame the characteristics of environment-oriented analysis of wine sustainability with the LCA approach. However, the LCA method has rapidly progressed in late years, and not all the most recent and challenging issues of this advance have been dealt with by the scholars concerned.

The worth of the present critical analysis lies in the systematic analysis and update of previous surveys on the wine LCA topic, all of them encompassing issues that belong to the conventional "attributional" LCA approach. Therefore, results of the analysis performed on the 81 studies suggest that other aspects still need to be assessed in the context of wine production or included in future analyses. These belong to methodological issues, which are currently neglected or only marginally treated, despite having been brought to the attention of the LCA community. They can be enumerated as follows:

i. Use of a consequential-LCA (C-LCA) perspective to enhance the evaluation of undesired or unexpected side-effects in the wine market.
ii. Assessment of biogenic carbon and temporal dynamics for carbon emission accounting to develop new characterisation factors for the wine LCIA profile.

With regard to (1), it is worth noting that current developments in LCA methods and databases are strongly focussed on the implementation of C-LCA tools, which aim at relating the effects of one or more choices by studying the environmental consequences of possible (future) changes between alternative product systems, and by modelling the causal relationships originating from the decision to change the output of the product (e.g. Ekvall and Weidema 2004; Earles and Halog 2011; Tillman 2000; Vázquez-Rowe et al. 2014a, b; Weidema 2003, 2006). Moreover, compared with current attributional approaches, C-LCA has been shown to provide some advantageous interpretation frameworks, which show decision-makers the side-effects of specific strategies for policy support (Zamagni et al. 2012). In other words, the C-LCA approach can address the kind of environmental assessment that analyses how biophysical flows of resources, emissions and products and their associated environmental burdens vary in response to changes in (marginal or structural) market implications in a specific life cycle beyond the foreground system (Ekvall and Andræ 2006; UNEP 2011; Vázquez-Rowe et al. 2014a, b).

This relatively novel perspective remains unexplored in wine LCA studies, probably because most authors have typically aimed at modelling the impact of actual (either more or less complex) case studies, rather than advancing pure methodological developments where wine is considered simply for demonstration purposes. It must also be taken into account that the evaluation of wine production with LCA and related tools (e.g. carbon footprint) has a relatively recent history (the first studies date from 2001), and even more recent is the diffusion of the C-LCA concept among practitioners and its effective implementation in LCA guidelines and tools such as databases and software (EC 2010; Ecoinvent 2013).

However, numerous techniques of C-LCA have been developed and tested, in particular for the agricultural production sector, among which are simplified approaches (Schmidt 2008) or more complex methodological combinations of life cycle tools with equilibrium models (Marvuglia et al. 2013; Vázquez-Rowe et al. 2014a, b). In fact, an environmental issue of great relevance and thus one of those most studied with C-LCA is land use, because it reflects the impact of new technologies or strategies in the agri-food sector (e.g. bioenergy production) and their relationship with other production sectors in the market. Therefore, the C-LCA approach represents fertile soil for trans-disciplinary studies in the wine LCA domain on the necessary and potentially hidden aspects of wine-related impacts: for example, the unexpected consequences associated with production cost and retail price changes, which can generate potential rebound effects on the market via a cascade (Binswanger 2001; Hertwich 2005; Rugani et al. 2013; Skuras and Vakrou 2002). With a C-LCA perspective, one could theoretically model such rebound effects with hybrid techniques, analysing the interactions between different actors in the wine life cycle, perhaps after technological modification (e.g. implementation of a new transportation strategy for bulk wine), and then assessing the indirect effects of possible modifications in economic segments other than the wine market (EC 2010; Rugani et al. 2013).

As regards biogenic carbon issues, (2), further in-depth observations could also be made. Hence, the handling of biogenic carbon balances in LCA of agri-forestry systems is of interest, especially in the sustainability and climate science community, as it directly relates to climate change issues. Accounting for CO_2 at each stage of the life cycle offers the advantage of allowing the dynamic modelling of emission and removal, making the analysis consistent with the "polluter pays" principle and the Kyoto rules, which imply that each GHG contribution (positive or negative) should be allocated to the causing agent (Rabl et al. 2007).

Besides the importance of assessing the specific contributions of GHG emissions generated through wine production, it is evident that wineries are starting to face the rebound effects caused by climate change itself on different appellations (Tate 2001; Mira de Orduña 2010). These effects are multiple, affecting different grape varieties in several ways, and in some cases generating opposite effects (Mira de Orduña 2010). More specifically, studies have proved the relationship of climate change with an advance in the harvest dates in several European appellations (Duchêne and Schneider 2005; Ganichot 2002; Jones and Davis 2000; Neman et al. 2001) and varying effects on the quality of the wine, such as higher sugar and alcohol concentrations owed to changes in temperature, climate or radiation (Canova et al. 2012; Jones and Davis 2000; Jones 2007; Jones et al. 2005; Mira de Orduña 2010). Effects of climate change on the wine quality may also have legal implications, in that in the near future modifications of local conditions may not allow the production of fine wines in the regions from which they have traditionally or legally come (Barriger 2011; Ramos et al. 2008). In addition, many studies have linked the alteration of wine phenology to the proliferation of forest fires attributed to the increased warming and aridity in certain areas (Tavşanoğlu and Úbeda 2011; Bento-Gonçalves et al. 2012). For example, oenological consequences described in the literature include the identification of smoke taints in wine (Anderson et al. 2008; Kennison et al. 2011; Simos 2008). Consequently, it appears that in years to come the expansion of these appellation areas may be strongly constrained by the loss of soil quality in neighbouring lands, which implies the reduced carbon stock potential of surrounding areas.

Various mitigating actions could be implemented even by small companies to reduce carbon release (Smyth and Russell 2009) or increase carbon sequestration (Smart 2010). The potential for adaptation and mitigation of climate change in the wine sector, which is highly sensitive to this global issue, does not seem to require substantial changes in the life cycle, such as changes in location, the trellis system (to shade vines with larger canopies), the pruning style and timing (to increase the size of the canopy and/or delay growth), the row orientation (to increase fruit protection from heat and/or sunburn), and irrigation management (if sufficient water is available) (Diffenbaugh et al. 2011). Moreover, it has been observed that additional carbon can be sequestered in the soil during the transition from conventional to organic systems (Venkat 2012). This implies that more farmers may be keen and/or incentivised to shift to organic production in the coming years, contributing to climate change mitigation with more effective tools. Soil management practices, such as residue incorporation and grassing, were also identified as the main factors affecting soil carbon sequestration (Bosco et al. 2013). Above all, pursuing a long-term CF management strategy is a great opportunity for winemakers to contribute directly to climate change mitigation actions at both local and global levels (Rugani et al. 2013).

However, strategies to improve the carbon budget of vineyards are still largely unknown, and an additional challenge will be to reach more consensus on how to assess the contribution from viticulture systems to the release of nitrous oxide, one of the most powerful GHGs (Schultz 2010). As previously mentioned (Sect. 3.3.4), the two stages of viticulture and winemaking are intrinsically related to biogenic carbon balance, being carbon sequestered during vine growth (e.g. Martin 1997; Poni et al. 2006; Soja et al. 2010) and released during the alcoholic fermentation of wine (Notarnicola et al. 2003), respectively. Within the studies evaluated here, some do not explicitly account for biogenic carbon trades and the stalk degradation in soil, because of the difficulties in obtaining a specific spatial estimate without a sampling campaign or validated models (Bosco et al. 2011). However, most authors do not probe the issue because they assume that the CF of wine should only consider fossil-based GHG sources (Barry 2011; Benedetto 2013; Notarnicola et al. 2003; Point et al. 2012; SAWIA 2004; Vázquez-Rowe et al. 2012b), and the commonly accepted principle is that fermentation should be considered negative because of the CO_2 that the vine sequesters (Greenhaigh et al. 2011). Moreover, CO_2 emissions from photosynthesis and must fermentation processes can be easily calculated, but scholars have not included them in the carbon balance because they are perceived as part of the short-term carbon cycle (e.g. CO_2 from wine fermentation, emissions from combustion or breakdown of vine pruning, etc.), as demonstrated in the application of the wine carbon calculation protocol by Pattara et al. (2012).

Nevertheless, from the review of studies that accounted for the contribution of biogenic carbon in their LCI (e.g. Soosay et al. 2012; Vázquez-Rowe et al. 2012b; Colman and Päster 2009; Zabalza et al. 2003), it emerges that biogenic CO_2 and removal activities generated by the carbon stock changes in biomass and soil, and the alcoholic fermentation, should not be neglected. Interestingly, Arzoumanidis et al. (2014b) have pointed to the need for introducing time-dependent carbon accounting in the wine LCA, in order to increase the accuracy of carbon balances for agricultural and bottling phases. Moreover, among the revisions for the new PAS 2050 guidelines (BSI 2011) is the inclusion of GHG emissions and removals from biogenic sources to demonstrate the relevance of considering CO_2 removals and biogenic carbon emissions. However, this change was made to bring the PAS in line with the approach taken in the GHG Protocol Product Standard (WRI and WBCSD 2011) and ISO 14067 (ISO 2013), assuming that biogenic carbon assessment is important for certain products associated with long-term carbon storage, such as perennial crops like vineyards, which can be expected to sequester more carbon than annual crops (CSWA 2009; Carlisle et al. 2009; Kroodsma and Field 2006; Freibauer et al. 2004). It is worth highlighting that the C pool in biomass is considerably smaller ($<1\,\%$ the size) than that in soil (Keightley 2011), and the corresponding vine biomass C pool is removed at the end of the vineyard production period (Bosco et al. 2013).

Future LCA studies in the wine sector will certainly benefit from the implementation of the above methodological progresses, specifically in relation to consequential LCA and biogenic carbon analysis. These could be used to outline a roadmap for more consensual sustainability assessment of wine production supply chains based on LCA, and possibly included in standardised tools or wine-LCA calculators.

3.3.9 Comparative Analysis

In the last few years, several impact assessment concepts have been developed beyond or grounded on LCA, and the environmental footprint concept has attracted increasing interest in both scientific and political communities (EU 2011). Methodological development is quite different, however. Evaluating "comparative analysis" can thus give an overview on the impact characterisation frameworks and models used in the wine LCA, showing the effectiveness of recent implementations across the most traditional and the newest impact categories, with a focus on local impacts and variability, integrated assessment and environmental sustainability analysis. Accordingly, the scoring approach described previously in Sect. 3.3 was useful to quantify the extent to which "Indicator(s)/method(s) other than LCIA" (key issue 6) are involved in the field of wine LCA (see key issue 6 in Table 3.6).

The majority of reviewed studies (73 %) falls within the first category (score = 1), where only conventional LCIA methods are applied (see Sect. 3.3.6 for an in-depth analysis).

In contrast, 14 % of reviewed studies fall within the second category (score = 2), where environmental assessment metric(s) other than those typically included in LCIA methods are applied: in particular, in the present literature review, such metrics are the hydrological water-balance model for measuring the water footprint (Herath et al. 2013a) and the Ecological Footprint (Niccolucci et al. 2008).

As regards the use of the EF for worldwide comparisons, surface measurements expressed in ha rather than gha (global hectares) showed that gha t^{-1} of wine was almost constant over time, this unit being unaffected by changes in yield (Niccolucci et al. 2008). Conversely, results expressed in ha t^{-1} varied over the period considered, demonstrating that local yield variations were accounted for (Niccolucci et al. 2008). A footprint measure reported in gha is globally consistent and can be compared between countries. However, it is unable to track specific changes in local resource management. Instead, actual hectares are an appropriate unit for analysing use and management of local natural resources, but cannot be used for worldwide comparisons.

EF and LCA are complementary in many respects, particularly because LCA has valuable potential for the validation of EF methodology and the development of instruments able to support decision-making both for companies and for public administrations (e.g. for spatial planning). In effect, LCA indicators traditionally applied in the wine sector like GWP, acidification potential, and eutrophication potential can estimate the load of environmental effects on soil, water and atmosphere during the life cycle phases; EF can support the evaluation of the ecosystem surfaces required to generate resources and absorb emissions associated with a unit of product. This information forms the basis for a coherent representation of the environmental profile of wine and the essential content for an environmental label of this consumption product, in either conventional or organic farming.

In the former case, authors have noticed "the grape growing as a land use and wine production as an industry do not have a deleterious impact on depletion of water resources in either region" (Herath et al. 2013a, p. 242). Interestingly, the conclusions of this work are that for agricultural-product WF to be meaningful, the natural variability in the production phase needs to be well accounted for. Given this variability in the impacts of water use on the local water resources, the authors recommend that WF should be assessed at a local level.

In contrast, Niccolucci et al. (2008) compared the EF of conventional and organic winemaking and concluded that the higher footprint of the conventional wine was essentially owed to the agricultural and packaging phases.

Finally, studies with an assigned score of three –*Application of LCIA + other (complementary) environmental assessment metric(s)* for comparative/combination purposes—account for 14 % of the total articles reviewed. In this context, an interesting comparison of different indicators of sustainability is that performed by Amienyo (2012), who ambitiously considered the Life Cycle Sustainability Assessment in the UK beverage sector by coupling LCA, Life Cycle Costings (+ value added analysis) and other specific social indicators (such as consumer health issues, employment and wages, intergenerational issues, child labour, forced labour, etc.). The analysis was conducted for five beverage categories, namely carbonated soft drinks, bottled water, beer, red wine, and spirits and liqueurs. For each beverage category, a standard procedure of LCA was followed by a focus on data quality, impact assessment and interpretation, GWP being the first impact category's indicator evaluated, followed by others such as Primary Energy Demand (PED), abiotic depletion (ADP), acidification (AP), eutrophication (EP), and toxicity indicators. It was observed that the combination of environmental and economic aspects facilitates the identification and comparison of environmental and economic hot-spots in the life cycle (Amienyo 2012).

The novelty of the approach implemented by Amienyo is its attempt to develop a Life Cycle Sustainability Assessment, which has never been proposed before in the wine sector. This is an extremely interesting methodological platform, both as regards the improvement of existing impact evaluation models (scope enlargement) and as regards wine companies (in particular the larger ones) willing to promote their products not only through environmental labels but open to consideration of other pillars related to the concept of sustainability (economy, environment, society).

Other indicators/environmental assessment metrics that have been used or coupled with LCI or LCIA methods are SOM changes (Bosco et al. 2013), ecodesign concepts (González-García et al. 2011a, b) and, related to the latter, the use of a process optimisation simulation (Jiménez et al. 2013), applied to determine impact in terms of the decisions made in the production process.

3.4 Lessons Learnt from LCA: Best Practices for Environmental Improvement in the Wine Sector

The implementation of LCA is oriented to identification of the most significant environmental impacts along the wine production chain. It reveals the "hotspots" of the whole system, in order to optimise the production steps and to support ecodesign strategies.

The review of the international LCA literature, discussed in this chapter, identified the following elements as the main hot-spots of the wine production chain:

- cultivation stage, mainly because of the use of pesticides and fertilisers;
- packaging, mainly because of the production of glass used for bottling;
- electric energy consumption in the winery;

- emission of VOC in the winery;
- distribution, because of fuel consumption in transportation processes.

The cultivation step contributes mostly on Ecotoxicity (ECT), Human Toxicity (HT), Eutrophication (NP) and Acidification (AP). The first two impact categories (ECT and HT) are strictly dependent on the use of pesticides, affecting water and soil toxicity and the human toxicity of workers in the field. The contribution on NP and AP essentially depends on the use of nitrogen and phosphate fertilisers. While NP is caused by water releases of phosphates and nitrates and to air emissions of NO_x and NH_3, AP is due to emissions of NO_x occurring during the fertilisers use.

The production of the glass bottle is one of the phases with the greatest impact in the wine life cycle, as highlighted Ardente et al. (2006). It especially affects energy consumption, Global Warming Potential (GWP), HT, and AP.

Vinification processes significantly contribute to the impact category of photochemical oxidation, because of the emissions of VOC during the alcoholic fermentation. Among these, the most problematic is ethyl alcohol, whose emission ranges between 43 and 71 g/hl in red wine (EPA 1995). Other impact categories affected by the vinification processes are those linked to electric energy production, but they have less impact compared with glass bottle production. In a winery, the stage with the highest energy consumption is the bottling, which accounts for about 60% of the total energy consumption, followed by the refrigeration phase.

Finally, the distribution phase of bottled wine is also relevant in the environmental profile of wine-related impacts when the winery and the retailer are at some distance from each other. Because the export of wine is increasingly by sea, the consumption of fossil fuels and the transportation means are elements that usually play a significant role in the generation of impacts such as GWP.

Another key issue is the relationship between technology and the quality of wine. Wine production is a complex activity in which technology plays the same important role as grape cultivation and winemaker skills. Although the raw materials are just grapes, yeast and some chemicals, the alternative production processes are highly variable and, as a result, the quality of output wines is highly variable as well. High quality wines have to add more technological steps to their production process and this results in a worsening of the environmental profile when assessed only on the basis of volume or mass (Notarnicola et al. 2010).

Despite the great variety of wines, most wine LCA studies, in particular those with comparative aims, consider as a functional unit a specific amount of product in litres or kilograms, without any reference to the main characteristics of products. This problem could be overcome via the use of other functional units, which could better represent the function of the system, such as a certain alcoholic degree or a certain hedonistic value (Notarnicola et al. 2010). The hedonistic value is an index, which measures the main characteristics of wine based on the traditional describers of the sensory feedback. Other scientifically more robust parameters could be considered in the definition of the functional unit, such as the total dry extract, the reducing sugars, the ash content, chloride and sulphate content, pH, free and total sulphur dioxide, chromatic properties such as luminosity and chromaticity, as defined by EC Regulation 2676/90 and its modifications which determine European

Community methods for the analysis of wines (EEC 1990). Notarnicola et al. (2010) have shown that with more technological production steps, the production of a high quality wine has a worse environmental performance if the comparison is made on the basis of volume or mass. If a different functional unit is considered, the results are completely inverted.

With regard to the vinification typology, it is very difficult to determine which is the most eco-friendly one (e.g. red or white wine). As also observed in Rugani et al. (2013), the variability of the impact associated with the same functional units of different wines worldwide is considerable (red vs. white, organic vs. conventional cultivation strategies). This means that it is extremely difficult to generalise and justify results only on the basis of wine typology; many other factors should be considered that potentially influence the impact associated with the FU. For example, if we consider the grape varieties of Aglianico for red wine and Chardonnay for white wine, the main difference is in their maturity stage, which, in Italy, corresponds to the end of August for Chardonnay and the middle of October for Aglianico. This difference implies that the Chardonnay viticulture needs eight pesticide treatments whereas the Aglianico needs ten of them; the consequence is a 20 % lower use of pesticides, diesel and lube oil in the case of Chardonnay. Nevertheless, the trivial amount of these inputs in the agricultural stage is counterbalanced by the different yields in the two vinifications. In fact, one 0.75 L bottle of red wine typically requires from 1.05 to 1.07 kg of grapes, and one of white wine about 1.2 kg (Notarnicola et al. 2003).

Even within the same vinification, there are technological steps, which increase the energy consumption and also the quality of the wine. In fact, the storage in barriques and the concentration of the must through reverse osmosis require greater resource consumption but, at the same time, they increase the quality of the resulting wines.

With regard to the above issues, below is a summary of the main guidelines to improve the energy and environmental performance of the wine sector.

3.4.1 Agricultural Stage

Integrated pest management and organic agriculture could be an option for the improvement of wine production environmental performance. However, as other studies have shown (Mattsson 1999), organic agriculture is not a better *a priori* solution than conventional agriculture. In the case of wine, the main problems are because of the great difference of yield, which is on average 40 % lower in the organic than in the conventional system, with consequent higher land use and energy and material consumption by the product unit (Nicoletti et al. 2001). Other problems are connected with the type of organic pesticides and fertilisers used: by its nature, manure is assimilated very slowly by plants, causing nitrogen compound emissions during its use; moreover, sulphur and copper sulphate have a more relevant impact in the production phase and a lower one during the use stage. A reduction in the use of these pesticides and consequent better environmental profile of the organic system should be targeted.

Moreover, the environmental profile of the organic production could be further improved by considering other environmental aspects, which cannot be assessed through an LCA; for example, organic farming increases biodiversity on a local scale, improves soil quality and increases the organic component of soils.

3.4.2 Winery

Energy efficiency The reviewed LCA studies show that one of the main impacts in the winery industry is electricity consumption. Improvements in energy efficiency or the use of locally-produced electricity (e.g. through installation of PV panels; Smyth and Russel 2009) can thus play an important role in reducing the energy and environmental impacts of the wine eco-profile.

The employment of plant and processes with high efficiency is also of paramount importance for (indirectly) decreasing energy consumption. The design of energy-efficient plants and the growing use of biotechnology in vinification are just two examples.

The use of biotechnologies is linked to the reduction of the energy consumption in the winemaking process in terms of the yeasts or enzymes used in grape treatments or wine refining to minimise the need for other treatments (Goode 2005).

Traditional filtration with fossil flours implies the problem of their disposal: consequently, new filtration technologies have been tested, e.g. the use of tangential filtration is a promising technology (Baker 2004).

Together with the above-mentioned practices, the implementation of an energy management system in accordance with ISO 50001 (ISO 2011) could result in better environmental and energy performance.

Airborne emissions Carbon dioxide represents the main air emission of the winery. In general, CO_2 is not taken into account in the analyses because it is linked to the natural carbon cycle (see Sect. 3.3.8). However, it is desirable to research for system solutions, which could enable its recovery in order to use it, for example, in carbonic maceration.

With the exception of CO_2, ethanol is the main compound emitted during alcoholic fermentation. Acetaldehyde, methyl alcohol, n-propyl alcohol, n-butyl alcohol, sec-butyl alcohol, isobutyl alcohol, isoamyl alcohol, and hydrogen sulphide are also emitted, but in much smaller quantities. In addition, a large number of other compounds are formed during the fermentation and ageing process as acetates, monoterpenes, higher alcohols, higher acids, aldehydes and ketones, and organosulphides (EPA 1995).

Fugitive ethanol emissions also occur during the screening of red wine, pressing of the pomace cap, ageing in barriques and the bottling process. In addition, small amounts of liquefied SO_2 are always added to the must prior to fermentation or to the wine after the fermentation is completed; SO_2 emissions can occur during these stages, but they are almost impossible to quantify.

Five potential emission control systems for VOC are available: carbon adsorption, water scrubbers, catalytic incineration, condensation, and temperature control, but all systems have their own disadvantages in terms of either low control efficiency or cost-effectiveness, or even overall applicability to the wide variety of wineries (EPA 1995). The only one, which has an emission abatement of about 98 %, is the wet scrubber but, like the other emission control systems, it is not currently used during winemaking because of its high cost. Starting from an ethanol emission of 55 g/hl of wine without abatement systems (EPA 1995), it is possible to reach the following values with the above-mentioned abatement systems: 4.6 g/hl with carbon adsorption, 13 g/hl with catalytic incineration, 0.67 g/hl with wet scrubbers.

The best practice would be the adoption of an abatement system of VOC emissions in the winery, which is compatible with the technology used; the choice of the abatement system must also be made taking into consideration the control of the operating costs.

Recovery of co-products The recovery and reuse of solid co-products—rasps, lees, marc—plays an important role in wines' eco-profile. In life cycle thinking it is possible to skip the burden of their disposal so, in industrial ecology terms, they become a raw material for new processes. The LCA approach allows us to assess the different eco-profiles because of the re-use/recycling of co-products and wastes, and to compare different environmental impacts rising from the above-mentioned options.

The best practice is the complete recovery of co-products and their use in other production chains. However, this should be further investigated in the future if more complex LCA studies are performed for the wine sector (such as those based on consequential LCA approaches; see Sect. 3.3.8), as the use of wine co-products outside the wine market or supply chain might not necessarily imply clean or impact-free recovery.

Wastewater treatment Winery activities represent a source of significant wastewater production, essentially because of the equipment used for cleaning operations and the losses during the processing of raw materials and product movement. Wastewater pollution has an organic and biodegradable nature, for the depuration of which it is possible to use an alternative process to the conventional one called "activated sludge" (Crittenden et al. 2005). Phyto-depuration makes use of the natural capacity of some aquatic plants to absorb substances contained in wastewater, or generated by the degrading action of microorganisms, through the roots (Kadlec and Wallace 2008). The plants that are generated by this process could be used as biomass for compost or energy production. On the other hand, activated sludge technology requires certain energy quantities and sludge needs to be appropriately treated before final disposal.

3.4.3 *Packaging*

High quality wines are stored in a glass bottle. The literature review shows that glass production is one of the highest contributors in terms of energy consumption and natural resources use.

The other products of wine packaging, such as cork, aluminium capsules and paper labels, generally show very low impacts in all the environmental categories.

As in other sectors, a possible solution is to replace the glass packaging with another material, and in fact, cardboard polylaminate has often been used. However, it is not appropriate for a high quality wine. Wine is a food product, which has no due date. The reason is that temporal evolution elevates the wine's quality. The use of polylaminate packaging entails a wine duration dependent on the duration of the packaging, imposing a restriction unrelated to the nature of the wine. Moreover, results of marketing research show that label design and bottle packaging are key factors in consumer choice (Barber et al. 2006; Lapsey and Moulton 2001). Therefore, glass is likely to continue to play an important role as packaging for wine.

The best practices should therefore be the use of the "design for environment or for recycling" techniques, in order to reduce the specific weight of the materials; the use of recycled materials may be another alternative.

Another approach to reducing environmental impact, practised mostly by companies in new wine-producing countries such as New Zealand (Dodds et al. 2013), is to export the wine in bulk and bottle it in the country of destination, in order to reduce the impact of transport.

Conclusions

A historical overview and an update on wine production opened the chapter, with a special focus on nutritional, cultural and functional aspects associated with the wine supply chain. The analysis proceeded with the presentation of a set of LCA-based methods and the road maps available so far to guide stakeholders (from academia, RDI and industry) in drawing up a life cycle study for wine. Discussion of current guidelines and good practices had the objective of highlighting existing consensus on an international and global scale, and showing the importance of future improvements to facilitate the process of harmonisation between definitions, concepts and approaches. An analysis of the different methods (both at a product and an organisation level) was performed regarding issues such as functional unit, system boundary, allocation and by-product, co-product and waste streams, use of resources and impact categories. A clear result that emerged was that there is no consensus on most of these issues amongst the methods examined.

A comprehensive critical analysis of LCA studies in the wine sector was conducted to ascertain fundamental methodological aspects related to goal and scope, system boundary, FU, data quality and availability, multi-functionality issues, inventory tools and impact assessment approaches, as well as results and research findings. This exercise helped to highlight the criticalities in the methodology and in the management of the wine supply chain and processes, pointing out the strengths and missing items, and generally providing useful insights and relevant recommendations for both LCA analysts and wine producers.

The main issues related to the environmental profile of wine are:

- FU: the majority of the analysed studies consider a mass or volume-related F.U., neglecting issues such as the quality of the product, namely a certain alcoholic level or a certain hedonistic value, especially in comparative studies;
- allocation: starting from the consideration that the production cycles of agricultural or agri-industrial have no more waste to dispose of, but by-products, it is necessary to identify an optimal strategy to deal with multifunctionality in the wine industry;
- the agricultural stage is one of those with the greatest impact on wine production; organic agriculture could be an option for the improvement of the wine production environmental performance, even if it is not an *a priori* better solution than conventional agriculture. Moreover, only few studies take into consideration the vineyard planting, an important factor to consider from an agronomic point of view and for its potential impact on GHG emissions.
- in the winery, improvements in energy efficiency can be achieved through the implementation of an energy control system, in accordance with, for example, ISO 50001:2011;
- consumption phase: not considered in most of the papers because of the negligible environmental impacts;
- glass production for packaging is one of the major contributors in terms of energy consumption and natural resource use: a possible solution is to replace glass packaging with another material or to use recycled glass;
- the provision of glass bottles, field-level emissions from fertilisers, and consumer transport are the life cycle stages proven to cause much of wine's total impact.
- Environmental management programmes that focus on these life cycle stages have a greater potential to result in substantial improvements to wine's environmental profile. In particular, continued research into the potential benefits of bulk transport, bulk packaging, alternative packaging materials, and bottle reuse systems may uncover important environmental improvement options for wine.

Future LCA studies in the wine sector will certainly benefit from the implementation of the above issues related to the sustainability assessment of wine and possibly included in standardised tools or wine-LCA calculators.

Acknowledgements The authors would like to thank Silvia Di Cesare and Federica Silveri for their valuable support during the critical analysis process.

References

Abbing, A. G. (2010). The sustainability performance of the South African-European wine supply Chain: Differences in sustainability from a scientific and actor perspective. Master thesis. Utrecht University.

ADEME. (2013). http://www.ademe.fr/internet/bilan_produit/login.asp. Accessed 12 Dec 2013.

Amienyo, D. (2012). Life cycle sustainability assessment in the UK beverage sector, M. Sc Thesis. Manchester, UK: The University of Manchester.

Anderson, K., Findlay, C., Fuentes, S., & Tyerman, S. (2008). Viticulture, wine and climate change. Garnaut Climate Change Review, University of Adelaide, Australia. http://www.garnautreview.org.au/ca25734e0016a131/WebObj/01-HViticulture/$File/01-H%20Viticulture.pdf. Accessed 10 Dec 2013.

Aranda A., Zabalza I., & Scarpellini, S. (2005). Economic and environmental analysis of the wine bottle production in spain by means of Life Cycle Assessment. *International Journal of Agricultural Resources, Governance and Ecology, 4*(2), 178–191.

Arcese G., Lucchetti M. C., & Martucci, O. (2012). Analysis of sustainability based on Life Cycle Assessment: An empirical study of wine production. *Journal of Environmental Science and Engineering, B 1,* 682–689.

Ardente, F., Beccali G., Cellura M., & Marvuglia, A. (2006). POEMS: A case study of an Italian wine-producing firm. *Environmental Management, 38*(3), 350–364.

Artaud-Wild, S. M., Connor, S. L., Sexton, G., et al. (1993). Differences in coronary mortality can be explained by differences in cholesterol and saturated fat intakes in 40 countries but not in France and Finland: A paradox. *Circulation, 88,* 2771–2779.

Arzoumanidis, I., Petti L., Raggi, A., & Zamagni, A. (2013a). Life Cycle Assessment (LCA) for the agri-food sector. In R. Salomone, M. T. Clasadonte, M. Proto, & A. Raggi (Eds.), Product-Oriented Environmental Management System (POEMS)—Improving sustainability and competitiveness in the agri-food chain with innovative environmental management tools (105–122). Dordrecht: Springer.

Arzoumanidis, I., Raggi, A., Petti, L., & Zamagni, A. (2013b). The implementation of simplified LCA in agri-food SMEs. In R. Salomone, M. T. Clasadonte, M. Proto, & A. Raggi (Eds.), Product-Oriented Environmental Management System (POEMS)—Improving sustainability and competitiveness in the agri-food chain with innovative environmental management tools (pp. 151–173). Dordrecht: Springer.

Arzoumanidis, I., Zamagni, A., Raggi, A., Petti, L., & Magazzeni, D. (2013c). A model of simplified LCA for agri-food SMEs. In R. Salomone, M. T. Clasadonte, M. Proto, & A. Raggi (Eds.), Product-Oriented Environmental Management System (POEMS)—Improving sustainability and competitiveness in the agri-food chain with innovative environmental management tools (pp. 123–150). Dordrecht: Springer.

Arzoumanidis, I., Raggi, A., & Petti, L. (2014a). Considerations when applying simplified LCA approaches in the wine sector. *Sustainability, 6*(8), 5018–5028.

Arzoumanidis, I., Fullana-i-Palmer, P., Raggi, A., Gazulla, C., Raugei, M., Benveniste, G., & Anglada, M. (2014b). Unresolved issues in the accounting of biogenic carbon exchanges in the wine sector. *Journal of Cleaner Production, 82,* 16–22.

Austin, G. A. (1985). *Alcohol in Western Society from Antiquity to 1800: A chronological history.* Santa Barbara: ABC—Clio.

Babor, T. (1986). *Alcohol: Customs and rituals.* New York: Chelsea House.

Baker, R. W. (2004). *Membrane technology and applications.* Wiley.

Barber, N., Almanza, B. A., & Donovan, J. R. (2006). Motivational factors of gender, income and age on selecting a bottle of wine. *International Journal of Wine Marketing, 18*(3), 218–232.

Barberini, M., Cecchini, F., Della Giovanpaola, M., Franchini, F., Mangani, F., Neri, P., Parmeggiani, G., & Tosi, M. (2004). Analisi del ciclo di vita del vino: Confronto tra la produzione convenzionale e quella biologica. Master thesis. Università di Urbino and ENEA, PROT—P135–033.

Barriger, L. (2011). Global warming and viticulture: the ability of wine regions to adapt in differing regulatory schemes. *Penn State Environmental Law Review, 19,* 311–332.

Barry, M. T. (2011). Life Cycle Assessment and the New Zealand wine industry: A tool to support continuous environmental improvement. Master Thesis. Massey University, Wellington, New Zealand.

Benedetto, G. (2013). The environmental impact of a Sardinian wine by partial Life Cycle Assessment. *Wine Economics and Policy, 2,* 33–41.

Bengoa, X., Maia de Souza, D., & Samson, R. (2009). Evaluating Tradeoffs between Material Supply, Lightweighting and Recyclability using Life Cycle Assessment—A Case Study on Wine Packaging. In: Proceedings of the conference. Life Cycle Assessment IX - toward the global life cycle economy. Boston, USA.

Bento-Gonçalves, A., Vieira, A., Úbeda, X., Martin, D. (2012). Fire and soils: Key concepts and recent advances. *Geoderma, 191,* 3–13.

BIER. (2010). Beverage industry sector guidance for greenhouse gas emissions reporting. Beverage industry environmental roundtable, Version 2.0.

BIER. (2012). Research on the carbon footprint of wine. Beverage industry environmental roundtable.

Binswanger, M. (2001). Technological progress and sustainable development: What about the rebound effect? *Ecological Econconomics, 36*(1), 119–132.

Bochu, J. L., Metayer, N., Bordet, C., & Gimaret, T. M. (2013). Development of carbon calculator to promote low carbon farming practices—Methodological guidelines (methods and formula), Deliverable to EC-JRC-IES by Solagro.

Bosco, S., Di Bene, C., Galli, M., Remorini, D., Massai, R., & Bonari, E. (2011). Greenhouse gas emissions in the agricultural phase of wine production in the Maremma rural district in Tuscany, Italy. *Italian Journal of Agronomy, 6*(15), 93–100.

Bosco, S., Di Bene, C., Galli, M., Remorini, D., Massai, R., & Bonari, E. (2013). Soil organic matter accounting in the carbon footprint analysis of the wine chain. *The International Journal of Life Cycle Assessment, 18*(5), 973–989.

BSI-British Standards Institution. (2011). PAS 2050:2011—Specification for the assessment of the life cycle greenhouse gas emissions of goods and services. British Standards Ed., London, UK. http://www.bsigroup.com/en/Standards-and-Publications/How-we-can-help-you/Professional-Standards-Service/PAS-2050/PAS-2050/. Accessed 11 March 2014.

Burja, C., & Burja, V. (2012). Decisions in Sustainable Viticulture using Life Cycle Assessment. *Journal of Environmental Protection and Ecology, 13*(3), 1570–1577.

Camilleri, C. S. (2009). The life cycle paradigm as an intrinsic component of supply network and value analysis: A contribution to stategic dialogue at the Yalumba wine company. Paper presented at the 6th Australian conference on life cycle assessment. Sustainability tools for a new climate. 16th–19th February, 2009. Melbourne, Australia.

Canova, L., Marini, G., Chernyavskaya, M., & Dallari, M. (2012). In vino qualitas: Climate change effects on Italian wine production. *Enometrica, 5*(1), 7–25.

Carballo Penela, A., García-Negro, M. C., & Quesada, J. L. D. (2009). A methodological proposal for corporate carbon footprint and its application to a wine-producing company in Galicia, Spain. *Sustainability, 1,* 302–318.

Carbonostics. (2013). http://www.carbonostics.com/. Accessed 12 Dec 2013.

Carlisle, E., Smart, D. R., Browde, J., & Arnold, A. (2009). Carbon footprints of vineyard operations. Pract. Winery Vineyard, Sep/Oct 2009. http://www.practicalwinery.com/sepoct09/carbon1.htm. Accessed 4 Feb 2014.

Carta, G. (2009). Evaluation of environmental sustainability of two italian wine productions Through the use of the Life Cycle Assessment (LCA) method. MSc thesis. University of Sassari, Italy (in Italian).

Catania, C. A., & La Mantia, F. P. (2006). Design and Eco-Design in Wine Industry (2006). In: A. Can Özcan, Elif Kocabiyik, & Zeynep Tuna Ultav, (eds.). Agrindustrial design—Proceedings of 1st international product and service design symposium and exhibition on agricultural industries: Olive oil, wine and design. 27th–29th April 2005, Izmir: Izmir university of economics, Department of industrial design.

CCaLC. (2013). http://www.ccalc.org.uk/. Accessed 12 Dec 2013.

Cecchini, F., Croce, E., Franchini, F., Mazzaracchio, P., & Neri, P. (2005a). Confronto della qualità ambientale di vini da produzione convenzionale e biologica attraverso l'LCA. Master Thesis. Alma Mater Studiorum Università di Bologna and ENEA, PROT—P135–077.

Cecchini, F., Neri, P., Franchini, F., Catto, G., Feoli, E., & Altobelli, A. (2005b). Analisi ambientale della filiera vitivinicola del Cabernet Sauvignon in Friuli Venezia Giulia secondo il metodo LCA. Master thesis. Università degli Studi di Trieste and ENEA, PROT—P135–080.

Cecchini, F., Montedonico, M., & Neri, P. (2006). Analisi del Ciclo di Vita—Life Cycle Assessment—di un vino Cabernet Sauvignon prodotto in Cile. Master Thesis. Alma Mater Studiorum Università di Bologna and ENEA, PROT—P135–101.

Cholette, S., & Venkat, K. (2009). The energy and carbon intensity of wine distribution: A study of logistical options for delivering wine to consumers. *Journal of Cleaner Production, 17,* 1401–141.

C.I.V. (Consorzio Interprovinciale Vini) (2008). Environmental product delaration—Bottled sparkling wine. *Grasparossa righi*. http://gryphon.environdec.com/data/files/6/7505/EPD%20S-P-00109%20ingl-2008-def.pdf. Accessed 21 Mar 2014.

Cleary, J. (2013). Life Cycle Assessments of wine and spirit packaging at the product and the municipal scale: A Toronto, Canada case study. *Journal of Cleaner Production, 44,* 143–151.

Colman, T., & Päster, P. (2009). Red, White, and 'Green': The cost of greenhouse gas emissions in the global wine trade. *Journal of Wine Research, 20*(1), 15–26.

Comandaru, I. M., Bârjoveanu, G., Peiu, N., Ene, S. A., & Teodosiu, C. (2012). Life Cycle Assessment of wine: Focus on water use impact assessment. *Environmental Engineering and Management Journal, 11*(3), 533–543.

Commission Regulation (EEC) No 2676/90. (1990). of 17 September 1990 determining community methods for the analysis of wines. (OJ L 272, 3.10.1990, p. 1).

Cozzani, I. (2005). Vino e salute, Aracne, 2005.

Criqui, M. H. & Ringel, B. L. (1994). Does diet or alcohol explain the French paradox? *Lancet, 344,* 1719–1723.

Christ, K. L., & Burritt, R. L. (2013). Critical environmental concerns in wine production: An integrative review. *Journal of Cleaner Production, 53,* 232–242.

Crittenden, J. C., Trussell, R. R., Hand, D. W., Howe, K. J., & Tchobanoglous, G. (2005). *Water treatment—Principles and design* (II ed.). Hoboken: Wiley.

CSWA-California Sustainable Winegrowing Alliance. (2009). Vineyard management practices and carbon footprints. http://www.sustainablewinegrowing.org/docs/Vineyards%20and%20GHGs%20Handout%20Final%20-%20May%202009.pdf. Accessed 2 Jan 2014.

De Camillis, C., Brandão, M., Zamagni, A., & Pennington, D. (Eds.). (2013). Sustainability assessment of future-oriented scenarios: A review of data modelling approaches in Life Cycle Assessment. Towards recommendations for policy making and business strategies. European Commission, Joint Research Centre, Institute for Environment and Sustainability. Publications Office of the European Union, Luxemburg.

Della Giovampaola, M., & Neri, P. (2004). Sangiovese "BIO" anche l'ambiente ci guadagna. Agricoltura, September, 38–39.

Diffenbaugh, N. S., White, M. A., Jones, G. V., & Ashfaq, M. (2011). Climate adaptation wedges: A case study of premium wine in the western United States. *Environmental Research Letters, 6*(2), 1–11.

Dillon, C. (2011). Waste management in the South African Wine Industry. Master thesis. Cape Wine Academy.

Dodds, R., Graci, S., Ko, S., & Walker, L. (2013). *What drives environmental sustainability in the New Zealand Wine Industry? An examination of driving factors and practices*. Emerald Group Publishing Limited.

Duchêne, E., & Schneider, C. (2005). Grapevine and climatic changes: A glance at the situation in Alsace. *Agronomy for Sustainable Development, 25*(1), 93–99.

Earles, J., & Halog, A. (2011). Consequential Life Cycle Assessment: A review. *The International Journal of Life Cycle Assessment, 16*(5), 445–453.

Ecoinvent. (2013). Swiss Centre for Life Cycle Inventories: ecoinvent database v3.0. http://www.ecoinvent.org/database/ecoinvent-version-3/. Accessed 12 Dec 2013.

EC-European Commission. (2010). International Reference Life Cycle Data System (ILCD) handbook e general guide for Life Cycle Assessment e detailed guidance. EUR 24708 EN, first ed. Publications Office of the European Union, Luxembourg, LU.

ECOSMES. (2013). http://www.ecosmes.net/everdee/login2. Accessed 12 Dec 2013.

Ekvall, T., & Andræ, A. S. G. (2006). Attributional and consequential environmental assessment of the shift to lead-free solders. *International Journal of Life Cycle Assessment, 11,* 344–353.

Ekvall, T., & Weidema, B. (2004). System boundaries and input data in consequential life cycle inventory analysis. *International Journal of Life Cycle Assessment, 9,* 161–171.

Ene, S. A., Teodosiu, C., Robu, B., & Volf, I. (2013). Water footprint assessment in the winemaking industry: A case study for a Romanian medium size production plant. *Journal of Cleaner Production, 43,* 122–135.

ENVIRONDEC. (2014). http://www.environdec.com/. Accessed 10 March 2014.

EPA. (1995). Emission factor documentation for AP-42, Section 9.12.2. Wines and Brandy.

EPD. (2006). Product Category Rules (PCR) for preparing an environmental product declara-tion (EPD) for Packaged Sparkling red, white and rosé wines. PCR 2006:03. The International EPD® System, Version 1.0.

EPD. (2013). Product Category Rules (PCR) for preparing an environmental product declara-tion (EPD) for UN CPC 24212– Wine of fresh Grapes, except sparkling wine; Grape must. PCR 2010:02. The International EPD® System, Version 1.03.

European Commission. (2011). Analysis of Existing Environmental Footprint Methodologies for Products and Organizations: Recommendations, Rationale, and Alignment. http://ec.europa.eu/environment/eussd/pdf/Deliverable.pdf. Accessed 4 Dec 2013.

Eveleth, R. (2013). Life Cycle Assessment: Organic and biodynamic wine. http://ssc.clickrally.com/wp-content/uploads/2010/06/Paper2.pdf. Accessed 24 Feb 2014.

Fearne, A., Soosay, C., Stinger, R., Umberger, W., Dent, B., Camilleri, C., Henderson, D., & Mugford, A. (2009). "Sustainable value chain analysis: A case study of South Australian wine", 14th Adelaide Thinker in Residence.

Food SCP RT. (2013). ENVIFOOD Protocol, Environmental Assessment of Food and Drink Protocol, European Food Sustainable Consumption and Production Round Table (SCP RT), Working Group 1, Brussels, Belgium.

Freibauer, A., Rounsevell, M. D. A., Smith, P., & Verhagen, J. (2004). Carbon sequestration in the agricultural soils of Europe. *Geoderma, 122*(1), 1–23.

Gabarell, X., Rives, J., Rieradevall, J., & Fernandez-Rodriguez, I. (n.d.). Life Cycle Assessment of the industrial production of natural cork stoppers for still wine. http://www.ecosostenible wine.com/files/P16_Gabarrell_Taps_Suro.pdf. Accessed 2 Mar 2014.

Ganichot, B. (2002). Evolution de la date des vendanges dans les Côtes du Rhône méridionales. In: 6emes Recontres Rhodaniennes (pp. 38–41). Orange, France: Institut Rhodanien.

Gazulla, C., Raugei, M., & Fullana-i-Palmer, P. (2010). Taking a life cycle look at crianza wine production in Spain: Where are the bottlenecks? *International Journal of Life Cycle Assessment, 15,* 330–337

Gonzalez, A., Klimchuk, A., & Martin, M. (2006). *Life Cycle Assessment of wine production process: Finding relevant process efficiency and comparison to eco-wine production*. Stockholm: Royal Institute of Technology.

González-García, S., Carles, M. G., Raúl García, L., Moreira, M. T., Gabarrell, X., Rieradevall i Pons, J., & Feijoo, G. (2011a). Assessing the global warming potential of wooden products from the furniture sector to improve their ecodesign. *Science of The Total Environment, 410–411,* 16–25.

González-García, S., Silva, F. J., Moreira, M. T., Pascual, R. C., Lozano, R. G., Gabarrell, X., Pons, J. R. I., & Feijoo, G. (2011b). Combined application of LCA and eco-design for the sustainable production of wood boxes for wine bottles storage. *International Journal of Life Cycle Assessment, 16*(3), 224–237.

Goode, J. (2005). *The science of wine: From vine to glass*. Octopus Publishing Group Ltd.
Greendelta (2011). openLCA case study: Conventional vs Organic Viticulture. www.openlca.org/documentation. Accessed 4 Feb 2014.
Greenhaigh, S., Mithraratne, N., Sinclair, R., Smith, S., McConachy, E., & Barber, A. (2011). GHG product accounting guidelines for the wine industry, Landcare Research contract report GHG0708, Ministry of Agriculture and Forestry, Wellington.
Guinée, J. B., Gorree, M., Heijungs, R., Huppes, G., Kleijn, R., Udo de Haes, H. A., Van der Voet, E., & Wrisberg, M. N. (2002). *Life Cycle Assessment. An operational Guide to ISO Standards* (Vol 1–3). Centre of Environmental Science Leiden University Ed, The Netherlands.
Hayes, P., & Battaglene, T. (2006). Regulatory response to climate change. Le Bulletin de L ' OIV. *Organisation Internationale de La Vigne et du Vigne, 79,* 697–708.
Herath, I., Green, S., Singh, R., Horne, D., van der Zijpp, S., & Clothier, B. (2013a). Water footprinting of agricultural products: A hydrological assessment for the water footprint of New Zealand's wines. *Journal of Cleaner Production, 41,* 232–243.
Herath, I., Green, S., Horne, D., Singh, R., McLaren, S., & Clothier, B. (2013b). Water footprinting of agricultural products: Evaluation of different protocols using a case study of New Zealand wine. *Journal of Cleaner Production, 44,* 159–167.
Hertwich, T. G. (2005). Consumption and the rebound effect: An industrial ecology perspective. *Journal of Industrial Ecology, 9*(1–2), 85–98.
IPCC (1997). Revised 1996 IPCC Guidelines for National Greenhouse Inventories. In J. T. Houghton, L. G. Melra Fllho, B. Um, K. Trdanton, I. Mamaty, Y. Bonduki, D. J. Griggs, & B. A. Callander (Eds). Prepared by the IPCC/OECD/IEA, Paris, France.
ISO. (2006). 14040: 2006. In Environmental management–Life Cycle Assessment–Principles and framework.
ISO. (2011). ISO 50001:2011, Energy management systems—Requirements with guidance for use.
ISO. (2013). *ISO 14067:2013—Greenhouse gases—Carbon footprint of products—Requirements and guidelines for quantification and communication*. Geneva: International Organisation for Standardisation.
Jiménez, E., Martínez, E., Blanco, J., Pérez, M., & Charmery, G. (2013). Methodological approach towards sustainability by integration of environmental impact in production system models through life cycle analysis: Application to the Rioja wine sector. Simulation 143–161.
Jones, G. V. (2007). Climate change: Observations, projections, and general implications for viticulture and wine production. Practical Winery and Vineyard, July/August, 44–64.
Jones, G. V., & Davis, R. E. (2000). Climate influences on grapevine phenology, grape composition, and wine production and quality for Bordeaux, France. *American Journal of Enology and Viticulture, 51*(3), 249–261.
Jones, G. V., White, M. A., Cooper, O. R., & Storchmann, K. (2005). Climate change and global wine quality. *Climatic Change, 73*(3), 319–343.
Kadlec, R. H., & Wallace, S. D. (2008). *Treatment Wetlands* (II ed.). CRC.
Kauhanen, J., Kaplan, G. A., Goldberg, D. E., Salonen, R., & Salonen, J. T. (1999). Pattern of alcohol drinking and progression of atherosclerosis. *Arteriosclerosis, Thrombosis, and Vascular Biology, 19,* 3001–3006.
Kavargiris, S. E., Mamolos, A. P., Tsatsarelis, C. A., Nikolaidou, A. E., & Kalburtji K. L. (2009). "Energy resources' utilization in organic and conventional vineyards: Energy flow, greenhouse gas emissions and biofuel production". *Biomass and Bioenergy, 33,* 1239–1250.
Keightley, K. E. (2011). Applying New Methods for Estimating in Vivo Vineyard Carbon Storage. *American Journal of Enology and Viticulture, 62*(2), 214–218.
Kennison, K., Ward, G., & Fisher, D. (2011). Key information on smoke effect in grapes and wine: What can be done to identify and reduce smoke effect in grape and wine production? Western Australia Agriculture Authority, Department of Agriculture and Food, Western Australia. http://www.awri.com.au/wp-content/uploads/smoke_effect_in_grapes_and_wine.pdf. Accessed 9 Mar 2014.

Kounina, A., Tatti, E., Humbert, S., Pfister, R., Pike, A., Ménard, J. F., Loerincik, Y., & Jolliet, O. (2012). The importance of considering product loss rates in Life Cycle Assessment: The example of closure systems for bottled wine. *Sustainability, 4,* 2673–2706.

Kroodsma, D. A., & Field, C. B. (2006). Carbon Sequestration in California agriculture, 1980–2000. *Ecololical Applications, 16*(5), 1975–1985.

Lapsey, J., & Moulton, K. (2001). *Successful wine marketing*. Springer.

Latunussa, C. E. L. (2011). Life Cycle Assessment (LCA) of Food and Beverage Packaging. Master thesis. Politecnico di Milano

Leger, A. S., Cochrane, A. L., & Moore, F. (1979). Factors associated with cardiac mortality in developed countries with particular reference to the consumption of wine. *Lancet, 313*(8124), 1017–1020.

Leonardi, R., Giacomazzi, M., Galesso, D., Vitali, M., & Picchioni, E. (2006). Product Category Rules (PCR) for preparing an environmental product declaration (EPD) for packaged sparkling red, white and rosé wines. PCR 2006:03.

Lucchetti, M. C., Martucci, O., & Arcese, G. (2010, September). Life Cycle Assessment of wine production. VII International Conference on Life Cycle Assessment in the agri-food sector, 22–24 September 2010, Bari, Italy, pp. 289–294.

Mann, J. C., Abramczyk, M. L., Andrews, M. R., Rothbart, J. A., Small, R. M. & Bailey, R. R. (2010, April). Sustainability at kluge estate vineyard and winery. Proceedings of the 2010 IEEE Systems and Information Engineering Design Symposium, Charlottesville, VA, USA.

Marenghi, M. (2005). Manuale di viticoltura: impianto, gestione e difesa del vigneto. Il Sole 24 Ore Edagricole.

Martin, D. (1997). The carbon dioxide budget of the New Zealand grape and wine industry. HortResearch Client Report No.97/161 commissioned by Wine Industry of New Zealand Ltd., The Horticulture and Food research Institute of New Zealand Ltd, Palmerston north, NZ. http://wineinf.nzwine.com/research_outputs.asp?id=3&cid=2&type=r. Accessed 19 Dec 2013.

Marvuglia, A., Benetto, E., Rege, S., & Jury, C. (2013). Modelling approaches for consequential Life-Cycle Assessment (C-LCA) of bioenergy: Critical review and proposed framework for biogas production. *Renewable and Sustainable Energy Reviews, 25,* 768–781.

Masoni, P., Sára, B., Bianconi, P., Marani, S., & Raggi, A. (2001). Adoption of Life Cycle Assessment in small- and medium-sized enterprises in Italy (Part I)—LCA case study in an Ital-ian medium-sized dairy company. *International Journal of Management Literature, 1,* 199–222.

Masoni, P., Sára, B., Scimìa, E., & Raggi, A. (2004). VerdEE—A tool for adoption of Life Cycle Assessment in small- and medium-sized enterprises in Italy. *Progress in Industrial Ecology, 1*(1/2/3), 203–228.

Mattila, T., Helin, T., Antikainen, R., Soimakallio, S., Pingoud, K., & Wessman, H. (2011). Land use in Life Cycle Assessment. The Finnish Environment, 24.

Mattila, T., Helin, T., & Antikainen, R. (2012a). Land use indicators in Life Cycle Assessment—A case study on beer production. *International Journal of Life Cycle Assessment, 17,* 277–286

Mattila, T., Leskinen, P., Soimakallio, S., & Sironen, S. (2012b). Uncertainty in environmentally conscious decision making: beer or wine? *International Journal of Life Cycle Assessment, 17,* 696–705.

Mattsson, B. (1999). Environmental Life Cycle Assessment (LCA) of Agricultural Food Product. Swedish Institute for Food and Biotechnology, SIK, Gothenburg.

Mira de Orduña, R. (2010). Climate change associated effects on grape and wine quality and production. *Food Research International, 43*(7), 1844–1855.

Morgante, A., Raggi, A., Petti, L., & Matteucci, P. (2004, September). Applicazione della procedura 'VerdEE' al ciclo di vita di un vino biologico abruzzese. Atti XXI Congresso Nazionale di Merceologia (su CD-ROM), Foggia.

Neman, R. R., White, M. A., Cayan, D. R., Jones, G. V., Running, S. W., Coughlan, J. C., & Peterson, D. L. (2001). Asymmetric warming over coastal California and its impact on the premium wine industry. *Climate Research, 19*(1), 25–34.

Neto, B., Dias, A. C., Machado, M. (2013). Life Cycle Assessment of the supply chain of a Portuguese wine: From viticulture to distribution. *International Journal of Life Cycle Assessment, 18,* 590–602.

Niccolucci, V., Galli, A., Kitzes, J., Pulselli, R. M., Borsa, S., & Marchettini, N. (2008). Ecological footprint analysis applied to the production of two Italian wines. *Agriculture, Ecosystems and Environment, 128,* 162–166.

Nicoletti, G. M., Notarnicola, B., & Tassielli, G. (2001). Comparison of conventional and organic wine. Proceedings of the International Conference LCA in Foods, Goteborg 26–27 April 2001.

Nigdikar, S. V., Williams, N. R., Griffin, B. A., et al. (1998). Consumption of red wine polyphenols reduces the susceptibility of low-density lipoproteins to oxidation in vivo. *The American Journal of Clinical Nutrition, 68,* 258–265.

Notarnicola, B., Tassielli, G., & Nicoletti, G. M. (2003). LCA of wine production. In B. Mattsonn & U. Sonesson (Eds.), Environmentally-friendly food production. Cambridge: Woodhead-Publishing.

Notarnicola, B., Tassielli, G., Renzulli, P., & Settanni, E. (2007, April). LCI of must enrichment by reverse osmosis pilot plant. Proc. 5th International Conference "LCA in Foods", Goteborg.

Notarnicola, B., Tassielli, G., Renzulli, P., & Settanni, E. (2008, November). Life Cycle Assessment of a pilot plant for the must enrichment by reverse osmosis. 6th International Conference on Life Cycle Assessment in the Agri-Food Sector, Zurich, Switzerland.

Notarnicola, B., Tassielli, G., & Settanni, E. (2010, September). Including more technology in the production of a quality wine: The importance of functional unit. Proceedings of VII International Conference on Life Cycle Assessment in the agri-food sector. Bari, 22–24th September 2010, vol. II, 235–240.

OIV. (1992). Products definition. http://www.oiv.int/oiv/info/endefinitionproduit?lang=en. Accessed 11 Mar 2014.

OIV. (2011). General principles of the OIV greenhouse gas accounting protocol for the vine and wine sector. Resolution OIV-CST 431–2011.

OIV. (2013). 2013 Global economic vitiviniculture data, Paris, October 2013. http://www.oiv.int/oiv/info/enconjoncture?lang=en. Accessed 13 Jan 2013.

OIV. (2014). http://www.oiv.int/oiv/cms/index?lang=en. Accessed 10 March 2014.

Patingre, J. F., Lorentzon, A., & De Caevel, B. (2010). Nordic Life Cycle Assessment Wine Package Study, Bio Intelligence Service. http://www.tetrapak.com/se/DocumentBank/LCA%20Nordic%20Wine%20comparative_August_10_with%20final%20statement.pdf. Accessed 3 Feb 2014.

Pattara, C., Cichelli, A., Civitarese, C., & Di Martino, M. (2012a). A comparison of carbon footprints in wine production: The case of two cooperative wineries in central Italy. *Bulletin de l'OIV, 85*(977–978–979), 307–316.

Pattara, C., Raggi, A., & Cichelli, A. (2012b). Life Cycle Assessment and carbon footprint in the wine supply-chain. *Environmental Management, 49,* 1247–1258.

PCR Development Initiative. (2013). Guidance for product category rule development. In W. Ingwersen & V. Subramanian (Eds.), *Product category rule guidance development initiative.* Version 1.0. http://www.pcrguidance.org. Accessed 29 Jan 2014.

Petti, L., Raggi, A., De Camillis, C., Matteucci, P., Sára, B., & Pagliuca, G. (2006). Life Cycle Approach in an organic winemaking firm: An Italian case-study. Proc. Fifth Australian Conference on Life Cycle Assessment, Melbourne, Australia, 22–24 November 2006. http://www.conference.alcas.asn.au/2006/Petti%20et%20al.pdf. Accessed 21 Feb 2014.

Petti, L., Ardente, F., Bosco, S., De Camillis, C., Masotti, P., Pattara, C., Raggi, A., & Tassielli, G. (2010a, April). Stato dell'arte della Life Cycle Assessment (LCA) nel comparto vitivinicolo. Proceedings of Scientific Convention of Rete Italiana LCA "La metodologia LCA: approccio proattivo per le tecnologie ambientali. Casi studio ed esperienze applicative", Padova, Italy.

Petti, L., Ardente, F., Bosco, S., De Camillis, C., Masotti, P., Pattara, C., Raggi, A., & Tassielli, G. (2010b, September). State of the art of Life Cycle Assessment (LCA) in the wine industry. 7th International Conference on LCA in the Agri-Food Sector, Bari, Italy, 22–24 September 2010, 493–498.

Pina, L., Dias, A., Neto, B., Arroja, L., & Quinteiro, P. (2011). The water footprint of wine production in portugal: A case study on vinho verde. Poster session proceedings of the 6th International Conference on Industrial Ecology (ISIE 2011 Conference), June 7–10, 2011, University of California, Berkeley (USA).

Pittock, B., Arthington, A., Booth, T., Cowell, P., Hennesy, K., Howden, M., Hughes, L., Jones, R., Lake, S., Lyne, V., McMichael, T., Mullet, T., Nicholls, N., Torok, S., & Woodruf, R. (2003). *Climate change: An Australian guide to the science and potential impacts* (p. 239). Canberra: Australian Greenhouse Oice.

Pizzigallo, A. C. I., Granai, C., & Borsa, S. (2008). The joint use of LCA and emergy evaluation for the analysis of two Italian wine farms. *Journal of Environmental Management, 86,* 396–406.

Point, E., Tyedmers, P., & Naugler, C. (2012). Life cycle environmental impacts of wine production and consumption in Nova Scotia, Canada. *Journal of Cleaner Production, 27,* 11–20.

Poni, S., Palliotti, A., & Bernizzoni, F. (2006). Calibration and evaluation of a STELLA software-based daily CO2 balance model in Vitis vinifera L. *Journal of the American Society for Horticultural Science, 131*(2), 273–283.

Rabl, A., Benoist, A., Bron, D., Peuportier, B., Spadaro, J. V., & Zoughaib, A. (2007). How to account for CO2 emissions from biomass in an LCA. *The International Journal of Life Cycle Assessment, 12*(5), 281.

Rallo, C. (2011). Applicazione della metodologia Life Cycle Assessment al Passito di Pantelleria. Master Thesis. Alma Mater Studiorum UniversitÃ di Bologna, Bologna, Italy.

Ramos, M. C., Jones, G. V., & Martínez-Casasnovas, J. A. (2008). Structure and trends in climate parameters affecting winegrape production in northeast Spain. *ClimateResearch, 38*(1), 1–15.

Reap, J., Roman, F., Duncan, S., & Bras, B. (2008). A survey of unresolved problems in Life Cycle Assessment. *The International Journal of Life Cycle Assessment, 13*(5), 374–388.

Reich-Weiser, C., Paster, P., Erickson, C., & Dornfeld, D. (2010). The Role of Transportation on the GHG Emissions of Wine. *Journal of Wine Research, 21*(2–3), 197–206.

Reynier, A. (2011). Manuel de viticulture: Guide technique du viticulteur Lavoisier.

Ribéreau-Gayon, J., & Peynaud, E. (1979) Trattato di scienza e tecnica enologica: La vinificazione, le trasformazioni del vino, Volume 3. Editions Aeb, 1980.

Rimm, E. B., Klatsky, A., Grobbee, D., et al. (1996). Review of moderate alcohol consumption and reduced risk of coronary heart disease: is the effect due to beer, wine or spirits? *British Medical Journal, 312,* 731–736.

Rives, J., Fernandez-Rodriguez, I., Rieradevall, J., & Gabarrell, X. (2011). Environmental analysis of the production of natural cork stoppers in southern Europe (Catalonia e Spain). *Journal of Cleaner Production, 19,* 259–271.

Rives, J., Fernandez-Rodriguez, I., Rieradevall, J., & Gabarrell, X. (2012). Environmental analysis of raw cork extraction in cork oak forests in southern Europe (Catalonia—Spain). *Journal of Environmental Management, 110,* 236–245.

Rives, J., Fernandez-Rodriguez, I., Rieradevall, J., & Gabarrell, X. (2013). Integrated environmental analysis of the main cork products in southern Europe (Catalonia—Spain). *Journal of Cleaner Production, 51,* 289–298.

Rugani, B., Niccolucci, V., Pulselli, R. M., & Tiezzi, E. (2009). A cradle-to-gate Life Cycle Assessment integrated with Emergy evaluation: sustainability analysis of an organic wine production. Proceedings of the SETAC Europe 19th annual meeting, "Protecting ecosystem health: Facing the challenge of a globally changing environment", Göteborg, Sweden, May 31steJune 4th, 2009, p. 274.

Rugani, B., Vázquez-Rowe, I., Benedetto, G., & Benetto, E. (2013). A comprehensive review of carbon footprint analysis as an extended environmental indicator in the wine sector. *Journal Cleaner Production, 54,* 61–77.

Ruggieri, L., Cadena, E., Martı´nez-Blanco, J., Gasol, C., Rieradevall, X., Gabarrell, X., Gea, T., Sort, X., & Sànchez, A. (2009). Recovery of organic wastes in the Spanish wine industry. Technical, economic and environmental analyses of the composting process. *Journal of Cleaner Production, 17,* 830–838.

Salomone, R., Clasadonte, M. T., Proto, M., Raggi, A., Arzoumanidis, I., Ioppolo, G., Lo Giudice, A., Malandrino, O., Matarazzo, A., Petti, L., Saija, G., Supino, S., & Zamagni, A. (2012). Improv-ing Life Cycle Management (LCM) tools for the food industry: A framework of Product-Oriented Environmental Management System (POEMS). 6th SETAC World Congress/22nd SETAC Europe Annual Meeting, Berlin.

Sanchez, S. T., Woods, J., Akhurst, M., Brander, M., O'Hare, M., Dawson, T. P., Edwards, R., Liska, A.J., & Malpas, R. (2012). Accounting for indirect land-use change in the Life Cycle Assessment of biofuel supply chains. *Journal of The Royal Society Interface, 9*(71), 1105–1119.

Sanchez Ramirez, P. K. (2011). A social LCA case study in the wine sector. Presentation, University G. d'Annunzio, Pescara.

SAWIA-South Australian Winc Industry Association. (2004). Australian Wine Industry State of the Environment 2003. South Australian Wine Industry Association Incorporated, Australian Government Department of Environment and Heritage and Winemakers' Federation of Australia, p. 40.

Saxe, H. (2010). LCA-based comparison of the climate footprint of beer vs. wine & spirits. Institute of Food and Resource Economics, Report no. 27.

Schlich, E. H. (2010). From Vineyard to Point of Sale: Allocation of Energy Use and CO2-Emission to entire Supply Chains of Wine. http://aic.ucdavis.edu/aaweconf/abstracts/Schlich.pdf. Accessed 20 Mar 2014.

Schmidt, J. (2008). System delimitation in agricultural consequential LCA—Outline of methodology and illustrative case study of wheat in Denmark. *International Journal of Life Cycle Assessment 13*(4), 350–64.

Schorb, A. (2000). Life cycle analysis for different methods of weed control in vine-yards. *Zeitschrift fur Pflanzenkrankheiten und Pflanzenschutz, 107*(17), 753–760.

Schultz, H. R. (2010). Climate change and viticulture: Research needs for facing the future. *Journal of Wine Research, 21*(2), 113–116.

Simos, C. (2008). The implications of smoke taint and management practices. *Australian Viticulture, 12*(1), 77–80.

Skuras, D., & Vakrou, A. (2002). Consumers' willingness to pay for origin labelled wine: A Greek case study. *British Food Journal, 104*(11), 898–912.

Smart, R. E. (2010). A Lump of Coal, a Bunch of Grapes &hellip. *Journal of Wine Research, 21*(2–3), 107–111.

Smyth, M., & Russell, J. (2009). From graft to bottle—Analysis of energy use in viticulture and wine production and the potential for solar renewable technologies. *Renewable and Sustainable Energy Reviews, 13*(8), 1985–1993.

Soja, G., Zehetner, F., Rampazzo-Todorovic, G., Schildberger, B., Hackl, K., Hofmann, R., Burger, E., & Omann, I. (2010). Wine production under climate change conditions: Mitigation and adaptation options from the vineyard to the sales booth. In building sustainable rural futures: The added value of systems approaches in times of change and uncertainty. In I. Darnhofer & M. Grötzer (Eds.), *University of natural resources and applied life sciences* (pp. 1368–1378). Vienna.

Soosay, C., Fearne, A., & Dent, B. (2012). Sustainable value chain analysis—A case study of Oxford Landing from "vine to dine". *Supply Chain Management: An International Journal, 17*(1), 68–77.

South Australian Wine Iindustry Association Incorpored (SAWIA). (2004). Australian wine industry state of the environment 2003. Report.

Sustainability in the Italian Viticulture. (2014). http://www.viticolturasostenibile.org/EN/Home.aspx. Accessed 10 March 2014.

Tate, A. B. (2001). Global warming's impact on wine. *Journal of Wine Research, 12*(2), 95–109.

Tavşanoğlu, Ç., & Úbeda, X. (2011). Fire and soils: Methodological issues and implications to management. *Environmental Research, 111*(2), 191–192.

Tillman, A. M. (2000). Significance of decision making for LCA methodology. *Environmental Impact Assessment Review, 20*(1), 113–123.

TSC. (2014). http://www.sustainabilityconsortium.org/. Accessed 10 March 2014.

UNEP. (2011). Global guidance principles for Life Cycle Assessment databases a basis for greener processes and products. "Shonan Guidance Principles". Life cycle initiative. SETAC and United Nations Environment Programme.

UNEP/SETAC. (2009) Guidelines for social Life Cycle Assessment of products. Life cycle initiative. SETAC and United Nations Environment Programme.

Vázquez-Rowe, I., Villanueva-Rey, P., Iribarren, D., Moreira, M. T., & Feijoo, G. (2012a). Joint life cycle assessment and data envelopment analysis of grape production for vinification in the Rías Baixas appellation (NW Spain). *Journal of Cleaner Production, 27,* 92–102.

Vázquez-Rowe, I., Villanueva-Rey, P., Iribarren, D., Moreira, M. T., & Feijoo, G. (2012b). Joint Life Cycle Assessment and data envelopment analysis of grape production for vinification in the Rías Baixas appellation (NW Spain). *Journal of Cleaner Production, 27,* 92–102.

Vázquez-Rowe, I., Rugani, B., & Benetto, E. (2013). Tapping carbon footprint variations in the European wine sector. *Journal of Cleaner Production, 43,* 146–155.

Vázquez-Rowe, I., Marvuglia, A., Rege, S., & Benetto, E. (2014a). Applying consequential LCA to support energy policy: Land use change effects of bioenergy production. *Science of The Total Environment, 472,* 78–89.

Vázquez-Rowe, I., Marvuglia, A., Flammang, K., Braun, C., Leopold, U., & Benetto, E. (2014b). The use of temporal dynamics for the automatic calculation of land use impacts in LCA using R programming environment. *The International Journal of Life Cycle Assessment, 19*(3), 500–516.

Venkat, K. (2012). Comparison of twelve organic and conventional farming systems: A life cycle greenhouse gas emissions perspective. *Journal of Sustainable Agriculture, 36,* 620–649.

Villanueva-Rey, P., Vázquez-Rowe, I., Moreira, M. T., & Feijoo, G. (2013). Comparative Life Cycle Assessment in the wine sector: Biodynamic vs. conventional viticulture activities in NW Spain. *Journal of Cleaner Production, 65,* 330–341.

Webb, L. B., Whetton, P. H., Barlow, E. W. R. (2007). Impact on Australian Viticulture from Greenhouse Induced Temperature Change. In A. Zerger & R. M. Argent (Eds.), *MODSIM 2005 International Congress on Modelling and Simulation* (pp. 1504–1510). Modelling and Simulation Society of Australia and New Zealand.

Weidema, B. P. (2003). Market information in Life Cycle Assessment. Environmental project no. 863, Copenhagen: Danish Environmental Protection Agency, DK. http://www2.mst.dk/udgiv/publications/2003/87-7972-991-6/pdf/87-7972-992-4.pdf. Accessed 7 Jan 2014.

Weidema, B. P. (2006). The integration of economic and social aspects in Life Cycle Impact Assessment. *The International Journal of Life Cycle Assessment, 11*(1), 89–96.

Weidema, B. P., & Wesnæs, M. S. (1996). Data quality management for life cycle inventories—An example of using data quality indicators. *Journal of Cleaner Production, 4*(3), 167–174.

Wine Institute. (2010). Greenhouse gas protocol and accounting tool. http://www.wineinstitute.org/ghgprotocol.

Woodward, L. J. (2010). Wine packaging: Alternatives to traditional glass. Dissertation, Cape Wine Academy, South Africa pp. 82.

WRAP-Waste & Resources Action Programme. (2007). The life cycle emissions of wine imported to the UK, Manufacturing/Final Report. Banbury. Accessed 12 Dec 2013.

WRI-World Resources Institute, WBCSD-World Business Council for Sustainable Development. (2011). GHG protocol—Product life cycle accounting and reporting standard. http://www.ghgprotocol.org/standards/product-standard. Accessed 12 Dec 2013.

Zabalza, I., Aranda, A., Scarpellini, S. (2003). Analysis and improvement of energy and environmental costs for small and medium enterprises in the wine sector. Proceedings of the 16th International Conference on Efficiency, Costs, Optimisation, Simulation and Environmental Impact of Energy Systems (ECOS 2003). June 30th–July 2nd, 2003, Copenhagen, Denmark.

Zamagni, A., Guinée, J., Heijungs, R., Masoni, P., & Raggi, A. (2012). Lights and shadows in consequential LCA. *The International Journal of Life Cycle Assessment, 17,* 904–918.

Chapter 4
Life Cycle Assessment in the Cereal and Derived Products Sector

Pietro A. Renzulli, Jacopo Bacenetti, Graziella Benedetto, Alessandra Fusi, Giuseppe Ioppolo, Monia Niero, Maria Proto, Roberta Salomone, Daniela Sica and Stefania Supino

Abstract This chapter discusses the application of life cycle assessment methodologies to rice, wheat, corn and some of their derived products. Cereal product systems are vital for the production of commodities of worldwide importance that entail particular environmental hot spots originating from their widespread use and from their particular nature. It is thus important for tools such as life cycle assessment (LCA) to be tailored to such cereal systems in order to be used as a means of identifying the negative environmental effects of cereal products and highlighting possible pathways to overall environmental improvement in such systems. Following a brief introduction to the cereal sector and supply chain, this chapter reviews some of the current cereal-based life cycle thinking literature, with a particular emphasis on LCA. Next, an analysis of the LCA methodological issues emerging from the literature review is carried out. The following section of the chapter discusses some practices and approaches that should be considered when performing cereal-based LCAs in order to achieve the best possible results. Conclusions are drawn in the final part of the chapter and some indications are given of the main hot spots in the cereal supply chain.

P. A. Renzulli (✉)
Ionian Department of Law, Economics and Environment, University of Bari, Via Lago Maggiore angolo via Ancona, 74121 Taranto, Italy
e-mail: pietro.renzulli@uniba.it

J. Bacenetti · A. Fusi
Department of Agricultural and Environmental Sciences, Production, Landscape, Agroenergy, University of Milan, Via Celoria 2, 20133 Milan, Italy
e-mail: jacopo.bacenetti@unimi.it

A. Fusi
e-mail: alessandra.fusi@unimi.it

G. Benedetto
Department of Science for Nature and Environmental Resources, University of Sassari, Via E. De Nicola, 1-07100 Sassari, Italy
e-mail: gbenedet@uniss.it

B. Notarnicola et al. (eds.), *Life Cycle Assessment in the Agri-food Sector*,
DOI 10.1007/978-3-319-11940-3_4

Keywords Cereal LCA · Agri-food sector · Cereal crop rotation · Footprint labels · Cereal product systems

4.1 Introduction

Cereal grains, the fruit of plants belonging to the grass family (Gramineae), represent the most important group of food crops produced throughout the world.

The agricultural revolution some 10,000 years ago made grains the major food raw material for humans (Diamond 2002). The global importance of cereal crops to the human diet and moreover their role in the recorded history of mankind and in agriculture cannot be overstated. Cereal crops are energy dense, containing 10,000–15,000 kJ/kg, about 10–20 times more energy than most succulent fruits and vegetables. Nutritionally, they are important sources of dietary protein, carbohydrates, the B complex of vitamins, vitamin E, iron, trace minerals and fibres (Cordain 1999).

Cereals occupy an important role as global commodity products, being bought almost immediately after harvest and sold on bulk markets. The growing and exporting of grains are vital for many countries of the world and account for significant contributions to their agricultural outputs.

The types of grain cultivated around the world depend on an array of environmental, cultural and economic factors, and the most critical environmental factors are temperature and water availability, which determine the crops grown in a given region. For instance, in regions where there is water availability, rice, and to some extent corn, tend to dominate. Neither corn nor rice can withstand frost and they

G. Ioppolo · R. Salomone
SEAM Department, University of Messina, Piazza Pugliatti 1, 98122
Messina, Italy
e-mail: giuseppe.ioppolo@unime.it

R. Salomone
e-mail: roberta.salomone@unime.it

M. Niero
QSA division, Department of Management Engineering, Technical University of Denmark Produktionstorvet 424 2800 Kgs. Lyngby, Denmark
e-mail: monni@dtu.dk

M. Proto · D. Sica · S. Supino
Department of Management & Information Technology (DISTRA-MIT),
University of Salerno, Via Giovanni Paolo II 132,
84084 Fisciano, Italy
e-mail: mariap@unisa.it

D. Sica
e-mail: dsica@unisa.it

S. Supino
e-mail: ssupino@unisa.it

must be grown in warm environments. Wheat, on the other hand, is grown in a wider variety of environments with a broad range of water availability and ambient temperatures; hence, it is widely produced in the temperate regions both in the winter and in the spring (Kirk-Othmer 1984). Consequently, climate change-related hazards and overall global warming as well as land use, water management, fertiliser and pesticide use, and food waste and losses (wastage) are critical factors affecting the productivity of cereal product systems.

The main challenge for such systems is not only to fulfil the need for more productive agricultural and food systems but also to make them more sustainable: in other words, producing 'more with less'. This task is made even more daunting by the combined effects of land degradation, over-extraction of groundwater, climate change, energy scarcity, the increase in the world population and the overall risk of species extinction. Specifically, cereal crops play a crucial role in agriculture intensification, characterized by increasing harvests, growing use of water resources and synthetic fertiliser and pesticide use beyond sustainable levels, all of which erode the sustainability of the platform upon which food production is based.

Cereal grain availability, environmental impacts and social issues are directly related not only to agronomic practices and water consumption in the cultivation phase, but also to the respective entire supply chain (Sect. 4.1.1). In this context, supply chain environmental analysis, with systemic use of life cycle assessment (LCA), is the central element in evaluating its 'goodness' and in proposing alternative configurations (Venkat 2007). Besides, a sustainable supply chain implies the management of material, information and capital flows aiming to achieve simultaneous balancing of economic, environmental and social goals, through cooperation between the actors involved and meeting customer needs (Seuring and Müller 2008).

Because of their central role in the world's agricultural production and in the human diet, both on the Italian and on the international level, rice, wheat, corn and some of their derived products will be the chief focus of this chapter, with the aim of analysing the approaches (Sect. 4.2) and methodological issues (Sect. 4.3) that need to be considered to complete a useful LCA of the cereal sector. Furthermore, since qualitative and quantitative methods and tools that are able to address environmental performance metrics will be fundamental in supporting policy makers, management strategies and operative decisions for the development of cereal food systems, in both industrialized and developing country contexts, some lessons learned for the optimised application of LCA to the cereal sector will also be highlighted (Sect. 4.4).

4.1.1 Introductory Scenario of the Cereal Supply Chain

The number of plant species nourishing humanity is extraordinarily limited. In fact, fewer than about 20 plant species provide 90 % of mankind's food supply (Cordain 1999), of which some cereals, such as rice, wheat and corn, represent a significant percentage in terms of both value and volume (Table 4.1).

Table 4.1 Top ten world crop productions (2011). (Source: FAO 2011a)

Commodity	Production	
	Int $ 1000	Gt
Rice, paddy	186,667,648	722,559.6
Wheat	84,281,536	701,395.3
Soybeans	65,903,601	262,037.6
Tomatoes	58,223,483	159,347.0
Sugar cane	56,903,836	1,800,377.6
Corn	55,478,433	885,289.9
Potatoes	49,681,577	373,158.3
Vegetables, fresh	45,936,531	268,833.8
Grapes	39,494,901	69,093.3
Apples	31,706,244	75,484.7
Total	*674,277,790*	*5,317,577.1*

Some cereals have been primary sources of nourishment for humans for thousands of years and today roughly half of the world's cropland is devoted to growing cereals. If we combine their direct intake (e.g. as cooked rice, bread, etc.) with their indirect consumption, in the form of non-vegetable foods like meat and milk (about 40 % of all grain is currently fed to livestock), cereals account for approximately two-thirds of all human calorie intake (Dyson 1999).

Developing countries depend more on cereal grains for their nutritional needs than the developed world (from 60 to 80 % of calories are derived directly from cereals in developing countries and approximately 30 % of calories in the developed world) (Awika 2011). In particular, in Europe, the average annual consumption of cereal grains is 131 kg per capita, wheat making up the majority (108 kg/capita/year), whereas in Asia, about half of the annual cereal consumption is rice. Wheat and rice are the most important cereals globally with respect to human nutrition, whereas corn is important especially in Central and South America and sorghum and millet are important in Africa (FAO 2011b). The world's population is predicted to exceed 9 billion people by 2050 and recent FAO estimates indicate that to meet the projected demand, global agricultural production will have to increase by 60 % from its 2005 to 2007 levels (FAO et al. 2013). This equates roughly to the additional production of around 1000 Mt of cereals and around 200 Mt of meat and fish per year by 2050 (FAO 2011c). These production gains are largely expected to come from increases in the productivity of crops, livestock and fisheries. However, unlike the 1960s' and 1970s' green revolution, our ability to reach these targets may be limited in the future by a scarcity of raw materials and energy resources.

The food supply chain varies greatly in relation to its location, the productive capacity of the producers and obviously the goods themselves (e.g. fresh or processed foods, etc.). In the business context, it may be considered as a complex network of chain actors, with many interdependencies and steps, encompassing different flows of materials, services and information. In Fig. 4.1, a simplified cereal supply chain is schematised. Moreover, the way in which cereals are handled, processed and transported throughout the entire chain influences not only the characteristics and prices of the products, but also several other issues.

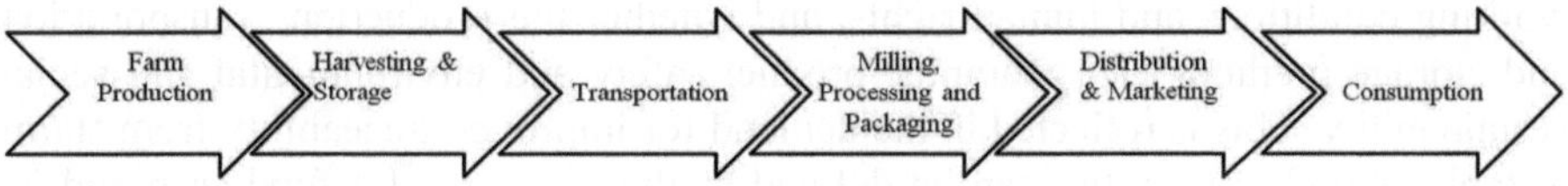

Fig. 4.1 The supply chain scheme for cereals and cereal-based products

Fig. 4.2 The main elements of sustainable cereals and cereal-based product supply chains

These issues are mainly related to the environmental and social dimensions, according to the triple bottom line approach (Elkington 1997), whereby performance is to be achieved in the economic, environmental and social dimensions (Fig. 4.2). The inclusion of environmental and social aspects in the analysis of cereal supply chains stems from two principal considerations. The first is related to market globalisation and to the growing role of multinational organisations, which has led to the lengthening of supply chains. In fact, the value provided to the customers derives from the complex aggregate of all the 'value added' along the entire supply chain. Secondly, the increasing pressure from developed countries has made necessary the close observance and monitoring of the sustainability approach of all the links in the supply chain (Editorial of Journal of Cleaner Production 2008). This is because consumers have become increasingly aware not only of the end-product consequences, but also of supply chain sustainability.

The need for better environmental performance will increase in forthcoming years, in terms of both the rising concern with national and international regulations and the ever-growing attention of end consumers to sustainability issues. Almost all products reach consumers through supply chain management and in the food sector each link of the supply chain affects the availability, affordability, diversity and overall nutritional quality of foods as well as their safety.

The result of the continuing sequence of food scandals and incidents, in almost every area of the world, has determined that food safety is currently considered the most important issue for all stakeholders. In fact, consumers' perceptions show a consolidated interest in the properties of the food they consume. The increasing need for transparent information has involved the entire cereal supply chain and is supported by several tools, most of which are based on the concept of traceability. In fact, with the globalisation of markets, consumers have become increasingly concerned about the origins of their food, the way in which agricultural land is used, working conditions and human rights, and whether the production, transportation and storage methods can guarantee product safety and environmental and social sustainability. This is reflected in the demand for improved traceability from 'farm to table'. Food traceability can be defined as the 'history' of a food crop and its subsequent transformations on its journey during its life cycle. Nowadays, traceability is becoming mandatory in many countries (the European Union, the United States and Japan), starting with some food products (Kraisintu and Zhang 2011). Legislation, protocols and quality assurance schemes perform different functions, but have in common that they all require compliance information to be recorded. The ability to collect this information and to use it to ensure product quality in 'real time' provides tangible benefits to the cereal supply chain. The latter can be especially complex since many different processing steps are taken in order to turn the coarse grain into a large spectrum of value-added products, ranging from meals to baby foods and pet foods, but also as a constituent of a large variety of other goods, such as beverages, drugs, hydrocolloids, biofuels, etc. Examples of the widespread food products deriving from cereal processing include: bread and pasta generally made, respectively, from wheat and durum wheat, couscous (a very fine grain cereal made from wheat), flour rice, crispy rice for breakfast, parboiled rice, corn meal, porridge, biscuits, snacks and many other derived products. For instance, the germ part of corn can also be refined to make a valuable vegetable oil or as a key ingredient in some margarines (Proto 1988). It is important to highlight that from the cereal cultivation stage, besides grains, representing the main product, there are many crop residues (such as cereal straw, corn stovers and rice hulls) that have the potential to be used as animal feed, bioenergy feed stocks and raw materials for a large number of both traditional and innovative industrial sectors.

4.1.2 Key Sustainability Aspects Associated with the Cereal Sector: Rice, Wheat, Corn and Derived Products

Generally, the first step in the cereal supply chain—after cultivation practices, harvesting, storage and transportation of cereal grains—is milling, a primary process that turns grain into flour. The cereal grain is composed of an embryo (the germ of the new plant), an endosperm (the starchy fraction), which accounts for about 80% of the bulk of grain, and a protective layer of the seed coat (the bran fraction). Milling is one of the most important steps for the cereal food sector, from which many kinds of white flour and several by-products derive. The flour obtained from cereals such as wheat, corn and rice is the main raw material for the production of a wide range of products, the processes of which are obviously very different.

The increased industrialisation of the food system has been accompanied by rapid integration of the various links in the cereal supply chain (Reardon and Timmer 2012) and among these transport and logistics have a relevant role.

Cereal supply chains generally span long distances and consequently require extensive use of fossil fuels to deliver goods to customers (wholesalers, retailers, final consumers). The global energy use and related atmospheric emissions of a cereal supply chain depend principally on the characteristics of the places of agricultural production, on the storage locations and on the markets to reach. In the traditional cereal systems, and overall in developing countries, consumers buy certain types of food (bulk grains, etc.) from small independent retailers (open markets, small shops, etc.), and others (processed and packaged foods) prevalently in stores and supermarkets. Conversely, a modern supply chain for cereals and derived products is a managed process, based on a combination of knowledge and skills, spanning biology and the social sciences of economics and laws, engineering and human behaviour, and more. All are set in an integrated framework that in the past was often based upon a vertical model whereby upstream and downstream activities were managed by one organization (FAO 2013a). However, the commodity nature of cereal grains promotes relationships that are transactional, in which parties are not even interested in establishing a close, long-term supply chain relationship.

As regards the supply chain of cereals and cereal-based products, the developed countries landscape is characterised by enormous complexity and it is better outlined as 'supply networks', in which some issues related to sustainability (for instance, climate change, resource depletion, food wastage) have a significant impact on the projected food supply and security.

Among the multiple paths to improving the production, food security and overall social and environmental performance of food, and particularly of the cereal sector, the minimisation of food loss (wastage) and waste appears to be a pivotal issue.

As mentioned in the first chapter, food is wasted throughout the food supply chain, from the initial agricultural production down to the final household consumption. The food losses in industrialised countries are as high as those in developing countries, but in the latter more than 40 % of the food losses occur at the post-harvest and processing levels, while in industrialised countries, more than 40 % of the food losses occur at the retail and consumer levels (Lipinski et al. 2013).

The FAO (2013a) estimates that each year roughly one-third in weight of all food produced for human consumption in the world is lost or wasted. In total, 54 % of the world's food wastage occurs 'upstream' during production, post-harvest handling and storage, while 46 % occurs 'downstream', at the processing, distribution and consumption stages. In terms of the measured calories of the various wasted foods, cereals are the largest source of wastage, representing more than half of the total (Fig. 4.3).

In particular, for cereals, wheat is the dominant crop supply in medium-and high-income countries, and the consumer phase involves the largest losses, which range from 40 to 50 % of the total cereal food waste. In low-income regions, rice is the dominant crop, especially in the highly populated region of South and Southeast Asia, where agricultural production and post-harvest handling and storage are stages in the food supply chain with relatively high food losses, as opposed to the distribution and consumption levels.

The worldwide cereal wastage is a twofold dimension: the first part is undoubtedly linked to ethical issues, because of the pervasive poverty of many people on the planet, and the second aspect is related to the need to avoid numerous environmental impacts (water pollution, atmospheric emissions, waste, etc.) deriving from the wastage in the multiple steps of supply chains. Indeed, cereal wastage represents

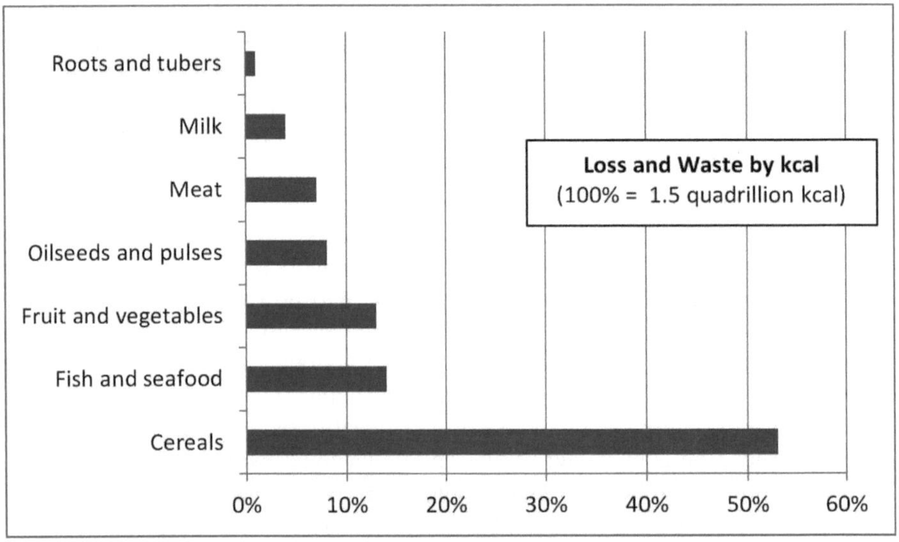

Fig. 4.3 The world's food wastage (2009). (Source: Adapted from FAO (2011b))

the main missed opportunity to improve global food security, on one hand, and to mitigate the environmental impacts and to implement more efficient resource use, on the other.

For the most important cereals (wheat, rice, corn), due the nature of the cereal-based product supply chains, it is difficult to assess the multitude of sustainability aspects in a systematic and coherent framework. Over the last decades, many studies have focused their analysis on specific practices in different phases of the supply chain tailored to a particular geographical context and often related to a single impact category. Many of these studies refer to the agricultural practices, which, as detailed in the following sections of the chapter, are often the most environmentally burdening phases of cereal-related supply chains. The use of fertilisers and pesticides is often responsible for such burdens, together with issues related to the use of the ever-decreasing available agricultural land (especially in Europe) and fossil fuel consumption. The wheat, rice and corn cultivation in the world accounts at present for some 60 % of global fertiliser use, and is expected still to account for just over half of fertiliser consumption by 2050 (Place and Meybeck 2013). Furthermore, agricultural practices are strongly site-specific, hence the impact deriving from the choice of a particular crop and the consumption of water and other resources are largely dependent on the characteristics of the production area.

It is clear that, in order to meet the increasing world future food demand, the cereal sector will increasingly face greater uncertainty and risks, both natural and economic. The first are linked to environmental damage resulting from the agricultural production system (externalities, such as biodiversity, soil loss, land degradation, GHG emissions, water pollution and solid waste production).

The economic challenges encompass principally the price volatility of both inputs and outputs.

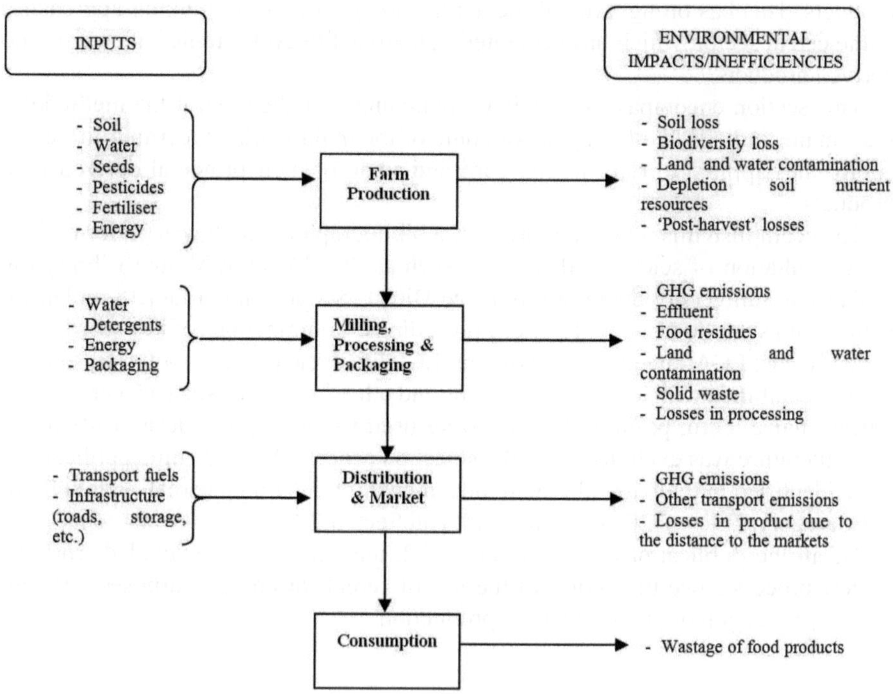

Fig. 4.4 The cereal and cereal-based product supply chain: an environmental perspective

An environmental perspective of the supply chain of cereals and cereal-based products is schematised in Fig. 4.4.

Progressing towards a more sustainable cereal supply chain requires an innovative management approach to enable measuring, assessment and monitoring capable of creating more efficient use of resources at every life cycle stage, from the farm to the consumer level. This necessary systemic vision, able to maximise the supply chain global performance, is the core concept of the life cycle thinking (LCT) framework. Taking an LCT perspective requires the development of new knowledge to look beyond the traditional vision in order to prioritise and set sustainability targets, improving the top-down and bottom-up cooperation along the supply chain.

4.2 Life Cycle Thinking Approaches Applied to the Production of Cereals and Derived Products: The State of the Art

As mentioned in the first chapter of the book, the sustainability of food products, including cereal-based ones, has become a main concern since a large part of the environmental burdens deriving from private consumption is attributable to such products. This has brought about the application of life cycle thinking approaches to the cereal sector, which has generated numerous life cycle studies of cereals and derived products.

This section encompasses a review, containing an indication of the methodologies, main findings and hot spots, of some of the main work concerning life cycle approaches applied to rice, wheat, corn and some of their principal derived food products.

To accomplish this reviewing process, a bibliographic search was performed via the consultation of scientific databases, such as the CASPUR Virtual Library (an Italian inter-university database), Science Direct, Scopus and Google Scholar, together with specific LCA conference proceedings. Combinations of key words such as 'life cycle, LCA, life cycle assessment, LCI, life cycle costing, footprint, sustainability' combined with logical expressions and other key words such as 'cereal, rice, wheat, maize, corn, pasta and bread' were used to identify the desired literature. Grey literature was excluded from the selection process. Seventy-nine publications were identified in total, together with documentation regarding cereal-related Product Category Rules (PCR) (see Table 4.2 and Sect. 4.2.6).

Of all the publications listed in Table 4.2, ten papers were excluded from the review process since they concern the use of cereal for energy purposes, thermoplastic production or other non-food production.

4.2.1 Classification of the Reviewed Life Cycle Thinking Approaches Applied to the Cereal Sector

The first studies regarding the application of life cycle approaches to the cereal sector date back to the beginning of the twenty-first century (e.g. Braschkat et al. 2003; Notarnicola and Nicoletti 2001; Petti et al. 2000). Since then, cereals have been studied intensively, in terms of their sustainability, via life cycle methodologies, confirming the topic as one of the major subjects of the last two international conferences on LCA in the agri-food sector held in Bari, Italy, in 2010 (Notarnicola et al. 2012a) and Saint Maló, France, in 2012 (van der Werf et al. 2013).

Specifically, such life cycle approaches have been extensively applied to the three most important cereals produced worldwide (rice, corn and wheat). As mentioned in the previous section, 69 studies, including LCAs regarding these three cereals, were reviewed: 65 % of them were published in scientific journals between 2000 and December 2013; 29 % of them are proceedings from conferences and workshops; and 5 % of them are research and/or project reports. The relevance of each paper to a specific cereal or derived product is illustrated in Fig. 4.5: rice (33 %), corn (11 %), wheat (38 %) and its derivatives (bread, 9 % and pasta, 9 %). Even if this review is certainly not exhaustive, it is undoubtedly a sound representation of life cycle studies in the cereal sector.

The scope of the reviewed studies can be broadly classified into four groups, which are summarized below.

Table 4.2 List of all the identified life cycle publications (prior to final screening and selection for review) concerning cereals or derived products

Reference	Main purpose	Methodology	Functional unit
Petti et al. (2000)	Applicative case study on pasta production	LCA	1 kg of pasta in primary and secondary packaging delivered to the company's direct customers
Notarnicola and Nicoletti (2001)	Comparative case study on pasta and couscous production	LCA	Quantity of product needed to produce a meal for 4 people
Kim and Dale (2002)	Applicative study of ethanol production from corn	LCA	1 kg of ethanol
Braschkat et al. (2003)	Comparative case study on bread with various cultivation methods	LCA	1 kg of bread ready for consumption
Brentrup et al. (2004)	Applicative case study on wheat	LCA	1 t of wheat grain
Narayanaswamy et al. (2004)	Applicative case study on bread, beer and canola cooking oil	LCA	Loaf of bread, 1 hL beer, 1 L cooking oil
Notarnicola et al. (2004)	Hybrid LCA approach applied to pasta production	LCA-IO	1 kg of pasta
Salomone and Ciraolo (2004)	Applicative case study on pasta production	LCA	1 t of dry packaged pasta
Breiling et al. (2005)	Rice-related GHG	Top-down approach CF	n.a.
Roy et al. (2005)	LCI rice processing	LCI	1 t of rice
Spatari et al. (2005)	Corn-derived ethanol automobiles	LCA	n.a.
Charles et al. (2006)	Applicative case study on wheat production system for bread making performed to optimise fertilisation	LCA	ha; 1 t of grain; 1 t of grain with constant quality (13 % of protein in dry grain)
Bevilacqua et al. (2007)	Applicative case study on pasta production	LCA	0.5 kg of packaged durum wheat pasta sold in the Italian market
Harada et al. (2007)	GHG of conventional puddling, no-puddling and no-tilling rice cultivation	CF	60 m^2 of land
Roy et al. (2007)	LCA+cost assessment of rice in Bangladesh	LCA	1 t of rice
Kim and Dale (2008a)	Effects of N fertiliser GHG from corn cultivation		1 kg of dry corn grain
Kim and Dale (2008b)	Ethanol derived from corn grain via dry milling	LCA	1 kg of ethanol

Table 4.2 (continued)

Reference	Main purpose	Methodology	Functional unit
Nemececk et al. (2008)	Comparison of cereal crop rotation with and without grain legume	LCA	Land, economic and energy-specific FUs
Pelletier et al. (2008)	Assessment of generic life cycle models of contemporary conventional and organic cultivation of cereals and other crops	LCA	1 t of each crop
Renouf et al. (2008)	Comparison of sugar production from corn sugarcane and beet	LCA	1 kg of monosaccharide
Schmidt (2008)	Illustrative framework for defining system boundaries in consequential agricultural LCA	LCA	1 kg of wheat demanded in Denmark
Yossapol and Nadsataporn (2008)	Applicative case study on rice production in Thailand	LCA	1 t of unmilled rice grain
Aldaya and Hoekstra (2010)	WF due to Italian consumption of pizza and pasta	WF	1 kg pasta, 725 g pizza
Blengini and Busto (2009)	Applicative case study of alternative agri-food chain management systems in northern Italy	LCA	1 kg of refined and packed rice
Kasmaprapruet et al. (2009)	Applicative case study on milled rice production in Thailand	LCA	1 kg of milled rice
Meisterling et al. (2009)	Case study of organic and conventional wheat	CF	0.67 kg of wheat flour
Roy et al. (2009a)	Review of LCA on some foods	LCA	n.a.
Roy et al. (2009b)	LCI of different forms of rice consumed in households in Japan	LCI	1 kg of final product or 1 MJ energy supplied by the final product
Kim et al. (2009)	Case study of corn grain and corn stover for ethanol in the US		1 kg of dry biomass
Vidal et al. (2009)	Case study of thermoplastic production from rice husk and other materials	LCA	1 kg of thermoplastics
Aldaya et al. (2010)	Case study of water footprint of wheat and rice (and cotton) in Asia	WF	m^3/ha
Biswas et al. (2010)	Comparative study on greenhouse gas emission due to the production of Australian wheat, meat and wool	LCA	1 kg of wheat
Hayashi et al. (2010)	LCI inventory DB study for crop production in Japan	LCI/IO	n.a.

Table 4.2 (continued)

Reference	Main purpose	Methodology	Functional unit
Kägi et al. (2010)	GHG rice—study on the use of confidence limits for communicating LCA results	CF	1 kg of rice
Khoo et al. (2010)	Calculation of GWP of beef, chicken, tofu, rice and tomatoes delivered to Singapore	LCA	1 kg of protein per food
Kissinger and Gottliebb (2010)	Profile of the environmental burden of production of grain in Israel	EF	1 t of grain
Kløverpris et al. (2010)	Methodological case study on wheat, with a focus on land use	LCI	1 t of wheat
Ruini and Marino (2010)	Applicative case study on durum wheat production	CF, EF, WF	kg of wheat considering the specific yield
Ruini et al. (2010)	Applicative case study on pasta production	EF	1 t of product
Seda et al. (2010)	Environmental assessment of different crops using different FUs	LCA	land area, yield end economic FUs
Settanni et al. (2010b)	Applicative study of a costing model to pasta	LCC	0.5 kg of pasta
Yoshikawa et al. (2010)	Case study of the CF of ecologically cultivated rice using sampling survey theory	CF	4 kg of polished rice
Wang et al. (2010)	Applicative study of rice production in China	LCA	1 t of rice
Williams et al. (2010)	Applicative and comparative study of wheat bread and other products	LCA	1 t of product
Berthoud et al. (2011)	Case study of freshwater eco-toxicity assessment due to wheat production	LCA (partial)	1 kg of wheat; 1 ha of land
Chapagain and Hoekstra (2011)	Applicative study regarding the WF of rice	WF	1 t of paddy rice
Espinoza-Orias et al. (2011)	CF of bread consumed in UK	CF	800 g loaf of bread
Ferng (2011)	Case study footprints on consumption of wheat and rice in Taiwan	EF	Hectare of land
Gan et al. (2011a)	Case study on field crop cultivation in semi-arid regions with crop rotation	CF	1 kg of wheat

Table 4.2 (continued)

Reference	Main purpose	Methodology	Functional unit
Gan et al. (2011b)	Case study on wheat cultivation with crop rotation in Canada	CF	1 kg of wheat
Lo Giudice et al. (2011)	Case study on Sicilian pasta production	LCI	1 kg of pasta
Nalley et al. (2011)	Comparative study on different agricultural techniques relative to corn, rice, wheat and other products	CF	1 acre of land
Prasara and Grant (2011)	Comparative study for energy production from rice husk	LCA	1000 t of rice husk
Röös et al. (2011)	Applicative case study on pasta production	CF	1 kg of pasta in paper packaging
Ruini (2011)	Environmental assessment and labelling of a multinational company	LCA, CF, WF	1 kg of product
Li Borrion et al. (2012)	Bioethanol production from wheat straw	LCA	1 kg of ethanol
Brankatschk and Finkbeiner (2012)	Comparison of different allocation methods for different agricultural products	LCA	n.a.
Van Stappen et al. (2012)	Scenario evaluation for food and non-food uses of cereal	LCA	Any useful output per hectare in an average year
Drocourt et al. (2012)	Applicative case study on rice production in a French region	LCA	1 t of rice
Fallahpour et al. (2012)	Comparative study of different farming systems for barley and wheat	LCA	1 t of cereal
Hokazono and Hayashi (2012)	Comparison of conventional and organic rice farming in Japan	LCA-IO	n.a.
Huang et al. (2012)	Applicative case study on wheat and maize	WF	1 kg of grain
Delivand et al. (2012)	Case study on the use of rice straw for ethanol production in Thailand	CF	1 t of dry rice straw
Kulak et al. (2012)	Case study of environmental improvements for bread production	LCA	1 kg of bread
Laurent et al. (2012)	Availability of data in existing LCI databases regarding cereals	LCA	n.a.

Table 4.2 (continued)

Reference	Main purpose	Methodology	Functional unit
Malça and Freire (2012)	Land use issues regarding wheat-based bioethanol	CF	1 MJ of fuel energy content
McConkey et al. (2012)	Applicative case study on wheat and maize	WF	1 t of grain
Muñoz et al. (2012)	Comparative study of conventional and organic wheat in Chile	LCA	1 t of wheat
Roer et al. (2012)	Applicative case study on wheat, barley and oats	LCA	1 kg of grain
Ruini et al. (2012)	Environmental assessment and labelling of a multinational company	LCA, CF, WF	1 kg of product
Shafie et al. (2012)	Case study of electricity generation in Malaysia from rice husk	LCA	1.5 MWh of electricity
Yoshikawa et al. (2012)	Comparison of different fertilisation techniques for rice production in Japan	CF	1 kg of rice
Eshun et al. (2013)	Case study of GHG emission in Ghana related to rice	CF	Land surface
Murphy and Kendall (2013)	Study of allocation issues for corn and stover production	LCI	1 ha of land
Ruini et al. (2013)	Comparison of different crop rotation systems	CF, WF, EF and others	1 t of durum wheat
Xu et al. (2013)	GHG study related to rice production in Chinese districts	CF	1 t of rice
Yan and Boies (2013)	Case study on uncertainty of GHG for wheat ethanol	CF	n.a.
Yoo et al. (2013)	Case study of water use issues for rice production in Korea	WF	n.a.
Zhang et al. (2013)	Effects of tillage on the soil organic carbon sequestration rate, C footprints and C sequestration in wheat–maize systems	CF	n.a.

LCA life cycle assessment, *LCIA* life cycle impact assessment, *LCI* life cycle inventory, *LCC* life cycle costing, *IO* input output, *CF* carbon footprint, *EF* environmental footprint, *WF* water footprint, n.a. not available

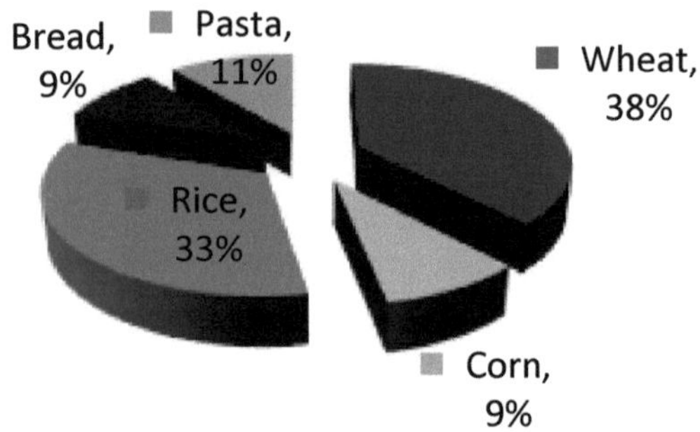

Fig. 4.5 Percentages of reviewed studies regarding crops/derived products

The *first group* of studies comprises papers (approximately 60 % of the total) that profile the environmental burden of a cereal/cereal product or compare different farming practices: Notarnicola and Nicoletti (2001) compared the life cycle of two foods (pasta and couscous); Roy et al. (2005) assessed the LCI of fresh parboiled and fresh rice produced by different production processes; Roy et al. (2009b) evaluated the life cycles of different forms of rice; Lo Giudice et al. (2011) focused on the LCI of pasta, taking into account all the different phases of the productive cycle; Aldaya and Hoekstra (2010) applied water footprint (WF) and LCA to pasta and pizza; Biswas et al. (2010) used LCA with the aim of calculating the GHG emissions for wheat, meat and wool; Röös et al. (2011) applied LCA to wheat and pasta; similarly, Salomone and Ciraolo (2004), Bevilacqua et al. (2007) and Röös et al. (2011) carried out an applicative case study on pasta; Berthoud et al. (2011) used the USEtox model to assess the share of the total freshwater ecotoxicity impact due to pesticide use and to identify active ingredients to replace these high-impact pesticides and estimate the effect of such substitution; Espinoza-Orias et al. (2011) adopted a simplified LCA vs. the carbon footprint (CF) calculation for bread consumed in the UK; Kulak et al. (2012) also studied bread with a focus on possible environmental improvements to its production; Fallahpour et al.'s (2012) study aimed to analyze the impact assessment of wheat and barley; Kasmaprapruet et al. (2009) applied LCA to milled rice; whilst Xu et al. (2013) and Yoo et al. (2013) calculated the carbon and water footprints of rice, respectively, from China and Korea.

This group of studies, which includes the analysis of the environmental impact of different agricultural practices, is very interesting since it provides useful information on the available choices for sustainable farming practices and deserves to be investigated further, as already pointed out by Benedetto et al. (2013). In particular, Braschkat et al. (2003) compared different industrial practices in the supply chain, whilst Brentrup et al. (2004) evaluated different N rates in wheat production. Kim et al. (2009) estimated the county-level environmental performance for continuous cultivation of corn grain and corn stover in various corn-growing locations in the Corn Belt states; two cropping systems were under investigation: corn produced for grain without collecting stover and corn produced for corn grain and corn stover harvesting. Nemececk et al. (2008) compared crop rotation with and without grain legumes; Kim and Dale (2008a) evaluated the global warming effects of N fertiliser application rates in the US using data at the county level; in their study, Meisterling et al. (2009) compared organic and conventional wheat practices; Ruini and Marino (2010), Ruini et al. (2013) compared durum wheat cultivation in two regions with different cropping characteristics and different kinds of rotation; Nalley et al. (2011) compared 57 different farming practices for cotton, rice, sorghum, soybeans and wheat; Hokazono and Hayashi (2012) used a multi-year comparative LCA of agricultural production systems with the aim of identifying the variability in environmental impacts during the conversion from conventional to organic farming; Muñoz et al. (2012) compared conventional and organic wheat crop systems; Kägi et al. (2010) carried out a comparison between conventional, organic and upland rice production; Charles et al. (2006) used an LCA approach for the optimisation of fertiliser use for wheat destined for bread production; Harada et al. (2007) studied GHG emission deriving from conventional puddling, non-puddling and no-tillage rice cultivation; and Gan et al. (2011a, b) studied the possibilities of reducing global warming effects due to wheat cultivation from the diversification of crop rotation. Yoshikawa et al. (2010) calculated the carbon footprint of ecologically cultivated rice in Japan using data from multiple producers in order to ensure the representativeness of the inventory results. In their other carbon footprinting study, Yoshikawa et al. (2012) studied the effects on GHG emissions of different fertilisation techniques for rice production. Finally, Zhang et al. (2013) identified carbon-friendly tillage systems for the North China Plain by evaluating the effects of different types of tillage on the sequestration rate of soil organic carbon for double-cropping cultivation systems based on wheat and maize.

The *second group* of studies focuses on comparing the environmental burden of different food products, e.g. Narayanaswamy et al. (2004) carried out an LCA case study for wheat-to-bread, barley-to-beer and canola-to-cooking oil with the objective of identifying the key environmental impacts along the food chain and assessing the relative contributions of pre-farm and farming to the total life cycle environmental impacts of products produced and consumed in Western Australia; Pelletier et al. (2008) generated a generic LC model of contemporary conventional and organic production systems in Canada in order to predict the 'cradle-to-farm-gate' cumulative energy demand for canola, corn, soy and wheat; Seda et al. (2010) analysed and compared the LCAs of wheat and corn, as well as horticultural crops, using different functional units and suggested the best alternative crop; Khoo et al. (2010) compared beef, chicken, soy-tofu, rice and tomato production; McConkey et al. (2012) applied the WF approach to compare the maize and wheat production processes; the LCA carried out by Renouf et al. (2008) compares different types of sugar production based on corn, sugarcane and beet; and Williams et al. (2010) evaluated the different burdens of bread wheat, oilseed rape and potatoes produced in various parts of the UK.

The *third group* includes studies that have adopted differing approaches to LCA or methodologies used in combination with LCA, or even different methodologies for the assessment of environmental impacts. It includes: Notarnicola et al. (2004), who applied LCA and IO-LCA to pasta production; Breiling et al. (2005), who used IO-LCA referred to GHG emissions for rice; Roy et al. (2007), who used LCA and cost assessment to determine the environmental load and production cost of rice in Bangladesh; Aldaya et al. (2010), who calculated a green and a blue WF for wheat, rice and cotton; Settanni et al. (2010b), who applied a novel costing model to pasta and LCC based on IOA to evaluate its consistency with LCA; Chapagain and Hoekstra (2011), who carried out a WF for rice production; Ferng (2011), who evaluated an environmental footprint (EF) in terms of crop land and energy land; Laurent et al. (2012), who analysed the available data in existing LCI databases regarding cereals and cereal products; Van Stappen et al. (2012), who carried out an environmental cereal LCA (attributional and consequential) together with a social LCA; Brankatschk and Finkbeiner (2012), who demonstrated the benefit deriving from the use of the cereal unit as a functional unit for a better allocation procedure in LCA studies of agricultural systems; and Murphy and Kendall (2013), who evaluated three different allocation methods for solving the problem of multi-functionality in the case of the production of corn grain.

The *fourth group* of studies concentrates its efforts on profiling the environmental burden of cereal production in a given area or on identifying the environmental hot spots in production systems' performance: e.g. Harada et al. (2007) estimated GHG emissions in northern Japan; Kissinger and Gottliebb (2010) assessed the ecological footprint for grain-based consumption in Israel (wheat, barley, maize); Huang et al. (2012) used WF and LCA for wheat and maize in China's main breadbasket basins; Drocourt et al. (2012) evaluated the environmental assessment of rice production in Camargue; Eshun et al. (2013) estimated the GHG emission and energy impact in rice production systems in Ghana; Schmidt (2008) focused his development of a framework for the definition of system boundaries in consequential LCA on Danish wheat production; and Yossapol and Nadsataporn (2008) carried out an LCA on rice produced in Thailand. Blengini and Busto (2009) applied the LCA methodology to rice production in northern Italy; Ruini and Marino (2010) calculated footprints for wheat productions in the south-west US and southern Italy; in their work, Hayashi et al. (2010) developed LCI data for Japanese crop production; similarly, Kløverpris et al. (2010) developed inventory data for land use deriving from wheat production in Brazil, China, Denmark and the USA; Muñoz et al. (2012) evaluated conventional and organic wheat crop systems in Chile; and Roer et al. (2012) assessed the life cycle environmental impact of grain production in central south-east Norway on a typical grain farm with a mix of barley, oat and spring wheat. The work of Ruini and Marino (2010) is an example of the second investigated issue of this group of studies, and is an application of the EF as a key performance indicator of a large-scale pasta producer.

In general, the LCA case studies have played a key role in supporting decision making in the cereal sector, but some authors highlight hot spots and methodological issues (some of the latter are discussed in detail in Sect. 4.3). For example, Schmidt (2008) presents a framework for defining system boundaries in consequential agricultural LCA. The framework is applied to an illustrative case study, i.e. the LCA of increased demand for wheat in Denmark. Different scenarios for meeting the increased demand for wheat show significant differences in emission levels as well as land use. The comparison of scenarios shows significant differences in the contribution to the included impact categories (climate change, eutrophication and land use). Therefore, the modelling of how increased demand can be met in an LCA appears to be crucial for the outcome of any study involving cereals.

Furthermore, most life cycle inventory data for crops do not include the ultimate (marginal) land use induced by crop consumption. Land use and land use change are usually considered at the inventory level in terms of land occupation and land transformation. Kløverpris et al. (2010) present, document and discuss a method that addresses this problem via its application to wheat consumption in Brazil, China, Denmark and the USA. The analysis shows that a combination of economic modelling, geographical data and agricultural statistics can resolve some of the obstacles to identifying ultimate or marginal land use changes when applying consequential LCA to crop production such as wheat.

Of all the LCA studies reviewed, only a few include pesticide diffusion models (Berthoud et al. 2011), nutrient balance models (Brentrup et al. 2004; Charles et al. 2006; Seda et al. 2010; Williams et al. 2010; Yoshikawa et al. 2012) and carbon storage accounting (Roer et al. 2012; Yoshikawa et al. 2012). On the other hand, the majority of the studies consider the emissions associated with fertiliser use (Berthoud et al. 2011; Blengini and Busto 2009; Brentrup et al. 2004; Charles et al. 2006; Espinoza-Orias et al. 2011; Fallahpour et al. 2012; Hokazono and Hayashi 2012; Kim et al. 2009; Meisterling et al. 2009; Muñoz et al. 2012; Murphy and Kendall 2013; Narayanaswamy et al. 2004; Nemececk et al. 2008; Pelletier et al. 2008; Renouf et al. 2008; Roer et al. 2012; Schmidt 2008; Seda et al. 2010; Yoshikawa et al. 2012). The most frequently applied method for calculating fertiliser emissions is the IPCC model, even though in some cases other methodologies were used, e.g. the DNDC model (Yoshikawa et al. 2012).

The most common hot spots identified when assessing the agricultural activities are fertiliser and pesticide production and use and fuel-related emissions (e.g. Braschkat et al. 2003; Roer et al. 2012; Williams et al. 2010). According to some authors (Braschkat et al. 2003; Pelletier et al. 2008), the adoption of organic cropping systems could improve the environmental profile of agricultural activities by lowering the overall impact, even though, as Blengini and Busto (2009) state, the lower grain yields obtained with organic systems could cancel this benefit.

4.2.2 LCA of Cereal Product Systems

Since over 78 % of the identified studies involve the implementation of classical environmental LCAs of cereal systems, involving multiple impact categories, the remaining part of this section reviews such LCA work regarding corn (Sect. 4.2.2.1), rice (Sect 4.2.2.2), wheat (Sect 4.2.2.3) and wheat-derived products (Sect 4.2.2.4). Following life cycle costing studies, simplified and hybrid LCA studies and footprints are discussed in detail in Sects. 4.2.4 and 4.2.5. Finally, Sect. 4.2.6 discusses cereal-related EPD labels and the relative PCRs.

4.2.2.1 Corn

Corn is an annual herbaceous plant that is widely cultivated throughout the world: over 170 million hectares are dedicated to corn cultivation (FAO 2013b). The United States produces 40 % of the world's harvest (273,832,130 t in 2012, according to the FAO), followed by China (208,130,000 t). This cereal is used both as human food and livestock feed and as a feedstock for the production of ethanol fuel; according to the RFA (2010), most ethanol produced in the United States is derived from corn grain.

Most of the studies regarding corn from a life cycle perspective consider this cereal as a feedstock for biodiesel production (e.g. Kim and Dale 2002, 2008b; Spatari et al. 2005). Only a few studies with the objective of evaluating the environmental performance of corn cultivation were identified and are discussed in this section.

Specifically, the major aims of the articles under study differed: while the study by Murphy and Kendall (2013) focused on the life cycle inventory for corn production, the goal of the study carried out by Kim et al. (2009) was to estimate the environmental performance of corn cultivation in various corn-growing locations in the Corn Belt states. Nalley et al. (2011) performed an environmental assessment of six of the largest row crops (among which is corn) produced in Arkansas, taking into consideration only the GHG emissions. Pelletier et al. (2008) compared different conventional and organic crops, including corn, in Canada. The inventory data used referred to the average agricultural practice specific to the area under study. All the studies under analysis used a cradle-to-farm-gate perspective.

As for the functional unit, both area units (i.e. 1 ha of corn and stover production (Murphy and Kendall 2013)) and mass-based units, such as 1 kg of dry biomass (Kim et al. 2009) or 1 kg of corn (Pelletier et al. 2008), were selected.

An important issue associated with LCA-oriented studies for multifunctional processes consists of the most appropriate choice of the allocation approach. In fact, as Murphy and Kendall (2013) demonstrate, the allocation method selected can heavily affect the results of the analysis. Along with corn grain, stover is also produced; as a by-product, it can be left on the field to maintain the soil condition, collected to be used as cattle fodder or harvested for biofuel production (Murphy and Kendall 2013). Different allocation approaches were performed in the studies analysed: no allocation, when corn stover is not collected (Kim et al. 2009), system expansion, when both corn grain and stover are harvested (Kim et al. 2009), and energy-based, economic allocation and subdivision, for the three different scenarios assessed by Murphy and Kendall (2013). The aim of this last study was in fact to explore these three allocation strategies for corn and stover, pointing out the advantages and disadvantages of each of them. The authors stated that 'value-based allocation methods, like energy and economic allocation, may be most appropriate when they reflect the goals of the production system. In addition, value-based methods are typically simple to apply thus may be more transparent'.

Another important issue when dealing with the environmental assessment of agricultural activities is represented by field emissions. Different models are available for estimating these emissions. Murphy and Kendall (2013) as well Pelletier et al. (2008) estimated N emissions from fertilisers according to the IPCC methodology

(IPCC 2006), while Kim et al. (2009) applied the DAYCENT model (Natural Resource Ecology Laboratory 2005), which is the daily time step version of the CENTURY model, a multi-compartmental ecosystem model. This model was also used to predict the carbon sequestration by soil.

The following impact categories were taken into account when assessing the environmental performance of corn cultivation: climate change, acidification, eutrophication, fossil energy (Kim et al. 2009; Murphy and Kendall 2013; Pelletier et al. 2008) and ozone layer depletion (Pelletier et al. 2008).

Regarding the most impacting materials identified, the studies under analysis show consistent results: the production and use of fertilisers generally dominate the total GHG and fossil energy impacts in conventional cropping systems. Pelletier et al. (2008) show that, for organic crop systems, the second major contributor after fertiliser emissions is fuel use.

According to Pelletier et al. (2008), the choice of the cultivation system (conventional or organic) affects the environmental performance of corn production. In their study, they show in fact that the organic crop production models generated consistently lower contributions to all the impact categories: this reduction was mainly due to the substitution of conventional nitrogen fertilisers with green manure.

4.2.2.2 Rice

Rice production is the second-largest cereal production worldwide, but in terms of dietary intake, rice is first in the world ranking, as the bulk of the world rice production is destined for food use, although some is used in domestic animal feeding. Rice is the primary staple for more than half the world's population, with Asia representing the largest producing and consuming region. In recent years, rice has also become an important staple food throughout Africa (FAO 2013b).

From an LCA perspective, most LCA studies on rice have Asia as the geographical boundary, i.e. Japan (Harada et al. 2007; Hokazono and Hayashi 2012; Roy et al. 2009b; Yoshikawa et al. 2010, 2009), Bangladesh (Roy et al. 2005, 2007), Thailand (Kasmaprapruet et al. 2009; Yossapol and Nadsataporn 2008) and China (Wang et al. 2010), followed by Europe, namely France (Drocourt et al. 2012) and Italy (Blengini and Busto 2009).

Some of these LCA studies are limited to the life cycle inventory level (Roy et al. 2005, 2009b); meanwhile, others report only the GHG emissions (Harada et al. 2007; Roy et al. 2009b).

The cultivation phase of rice emerged as the hot spot in the life cycle of rice (Kasmaprapruet et al. 2009). Rice is present in many varieties, e.g. brown, partially milled, well-milled, germinated brown and parboiled (i.e. rice that has been boiled in the husk), which differ in their production process and therefore also in their environmental impacts, as already discussed by Roy et al. (2009a) in their review paper of LCA of food products. When parboiled rice was compared with non-parboiled rice, the latter showed lower environmental loads (Roy et al. 2005), whilst the partially milled rice (milling 2%) was found to be the most environmentally friendly rice by Roy et al. (2009b). Apart from the reduction in the environmental impacts, the choice of different types of rice has implications for their nutritional context, e.g. the partially milled rice leads to the retention of some of the nutrients that are beneficial to human health (Roy et al. 2009b), and parboiling improves the milling yield, storability and nutritional content (Roy et al. 2007). However, this kind of comparison can be misleading as the different varieties have different tastes and require different amounts of water during cooking. As mentioned by Roy et al. (2009b), to gain a certain amount of energy from cooked rice, greater amounts of parboiled rice need to be consumed compared with the well-milled rice because of the higher volume expansion ratio. The environmental load is dependent not only on the form of the rice but also on the packaging used; as shown by Roy et al. (2009b), the paper bag packaging option seems to be preferable to the polyethylene bag option. Further differences are connected with the cultivation techniques, e.g. organic farming or upland farming (upland rice is rice cultivated without submersion and grown under a reduced water regime). The study performed by Blengini and Busto (2009) shows that organic and upland farming have the potential to decrease the impact per unit of cultivated area. However, due to the lower grain yields, the environmental benefits per kg of the final products are greatly reduced in the case of upland rice production and almost cancelled for organic rice. The comparison between conventional and organic farming is a delicate issue, the results of which can be biased by the assumption that the year-to-year variations in agricultural production are negligible. However, as demonstrated by Hokazono and Hayashi (2012), it is necessary to investigate the variability in environmental impacts during the conversion period, because the performance of organic farming in the conversion process from conventional farming is unstable. Although the environmental impacts of organic rice production are higher than those of conventional rice production on average, they decrease to the same level as conventional farming in the last phase of conversion. Other options to decrease the environmental impacts of rice production refer to the use of alternative types of fertiliser. Yoshikawa et al. (2012) compared two types of cultivation: reduced chemical fertiliser use and green manure use. The results show that the utilisation of green manure reduces the production costs and the impact due to energy consumption and eutrophication, though it increases the farmer's labour time and GHG emissions. Furthermore, in order to reduce the total impact of rice production, improved water management would provide a significant benefit for green manure use.

One peculiar aspect of the rice life cycle is connected with water management practices, mainly due to long submersion times, which lead to the anaerobic decomposition of organic matter and the consequent methane production, which determine the GHG emissions (Blengini and Busto 2009; Drocourt et al. 2012; Harada et al. 2007).

4.2.2.3 Wheat

The LCA studies performed on wheat can be distinguished into two main categories: studies addressing wheat as cereal, without indication of its final use

(Berthoud et al. 2011; Brentrup et al. 2004; Fallahpour et al. 2012; Roer et al. 2012; Schmidt 2008), and studies of wheat used for bread production (Charles et al. 2006; Meisterling et al. 2009; Williams et al. 2010). One key issue in the LCA of wheat is the assessment of the impact of the fertilization rate on the final results, particularly nitrogen (N) fertiliser (Brentrup et al. 2004; Charles et al. 2006). However, as concluded by Berthoud et al. (2011), pesticides and their effects on the ecosystems should not be neglected, as they strongly contribute to freshwater ecotoxicity impacts.

A further main topic for cereal LCAs—and wheat LCA in particular—is the comparison between different farming techniques, i.e. conventional vs. organic farming (Meisterling et al. 2009) or irrigated vs. rain-fed farming (Fallahpour et al. 2012). In the case of climate change, Meisterling et al. (2009) show that when conventional and organic wheat are transported the same distance to market, the organic wheat system produces less CO_2-eq per functional unit than the conventional wheat system. The shipping distance of the wheat, as well as the transport mode of the finished product, could cancel out or enhance the advantage of the organic wheat. With regard to the irrigation issue, Fallahpour et al. (2012) show that under low consumption of N fertiliser, the environmental impacts of rain-fed wheat are lower than those of irrigated wheat.

When other impact categories are included in the assessment, similar trade-offs can be expected between impact categories, as well as different results according to the different FUs chosen for the assessment. This is the reason why some studies (Charles et al. 2006; Roer et al. 2012) include a sensitivity analysis with different functional units, mainly mass (1 kg of dry matter) or area (1 ha of cultivated land). Defining the functional unit in terms of mass is not always a good measure of the quality of the food produced; the energy (MJ) and protein content (kg) can be of greater interest (Roer et al. 2012). To compare different systems of production managed with different fertilisation intensities, it is necessary to consider both the yield and the quality of the product. Assessment of the wheat production system shows that increased fertilisation needs a sufficient increase in yield to justify the additional emissions (Charles et al. 2006). Finally, in order to reveal the importance of system boundaries, attention should be paid to the inclusion/exclusion of factors in LCA studies, such as machinery manufacturing, buildings, pesticide production and use, humus mineralisation and nitrous oxide loss from the use of mineral fertiliser, as shown by Roer et al. (2012).

4.2.2.4 Wheat Products

Pasta and bread have been the object of various LCA analyses. One key element when assessing these processed products is the system boundary selection: the majority of the studies adopted a cradle-to-grave approach, including all the life cycle phases up to disposal in the analysis (Bevilacqua et al. 2007; Espinoza-Orias et al. 2011; Notarnicola and Nicoletti 2001; Notarnicola et al. 2004; Salomone and Ciraolo 2004). Taking into account the whole life cycle of a product appears to be an important element since, in some cases, the last stages of the product life cycle were found to be not negligible. In fact, besides the cultivation phase, which resulted as being determinant in all the studies carried out on pasta and bread, other stages of the life cycle, such as distribution and use, indicated 'environmental importance'. While for bread the impact of the consumption phase resulted as significant depending on the consumer's behaviour (if bread is refrigerated or toasted) (Espinoza-Orias et al. 2011), the use phase associated with pasta appeared to be relevant in terms of energy consumption and related impacts (Bevilacqua et al. 2007; Ruini et al. 2013). In some studies (i.e. Kulak et al. 2012; Salomone and Ciraolo 2004), the production (pasta production and bread baking) and distribution phases (Kulak et al. 2012) were also found to be critical.

A comparative approach was used in different studies: Notarnicola and Nicoletti (2001) assessed two different wheat-derived products (pasta and couscous), Bevilacqua et al. (2007) compared the results obtained with alternative production systems designed to reduce the environmental impact of pasta (conventional vs. organic crop systems, plastic vs. cardboard packaging), while Braschkat et al. (2003) analysed eight different scenarios for bread production. When considering the use of organic wheat for pasta or bread production, lower impacts were obtained, but more land area was required (Braschkat et al. 2003). Different milling and baking technologies were also assessed by Braschkat et al. (2003), revealing that home-made bread has a greater overall impact when compared with industrial bread.

All the functional units selected are based on mass, i.e. 1 kg of bread (Braschkat et al. 2003; Kulak et al. 2012), 1 kg of packaged pasta or the amount of pasta needed to prepare four portions (Notarnicola and Nicoletti 2001).

The typical impact categories taken into account in the studies regarding wheat products are global warming, acidification, eutrophication, ozone layer depletion, eco-toxicity and abiotic depletion. In some of the studies related to pasta, the normalisation phase of the results was carried out. The normalised results reveal the most affected impact categories, i.e. land use and fossil fuel, followed by respiratory inorganics and climate change, according to Bevilacqua et al. (2007).

4.2.3 Life Cycle Costing (LCC)

Evaluating the costs of a product system from a life cycle perspective is a task performed with the general intention of evaluating possible new investments in a supply chain (SC) or with the aim of optimising the existing resources and reducing the costs along the whole SC, including the consumption and end-of-life phases. Since the underlying framework of LCC is similar to that of LCA, such costing methods are also implemented as a means of evaluating the environmental costs of a product system or the most cost-effective method of making environmental improvements to it. In this respect, typically, LCC approaches combine some discounted cash flow analysis with LCA. However, in order to apply LCC effectively in the same holistic manner as LCA to the entire life cycle of non-durable products such as cereal products, input–output analysis (IOA) based approaches (Settanni et al. 2010a) need to be implemented from a microeconomic point of view. The application of such approaches allows, from a supply chain management (SCM) perspective, both the evaluation of the economic performance of an SC and the inclusion of considerations for environmental concerns. This input–output cost accounting methodology, fundamentally different from typical costing activities, has not yet become as mainstream as pure environmental LCA, and is thus still an object of academic research.

In Settanni et al. (2010b), a novel IOA costing model is applied to the fresh pasta supply chain. Here the authors address the problem of representing a southern Italian pasta factory as a series of interacting processes, including those with the suppliers and customers, and then transforming such relationships into financial transactions via matrix operations. This kind of modelling involving reciprocal interdependences, in terms of interconnected material flows, among the processes of an SC allows the management of inter-organisational cost issues. Specifically, such an approach permits the assessment of the activity levels along the pasta SC, together with the expected required resources, the related environmental burdens, and, subsequently, the internal production costs along such an SC. Even though the purpose of the paper is that of demonstrating the effectiveness of the approach, which in this case, for simplification purposes, excludes the agricultural stage, direct process and unit product costs were obtained, for the various types of pasta produced, and were simultaneously combined with the relative environmental burdens calculated with an LCA based on the same inventory data structure. Furthermore, the model is able to indicate the costs of inefficiencies along the different stages of the production process. To increase the robustness of the method, a non-deterministic analysis was performed via the use of uncertainties related to the main technical–economical parameters used for the study; this allowed the unit costs to be turned into calculated probability distributions.

The reviewed cereal based literature also encompasses work concerning costing activities, such as for example Roy et al. (2007), carried out in parallel with an LCA study, without a common integrated life cycle framework, using standalone calculations. In this case, the authors used the LCA methodology to determine the environmental load of different rice production processes in Bangladesh together with estimating the production cost of the rice in order to aid the decision-making processes employed for the identification of potential improvements of a product, a process or an activity.

Specifically, the production costs of milled and head rice were calculated, both per unit mass and energy. The results indicate that the production cost of untreated rice (per unit mass or energy) is higher than that of parboiled rice for the head rice option. However, the production cost of parboiled rice is found to be higher than the untreated rice for the milled rice option (probably due to the difference in rice yield and energy consumption in the production processes). The analysis of the production costs per tonne of rice (ranging from US$ 135.9/t to US$ 145.5/t) indicates that milled rice would be acceptable for the local consumers, in economic terms, whilst the untreated rice would be the best choice for sustainable consumption.

4.2.4 Simplified Life Cycle Assessment (S-LCA) and Hybrid Methods

Guinée et al. (2002) define simplified LCA as: '…a simplified variety of detailed LCA conducted according to guidelines not in full compliance with the ISO 14,040 standards and representative of studies typically requiring from 1 to 20 person-days of work'. Such a type of LCA is usually implemented due to time and/or cost constraints and typically leads to an indication of the main environmental criticalities of the product system analysed as opposed to a reliable quantification of the various burdens occurring during its life cycle. The simplification can occur at the inventory level and/or during the impact assessment phase of the study.

Simplified LCI, when based on process analysis, can be achieved via modelling simplification and data collection simplification strategies. Modelling simplification involves the cut-off of life cycle sub-phases or the removal of smaller elementary flows of the product system, or the modelling of a series of processes as a unique process. Data simplification processes usually involve the use of generalised non-detailed or non-specific databases in order to overcome the difficulties (time requirements, costs, confidential nature) of data collection.

Another approach to the simplification of the inventory phase is IO LCI, which uses IO tables (Suh and Huppes 2005). Such an approach is successful if sufficiently detailed applicable sectorial environmental data exist for the desired country.

LCIA simplification generally involves the exclusion of certain indicators or the aggregation of some of these into a unique new one, for example the cumulative energy demand indicator (Huijbregts et al. 2010).

Hybrid LCA methods can involve the integration of information from IO accounts coupled with process-specific data not as a means of simplification, but rather as a means of avoiding cut-off or truncation errors and hence making the study more complete.

In such a context, Notarnicola et al. (2004) evaluated how the conjunct adoption of a typical LCA approach and an IOA-based one can help improve the inventory set-up of the pasta sector. Furthermore, the study tried to quantify the hybrid approaches in order to improve the overall results. The IO-LCA methods are based on the utilisation of environmental matrices and input–output tables developed in America and in Europe. A comparison of the ISO 14,040 LCA results with those of the IO-LCAs highlights the capability of the IO approach to avoid truncation errors from cradle to gate and the capability of avoiding closed loops. At the same time, the results highlight the problem of gate-to-grave truncation of the IO approaches due to the nature of the input–output tables, of which the European ones seem to be less detailed than the US ones. Hybridising the approach via conjunct adoption of the above-mentioned approaches tended to obtain more detailed results, in particular for the impacts due to fertilisers and pesticides, for which the traditional LCA approach lacks specific data. The authors conclude that, for the pasta case study, IO-LCA approaches should not be used as standalone simplified methods but instead should be used in a hybrid approach with traditional LCAs, keeping in mind that the quantity and quality of the available data will determine the level and the type of adopted combination of approaches.

Similar conclusions are explicated by Hayashi et al. (2010). In this study, when comparing their LCI database regarding crops (including rice—conventional and non-conventional) with those derived from Japanese IO tables, the authors express the need to be cautious when validating results derived from IOA due to the coarseness of such a method for assessing agricultural technologies.

Other IO approaches have also been identified in the literature regarding the life cycle of rice. Specifically, Hokazono and Hayashi (2012), in their study concerning the change in environmental impact during the conversion from conventional to organic rice farming, due to the lack of data regarding duck rice farming, implemented IO analysis, via Japanese input tables, to build the inventories describing such a type of farming. Breiling et al. (2005), when evaluating rice-related greenhouse gases in Japan, used data from IO tables to track primary and secondary CO_2 emissions and detailed the secondary emissions using the IO tables of different Japanese prefectures. The main findings of this work are presented in Sect 4.2.5.

4.2.5 The Carbon, Water and Ecological Footprints

4.2.5.1 The Ecological Footprint (EF)

The ecological footprint (EF) is nowadays one of the most widespread indicators used for assessing the sustainability of humanity's demands on nature. Over the years, the assessment of the EF has been increasingly applied to food products. Regarding cereal production and cereal-derived foods, most of the studies concern wheat and pasta production.

The EF has been used as one environmental assessment method (together with the carbon footprint and water footprint) for the Environmental Product Declaration (EPD) of several cereal-based products from the Barilla company (see also Sect 4.2.6). Barilla, the largest world producer of pasta, was the first private company to develop a system based on the International EPD PCR (Product Category Rules) to certify the results of its LCAs, not only in terms of its carbon footprint, but also in terms of its water and ecological footprints. Ruini and Marino (2010) assessed the EF for wheat cultivation in southern Italy; considering a grain yield of 3.2 t/ha, they assessed the EF of wheat as being equal to 9.5 global m^2/kg: this value is considerably higher than the value assessed for the south-western USA (4.7 global m^2/kg). Ruini and Marino (2010) evaluated the EF of semolina pasta made by Barilla considering a cradle-to-factory-gate boundary. The objectives of this work were to quantify the environmental appropriation of each phase of pasta production, including the phase of final consumption. The total footprint for dry durum semolina pasta is 1.63 global ha/t of the final product (16.2 global m^2/kg) at the platform (this result regards the part of the productive chain from cradle to gate). The largest contribution, 77.6 %, is due to durum wheat cultivation, followed by packaging (14.4 %), while other industrial processes, such as milling, pasta production and transport, usually associated with heavier pressure on ecosystems, are far less land-intensive, covering on the whole only 8 %. In another study, Ruini (2011) calculated for semolina pasta an EF equal to 10.9 global m^2/kg of pasta considering durum wheat cultivation (84.4 %), milling (1.0 %), packaging (5.5 %), pasta production (7.3 %) and distribution (1.8 %). In this study, cooking was excluded from the system boundary, but the authors estimated for this process an EF ranging from 2 global m^2/kg of pasta (when carried out using natural gas) to 6 global m^2/kg of pasta (when carried out using electricity). In the same study, the authors also report an EF for rice equal to 14 global m^2/kg of rice.

According to Cerutti et al. (2012), the agricultural phase accounts for almost 49 % of the whole product EF, the industrial phase (which includes elaboration, packaging and distribution) accounts for 9 % and the consumer phase (which includes the impact of cooking) accounts for 42 % of the whole product's EF.

Ferng (2011) measured Taiwan's rice and wheat consumption footprints in terms of cropland and energy land from 1989 to 2008 and identified the cropland location by source country. The results of this study indicate that Taiwan has continuously enlarged and dispersed the cropland for its rice and wheat consumption footprints in foreign countries, but it has decreased its footprint in domestic territories.

Kissinger and Gottliebb (2010) analysed the ecological footprint of grain-based consumption in the state of Israel during the last two decades. They found that most of Israel's grain footprint falls on North America, followed by the Black Sea region. The study also shows that while the overall consumption of grain products has increased throughout the research period, the size of the footprint has been dropping in recent years as a consequence of changing sources of supply and grain composition.

4.2.5.2 The Carbon Footprint (CF)

Food systems include agricultural phases as well as transport, processing and disposal and are among the main contributors of anthropogenic GHG emissions. Over the years, considering the importance of GHG emissions for the climate change impact category, the need to account for the emissions associated with the agricultural sector has become increasingly relevant. Consequently, the carbon footprint (CF) has become one of the key indicators of environmental sustainability aiming to identify the hot spots and stimulate emission reduction.

With regard to the agricultural sector, during the last 10 years, several studies focusing on the evaluation of the CF of different cereals have been carried out. Cereals represent one of the most important agricultural commodities and their cultivation is widespread worldwide in developed as well as in developing countries. Although the final destination of cereals can differ, their cultivation practices are quite standardised in the different geographical areas and involve significant GHG emissions.

Kim and Dale (2008a) evaluated the impact of nitrogen fertilisation on the CF of maize production in the USA (Corn Belt states). Depending on N application, the

CF ranges from 227 to 518 kg CO_2eq/t and N_2O is responsible for between 31 and 59% of the overall GHG emissions, whilst 63–97 kg CO_2eq per tonne of dry corn grain are associated with nitrogen fertilisation. Biswas et al. (2010) studied the CF of the main products of Australian agriculture. The CF for wheat is 0.40 kg CO_2eq/kg, but soil tillage as well as soil carbon sequestration were not taken into account in this study. Meisterling et al. (2009) assessed the CF of organic and conventional wheat in the USA; lower values are reported for organic agriculture (160 kg CO_2eq/t) than for conventional cultivation (190 kg CO_2eq/t) and N_2O emissions are the main GHG sources. Seed transport for 420 km involves an emission of GHG equal to 30 kg CO_2eq/t.

Ruini and Marino (2010) evaluated the CF for wheat cultivation in southern Italy. The CF is equal to 610 kg CO_2eq/t; the bigger contributions are caused by the production and use of fertilisers causing principally nitrous oxide releases. Diesel use is also an important contributor to the total. The CF of wheat cultivated in different cropping systems was evaluated by Gan et al. (2011b) for semi-arid climatic conditions in Canada. When wheat cultivation is followed by the cultivation of another cereal (on the same land area), the CF is higher (460 kg CO_2eq/t) than crop systems with legumes (200 kg CO_2eq/t) or canola (301 kg CO_2eq/t). The main hot spots identified are the production and application of N fertilisers, which account for about 57–65% of the CF, and crop residue decomposition (16–30%). Besides the choice of different cropping systems, in this study the other strategies and practices evaluated for lowering the CF include an improvement of N use efficiency, the increment of the Harvest Index and the improvement of crop residue management in farming systems. With the correct combination of these strategies, a CF reduction varying from 25 to 34%, depending on the soil conditions, can be achieved. A second study, aimed at evaluating the impact of different cropping systems on the CF of durum wheat, was conducted by Gan et al. (2011a) under the same climatic conditions. The total GHG emissions from the decomposition of crop residues along with various production inputs were used for estimating the CF. On average, the emissions from the decomposition of crop straw and roots accounted for 25% of the CF, those from the production, transportation, storage and delivery of fertilisers and pesticides to farm gates and their applications accounted for 43% of the CF and emissions from farming operations accounted for 32% of the total. Regarding the impact on the CF of different cropping systems, the authors report that durum wheat: (1) preceded in the previous year by an oilseed crop had a CF of 0.33 kg CO_2eq/kg of grain (7% lower than durum in a cereal–cereal–durum system); (2) preceded by a biological N-fixing crop the previous year lowered its CF by 17% compared with durum preceded by a cereal crop; (3) produced in a pulse–pulse–durum system had a CF of 0.27 kg CO_2eq/kg (34% lower than durum grown in cereal–cereal–durum systems). In addition, Ruini et al. (2013) assessed the CF of durum wheat cultivated after different crops. The trend is similar to the one highlighted by Gan et al. (2011a): the CF is higher when the wheat follows another cereal (580 kg CO_2eq/t), while the cultivation of durum wheat after vegetable (405 kg CO_2eq/t) or leguminous cultivation (380 kg CO_2eq/t) contributes to reducing the GHG significantly.

A CF study of bread was carried out by Espinoza-Orias et al. (2011) in the United Kingdom. In particular, the authors assessed the CF for a standard 800 g loaf of sliced bread, made of wheat flour on an industrial scale and consumed at home. Specifically, the CF of bread depends on the thickness of the slices, packaging and types of flour. For example, the CF ranges from 1.11 kg CO_2eq/loaf for wholemeal bread cut into thick slices to 1.24 kg CO_2eq/loaf for white bread cut into medium slices. For bread packaged in plastic bags, the results range from 0.98 kg CO_2eq/loaf for thick-sliced wholemeal bread to 1.10 kg CO_2eq/loaf for medium-sliced white bread.

A CF evaluation for durum wheat semolina dried pasta produced in Italy and packaged in paperboard boxes was carried out by Ruini and Marino (2010); a CF value of 1.284 kg CO_2eq for 500 g of pasta is reported. The main contributions to the CF are cooking (39 %), wheat cultivation (36 %), pasta production (13 %), grain milling (5 %), packaging (4 %) and transport (3 %). Röös et al. (2011) assessed the CF for Swedish pasta; they report, for wheat at the farm gate, CF values varying between 0.25 and 0.47 kg CO_2eq/kg wheat. The mean CF of 1 kg of Swedish pasta is 0.50 kg CO_2eq/kg (0.31 kg CO_2eq/kg wheat before the milling process). The main contributing processes are N_2O from soil (74 %) and fertiliser production and application (21 %).

When compared with wheat, only a few rice studies on CF evaluation have been carried out. Xu et al. (2013) assessed the CF for the five main rice-growing regions in China. In this study, the material and energy consumptions were estimated for these five regions using governmental statistical data, industrial standards and relevant technical data. The CF of rice production ranges from 2504 to 1344.92 kg CO_2eq/t. As possible mitigation strategies, the reduction of urea applications and intermittent irrigation are proposed in the paper. The CF of milled rice produced in Thailand and imported into Singapore was evaluated by Khoo et al. (2010). In this study, the authors compared different food products (beef, tofu, tomatoes and rice) in terms of protein content. The CF for milled rice is 219 kg CO_2eq/kg of protein; the methane emissions from paddy fields represent the main emission source of GHG.

Nalley et al. (2011) estimated the GHG emissions of the six largest row crops (corn, cotton, rice, sorghum, soybeans and wheat) produced in Arkansas using 57 different production practices. The CF estimation was carried out using a cradle-to-farm-gate LCA on a county-by-county basis. For rice, the CF value ranges from 2250 kg CO_2/ha (with conventional seeding and cultivation carried out in clay soils) to 2082 kg CO_2/ha (with no tillage and water seeded). For corn, the CF ranges from 640 kg CO_2/ha in furrow clay soil to 533 kg CO_2/ha in loamy soil, while for wheat, the CF shows a higher value (318 kg CO_2/ha) when the cultivation takes place after rice and a lower value (269 kg CO_2/ha) in sandy or silt soil after the cultivation of other cereals. Eshun et al. (2013) assessed a rice CF equal to 477 kg CO_2/ha in Ghana with 'cradle-to-national-retailer' system boundaries.

Furthermore, Yoshikawa et al. (2010, 2012) evaluated the CF of rice in Japan. In the first study, the carbon footprint of ecologically cultivated rice was evaluated. The functional unit in this study is a 4 kg package of polished rice. The system boundary includes raw material production, rice polishing, distribution and retailing, rice cooking and waste treatment. Environmental loads related to durables (agricultural equipment, facilities, cooking equipment, etc.) are not included because of uncertainty regarding their durability. The results show that the carbon footprint of rice is 7.7 kg CO_2eq/package (for 4 kg of polished rice, which amounts to 1.925 kg CO_2eq/kg). About 65 % of the emissions are related to the raw material production stage; almost all the emissions derive from agricultural production. CH_4 emission from paddy fields is caused by anaerobic fermentation and accounts for 50 % of the CF from agricultural production. Besides CH_4 emission, the emission of GHGs from fertiliser, energy and the transportation of input materials each accounted for more than 5 % of the CF in the agricultural phase. In the second study, the assessment was carried out considering two types of cultivation: one with reduced use of chemical fertiliser and another in which green manure is utilised. The assessment is carried out considering 'cradle-to-factory-gate' boundaries and the FU is the mass of white rice. The CF is 2.25 kg CO_2eq/kg for polished rice with cultivation that involves chemical fertiliser reduction (73, 12 and 8 % of the life cycle GHG emissions are due to field emissions—CH_4, N_2O, fertiliser production, and fuel and electricity, respectively), while the CF is equal to 4.89 kg CO_2eq/kg for polished rice when cultivation is carried out using green manure (the CH_4 emissions—2.8 times higher than rice cultivation with mineral fertilisers—represent about 80 % of the global GHG emissions). The impact of different forms of tillage management on the rice CF in Japan was studied by Harada et al. (2007). In particular, scenarios for conventional puddling and no-tilling rice cultivation were compared. The CF from the no-tilling field is 1741 kg CO_2/ha, lower than that from the conventional puddling field. In conclusion, considering that the fuel consumption is also lower, the authors state that no-tilling rice cultivation has the potential to save 1783 kg CO_2/ha from paddy fields. Breiling et al. (2005), for Japan, estimated the CF of rice considering not only direct rice-related GHG emissions but also GHGs hidden in the other categories, primarily energy, industry and waste. The study highlighted that since 1990 the GHG emissions in rice production have been reduced, but this reduction has been offset by the increase in other sectors.

4.2.5.3 The Water Footprint (WF)

Cereal cultivation, in particular maize and rice, involves the consumption of high water volumes. Irrigation is essential to reach good production levels and it can help to stabilise the yields. Moreover, in the case of flooded paddies, water plays an important role in thermoregulation, allowing cultivation in temperate areas characterised by a cold spring as well. For spring wheat cultivation, irrigation is not usually carried out, but water is needed for the processing operations and, in particular, for pasta production. In recent decades, the water availability, for agricultural use as well as for industrial processes, has decreased due to climate change. Consequently, the importance of water footprint (WF) assessment has greatly increased and several studies have been carried out for the evaluation of water consumption during the life cycle of food products, as well as for the identification of the hot spot processes. Generally, for WF assessments for which only the cultivation system is considered ('from-cradle-to-farm-gate' boundary), more attention has been paid to rice than to maize and wheat.

Aldaya et al. (2010) carried out a WF evaluation for rice cultivation in Asia and they obtained a WF equal to 2600 m^3/t in Kazakhstan, 3500 m^3/t in Uzbekistan and 4000 m^3/ha in Tajikistan; lower WF values refer to clay soils and arid conditions. Chapagain and Hoekstra (2011) assessed the WF for the 13 major rice-producing countries. Although an average value of 1325 m^3/t was attained, the study highlights pronounced differences between different countries: rice production in Pakistan shows the highest WF (2874 m^3/t), followed by India (2020 m^3/t) and Thailand (1617 m^3/t), while the lowest values are reported for Vietnam (638 m^3/t), Japan (802 m^3/t) and the USA (829 m^3/t). Besides big differences in WF values, the share of green, blue and grey water also varies greatly; in India, Indonesia, Vietnam, Thailand, Myanmar and the Philippines, the green water fraction is substantially larger than the blue one, whereas in the USA and Pakistan, the blue water footprint is four times higher than the green component. The WF of rice cultivated in Asia has also been evaluated by Yoo et al. (2013). Specifically, this study refers to rice cultivation in South Korea. The WF of rice is 844.5 m^3/t, and green, blue and grey water accounts for 294.5, 501.6 and 48.4 m^3/t, respectively.

McConkey et al. (2012) evaluated the WF for maize and wheat cultivation in Canada. In semi-arid conditions, maize cultivation with irrigation has a WF equal to 3310 m^3/t. In these areas, the spring wheat shows a WF ranging from 4110 m^3/t, with irrigation, to 19,200 m^3/t when cultivation takes place on summer fallow without irrigation. In sub-humid and humid areas, the maize is not irrigated and it shows a WF equal to 5540 m^3/t (sub-humid conditions) and 7290 m^3/t (humid conditions), while the WF values for spring wheat range from 10,500 to 19,600 m^3/t. Generally, the grey water, computed following the Canadian environmental objectives in terms of P concentration, represents about 80 % of the total WF. Huang et al. (2012) compared the WF of wheat and maize in China's main breadbasket basins. The authors report remarkable differences for wheat cultivated in the different regions (from 1262 to 31 m^3/t). The water footprints of maize range from 35 to 515 m^3/t. Ruini and Marino (2010) compared the WF of durum wheat cultivation in Italy with that of other countries. For Italy, the WF values are 450, 920 and 1100 m^3/t, respectively, for northern, central and southern regions. These values are higher than the WF for durum wheat in France (450 m^3/t), but they are lower than the ones obtained for other European countries (Spain 1400 m^3/t, Turkey 1520 m^3/t and Greece 1220 m^3/t) as well as for the northern USA (2230 m^3/t) and Australia (2750 m^3/t). For pasta, the authors report a WF equal to 0.7 m^3 per 500 g of product.

Compared with rice, few wheat WF studies, relative to the cultivation phase, have been carried out, but, unlike the case of rice WFs, some papers analyse in particular detail the processing steps needed for derived products, such as pasta and bread. Aldaya and Hoekstra (2010) analysed the WF related to pasta and pizza margherita. They report a WF equal to 1574 m^3/t for durum wheat (748, 525 and 301 m^3/t, respectively, for green, blue and grey water). For pasta, the durum wheat grains need to be processed into flour. Considering a semolina yield of 72 % (the rest consists of the wheat bran and germ) and that the semolina constitutes 88 % of the total value of the separate products, the authors calculated the WF of semolina to be 1924 m^3/t and that of pasta (assuming that it is made from semolina (1 kg), water (0.5 dm^3) and salt) to be 1924 dm^3/kg. The authors also report a WF of bread wheat of 786 m^3/t and a WF of bread wheat flour of 961 m^3/t. Finally, the reported WF of pizza margherita is 1216 dm^3/kg.

4.2.6 Product Category Rules (PCRs) and Environmental Product Declarations (EPDs)

The European Union (EU) promotes environmental strategies and policies oriented towards the development of a European market characterised by an exchange of greener products. As such, as illustrated in the first chapter, one of the most important actions is the development of life cycle assessment (LCA) based environmental labels, on one hand stimulating producers to improve the environmental performance of their products and on the other allowing consumers, with their choices, to privilege the market for more ecological products.

One of the most interesting types of LCA-based environmental labels is the Environmental Product Declarations (EPDs) and their relative Product Category Rules (PCRs). They are considered complementary to the general requirements of the EPD programmes and they form the basis for third-party verification of LCA studies on the products and the related statements.

The current EDP systems in the agri-food sector, which use the type III programme according to the requirements of the ISO standard 14025:2006, are the International EPD® System (EU), EPD Norge (EU), Earthsure® (USA), the Sustainability Measurement and Reporting System (SMRS) (the USA) and the Ecoleaf environmental label (Japan).

Focusing on the cereal sector, due to its wide variety, the field of application of this chapter is restricted to three main cereals: wheat, rice and corn. Among these, wheat is characterised by a higher number of PCRs and EPDs because, even though it has quite a limited area of production interest in the world, there is a growing market demand in new geographical areas, especially for wheat-derived products like pasta and bread. The following paragraphs illustrate the PCRs published in the section on food products, basic module grain mill products, starches and starch products and other food products, with their related EPDs, with regard to the International EPD® System (Environdec 2014), which is the most widespread scheme among the ones mentioned above:

1. PCR 2010:01 (CPC 2371): Uncooked pasta, not stuffed or otherwise prepared. The EPDs based on this PCR are:
   - developed by the company Lantmännen—Kungsörnen spaghetti, Kungsörnen Macaronis 'Gammaldags Idealmakaroner', Kungsörnen Ideal Macaroni in bulk packs, Kungsörnen pasta in bulk packs, Kungsörnen wholegrain pasta in bulk packs, Kungsörnen white fibre in bulk packs;

  - developed by Barilla—dry semolina pasta from durum wheat EPD, which covers the classic semolina pasta cuts (spaghetti, penne, fusilli, etc.), piccolini (miniatures of classic semolina cuts) and specialità (reginette, orecchiette, ruote, etc.);
  - developed by De Cecco—pasta di semola De Cecco EPD, which includes a traditional range manufactured in cello bags (spaghetti, penne, fusilli, etc.) and the specialities (farfalle, zita, etc.);
  - developed by Sgamboro—Pasta Sgamboro Etichetta Gialla.

2. PCR 2011:07 (CPC 2372): Pasta, cooked, stuffed or otherwise prepared; couscous—for this PCR no EPDs have been actually published;
3. PCR 2012:06 (CPC 234): Bakery products—this PCR incorporates the Product Category Rules 2010:05 (CPC 2349): Bread and other bakers' wares valid until 9 March 2013 and the Product Category Rules 2010:06 (CPC 2343): Pastry goods and cakes valid until 9 March 2013. Several EPDs, based on the abovementioned deregistered PCRs, are valid until the end of the year 2014, and they are not presented because they are not relevant to the research topics. All the products in the CPC Group 234 'Bakery products' are included in this PCR, especially the following classes:
   - CPC 2341: Crispbread; rusks, toasted bread and similar toasted products
   - CPC 2342: Gingerbread and the like; sweet biscuits; waffles and wafers
   - CPC 2343: Pastry goods and cakes, fresh or preserved
   - CPC 2349: Other bread and other bakers' wares

   The EPDs based on this PCR are:
   - developed by Barilla—Mulino Bianco Pan Bauletto Bianco, Mulino Bianco Fette Biscottate dorate, integrali, malto d'orzo e cereali, Muli-no Bianco Tarallucci, Mulino Bianco Girotondi, Mulino Bianco Bat-ticuori, Cracker Gran Pavesi, Ringo Pavesi, Mulino Bianco Flauti, Mulino Bianco Plumcake, Petit Pavesi, Mulino Bianco Pagnotta di Gran Duro, Mulino Bianco PanCarrè, Grancereale classico alla frutta, Harrys American Sandwich Complet, Harrys American Sandwich Nature, Harrys Brioche Tranchée, Harrys Extra Moelleux, Mulino Bianco Granetti, Mulino Bianco Saccottini, Mulino Bianco Michetti, Mulino Bianco Pan Goccioli, Pan di Stelle, Mulino Bianco Cracker salati e non salati;
   - developed by Lantmännen—Kungsörnen pancake with diced pork, Kungsörnen potato pancakes (raggmunk), Lantmännen batter pudding (ugnspannkaka).

4. PCR 2013:04 (CPC 231): Grain mill products—this PCR covers products belonging to the UN CPC Group 231 'Grain mill products' and replaces PCR 2010:03 (CPC 2313): Groats, meal and pellets of wheat and other cereals, which expired on 29 April 2013.

   This group includes the following CPC classes:
   - 2311—Wheat and meslin flour
   - 2312—Other cereal flours
   - 2313—Groats, meal and pellets of wheat and other cereals

- 2314—Other cereal grain products (including cornflakes)
- 2316—Rice, semi-or wholly milled, or husked
- 2317—Other vegetable flours and meals
- 2318—Mixes and doughs for the preparation of bakers' wares

The EPDs based on this PCR (CPC 2313) are:

- developed by the company Lantmännen—Axa oatmeal 'Havre Gryn', Kungsörnen plain flour, Kungsörnen oat berry, Kungsörnen pearled barley, Kungsörnen wheat berry, Kungsörnen pearled barley in bulk packs, Kungsörnen wheat flour with whole grain, Kungsörnen graham flour, Kungsörnen plain flour in bulk packages, Axa oatmeal in bulk packs.

The above-mentioned EPD list highlights the dominance of the Lantmännen and Barilla companies as the main EPD developers. Lantmännen, part of the Lantmännen Cerealia team, develops, manufactures and markets mainly cereal-based products under strong brands such as AXA, Golden Eagle, Home, Gyllenhammar, Gooh, GoGreen, Soups, Amo, Kornkammeret and Regal. The range consists mainly of breakfast foods, flour, flour mixes, pasta, pancakes, beans/lentils and dishes that are sold in grocery stores in northern Europe. In addition, Barilla occupies a representative position in terms of experience in EPD development: it is the first private company to have developed an EPD Process System. This company is one of the top Italian food groups, producing more than 100 products in about 50 plants around the world. The company has been using the LCA methodology for more than a decade. Since 2008, life cycle thinking has made its way into the company strategy, as an instrument to study the production chain thoroughly and localise the most substantial environmental impacts. Moreover, Barilla, at this moment, developing 56 % of the above-mentioned EPDs, could be considered as a guide in this field.

Indeed, Barilla defines a common system process according to its experience, the 'funnel process', which, in three main steps, represents an internal standard to develop PCR–EPDs, gathering, aggregating, analysing and processing the data, to reduce them to more manageable results. Specifically, such steps are:

- data collection and management—the identification and gathering of product-specific information regarding the product recipe (the amount of food raw materials per unit of product), the bill of materials packaging list, the production plants in which the product is manufactured and the related production volume, logistic distribution data for the finished product and other relevant environmental aspects;
- data processing—elaboration of the product system using the LCA database distinguished in data module groups (raw materials, packaging raw materials, energy, plants and transport). This step occupies a central role in PCR–EPD elaboration because PCRs describing the product category include requirements for the LCA that provides the basis for an EPD: the functional unit, system boundaries, cut-off rules, allocation rules, data quality requirements and indicators. All the data modules are internally verified and are ready to be used for EPD purposes;
- result management—the product group model calculation tool carries out a collection of the results in a specific LCA data sheet, which is developed for each product group in a specific fashion following the PCR and is internally vetted.

The reliability of the EPDs is ensured by several verification levels, carried out internally by the Data Assessor and Process Assessor and externally by the Verification Body. Internal verification is applied in all three steps in a continuative way, in order to verify the LCA calculation and maintain conformity. Indeed, internal assessments, at planned intervals, are conducted to determine the reliability, relevance and independence of the EPD.

The Verification Body, an external auditing body, represents an accredited body certified for auditing management systems that verifies the entire EPD process system.

The use of the Barilla EPD Process System has shortened the EPD publication time, which now lasts about 8–10 weeks.

The EPD represents an environmental success action. From the EPD process results (i.e. by LCA calculation of semolina pasta), Barilla has achieved a mix of environmental objectives; the company has obtained a reduction in GHG emissions by acting on the phases of cultivation of durum wheat and pasta production. The result is due to a combination of actions:

- the rotation of cultivation and the careful use of fertilisers, changing the production rules, avoiding −55 % of the GHG emissions (390 kg of GHG per tonne of produced durum wheat);
- the increase in the proportion of recyclable packaging from 92 to 95 %;
- the rationalisation of logistics with the optimisation of the transport saturation avoiding—8 % of GHG emissions;
- the reduction of water recommended for cooking, from 1 to 0.8 L of water per 100 g of pasta, avoiding—5 % GHG emissions;
- the improvement in the efficiency of the energy management systems in factories, introducing a CHP (combined heat and power) plant in pasta production, using renewable energy, avoiding −13 % of GHG emissions.

In conclusion, the analysed PCRs do not present relevant differences; given the amount of information, the following list summarises the main differences:

- in the functional unit section, PCR 2012:06 (CPC 234) defines 1 kg of product (as do the others), but the packaging weight is not included in the kg of product;
- in the core process section, only PCR 2010:01 (CPC 2371) is limited to the product production (pasta in this case);
- in the downstream processes section, not only the PCRs include the distribution (e.g. the PCR 2010:01 (CPC 2371));
- in the allocation section, all the PCRs allow partitioning with the allocation by mass, but PCR 2010:01 (CPC 2371) underlines that the products that are not compliant with the quality requirements and are destined for other chains (such as animal food) must be considered waste.

Instead, all the above-mentioned PCRs, in order to communicate and compare the environmental performances of different products, in the additional environmental information section (Sect. 10.4) and annex 1–2, include in the LCA report some additional optional indicators, widening the EPD scope (e.g. in the case of Barilla) in terms of the ecological footprint, carbon footprint and virtual water content (e.g. PCR 2010:01 (CPC 2371) introduces the ecological footprint, the water footprint, land use, land use change and forestry, marine water eutrophication and aquatic eco-toxicity indicators). However, comparability remains a critical factor for the EPDs (Schau and Fet 2008); hence, it is best to compare only products that are similar to each other and that are within the same class included in each PCR. Finally, on the basis of the foregoing analysis, the LCA–PCR link seems to be a critical aspect that still needs a wide and deep action for harmonisation from a global perspective.

4.3 Review of the Methodological Issues in the Cereal Sector

4.3.1 Definition of the Functional Unit (FU)

The main applications of LCA in the cereal sector (maize, wheat and rice) have been devoted to different goals: identifying the environmental hot spots in production systems' performance, profiling the environmental burden of production in a given area, comparing the environmental burden of different food products and different farming practices, as well as evaluating the environmental properties of a supply chain.

As a consequence, different LCA studies of the same product systems can have different functional unit (FU) definitions, making the choice of the FU very controversial (Reap et al. 2008). In fact, as mentioned in the first chapter, this can lead to different or even contrasting LCA results. Ideally, an LCA with multiple FUs can provide a better picture of the sustainability of the product systems under assessment and at the same time make the study more comparable with others.

In most of the studies reviewed in the current chapter, the FU is based on mass: about 31 % of the reviewed papers use 1 kg of the investigated product at different stages of the value chain. For example, for the agricultural stage, 1 kg of dry corn grain or milled rice at the mill gate is considered. If the LCA considers the final consumption stage, the FU can be that of 1 kg of pasta, in primary and secondary packaging, delivered to customers, 1 kg of bread ready for consumption, 1 kg of refined rice packed and delivered to the supermarket, 1 kg of short pasta, 1 kg of pasta or a 725 g pizza margherita.

In some cases, the FU includes quality aspects, such as the energy content of the final product. Roy et al. (2009b) express the FU in MJ of energy supplied by different forms of cooked rice to enable the comparison among them. In such a context, other FUs are the protein content per food (Khoo et al. 2010), the content of glucose and fructose (Renouf et al. 2008) or dry biomass production (Kim and Dale 2002).

Over 14 % of the reviewed studies use mass as the FU but refer to metric tons (Brentrup et al. 2004; Drocourt et al. 2012; Fallahpour et al. 2012; Kissinger and Gottliebb 2010; Kløverpris et al. 2010; McConkey et al. 2012; Muñoz et al. 2012;

Roy et al. 2005, 2007; Ruini and Marino 2010; Salomone and Ciraolo 2004; Williams et al. 2010). Some studies, mainly dealing with derived cereal products, use different FUs: 1 loaf for bread; 1 hl for beer; 1 L for canola cooking oil (Narayanaswamy et al. 2004); 1 t of grain; or 1 t of grain with constant quality (13 % protein in dry grain (Charles et al. 2006)). Nemececk et al. (2008) define three FUs, one for a land management function (i.e. hectares/years), one for a financial function (gross margin 1 in €) and finally one for a productive function (MJ gross energy of the product).

Another commonly used FU is based on land surface area: Van Stappen et al. (2012) adopt any useful output per hectare in an average year and illustrate the competition for land between food and non-food products; Murphy and Kendall (2013) choose 1 ha of corn and stover production; Eshun et al. (2013) and Ferng (2011) use 1 ha; Seda et al. (2010) use land surface area together with yield and economic benefit; Harada et al. (2007) use 60 m^2; and Nalley et al. (2011) use acres.

Some studies perform the LCA according to different FUs. Seda et al. (2010) choose the land area (ha), yield (tonnes) and economic benefit (€) as a basis for comparing the environmental impacts of cereals and horticultural crops. The land area functional unit provides explicit information on the intensity of use of agricultural inputs. Yield as an FU is a reflection of agricultural activity as a producer of market goods, and it can be used to evaluate the effect of cultivation techniques on yield (e.g. different rates of fertilisation). The study also includes a cost–benefit analysis in order to define the eco-efficiency concept better, which is the management philosophy encouraging business to search for environmental improvements that obtain parallel economic benefits. In general, when using land surface area, the impacts of horticultural products are higher. The cost–benefit analysis reveals that the economic benefit of the horticultural crop alternative is seven times higher than that of cereals. From this case study, it can be concluded that horticultural crops would be a suitable choice based on productivity and economic terms (the weight of the product or the economic benefit were used as functional units). The differences could be attributed to the higher yield and retail prices of horticultural crops in comparison with cereals. On the contrary, when land area is used as an FU, the cereal crops tend to be more sustainable.

Roer et al. (2012) illustrate the importance of carefully selecting the functional unit, choosing 1 kg of cereal dry matter as the FU and then performing a sensitivity analysis in terms of energy content (MJ), protein content (kg) and area occupied (ha). Even though the choice of the FU does not change the overall rating of the included cereals, the relative differences change. Furthermore, despite the widespread use of kg dry matter, this unit is not always a good measure of the quality of the food produced; the energy and protein content can indeed, as reviewed above, be more appropriate. The situation changes completely if area is used as the functional unit, as in this case more intense management per ha can overshadow the higher productivity. This factor underlines the need to be very specific regarding the motivation and scope of the study and thus the selection of the functional unit, which should be goal-driven. For example, in the case of the selection of alternative agricultural production systems, Hayashi (2013) recommends the definition of decision criteria rather than trying to make decisions on the basis of multiple functional units. The author shows that a comparison based on the functional units is not fair because the product information (yield) is not contained in impacts per area unit. When decision criteria are introduced, two aspects need to be considered: impacts per area unit (which have to be minimised) and yield per area unit (to be maximised). The ratio of the former to the latter leads to the definition of impacts per product unit, which can be recognised as an integrated upper-level criterion (Hayashi 2013).

4.3.2 System Boundaries and Cut-Off Criteria and End-of-Life Aspects

Among the reviewed studies, more than 61 % explicitly specify the system boundaries. To outline better the definition of the system boundaries of cereal-related life cycle studies, the reviewed studies have been subdivided into five categories, according to the cereal considered: wheat, rice, maize, derived products and a combination of these.

The authors who have dealt with wheat as a case study all start at the wheat cultivation phase, apart from Röös et al. (2011), who also consider the seed planting stage. Charles et al. (2006), Muñoz et al. (2012), Ruini and Marino (2010), Ruini et al. (2013) and Schmidt (2008) stop at the stage of cultivation, while Bevilacqua et al. (2007) and Salomone and Ciraolo (2004) consider all the stages of the production cycle (processing, packaging, transportation, distribution, consumption and disposal), and Röös et al. (2011) stop at the distributive stage. Brentrup et al. (2004) and Meisterling et al. (2009) consider the extraction of raw materials, production and transportation of input (fertiliser, pesticides, machinery, fuel).

Only for 56 % of the studies on rice it is possible to analyse the process coverage, and the definition of the system boundaries, when specified, is very different: for example, Blengini and Busto (2009) consider agricultural processes, drying and storing, and refining and packaging; Roy et al. (2009b) include the cultivation, processing and distribution of rice produced and consumed; Khoo et al. (2010) take into account land use, cultivation, harvesting, milling, drying, refining and storage, and transportation to a national retailer; and Hokazono and Hayashi (2012) define the boundaries in a generic manner: 'all farm-level and upstream processes of rice production in the paddy fields'.

Studies that analyse wheat-derived products (mainly pasta and bread) have a greater process coverage and include all the stages of production, packaging, transportation, distribution and consumption of the processed product, including planting and cultivation, and in some cases they also consider the stage of waste disposal (Bevilacqua et al. 2007; Espinoza-Orias et al. 2011; Salomone and Ciraolo 2004).

The studies on corn mainly focus on the agricultural phase and consider farming operations (such as soil tillage, seedbed preparation, sowing, fertiliser and pesticide application, harvesting, collection and collection of stover). Regarding the studies focusing on the analysis of a combination of products, out of a total of 19 studies, 5 do not indicate the system boundaries; of the 14 remaining, some consider only the cultivation phase (Brankatschk and Finkbeiner 2012; Fallahpour et al. 2012; Gan et al. 2011b; Pelletier et al. 2008; Zhang et al. 2013). Nalley et al. (2011), in addition to farming, also consider the production of inputs; others (Huang et al. 2012; McConkey et al. 2012; Roer et al. 2012; Seda et al. 2010) consider only plantation and cultivation; and Biswas et al. (2010) and Narayanaswamy et al. (2004) also consider storage, processing, retail, consumption and transport.

In terms of the overall main boundaries and the cut-off of some stages of the life cycle under analysis, 40 % of the studies reviewed in this chapter implement 'cradle to farm gate' as the main boundaries; about 9 % consider 'cradle to consumption'; over 6 % 'cradle to grave including end of life'; 5 % 'cradle to factory gate (end of transformation)'; 4 % 'cradle to national retailer'; and two studies chose 'cradle to international retailer'. Breiling et al. (2005) delimit the analysis 'from land preparation to harvesting', whilst Settanni et al. (2010b) delimit the analysis by 'entry gate to output gate of factory'. Geographical boundaries are specified in over 65 % of the reviewed studies, and time boundaries in only 28 %. The choice of geographical boundaries should be consistent with the system boundaries of unit process data sets, because of the critical issues that may arise due to the fact that some phases are carried out within the geographic boundaries of the country indicated, while other phases (e.g. sales, use, treatment and waste) can take place in other areas.

The inclusion or exclusion of process units in the system boundaries is a subjective choice, which can be relevant to the outcomes of an LCA, even in the case of cradle-to-farm-gate analysis. The lack of data, which is one of the constraints of LCA, often contributes to spreading the tendency towards simplification (see Sect. 4.2.4), e.g. excluding the contribution of some inputs, such as capital goods (machinery and buildings), which can, in certain circumstances, contribute significantly to the total impact of the production systems (Blengini and Busto 2009; Roer et al. 2012). Furthermore, the contribution of the production and use of pesticides, mineralisation in humus and nitrogen oxides from mineral fertilisers is usually neglected, with consequent underestimation of the actual total environmental impact, but as demonstrated by Roer et al. (2012), all of this has a significant environmental impact in the cereal sector.

The end of life is a relevant step in the life cycle of agricultural products, but in the specific case of cereals and their derivatives, it assumes a marginal role, as cereals are usually used for human consumption or as raw materials for the manufacturing of other products. The end of life is considered in the case of packaging materials, e.g. rice (Kägi et al. 2010), bread (Espinoza-Orias et al. 2011) or, in the case of the Product Category Rules, e.g. bakery products, cooked and uncooked pasta (see Sect. 4.2.6).

Most of the studies investigated in this chapter adopt an attributional LCA: in this case, the life cycle of the system is modelled as it is and the principal system boundaries and included life cycle stages can be derived from the goal and scope of the study. When consequential modelling is considered, processes of other systems (other than those specifically assessed) are to be included in the system boundary of the analysed system. Schmidt (2008) presents a framework for defining system boundaries in consequential agricultural LCA using wheat production as a case study. He also argues that the proposed methodology contributes to increased completeness of the identification of the processes actually affected.

4.3.3 Criteria for the Allocation of Multifunctional Processes

Allocation is a crucial issue in LCA studies, because the uncertainty of LCA results is largely dependent on the methodological choices related to allocation criteria (Brankatschk and Finkbeiner 2012; Curran 2008; Gnansounou et al. 2009; Kim and Dale 2002). In the cereal production sector, the allocation problem is particularly relevant because this type of production almost always implies cultivation systems that produce multiple products (co-products or by-products) by rotating crops or processing diverse parts of a plant for different uses (e.g. for food or energy purposes).

By analysing the scientific papers included in the state-of-the-art analysis for the cereal sector presented in Sect. 4.2 (see Table 4.2), some elements regarding allocation methodologies and criteria can be highlighted.

First of all, it should be highlighted that three papers specifically investigate the allocation issue, presenting different criteria to face the problem with applicative examples in the cereal sector:

- In Kim and Dale (2002), the authors focus on the ethanol production system from corn grain and present a study in which allocation is avoided through system expansion. In order to avoid the allocation procedures completely, five mutually interdependent product systems were required (ethanol production from corn dry milling and corn wet milling, corn grain production, soybean products from soybean milling, urea production). The system expansion approach is equivalent to assuming that the environmental burdens associated with ethanol from dry milling are equal to those associated with ethanol from wet milling. This approach is interesting because it can be used to compare the environmental burdens associated with ethanol with those associated with petroleum-based fuel as well. However, the proposed approach would not work for an LCA study aiming to compare the environmental burdens between different ethanol production technologies;
- In Brankatschk and Finkbeiner (2012), the authors demonstrate the benefits deriving from the use of the 'cereal unit' (CU) as an allocation procedure in LCA studies of agricultural systems, presenting a comparison of different allocation methods (cereal unit, mass allocation, energy allocation and economic allocation) for different agricultural products (barley, soybeans, sugar beet plant, wheat plant, sunflower plant, rapeseed plant and rape seeds). The CU is a common denominator that could be used for evaluating agricultural products and by-products based on the feeding value of agricultural products. The results highlight that the application of CU allocation could reduce the variability and potential bias in the LCA results of agricultural systems;

- Murphy and Kendall (2013) explored three different approaches to allocation for corn and stover: economic allocation, energy-based allocation and a subdivision approach, which assigns to stover only those additional activities caused by its harvest. For most indicators, subdivision produces impacts approximately equal to those of economic allocation. Both economic allocation and subdivision assign lower impacts to stover than energy allocation. No definitive conclusions on which allocation criteria to be preferred are defined in the paper, but the authors argue that in the long term, once commercial production systems and associated markets are established, economic allocation may be preferable, while for current LCAs of stover production, the most reasonable approach could be that of using a range of values based on multiple allocation methods.

Of the remaining scientific articles (taking into account only the case studies of cereals used for food purposes), it should be noted that only eight of them clearly report whether allocation was applied, also specifying the methods used (for which the economic allocation is the most frequent); how allocation was treated in these studies is briefly presented in the following paragraphs.

In Blengini and Busto (2009), the LCA methodology is applied to the rice production system, from the paddy field to the supermarket. Rice production generates different marketable products and by-products (refined rice, broken grains, rice flour, husk straw, etc.) for which the allocation of burdens to the co-products was based on relative economic value, as suggested by Williams et al. (2005) (which is the same method as that used by Williams et al. (2010), as reported below and in footnote 1).

Kasmaprapruet et al. (2009) present an LCA analysis of milled rice production, from rice cultivation to the mill. The allocation step was performed based on economic allocation from which resulted the following allocation of environmental burdens: 51 % to milled rice, 27 % to broken rice, 20 % to rice bran and 2 % to rice husk.

Biswas et al. (2010), using LCA methodology, compared the life cycle global warming potential of three important Australian agricultural productions (wheat, meat and wool), including two major life cycle stages: pre-farm and on-farm. In order to calculate the inputs and outputs of the co-products, the authors chose to apply an economic allocation method in which the allocation factors to partition the greenhouse emissions to the various products (wool, sheep meat and wheat) are derived using the ratio of market value for those products; the method used was based on Guinée et al. (2004).

Hokazono and Hayashi (2012) present an LCA of three rice production systems in Japan: organic, environmentally friendly and conventional. The allocation procedure was applied to brown rice and rice ducks (a by-product of paddy fields sold as poultry in small markets at relatively high prices) following economic allocation criteria (approximately 10 % of the impact was allocated to rice ducks, which varied from 8.1 to 10.4 % depending on the rice yields). Allocation was also conducted between white rice and rice bran (both obtained by polishing brown rice), again using economic criteria (99.6 and 0.4 % of the impact were allocated, respectively, to white rice and rice bran since the economic value of rice bran is much lower than that of white rice).

Espinoza-Orias et al. (2011) estimated the carbon footprint of bread produced and consumed in the UK. In this case, allocation problems arise in the wheat milling stage, which co-produces flour, wheat germ and bran. The authors decided, in the absence of data to perform system expansion, to face allocation using an economic value approach (suggested both by PAS 2050 and by ISO 14,044). The result is that GHG emissions deriving from the wheat milling stage were allocated 88 % to white flour, 92.5 % to wholemeal flour and 90 % to brown flour.

Williams et al. (2010) describe the production burdens of three organic and non-organic arable crops (bread wheat, oilseed rape and potatoes). For the specific case of wheat (which is grown for bread making), the burdens were allocated between the bread and the feed fractions according to their economic value.[1]

In Notarnicola and Nicoletti (2001), a comparative LCA between pasta and couscous is presented. Two different allocations were performed: one referring to the stage of agricultural production, from which grain and straw are obtained, and the other related to semolina production, from which flour, bran and fodder grain are obtained. The allocations were made with a combination of economic and mass criteria by applying the following formula:

$$A = \frac{q_i \text{ x } p_{ui}}{\sum q_i \text{ x } p_{ui}}$$

where

A = economic factor of allocation;
q_i = mass allocation factor;
p_{ui} = relative price.

Also in Notarnicola et al. (2004), the allocation problems related to the co-production of durum wheat and straw or semolina, pollard, millfeed, screenings and germ were solved on the basis of the relative quantities and marker prices.

It should be pointed out that the Product Category Rules (PCRs) published for this sector also suggest different allocation rules depending on the specific type of product, in particular:

- the PCRs on 'grain mill products', 'bakery products', 'pasta, cooked, stuffed or otherwise prepared and couscous' and 'uncooked pasta, not stuffed or otherwise prepared' suggest the use of mass allocation when the inputs and outputs of the system should be partitioned between the different products or functions;

[1] 'The total burdens of producing grain and straw are: $T = H + (1 - p_s)I + p_s B + D$. Then the burden allocated to grain is: $G^* = (H + I)\,(Y_g / (Y_g + v_s p_s Y_s)) + D$, and the burden allocated to straw is: $S^* = \frac{(H + I)(v_s p_s Y_s)}{(Y_g + v_s p_s Y_s)} + p_s(B - I)$ where H is the vector of burdens of producing grain up to the end of combine harvesting per hectare, I is the vector of burdens of chopping for incorporation for all straw produced, D is the vector of burdens of drying and storage of grain, B is the vector of straw baling burdens for all straw produced, p_s is the proportion of straw baled and harvested, Y_g is the net yield of grain per hectare at standard DM content, Y_s is the yield of straw per hectare (whether harvested or not) at standard DM content, and v_s is the relative value of the straw prior to baling versus the grain, typically 0.05' (Williams et al. 2010).

- the PCRs on 'bread and other bakers' wares', 'groats, meal and pellets of wheat and other cereals' and 'pastry goods and cakes' report that 'allocation between different products and co-products shall be based on economical allocation'.

To avoid allocation, as recommended by the ISO requirements, system subdivision or system expansion should be implemented when possible. System subdivision means dividing the unit process to be allocated into two or more sub-processes and then collecting the input and output data related to these sub-processes (ISO 2006), while system expansion means that the system boundary is expanded in order to include the displacement of substitute products (the co-products) in the market, which will generate environmental credits due to the avoided production of displaced products (Ekval and Weidema 2004).

Of the papers reviewed in this chapter, system subdivision was implemented by Murphy and Kendall (2013) (as reported above), while system expansion was specifically mentioned only by Renouf et al. (2008), in which an LCA of sugarcane production and processing in Australia is presented and this system is then compared with other sugar-producing crops (US corn and UK sugar beet). Among the various conclusive remarks stressed by the authors of this case study, it should be noted that they state that a crop's agronomic characteristics can influence its environmental performance and one of the main characteristics is the nature and quantities of co-products deriving from crops, which can displace other products in the markets, giving environmental credits.

All the other case studies included in the state-of-the-art analysis do not mention allocation criteria in any way, except for two cases (Petti et al. 2000; Schmidt 2008), in which the authors specify that, for simplification reasons, all the burdens were allocated to the main crop/product, and one paper (Nemececk et al. 2008), in which it is specified that only allocation for shared infrastructure (machinery and buildings) was performed (in particular following the procedures described by Nemecek and Erzinger (2005) and Nemecek and Baumgartner (2006).

4.3.4 Data Availability and Quality

Data availability and data quality are one of the main problems that LCA practitioners face when developing an LCA study; the significance of the problem is also demonstrated by the fact that the Society of Environmental Toxicology and Chemistry (SETAC) has set up a working group on this specific topic (SETAC LCA Working Group on Data Availability and Data Quality—Bretz 1998) and many other initiatives, at the national and international levels, have been initiated to deal with this issue.

LCI data availability is particularly significant in some specific industrial and productive sectors, such as agri-food, in which there is still a lack of complete and reliable data for many processes and kinds of food.

Concerning the cereal sector, it can be highlighted that, starting from the state-of-the-art analysis for the cereal sector presented in Sect. 4.2 (see Table 4.2), only three papers specifically investigate the problem of data availability:

- In Notarnicola et al. (2004), LCA and IO-LCA are applied to the pasta life cycle in order to verify whether the adoption of these two tools could improve the quality of the inventory set-up. In particular, in the study, two IO-LCA approaches are considered: the Economic Input–Output Life Cycle Assessment (EIOLCA), developed in the US, and the Missing Inventory Estimation Tool (MIET), developed in the Netherlands; the input–output tables of EIO-LCA are substantially different in nature from LCA, while MIETs are more similar in structure to LCA. The results show that hybrid approaches (involving the integration of IO with LCA and vice versa) may resolve the truncation error problems of LCA, together with the closed-loop incompleteness issues. The study also highlighted that, in general, IO-LCA approaches should not be used to carry out streamlining LCA, but to make the LCA set-up and the LCA results more comprehensive as well as to make them less site-independent, keeping in mind data quality and quantity;
- In Hayashi et al. (2010), a life cycle inventory (LCI) database for crop production in Japan (the NARO LCI database) is presented. The database was developed using modularisation techniques; SimaPro 7.2 was utilised for database construction and management, and Ecoinvent 2.1 was used as the basis of the development. The database includes inventories for paddy and upland field crops, agricultural work, fertilisers, pesticides and agricultural machinery;
- In Laurent et al. (2012), a summary of the results of an analysis aimed at assessing the available data in existing LCI databases regarding cereals (wheat, barley, maize, sorghum, rice and rye) and cereal-containing products is presented. The analysis was conducted on ten French and international databases, eight of which include cereal-related data (Ecoinvent, DiaTerre, LCA Food, Bilan Carbone®, AUSLCI, CPM Database, USLCI and Agri-Footprint), while two (Probas and BUWAL 250) do not include cereal-related data. The analysis highlights that the Ecoinvent Database is by far the most complete database, with Swiss and European data for agricultural raw materials, inputs and processes. Data about some cereal-based finished products can be found in the LCA Food Database (wheat, bread, pastries, oat flakes) and in the French Bilan Carbone® database. However, very few data can be found in the databases about agricultural processes, food industry processes, storage or mass-market retailing. Only the Ecoinvent Database and the LCA Food Database provide specific geographic data: Swiss data in Ecoinvent and Danish data in the LCA Food Database. The study also raises the issue of methodological comparability: all the databases set their own hypotheses and methodological rules (allocation, cut-off rules, etc.) and major differences can be found between data from different databases. This variability makes it difficult to implement environmental labelling of cereal-based products with sufficient accuracy and comparability.

Concerning data quality, it is interesting to stress that different papers applied some form of data quality check: sensitivity analysis was applied in seven cases (Charles et al. 2006; Kägi et al. 2010; Kim et al. 2009; Kløverpris et al. 2010; Narayanaswamy et al. 2004; Nemececk et al. 2008; Pelletier et al. 2008); completeness and consistency checks in one case (Narayanaswamy et al. 2004); uncertainty checks in one case (Röös et al. 2011); and comparisons with other studies and/or LCI databases in five papers (Aldaya et al. 2010; Espinoza-Orias et al. 2011; Harada et al. 2007; Hayashi et al. 2010; Williams et al. 2010). This indicates that data quality checks are increasingly gaining importance in LCA practice since they can be used to verify the reliability of uncertainty data and can assess more carefully the kind of influence such data can have on the final results.

4.3.5 *Life Cycle Impact Assessment (LCIA)*

The LCIA is a crucial phase of an LCA in which large quantities of data regarding natural resource use and emissions are transformed into useful information for the evaluation of the product system under analysis in terms of impacts on human health and on the environment. Unlike traditional risk assessment analysis, LCIA is not site-or emission-specific nor time-dependent (Margni and Curran 2012). However, the nature of food products, including cereals and derived products, and at times the type of food LCA study, is such that site-specific data must be considered (Notarnicola et al. 2012) in order to assess the potential impacts properly. In fact, as pointed out in the first chapter of the present book, especially for the agricultural phase of a food LCA, site specificity can greatly influence the results of the impact assessment. For example, the pedoclimatic conditions can heavily influence the impact deriving from the use of fertiliser and pesticides or the water use impact category.

The evolution of LCIA methods over the last decades has brought about numerous models, mainly involving a combination of midpoint and endpoint modelling, with numerous characterisation models that can potentially generate different results. The most widely used LCIA methods for cereal LCA are CML (Muñoz et al. 2012; Narayanaswamy et al. 2004; Nemececk et al. 2008; Notarnicola et al. 2004; Pelletier et al. 2008; Salomone and Ciraolo 2004; Williams et al. 2010), Ecoindicator (Bevilacqua et al. 2007; Petti et al. 2000; Renouf et al. 2008), EDIP (Nemececk et al. 2008; Schmidt 2008), ReCiPe (Roer et al. 2012), Impact (Drocourt et al. 2012) and LIME (Yoshikawa et al. 2012). Furthermore, as pointed out by Margni and Curran (2012), the rapid and fervent development of methodologies indicates that LCIA has not yet reached a stable and generally accepted standard; hence, methodologies that are older than 10 years may not reflect the state of the art and may entail methodological weaknesses that have been resolved with more recent methodologies. Thus, when consulting cereal-related LCA results dating back at least 10 years or when using one of the older methodologies, the results should be carefully analysed and if possible compared with similar results obtained with a more recent LCIA methodology implementation. However, despite this fervid development and the ongoing discussion, there are still some LCA studies that do not report the LCIA method used, therefore preventing any sort of comparison or relative assessment.

As far as specific impact categories are concerned, the IPCC (Intergovernmental Panel on Climate Change) and the WGO (World Meteorological Organization) models are the only internationally accepted ones commonly used for the GWP (global warming potential) and ODP (ozone depletion potential) assessment in all the methodologies. This is reflected in the explicit use of such methodologies in LCA regarding cereals, with particular regard to the GWP and the use of the carbon footprint methodology (e.g. Drocourt et al. 2012; Eshun et al. 2013; Espinoza-Orias et al. 2011; Gan et al. 2011a, b; Murphy and Kendall 2013; Nalley et al. 2011; Ruini et al. 2013). For other indicators, there are multiple characterisation models, not all unanimously accepted and each with limitations that inevitably will produce variability among LCA results regarding similar products. A list of LCIA methods, identified as the best among the existing characterisation models, was provided for each impact category in the context of the ILCD Handbook (EC-JRC 2011). If the identified model was judged of sufficient quality, it was recommended and the list of the recommended models for each impact category was provided by Hauschild et al. (2013). Some examples of applications of these recommendations are already present in the cereal sector, e.g. with regard to Usetox for toxicity-related impacts (Berthoud et al. 2011). Furthermore, all LCIA models assume that the emissions (with the exception of those relative to global warming and ozone depletion) occur in the country where the methodology was developed, which is not necessarily true and may need to be accounted for in the interpretation phase of the LCA study.

In order to deal with the above-mentioned site specificity, characterisation methodologies (e.g. IMPACT World+ 2014) are being developed in order to address the regionalisation of impact categories. Furthermore, software producers, of products such as Ecoinvent (2014), are moving towards more regionalised data sets (when applicable), but such effort is limited to data regarding different macro geographical regions (countries or areas of continents). In reality, data sets of a specific region of a country can produce LCIA results that differ considerably from the average national impact values; see for example Laurent et al. (2012) in the cereal sector.

Water use and land use issues related to cereal crops are undoubtedly affecting the ecosystem worldwide. This is particularly true for cereal crops such as wheat or rice that are used for the production of staple foods in many countries. The impact assessment for such impact categories is by no means standardised (Notarnicola et al. 2012) but should nonetheless, if possible, be included in order to improve the overall quality of the LCA of cereal or derived products (e.g. Kløverpris et al. (2010) for land use changes in wheat production). However, most of the cereal-related LCA studies include land use and land use change in terms of land occupation and land transformation, therefore neglecting a proper impact assessment (e.g. Bevilacqua et al. 2007; Braschkat et al. 2003; Brentrup et al. 2004; Charles et al. 2006; Drocourt et al. 2012; Kulak et al. 2012; Ruini et al. 2013; Schmidt 2008; Williams et al. 2010).

Of the studies reviewed in the previous sections of this chapter, many include pesticide and fertiliser production, but very few actually include the modelling of their diffusion in the environment. In most cases, it is assumed that all the pesticide or fertiliser is absorbed by the cereal plant. In reality, the pedoclimatic conditions and farming practices can strongly influence how much of these chemicals are transferred to the environment. Some studies adopt the PestLCI model by Birkved and Hauschild (2006) and Dijkman et al. (2012), which estimates pesticide emissions to air, surface water and groundwater for use in life cycle inventory (LCI) modelling of field applications, e.g. Berthoud et al. (2011); however, such a method is rather complicated and requires large quantities of data for its correct implementation. If the data are not available, there is a risk of basing the evaluation of the pesticide diffusion on too many assumptions, hence making the modelling ineffective. Furthermore, the results obtained with the PestLCI model need to be applied in combination with characterisation factors obtained from emission route-specific impact assessment models, such as USEtox.

Finally, LCIA includes options for the normalisation, grouping and weighting of impacts. The implementation of such approaches is subjective (e.g. weighting factors may be based on economic, political or environmental considerations) and can make the results of the LCA inapplicable to product systems of the same nature originating from different geographical areas. Therefore, in the cereal sector, weighting is only rarely implemented (Bevilacqua et al. 2007; Brentrup et al. 2004; Fallahpour et al. 2012; Notarnicola and Nicoletti 2001, 2004 Petti et al. 2000).

4.3.6 Interpretation and Comparison of the Results

There are many tools to assess the robustness of an LCA: one of these is the completeness check for both process coverage and I/O coverage (e.g. all the included material or energy input and emission associated with the system under analysis). Such coverage is seldom complete and complicated by the high variability of the system boundaries of cereal-related LCAs, even within the same LCA approach (cradle to gate or cradle to grave), which makes the comparison between different studies analysing the same product a harsh task. In this regard, Schau and Fet (2008) stress the need for a set of rules to determine the system boundaries for different product categories so that a comparison of the environmental impacts of different batches of products can be possible.

Most cereal-related studies have an interpretation phase that entails a description of significant related aspects and a discussion on the limitations and recommendations. However, not all of the analysed studies carried out a sensitivity or an uncertainty analysis to test the extent to which the results are affected by specific methodological choices.

In accordance with ISO 14,044, the sensitivity and uncertainty analysis should be based on those model choices known to have a major influence on the results of the study, such as (Guinée et al. 2002):

- allocation rules: examples for the cereal sector could be economic vs. energetic vs. mass allocation
- boundary setting: examples could be the inclusion or exclusion of the transport of agricultural inputs, the production of agricultural machinery, etc.
- process data: examples could be the variations of the type of fertilisers used, pesticides, fertiliser emissions, etc.

- cut-off criteria: changing the cut-off rules (the boundary between processes that are relevant and irrelevant to the product system)
- characterisation method: alternative characterisation methods, which could be adopted instead of the baseline method
- normalisation data and weighting method (if carried out)

Once one or more variables from the above list have been selected, the changes produced by their variation in the LCA results should be analysed.

Within the cereal and derived products context, several authors have implemented a sensitivity analysis in their studies (Blengini and Busto 2009; Charles et al. 2006; Drocourt et al. 2012; Espinoza-Orias et al. 2011; Kägi et al. 2010; Kløverpris et al. 2010; Meisterling et al. 2009; Roer et al. 2012; Yoshikawa et al. 2012).

Specifically, for such analyses, the following parameters were taken into account: field emissions (Blengini and Busto 2009; Charles et al. 2006; Kägi et al. 2010; Roer et al. 2012), allocation criteria (Blengini and Busto 2009), transportation distance and yields (Drocourt et al. 2012; Meisterling et al. 2009) and water requirements (Blengini and Busto 2009). Regarding field emissions from fertiliser use, two alternatives are available: either varying the emissions factor within the range provided in the model (i.e. the IPCC method), as performed by Roer et al. (2012), or using different models to assess the emissions (Charles et al. 2006). Espinoza-Orias et al. (2011), who determined the carbon footprint of bread, opted for a non-agricultural variable, carrying out a sensitivity analysis with different percentages of bread waste (from 10 to 30 %).

As for the results of the sensitivity analysis (expressed in terms of percentage variations with respect to the results obtained), the following considerations can be drawn: the range of variation in field emissions (N_2O and, in the case of rice cultivation, also CH_4) affected climate change from 11 (Roer et al. 2012) up to 36 % (Blengini and Busto 2009) and photochemical oxidant, terrestrial acidification and particular matter formation from 32 up to 53 % (NH_3 and NO_x) (Roer et al. 2012). In addition, Drocourt et al. (2012) and Kägi et al. (2010) underline the importance of the field emission parameters on LCA results. In fact, when it comes to direct field emissions, Kägi et al. (2010) found that the results varied from 15 (upland rice) up to 31 % (organic rice).

According to Blengini and Busto (2009), in their study on rice, the maximum variation caused by allocating burdens to straw was − 10 % (for eutrophication potential, photochemical ozone creation potential and water use), while the change in water for irrigation affected the total water requirement by ± 27 %. According to Murphy and Kendall (2013), the allocation approach could have a great influence on LCA results. In fact, in their study on corn, they state that, for most indicators, the subdivision approach produces impacts approximately equal to those of economic allocation and both economic allocation and subdivision assign significantly less impact to stover than energy allocation. Finally, in their study on bread, Espinoza-Orias et al. (2011), by varying the percentage of waste, calculated a variation of 10–12 % in GHG emissions.

In conclusion, in order to evaluate the robustness of an LCA model, a sensitivity analysis is most effective. Normally, the most uncertain parameters have to be taken into consideration to run a sensitivity test. These variables are often associated with field emissions when agricultural activities, such as cereal cultivation, are evaluated. These emissions, in fact, can strongly affect the results of an LCA study. Allocation criteria are also likely to influence the results of a study and should therefore also be carefully considered in this step of the analysis.

4.4 Some Lessons Learned from the Application of Life Cycle Assessment in the Cereal Sector

In this section, some indications are given of the best possible application of the LCA methodology based on the issues discussed in the previous sections. Such indications are by no means exhaustive or absolute but should be considered, whenever possible, when performing an LCA of a cereal product system, in order to fulfil the scope of the study and achieve the best possible results.

4.4.1 Goal and Scope Definition

The first phase of an LCA study consists of defining the goal and the scope, which aim to provide a description of the product system.

According to the ISO standard (ISO 14,040 2006), the goal of the study should define the application and the reason for carrying out the study, the intended audience and whether the results are intended to be disclosed to the public. The scope should clearly describe the system of the studied product or process and its boundaries, the system functions, the functional unit and reference flow, the environmental impact assessment methodology applied, the data requirements and finally the assumptions and limitations.

The goals of the cereal LCA studies reviewed in this chapter were different. The majority of them aimed to profile the environmental burden of a cereal, in order to identify the environmental hot spots of the system investigated. Other studies privileged a comparative approach, aiming to evaluate different farming and industrial practices or different cereal products. Investigating different agricultural practices (which includes evaluating different N rates used in the fertilisation phase or comparing organic and conventional systems), as well as industrial alternatives, could be very useful since it provides support in choosing the most productive approach in terms of environmental sustainability. From this perspective, effort should be made to identify the best agricultural techniques to put in place for cereal cultivation, consistent with the geographical specificities, in order to achieve a lower environmental burden. The same conclusion could be drawn for the industrial practices to be pursued along the supply chain of cereal-derived products.

4.4.2 The Functional Unit

The identification of an FU is the core of any LCA, providing a reference unit to which the inventory data are normalised. As already mentioned, the results of an LCA are strongly dependent on the FU chosen and this introduces a kind of uncertainty. Particularly in the cereal studies field, a 'one-size-fits-all' solution cannot be envisaged.

In the wide array of LCA studies found in the literature regarding cereals and cereal-based products, the selected FUs vary to a great extent, reflecting the significant differences in the characteristics of goods. Such differences concern not only the type of cereal (wheat, corn, rice) but also the kind of product deriving from a specific step in its supply chain (e.g. grains, flours, pasta, derived products, etc.). At times, the choice of FU is based on the mass or volume. This is often inadequate because the qualitative characteristics of grain can differ widely, as can those of cereal-based products. Most differences derive from complex production processes, which entail different technologies. Thus, cereal-derived products are not always comparable, even when belonging to the same cereal. Therefore, it may be useful to choose a set of multiple FUs based on mass, volume, cultivated area, economic value and qualitative characteristics, such as nutritional content. Charles et al. (2006), for example, in order to compare properly different wheat cultivation systems for bread making, included a quality parameter in the selected FU (13 % of protein in dry grain). The inclusion of quality parameters could in fact be an important aspect when it comes to comparing different systems within the same study or different studies assessing the same product.

Furthermore, since the main goal of an LCA is that of supporting market operators in their decisions, it is crucial to identify an appropriate FU—or, better, several FUs (Notarnicola et al. 2012)—in order to increase the relevance and impact of LCA sustainability information. A suggested tailored range of FUs for the cereal supply chain is schematised in Fig. 4.6.

4.4.3 System Boundaries

The definition of the system boundaries is a crucial step in the scope definition of every LCA study, but in the specific case of cereals, some aspects need to be considered. The recommendations are different according to the life cycle inventory modelling techniques used, i.e. whether it is an attributional or consequential LCA.

As demonstrated by Roer et al. (2012), in the case of attributional LCA, the system boundaries should be set in such a way that important processes for the environmental impact caused by food production are not excluded. In most cereal LCA studies, the manufacturing of machinery, buildings, humus mineralisation, production and the use of pesticides and/or nitrogen oxide loss due to the use of mineral fertiliser are excluded, but Roer et al. (2012) show that all these factors have a significant environmental impact, but with different contributions according to the

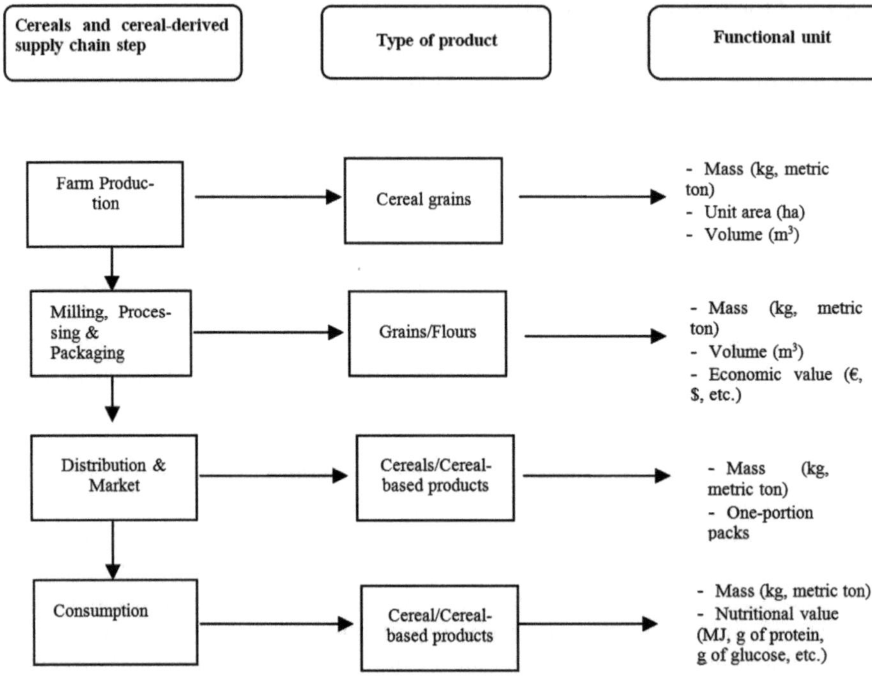

Fig. 4.6 Possible FUs related to cereal and cereal-derived product systems

impact categories considered. Therefore, the inclusion of these factors should be considered according to the impact categories included in the LCA study:

- the influence of the production of machinery was found to be relevant to all the impact categories, mostly the toxicity ones;
- the production of buildings had only a minor influence on the total environmental impact, although excluding buildings decreased the impact of the climate change impact category by 6–7 %;
- humus mineralisation only had an impact on climate change, and the exclusion of this factor decreased the carbon content by 2–30 %;
- the exclusion of the production and use of pesticides determined the highest reductions for the ecotoxicity impact categories;
- the exclusion of NO_X loss from the use of mineral fertiliser gave high reductions for particulate matter formation (65–71 %), photochemical ozone formation (83–82 %) and terrestrial acidification (68–75 %).

If all the above-mentioned factors are excluded simultaneously, the climate change impact is reduced by more than 40 %, with a consequent underestimation of the actual total environmental impact and increased difficulties in comparing different studies.

In the case of consequential LCA, the processes of further systems in addition to the one analysed need to be included in the system boundary of the analysed system in order to take into account the effects of an increased demand for cereals in one region. Schmidt (2008) presents a framework for defining system boundaries in consequential agricultural LCA, through the definition of different scenarios describing how the increased demand for wheat can be met. The comparison of scenarios shows significant differences in the contribution to the included impact categories (climate change, eutrophication and land use).

4.4.4 Availability and Quality of Data

The availability and quality of data issues of the cereal sector overlap with the problems of the more general agri-food sector. In particular, the former are connected to the lack of availability of data in the agricultural phase, such as the production of some fertilisers, herbicides and pesticides, the dispersion of chemical compounds into the environment (air, water and soil), the balance of CO_2 emissions, etc.

According to ISO 14,044, the quality of data should be carefully evaluated considering its time-related, geographical and technological coverage, and any information should be precise, complete, representative, consistent and reproducible, paying a high degree of attention to the source of the data and the uncertainty of the information.

The quality of data is strictly related to the availability of primary data, so it can be generally suggested to use literature data for the background system and primary specific data for the foreground. Estimations are very frequently not accurate, so, in principle, they should be avoided. Nonetheless, this is not always possible and provided that secondary data are derived from careful source selection and estimations, they can be used to obtain reasonably accurate LCA results. For example, in their study, Espinoza-Orias et al. (2011) performed a CF assessment of bread by following the PAS 2050 methodology using primary data and then performed the same study using secondary data, attaining similar results. Of course, in the case of the use of secondary data, all assumptions and estimations should be clearly declared and fully explained in order to avoid incomparability among different case studies. Furthermore, even if primary or specific data are available, a statistical approach to data collection and its evaluation should be used whenever possible (e.g. the use of confidence intervals or variance analysis). This implies taking, whenever possible, multiple measurements and readings of primary data in order to check the representativeness of the sampled data and also to verify that these data sets originate from the same stochastic distributions. Such approaches are implemented in very few LCA studies (e.g. Harada et al. 2007; Williams et al. 2010; Yoshikawa et al. 2010) and their importance is illustrated in Kägi et al.'s (2010) study on rice CF. Finally, considering the variability of data affecting the sector caused by the lack of specific and/or primary data, a consistency check of the data quality should be carried out, whenever possible, in order to evaluate how the different choices affected the results.

4.4.5 Allocation Methods

As highlighted in Sect. 4.3.3, in the cereal production sector, the allocation problem is particularly important because this type of production almost always implies cultivation systems that produce multiple products or imply the processing of diverse parts of a plant for different uses.

As suggested by ISO 14,044, allocation should be avoided through system expansion to include the additional functions related to the co-products or dividing the unit process to be allocated into two or more sub-processes and collecting the input and output data related to these sub-processes. The choice of system expansion or of subdivision of the processes in which the allocation problem occurs should always be preferred to avoid data distortion, but when allocation cannot be avoided, the inputs and outputs of the system should be partitioned between its different products or functions in a way that reflects the physical relationships between them.

In the cereal sector, the literature analysis highlighted that, when allocation was carried out, the most common solution was to use economic allocation methods, although other methods were also experienced, such as mass allocation, the cereal unit allocation (Brankatschk and Finkbeiner 2012), energy-based allocation or a combination of economic and mass criteria. In addition, the PCRs published for this sector suggest different allocation rules depending on the specific type of product; in particular, of the four PCRs actually published for this sector, three suggest using mass allocation while the other suggests economic allocation. Moreover, as highlighted in Sect. 4.2.6, the products that are not compliant with the quality requirements and/or are destined for other chains (such as animal food) must be considered as waste (PCR 2010:01 (CPC 2371)).

In the opinion of the authors of this chapter, and also in line with the suggestions carried out for other agri-food sectors (see for example the chapter on the olive oil sector), when allocation cannot be avoided, the allocation procedures in the cereal sector (which mainly involve by-products), should take into account both the mass and the economic value of the by-products in order to balance the huge quantities of by-products obtained with their low economic value.

4.4.6 Life Cycle Impact Assessment (LCIA)

The scope of an LCA study in the cereal sector can be varied: evaluating/identifying environmental hot spots, comparing different options, obtaining some kind of sustainability label, for marketing purposes, or other aims. Hence, such a scope will determine whether the LCIA phase of the study should consider optional elements, such as normalisation, grouping and weighting of the results. For example, if the scope of the study is to compare LCAs of the same cereal product system, based on the approaches used by the other studies with which one wishes to compare the results, it may or may not be appropriate to exclude these optional steps of the impact assessment phase. As mentioned in Sect. 4.3.5, a cereal-based LCIA will have to deal with particular site-and time-dependent issues when addressing the impacts deriving from fertilisers, pesticide application and land and water use. Specifically, many of the cereal-based LCAs reviewed in this chapter do not specify (or simply completely ignore) how to deal with the modelling of the impacts deriving from the diffusion of pesticides in the environment and the balance of nutrients.

Fertiliser applications have been identified as one of the main emission sources, in particular for impact categories such as acidification and eutrophication. However, the importance of these emission sources can vary greatly in different cultivation areas because of the different climatic conditions and cultivation practices. In particular, the emissions of nitrous oxide (N_2O), ammonia (NH_3) and methane (CH_4) from the soil depend firstly on:

- The amount of nitrogen applied and the type of fertiliser. The application of organic fertilisers involves higher emissions of ammonia and methane compared with the application of the same amount of nitrogen with mineral fertilisers;
- The temperature, wind, soil and water content at the moment of fertiliser application (and during the days following the application);
- The method of application. For example, for organic fertilisers, the injection into the soil strongly reduces the ammonia emissions and a similar effect can be achieved with fast soil incorporation after the spreading. In such a context, Carozzi et al. (2013) report strong reductions of NH_3 emissions if spreading is carried out with specific techniques (higher than 80 % when the organic fertiliser is injected into the soil or is quickly incorporated into the soil).

In the absence of primary data regarding these emissions, some methods have been developed for their assessment. The method proposed by the IPCC is one of the most utilised due to its simplicity. Nevertheless, others methods have also been developed. For example, Brentrup et al. (2000) assessed the emission of NH_3, NO_3 and N_2O not only considering the amount of N applied but also taking into account the timing of application, the climatic conditions (temperature, wind and rain) and the soil conditions (water content, structure, texture and field capacity). When this information is available, this method should be utilised instead of the IPCC's method. In particular, when the analysis is focused on the agricultural step of the production system, simplification should be avoided because it will not be possible to evaluate the differences linked to the use of different fertilisers or to spreading with different spreading machines.

Furthermore, when the LCA regards cultivation carried out in specific conditions (for example in greenhouse or soilless cultivation or with high surface irrigation), detailed methods must be utilized for the assessment of fertiliser-related emissions (see the chapter on the fruit sector).

Various models are available for the quantification of the emissions from pesticide use (e.g. Audsley et al. 2003; Birkved and Hauschild 2006; Dijkman et al. 2012) and should be carefully chosen based on the complexity and age of the model and the specific data available for its implementation. Furthermore, due to the non-univocal definition of technosphere (anthropogenic system) and ecosphere (environment) in the case of agricultural soil, different approaches can be used to account for the impact from pesticide use with a subsequent effect on the impact assessment step. An operative framework for toxicological assessments of pesticides is proposed by van Zelm et al. (2013), who defined a procedure to help LCA practitioners to gather the right data and use the proper models to include all the relevant emission and exposure routes where possible. Furthermore, very few cereal-based LCAs deal with the impacts deriving from land and water use. Even though no standardised methodologies have been adopted for such impact categories, if site-specific data are available, the implementation of one of the many methods that are described in the literature (Notarnicola et al. 2012), provided that the relative assumptions made are clearly stated, can give an indication of the effects of land and water use that are undoubtedly responsible for a large part of the overall impacts attributable to cereal product systems. For example, a method for identifying ultimate or marginal land use changes when studying crop consumption via LCA is proposed by Kløverpris et al. (2010). Furthermore, in such a respect, LCIA methods that deal with site specificity are being developed (e.g. IMPACT World+ 2014) and should be used whenever possible to improve the overall results of the LCA.

Finally, as highlighted in the literature review (Sect. 4.2—e.g. Braschkat et al. 2003; Pelletier et al. 2008), organic cultivation (e.g. duck rice) and certain agricultural practices (e.g. reduced tillage) can have a beneficial effect on the sustainability of cereal product systems. In such a context, with regard to the climate change impact category, the LCIA phase should carefully account for any sequestration or specific emissions of biogenic greenhouse gases, which in many studies are erroneously assumed to generate an overall GHG balance of zero.

4.4.7 Interpretation

Generally the main recommendation for the interpretation of the LCA results is to perform combined sensitivity and uncertainty analysis in order to test the influence of the variability of the input data and the uncertainty connected with subjective choices on the final outcomes of the LCIA results.

In the reviewed cereal-related LCA studies, the most important issues with regard to the sensitivity check emerged as being linked to the cultivation phase with reference to the emissions related to the use of input (mainly N emissions from fertiliser use); the yield levels; the comparative analyses; and the transport and waste phases. Some authors also stress the need to perform a sensitivity check for methodological issues, such as the choice of the functional unit or system boundaries; the methodological exploration of LCI modelling of land use; and to investigate the effect of uncertainty due to the level of precision of the collected input data, the variations in climate and farm practice schedules.

Concerning uncertainty analysis, a practical solution for testing the model outputs according to the variability of inventory data is suggested by Niero et al. (2012). The approach is based on a combined qualitative and quantitative analysis, implemented through a qualitative assessment by data quality indicators and a quantitative analysis through the Monte Carlo sampling technique. If empirical data are available for calculating the uncertainty distribution, this should be the preferred option, instead of using expert judgement to make qualified estimates.

With these combined tools, the conclusions of the LCA study can be strengthened and the robustness and transparency of the study can be improved.

Conclusions

Cereals and their derived products represent agricultural commodities of worldwide importance, with particular environmental hot spots originating from their widespread use and from their particular nature. The review illustrated in this chapter of the life cycle approaches related to cereals has highlighted that the agricultural phase is in most cases the one responsible for a larger share of the impacts of such product systems. Specifically, fertiliser and pesticide production and use and fuel-related emissions seem to be a common source of impact. Fuel use is responsible for a large contribution to the energy demand and acidification. Fertilisation and pesticide usage are also responsible for a large quota of the overall energy use during the life cycle of cereal-based products and hence are also responsible for the production of GHG. Such energy demand, together with ozone-depleting and acidifying emissions typical of intensive agricultural systems, are the reason for the lower impact of alternative organic types of cultivation that generally do not involve the use of pesticides and avoid the production of fertilisers, including nitrogen-based ones. However, even though organic agricultural approaches can potentially lower the overall impact of cropping systems, the lower fertiliser use and relative energy use in such systems can at times be counterbalanced by larger energy use for fieldwork and lower yields, which in turn lead to overall greater land occupation needed for the cereal production.

The growing number of CF studies highlights an emphasis on the study of the effects of cereal systems on climate change. In this context, rice differentiates itself from corn and wheat since there are many types of rice and production processes, all of which are responsible not only for the above-mentioned major contributions deriving from the production and use of fertilisers but also for the contribution of methanogenesis occurring in the waterlogged soil. Accordingly, there still seems to be no unique methodology for the reduction of the impacts of crop production, and the results from the studies vary substantially; however, the literature indicates that careful use of organic approaches and controlled water use and puddling methods need to be considered in order to reduce both WFs and CFs. What can certainly improve the overall sustainability of cereal production is the use of correct crop rotation by following the cereal cultivation, whenever possible, with legume or vegetable cultivation.

The literature review has also highlighted that the user behaviour when dealing with cereal-based products, for example in terms of transport distance and typology for the purchase of such products or the disposal of waste, can heavily influence the overall environmental sustainability of such product systems. There is thus a need to inform customers better and enhance their awareness of the possibilities of contributing to more sustainable cereal product systems with particular reference to the end of life of the product, which is often one of the least-studied LCA phases.

Overall, the implementation of LCA approaches, at an institutional level (both in developed and in developing countries), at the large corporate firm and SME levels, has increased the environmental consciousness of the people involved in the cereal sector, including users and customers, with an overall reduction of the burdens deriving from such product systems. There is in fact growing use of LCA in the cereal sector for the obtaining of environmental labels (e.g. EPDs). However, there is still a need for a better understanding of the difficulties that can be encountered when performing an LCA of a cereal product system in order to gain the best possible results, which can be used to improve the sustainability of the system. Some of these methodological aspects have been discussed in this chapter, such as the site and time dependency of pesticide diffusion modelling, the need for a deeper analysis and a standardised methodology for calculating the effects of land use on the quality of soil and biodiversity and the need for better quantification and qualification of water use. Among the reviewed studies, the system boundary and functional unit definition appear to be critical stages of the LCAs, in which it may be necessary to use one or more FUs, inevitably using allocation for environmental burden partitioning, and in which certain assumptions have to be made in order to progress with the overall assessment and overcome the lack of data and time or cost issues. The sources of information at the base of these assumptions are not always accurate; hence, it is important to explicate them carefully and evaluate the representativeness of the results and their variability in order to produce useful cereal LCA work that can help understand and improve the sustainability of such product systems.

References

Aldaya, M. M., & Hoekstra, A. Y. (2010). The water needed for Italians to eat pasta and pizza. *Agricultural Systems, 103*(6), 351–360.

Aldaya, M. M., Muñoz, G., & Hoekstra, A. Y. (2010). Water footprint of cotton, wheat and rice production in Central Asia. Value of Water Research Report Series No.41, UNESCO-IHE.

Audsley, E., Alber, S., Clift, R., Cowell, S., Crettaz, P., Gaillard, G., Hausheer, J., Jolliet, O., Kleijn, R., Mortensen, B., Pearce, D., Roger, E., Teulon, H., Weidema, B. and Van Zeijts, H. (2003). Harmonisation of Environmental Life Cycle Assessment for Agriculture. Final Report Concerted Action AIR3-CT94-2028. European Commission, Silsoe, United Kingdom, p. 70.

Awika, J. M. (2011). Major cereal grains production and use around the world. In J. M. Awika, V. Piironen, & S. Bean (Eds.), *Cereal science: Implications to food processing and health promotion* (pp. 1–13). Washington, DC: ACS Symposium Series, American Chemical Society.

Benedetto, G., Petti, L., & Raggi, A. (2013, November). *The potential of the LCT approaches in supporting economic choices at company level in the agri-food sector*. Poster paper presented during the19th SETAC LCA Case Study Symposium, Rome, Italy.

Berthoud, A., Maupu, P., Hue, C., & Poupart, A. (2011). Assessing freshwater ecotoxicity of agricultural products assessing freshwater ecotoxicity of agricultural products in life cycle assessment (LCA): A case study of wheat using French agricultural practices databases and USEtox model. *The International Journal of Life Cycle Assessment, 16,* 841–847.

Bevilacqua, M., Braglia, M., Carmignani, G., & Zammori, F. A. (2007). Life cycle assessment of pasta production in Italy. *Journal of Food Quality, 30,* 932–952.

Birkved, M., & Hauschild, M. Z. (2006) PestLCI—A model for estimating field emissions of pesticides in agricultural LCA. *Ecological Modelling, 198,* 433–451

Biswas, W. K., Graham, J., Kelly, K., & John, M. B. (2010). Global warming contributions from wheat, sheep meat and wool production in Victoria, Australia e a life cycle assessment. *Journal of Cleaner Production, 18,* 1386–1392.

Blengini, G. A., & Busto, M. (2009). The life cycle of rice: LCA of alternative agri-food chain management systems in Vercelli, Italy. *Journal of Environmental Management, 90,* 1512–1522.

Brankatschk, G., & Finkbeiner, M. (2012, October). *Allocation challenges in agricultural life cycle assessments and the Cereal Unit allocation procedure as a potential solution.* Paper presented at the 8th International Conference on LCA in the Agri-Food Sector, Saint-Malò, France.

Braschkat, J., Patyk, A., Quirin, M., & Reinhardt, G. A. (2003, October). *Life cycle assessment of bread production—a comparison of eight different scenarios.* Paper presented at the 4th Conference on LCA in the Agri-Food Sector, Bygholm, Denmark.

Breiling, M., Hashimoto, S., Sato, Y., & Ahamer, G. (2005). Rice-related greenhouse gases in Japan, variations in scale and time and significance for the Kyoto Protocol. *Paddy and Water Environment, 3*(1), 39–46.

Brentrup, F., Küsters, J., Lammel, J., Kuhlmann, H., (2000). Methods to estimate on-field nitrogen emissions from crop production as an input to LCA studies in the agricultur*al* sector. *Int. J. Life Cycle Assess.,* 5, 349–357.

Brentrup, F., Küsters, J., Lammela, J., Barraclough, P., & Kuhlmann, H. (2004). Environmental impact assessment of agricultural production systems using the life cycle assessment (LCA) methodology II. The application to N fertilizer use in winter wheat production systems. *European Journal of Agronomy, 20,* 265–279.

Bretz, R. (1998). SETAC LCA workgroup: Data availability and data quality. *The International Journal of Life Cycle Assessment, 3*(3), 121–123.

Carozzi, M., Ferrara, R. M., Rana, G., & Acutis, M. (2013). Evaluation of mitigation strategies to reduce ammonia losses from slurry fertilisation on arable lands. *Science of the Total Environment, 449,* 126–133.

Cerutti, A. K., Beccaro, G., Bagliani, M., Contu, S., Donno, D., & Bounous, G. (2012). Ecological footprint applied in agro-ecosystems: Methods and case studies, the functioning of ecosystems, Prof. Mahamane Ali (Ed.), ISBN: 978-953-51-0573-2, In Tech, doi:10.5772/35966. http://www.intechopen.com/books/the-functioning-of-ecosystems/ecological-footprint-applied-in-agro-ecosystems-methods-and-case-studies.

Chapagain, A. K., & Hoekstra, A. Y. (2011). The blue, green and grey water footprint of rice from production and consumption perspectives. *Ecological Economics, 70,* 749–758.

Charles, R., Jolliet, O., Gaillard, G., & Pellet, D. (2006). Environmental analysis of intensity level in wheat crop production using life cycle assessment. *Agriculture, Ecosystems and Environment, 113,* 216–225.

Cordain, L. (1999). Cereal grains: Humanity's double-edged sword. *World Review Nutrition Dietetics Basel, Karger, 84,* 19–73.

Curran, M. A. (2008). *Development of life cycle assessment methodology: A focus on co-product allocation.* Rotterdam: Erasmus University.

Delivand, M. K., Barz, M., Gheewala, S. H., & Sajjakulnukit, B. (2012). Environmental and socio-economic feasibility assessment of rice straw conversion to power and ethanol in Thailand. *Journal of Cleaner Production, 37,* 29–41.

Diamond, J. (2002). Evolution, consequences and future of plant and animal domestication. *Nature, 418,* 700–707. doi:10.1038/nature01019.

Dijkman, T. J., Birkved, M., & Hauschild, M. Z. (2012) PestLCI 2.0: A second generation model for estimating emissions of pesticides from arable land in LCA. *The International Journal of Life Cycle Assessment, 17,* 973–986.

Drocourt, A., Mervant, Y., Milhau, F., Chinal, M., & Hélias, A. (2012, October). *Environmental assessment of rice production in Camargue, France*. Paper presented at the 8th Conference on LCA in the Agri-Food Sector, Saint-Malò, France).

Dyson, T. (1999). World food trends and prospects to 2025. *Proceedings of the National Academy of Sciences United Sates of America, 96,* 5929–5936. doi:10.1073/pnas.96.11.5929

Ecoinvent—version 3. http://www.ecoinvent.ch/. Accessed 10 Jan 2014.

Editorial of Journal of Cleaner Production. (2008). Sustainability and supply chain management: An introduction to the special issue. *Journal of Cleaner Production, 16,* 1545–1551. doi:10.1016/j.jclepro.2008.02.002.

Ekval, T., & Weidema, B. P. (2004). System boundaries and input data in consequential life cycle inventory analysis. *The International Journal of Life Cycle Assessment, 9,* 161–71

Elkington, J. (1997). Cannibals with Forks: The Triple Bottom Line of 21st Century Business. Oxford: Capstone Publishing Ltd.

Environdec. (2014). http://www.environdec.com/PCR/Pcr-Search/?Category=6192/. Accessed 13 Jan 2014.

Eshun, J. F., Apori, S. O., & Wereko, E. (2013). Greenhouse gaseous emission and energy analysis in rice production systems in Ghana. *African Crop Science Journal, 21*(2), 119–125.

Espinoza-Orias, N., Stichnothe, H., & Azapagic, A. (2011). The carbon footprint of bread. *The International Journal of Life Cycle Assessment, 16,* 351–365.

Fallahpour, F., Aminghafouri A., Ghalegolab Behbahani, A., & Bannayan M. (2012). The environmental impact assessment of wheat and barley production by using life cycle assessment (LCA) methodology. *Environment Development and Sustainability, 14,* 979–992.

FAO. (2011a). FAOSTAT statistical database. http://faostat.fao.org/.

FAO. (2011b). Save and grow. A policy maker's guide to the sustainable intensification of smallholder crop production. http://www.fao.org/docrep/014/i2215e/i2215e.pdf. Accessed 15 Nov 2013.

FAO. (2011c). Energy-smart food for people and climate. Issue paper. http://www.fao.org/docrep/014/i2454e/i2454e00.pdf. Accessed 25 Nov 2013.

FAO. (2013a). Food wastage footprint. Impacts on natural resources. Summary reports. http://www.fao.org/docrep/018/i3347e/i3347e.pdf. Accessed 10 Dec 2013.

FAO. (2013b). FAO statistical yearbook 2013 world food and agriculture. Food and Agriculture Organization of the United Nations, Rome. ISBN 978-92-5107396-4.

FAO, IFAD, & WFP. (2013). The State of Food Insecurity in the World 2013. The multiple dimensions of food security. Rome, FAO.

Ferng, J. J. (2011). Measuring and locating footprints: A case study of Taiwan's rice and wheat consumption footprint. *Ecological Economics, 71,* 191–201.

Gan, Y., Liang, C., Hamel, C., Cutforth, H., & Wang, H., (2011a). Strategies for reducing the carbon footprint of field crops for semiarid areas. A review. *Agronomy for Sustainable Development, 31,* 643–656.

Gan, Y., Liang, C., Wang, X., & McConkey, B. (2011b). Lowering carbon footprint of durum wheat by diversifying cropping systems. *Field Crops Research, 122,* 199–206.

Gnansounou, E., Dauriat, A., Villegas, J., & Panichelli, L. (2009). Life cycle assessment of biofuels: Energy and greenhouse gas balances. *Bioresource Technology, 100,* 4919–4930.

Guinée, J. B., Gorée, M., Heijungs, R., Huppes, G., Kleijn, R., Koning, A., de Oers, L., van Wegener Sleeswijk, A., Suh, S., Udo de Haes, H. A., Bruijn, H., de Duin, R., & van Huijbregts, M. A. J. (2002). *Handbook on life cycle assessment. Operational guide to the ISO standards.* Kluwer Academic, Dordrecht.

Guinée, J. B., Heijungs, R., & Huppes, G. (2004). Economic allocation: Examples and derived decision tree. *The International Journal of Life Cycle Assessment, 9*(1), 23–33.

Harada, H., Kobayashi, H., & Shindo, H. (2007). Reduction in greenhouse gas emissions by no-tilling rice cultivation in Hachirogata polder, northern Japan: Life-cycle inventory analysis. *Soil Science and Plant Nutrition, 53,* 668–677.

Hauschild, M. Z., Goedkoop, M., Guinée, J., Heijungs, R., Huijbregts, M., Jolliet, O., Margni, M., De Schryver, A., Humbert, S., Laurent, A., Sala, S., & Pant, R. (2013). Identifying best existing practice for characterization modeling in life cycle impact assessment. *The International Journal of Life Cycle Assessment, 18,* 683–697.

Hayashi, K., Uchida, S., Hokazono, S., & Sato, M. (2010, September). Modelling life cycle inventories for crop production in Japan: Development of the NARO LCI database. Paper presented at the VII International Conference on Life Cycle Assessment in the Agri-Food Sector, Bari, Italy.

Hayashi, K. (2013). Practical recommendations for supporting agricultural decisions through life cycle assessment based on two alternative views of crop production: the example of organic conversion. *Int. J. Life Cycle Assess.,* 18, 331–339.

Hokazono, S., & Hayashi, K. (2012). Variability in environmental impacts during conversion from conventional to organic farming: A comparison among three rice production systems in Japan. *Journal of Clearner Production, 28,* 101–112.

Huang, J., Ridoutt, B. G., Huang, F., & Chen, F. (2012, October). Water footprints of wheat and maize: Comparison between China's main breadbasket basins. Paper presented at the 8th International Conference on LCA in the Agri-Food Sector, Saint-Malò, France.

Huijbregts, M. A. J., Hellweg, S., Frischknecht, R., Hendriks, H. W. M., Hungerbuhler, K., & Hendriks, A. J. (2010). Cumulative energy demand as predictor for the environmental burden of commodity production. *Environmental Science and Technology, 44*(6), 2189–2196.

Impact World. http://www.impactworldplus.org/en/presentation.php. Accessed 10 Jan 2014.

IPCC. (2006) IPCC guidelines for national greenhouse gas inventories, vol. 4.

ISO. (2006). 14025: Environmental labelling and declarations—Type III environmental declarations—Principles and procedures. International Organization for Standardization.

JRC-IES-European Commission. (2011). Recommendations based on existing environmental impact assessment models and factors for life cycle assessment in European context. ILCD Handbook—International Reference Life Cycle Data System. . http://lct.jrc.ec.europa.eu/assessment/assessment/projects#consultation_impact. Accessed 3 Jan 2014.

Kägi, T., Wettstein, D., & Dinkel, F. (2010, September). *Comparing rice products: Confidence intervals as a solution to avoid wrong conclusions in communicating carbon footprints.* Paper presented at the VII International Conference on Life Cycle Assessment in the Agri-Food Sector, Bari, Italy.

Kasmaprapruet, S., Paengjuntuek, W., Saikhwan, P., & Phungrassami, H. (2009). Life cycle assessment of milled rice production: Case study in Thailand. *European Journal of Scientific Research, 30*(2), 195–203.

Khoo, H. H., Wong, J. L., & Tan, R. B. H. (2010, September). *An LCA approach for evaluating the Global Warming Potential of various food types imported to Singapore.* Paper presented at the VII International Conference on Life Cycle Assessment in the Agri-Food Sector, Bari, Italy.

Kim, S., & Dale, B. E. (2002). Allocation procedure in ethanol production system from corn grain i. system expansion. *The International Journal of Life Cycle Assessment, 7*(4), 237–243.

Kim, S., & Dale, B. E. (2008a). Effects of nitrogen fertilizer application on greenhouse gas emissions and economics of corn production. *Environmental Science and Technology, 42,* 6028–6033.

Kim, S., & Dale, B. E. (2008b). Life cycle assessment of fuel ethanol derived from corn grain via dry milling. *Bioresource Technology, 99*(12), 5250–5260.

Kim, S., Dale, B. E., & Jenkins, R. (2009) Life cycle assessment of corn grain and corn stover in the US. *The International Journal of Life Cycle Assessment, 14,* 160–174.

Kirk-Othmer (1984). Wheat and other cereal grains. In *Encyclopedia of Chemical technology* (Vol. 24). New York: Wiley.

Kissinger, M., & Gottliebb, D. (2010). Place oriented ecological footprint analysis—The case of Israel's grain supply. *Ecological Economics, 69*(8), 1639–1645.

Kløverpris, J. H., Baltzer, K., & Nielsen, P. H. (2010). Life cycle inventory modelling of land use induced by crop consumption. Part 2: Example of wheat consumption in Brazil, China, Denmark and the USA. *The International Journal of Life Cycle Assessment, 15,* 90–103.

Kraisintu, K., & Zhang, T. (2011). The role of traceability in sustainable supply chain management. Master of Science Thesis Report No. E2011:085. Chalmers University of Technology. Gothenburg, Sweden. http://publications.lib.chalmers.se/records/fulltext/146242.pdf. Accessed 3 Dec 2013.

Kulak, M., Nemecek, T., Frossard, E., & Gaillard, G. (2012, October). *Ecodesign opportunities for a farmer's bread. Two case studies from north-western France*. Paper presented at the 8th Conference on LCA in the Agri-Food Sector, Saint-Malò, France.

Laurent, E., Bellino, R., & Le Pochat, S. (2012, October). *Availability and completeness of LCI of cereal products and related databases*. Paper presented at the 8th Conference on LCA in the Agri-Food Sector, Saint-Malò, France.

Li Borrion, A., McManus, M. C., & Hammond, G. P. (2012). Environmental life cycle assessment of bioethanol production from wheat straw. *Biomass and Bioenergy, 47,* 9–19.

Lipinski, B., Hanson, C., Lomax, J., Kitinoja L., Waite, R., & Searchinger, T. (2013). *Reducing food loss and waste. Working paper*. World Resources Institute. http://www.wri.org/sites/default/files/reducing_food_loss_and_waste.pdf. Accessed 3 Dec 2013.

Lo Giudice, A., Clasadonte, M. T., & Matarazzo, A. (2011). LCI preliminary results in the Sicilian durum wheat pasta chain production. *Journal of Commodity Science, Technology and Quality, 50*(I), 65–79.

Malça, J., & Freire, F. (2012). Addressing land use change and uncertainty in the life-cycle assessment of wheat-based bioethanol. *Energy, 45*(1), 519–527.

Margni, M., & Curran, M. A. (2012). Life cycle impact assessment. In M. A. Curran (Ed.), *Life cycle assessment handbook: A guide for environmentally sustainable products* (pp. 67–104). USA: Wiley.

McConkey, B., Chen, C., Maxime, D., Kulshrestha, S., Vergé, X., & Desjardins, R. (2012, October). *Green, blue, and grey water use of Canadian wheat and maize*. Paper presented at the 8th International Conference on LCA in the Agri-Food Sector, Saint-Malò, France.

Meisterling, K., Samaras, C., & Schweizer, V. (2009). Decisions to reduce greenhouse gases from agriculture and product transport: LCA case study of organic and conventional wheat. *Journal of Cleaner Production, 17,* 222–230.

Muñoz, E., Montalba, R., & Herrera, J. (2012, November). *Evaluation of two styles of Chilean wheat by means of LCA methodology*. Paper presented at the 2nd anviR LCA Conference, Lille, France.

Murphy, C., & Kendall, A. (2013). Life cycle inventory development for corn and stover production systems under different allocation methods. *Biomass and Bioenergy, 58,* 67–75.

Nalley, L., Popp, M., & Fortin, C. (2011). The impact of reducing greenhouse gas emissions in crop agriculture: A spatial and production-level analysis. *Agricultural and Resource Economics Review, 40*(1), 63–80.

Narayanaswamy, V., Altham, V., Van Berkel, R., & McGregor, M. (2004). Environmental Life Cycle Assessment (LCA) Case Studies for Western Australian Grain Products, Project Funded by Grain Research & Development Corporation, Curtin University of Technology, Perth, Western Australia.

Natural Resource Ecology Laboratory. (2005). Century soil organic matter model: User's guide and reference, Colorado State University. http://www.nrel.colostate.edu/projects/century5/reference/index.htm.

Nemecek, T., & Baumgartner, D. (2006). Environmental impacts of introducing grain legumes into European crop rotations and pig feed formulas. Concerted Action GL-Pro, Final report WP4. Agroscope Reckenholz-T¨anikon Research Station ART, p. 63. http://www.art.admin.ch/themen/00617/00789/index.html?lang=En-US.

Nemecek, T., & Erzinger, S. (2005). Modelling representative life cycle inventories for Swiss arable crops. *The International Journal of Life Cycle Assessment, 10,* 68–76.

Nemececk, T., von Richthofen, J. S., Dubois, G., Casta, P., Charles, R., & Hubert, P. (2008). Environmental impacts of introducing grain legumes into European crop rotations. *European Journal of Agronomy, 28,* 380–393.

Niero, M., Manzardo, A., Toniolo, S., Zuliani, F., & Scipioni, A. (2012, October). *Uncertainty analysis in a comparative LCA between organic and conventional farming of soybean and barley.* Paper presented at the 8th International Conference on LCA in the Agri-Food Sector, Saint-Malò, France.

Notarnicola, B., & Nicoletti, G. M. (2001). Life cycle assessment of pasta and couscous. *Tecnica Molitoria, 52*(1), 19–27, 33.

Notarnicola, B., Mongelli, I., Tassielli, G., & Nicoletti, G. M. (2004). Environmental input-output analysis and hybrid approaches to improve the setup of the pasta life cycle inventory. *Journal of Commodity Science, 43*(II), 59–86.

Notarnicola, B., Tassielli, G., & Renzulli, P. A. (2012). Modeling the agri-food industry with life cycle assessment. In M. A. Curran (Ed.), *Life cycle assessment handbook: A guide for environmentally sustainable products* (pp. 159–183), Wiley.

Notarnicola, B., Hayashi, K., Curran, M. A. & Huisingh, D. (2012a). Progress in working towards a more sustainable agri-food industry. *Journal of Cleaner Production,* 28, *1–8*

Pelletier, N., Arsenault, N., & Tyedmers, P. (2008). Scenario modelling potential eco-efficiency gains from a transition to organic agriculture: life cycle perspectives on Canadian canola, corn, soy, and wheat production. *Environmental Management, 42,* 989–1001.

Petti, L., Raggi, A., Pagliuca, G., & Ancona, F. (2000, May). *LCA implementation to pasta products: A case study.* Paper presented 3rd SETAC World Congress, Brighton, UK.

Place, F., & Meybeck, A. (Eds.). (2013, April). *Food security and sustainable resource use—what are the resource challenges to food security*? Background paper for the conference Food Security Futures: Research Priorities for the 21st Century, Dublin, Ireland.

Prasara A. J., & Grant, T. (2011). Comparative life cycle assessment of uses of rice husk for energy purposes. *The International Journal of Life Cycle Assessment, 16*(6), 493–502.

Proto, M. (1988). *Evoluzione e stato attuale della produzione e csumi di margarina. Industrie Alimentari 17*(4), 640–648.

Reardon, T., & Timmer, C. P. (2012). The economics of food system revolution. *Annual Review of Resource and Economics, 4,* 225–264. doi:10.1146/annurev.resource.050708.144147.

Reap, J., Roman, F., Duncan, S., Bras, B., (2008). A survey of unresolved problems in life cycle assessment. *Int. J. Life Cycle Assess.,* 13(5), 374–388.

Renewable Fuels Association (RFA). (2010). Ethanol Industry Outlook: Climate of Opportunity. http://www.ethanolrfa.org/. last visited 16/12/2013. Accessed 14 Dec 2013.

Renouf, M. A., Wegener, M. K., & Nielsen, L. K. (2008). An environmental life cycle assessment comparing Australian sugarcane with US corn and UK sugar beet as producers of sugars for fermentation. *Biomass and Bioenergy, 32,* 1144–1155.

Roer, A. G., Korsaeth, A., Henriksen, T.M., Michelsen, O., & Strømman, A. H. (2012). The influence of system boundaries on life cycle assessment of grain production in central southeast Norway. *Agricultural Systems, 111,* 75–84.

Röös, E., Sundberg, C., & Hansson, P. A. (2011). Uncertainties in the carbon footprint of refined wheat products: A case study on Swedish pasta. *The International Journal of Life Cycle Assessment, 16*(4), 338–350.

Roy, P., Shimizu, N., & Kimura, T. (2005). Life cycle inventory analysis of rice produced by local processes. *Journal of the Japanese Society of Agricultural Machinery, 67*(1), 61–67.

Roy, P., Shimizu, N., Okadome, H., Shiina, T., & Kimura, T. (2007). Life cycle of rice: Challenges and choices for Bangladesh. *Journal of Food Engineering, 79,* 1250–1255.

Roy, P., Nei, D., Orikasa, T., Xu, Q., Okadome, H., Nakamura, N. & Shiina, T. (2009a). A review of life cycle assessment (LCA) on some food products. *Journal of Food Engineering,* 90(1), 1–10.

Roy, P., Ijiri, T., Nei, D., Orikasa, T., Okadome, H., Nakamura, N. & Shiina, T. (2009b). Life cycle inventory (LCI) of different forms of rice consumed in households in Japan. *Journal of Food Engineering,* 91, 49–55.

Ruini, L. (2011, Giugno). *LCA & EPD: dall'Analisi all'Azione*. Paper presented at the Conference of the Rete Italiana LCA, Rome, Italy.

Ruini, L., & Marino, M. (2010, September). *An overview of environmental indicators aimed to an easy life cycle assessment results presentation by an important food supplier: The example of durum wheat cultivation*. Paper presented at the VII International Conference on Life Cycle Assessment in the Agri-Food Sector, Bari, Italy.

Ruini L., Campra L., Marino, M., Filareto A., Bagliani M., & Contu, S. (2010, June). *Ecological footprint as large-scale pasta producer Key Performance Indicators (KPI). The State of the Art in Ecological Footprint Theory and Applications*. Paper presented at FOOTPRINT FORUM 2010 Academic Conference, Siena, Italy.

Ruini, L., Marchelli, L., Marino, M., & Filareto, A. (2012, October). *Barilla EPD process system to increase reliability, comparability and communicability of LCA studies*. Paper presented at the 8th International Conference on LCA in the Agri-Food Sector, Saint-Malò, France.

Ruini, L., Ferrari, E., Meriggi, P., Marino, M., & Sessa, F. (2013, May). *Increasing the sustainability of pasta production through a life cycle assessment approach*. Paper presented at the 4th International Workshop Advances in cleaner production, São Paulo, Brazil.

Salomone, R., & Ciraolo, L. (2004, September). Applicazione della metodologia Life Cycle Assessment alla produzione di pasta: alcuni risultati preliminari. Paper presented at the XXI Congresso Scienze Merceologiche Risorse naturali e sviluppo economico-sociale. Foggia, Italy.

Schau, E. M., & Fet, A. M. (2008). LCA studies of food products as background for environmental product declarations. *International Journal of Life Cycle Assessment, 13*(3), 255–264.

Schmidt, J. H. (2008). System delimitation in agricultural consequential LCA—Outline of methodology and illustrative case study of wheat in Denmark. *The International Journal of Life Cycle Assessment, 13,* 350–364.

Seda, M., Anton, A., & Munoz, P. (2010, September). *Analysing the influence of the functional unit in agricultural LCA*. Paper presented at the VII Conference on LCA in the Agri-Food Sector, Bari, Italy, 85–90.

Settanni, E., Tassielli, G., & Notarnicola, B. (2010a). Combining LCA of food products with economic tools. In U. Sonesson, J. Berlin & F. Ziegler (Eds.), *Environmental assessment and management in the food industry*, Woodhead.

Settanni, E., Notarnicola, B., & Tasselli, G. (2010b, September). *Application of a costing model consistent with LCA to the production of pasta in a small-sized firm*. Paper presented at the VII International Conference on Life Cycle Assessment in the Agri-Food Sector, Bari, Italy.

Seuring, S., & Müller, M. (2008). From a literature review to a conceptual framework for sustainable supply chain management [Electronic version]. *Journal of Cleaner Production, 16,* 1699–1710.

Shafie, S. M., Mahlia, T. M. I., Masjuki, H. H., & Rismanchi, B. (2012). Life cycle assessment (LCA) of electricity generation from rice husk in Malaysia. *Energy Procedia, 14,* 499–504.

Spatari, S., Zhang, Y., & MacLean, H. L. (2005). Life Cycle Assessment of switchgrass- and corn stover-derived ethanol-fueled automobiles. *Environmental Science & Technology, 39,* 9750–9758.

Suh, S., & Huppes, G. (2005). Methods for Life Cycle Inventory of a product. *Journal of Cleaner Production, 13,* 687–697.

Van der Werf, H. M. G., Corson, M. S. C., & Wilfart, A. (2013). LCA Food 2012—Towards sustainable food systems. *The International Journal of Life Cycle Assessment, 18*, 1180–1183.

Van Stappen, F., Delcour, A., Gheysens, S., Decruyenaere, V., Stilmant, D., Burny, P., Rabier, F., & Goffart, J-P. (2012, October). *Assessment of existing and potential cereal food and non-food uses by combining E-LCA and S-LCA*. Paper presented at the 8th International Conference on LCA in the Agri-Food Sector, Saint-Malò, France.

Venkat, K. (2007). *Analyzing and optimizing the environmental performance of supply chains*. ACEEE Summer Study on Energy Efficiency in Industry T. Surya Technologies, Inc.

Vidal, R., Martínez, P., & Garraín, D. (2009). Life cycle assessment of composite materials made of recycled thermoplastics combined with rice husks and cotton linters. *The International Journal of Life Cycle Assessment, 14*(1), 73–82.

Wang, M., Xia, X., Zhang, Q., & Liu, J. (2010). Life cycle assessment of a rice production system in Taihu region, China. *The International Journal of Sustainable Development & World Ecology, 17*(2), 157–161.

Williams, A.G., Audsley, E., & Sandars, D. L. (2005). Final Report to Defra on Project IS0205: Determining the environmental burdens and resource use in the production of agricultural and horticultural Commodities. Department of Environment, Food, and Rural Affairs (Defra), London. http://www2.defra.gov.uk/research/Project_Data/More.asp%3FI%3DIS0205†%26M%3DKWS%26V%3DERM.

Williams, A. G., Audsley, E., & Sandars, D. L. (2010). Environmental burdens of producing bread wheat, oilseed rape and potatoes in England and Wales using simulation and system modelling. *The International Journal of Life Cycle Assessment, 15,* 855–868.

Xu, X., Zhang, B., Liu, Y., Xue, Y., & Binsheng, D. (2013). Carbon footprints of rice production in five typical rice districts in China. *Acta Ecologica Sinica, 33*(4), 227–232.

Yan, X., & Boies, A. M. (2013). Quantifying the uncertainties in life cycle greenhouse gas emissions for UK wheat ethanol. *Environmental Research Letters, 8,* 015024.

Yoo, S.H., Choi, J.Y., Lee, S.H., & Kim, T. (2013). Estimating water footprint of paddy rice in Korea. *Paddy and Water Environment, 12*(1), 43–54.

Yoshikawa, N., Ikeda, T., Amano, K., & Shimada, K. (2010, September). Carbon footprint estimation and data sampling method: a case study of ecologically cultivated rice produced in Japan. Paper presented at the VII International Conference on Life Cycle Assessment in the Agri-Food Sector, Bari, Italy.

Yoshikawa, N., Ikeda, T., Amano, K., & Fumoto, T. (2012, November). *Life-cycle assessment of ecologically cultivated rice applying DNDC-Rice model.* Paper presented at the 10th International Conference on EcoBalance 'Challenges and Solutions for Sustainable Society', Yokohama, Japan.

Yossapol, C., & Nadsataporn, H. (2008, November). *Life Cycle Assessment of rice production in Thailand.* Paper presented at the 6th Conference on LCA in the Agri-Food Sector, Zurich, Switzerland.

van Zelm, R., Larrey-Lassalle, P., & Roux, P. (2013). Bridging the gap between life cycle inventory and impact assessment for toxicological assessments of pesticides used in crop production. *Chemosphere.* doi:http://dx.doi.org/10.1016/j.chemosphere.2013.11.037.

Zhang, M.-Y., Wang, F. -J., Chena, F., Malemela, M. P., & Zhang, H. -L. (2013). Comparison of three tillage systems in the wheat-maize system on carbon sequestration in the North China Plain. *Journal of Cleaner Production, 54,* 101–107.

Chapter 5
Life Cycle Assessment in the Livestock and Derived Edible Products Sector

Rocco Roma, Sara Corrado, Annalisa De Boni, Maria Bonaventura Forleo, Valentina Fantin, Michele Moretti, Nadia Palmieri, Andrea Vitali and Camillo De Camillis

Abstract The livestock production sector represents more than 40% of the economic value of EU primary productions. This sector consists of a huge diversity of processes and techniques depending on the animal species and the final products. Because of these differences, livestock productions are associated with several adverse effects on the environment, especially in the breeding phases and feeding composition and management; moreover, in terms of raising awareness of the environmental implications of livestock productions, LCA applications are of increasing importance for systematic assessment of the environmental burdens connected with this sector. After an overview of the structural and economic characteristics of the most significant livestock supply chain and its main environmental problems, we provide a description of the international state of the art of LCA implementations for livestock. Methodological problems connected with the application of LCA are investigated, starting with the critical analysis of international papers and the few Italian papers in the scientific literature. Finally, the best practices regarding LCA methodology implementation are proposed, in order to improve results and manage the methodological problems identified.

Keyword Livestock sector · Life cycle assessment · Life cycle costing · Environmental product declaration · Footprint labels

R. Roma (✉) · A. De Boni · M. Moretti
Department of Agro-Environmental Science, University of Bari, Via Orabona 4, 70126 Bari, Italy
e-mail: rocco.roma@uniba.it

A. De Boni
e-mail: annalisa.deboni@uniba.it

M. Moretti
e-mail: michele.moretti@uniba.it

B. Notarnicola et al. (eds.), *Life Cycle Assessment in the Agri-food Sector*,
DOI 10.1007/978-3-319-11940-3_5

5.1 Introduction

The global livestock industry accounts for almost 40 % of agricultural GDP (Steinfeld et al. 2006) and the global meat production is projected to double by 2050 following the increase in meat demand (FAO 2006). At the same time, the FAO (2006) states that the livestock sector is one of the most critical sectors in terms of environmental problems such as climate change, water and air pollution, land degradation, and biodiversity loss (Steinfeld et al. 2006). Livestock systems occupy about 30 % of free terrestrial surface area and with a value of at least $ 1.4 trillion they are a major global asset. In developed countries production and consumption of livestock products are now growing slowly, albeit at high levels of production, accounting for 53 % of their agricultural GDP (World Bank 2009). Impacts may vary significantly depending on the supply chain in question (meat and dairy, pigs, sheep, goats or chickens) and the practices and techniques employed. In recent years, this sector has received particular attention, and has been the subject of a number of studies since it was defined as one of the productive sectors with the highest environmental impacts (Steinfeld et al. 2006; Weidema et al. 2009). This is because of 75–90 % of the energy consumed by the animal in its diet is then used for body maintenance or lost in manure and by-products such as skin and bones. Livestock competes with the other productive sectors for the use of scarce resources such as land, water and energy, and, according to the Food and Agriculture Organization (FAO), it is responsible

S. Corrado
Institute of Environmental and Chemical Agriculture, Università cattolica del Sacro Cuore, Via Emilia Parmense 84, 29122 Piacenza, Italy
e-mail: sara.corrado@unicatt.it

M. B. Forleo · N. Palmieri
University of Molise, Via F. De Sanctis, 86100 Campobasso, Italy
e-mail: forleo@unimol.it

N. Palmieri
e-mail: nadia.palmieri@unimol.it

V. Fantin
LCA and Ecodesign Laboratory-ENEA (Italian National Agency for New Technologies, Energy and Sustainable Economic Development), Via Martiri di Monte Sole 4, 40129 Bologna, Italy
e-mail: valentina.fantin@enea.it

A. Vitali
DAFNE Department, University of Tuscia, Via S. Camillo de Lellis, 01100 Viterbo, Italy
e-mail: mail@andreavitali.it

C. De Camillis
Food and Agriculture Organization of the United Nations (FAO), Agriculture and Consumer Protection Department, Room C 535, Viale delle Terme di Caracalla, 00153 Rome, Italy
e-mail: Camillo.DeCamillis@fao.org

European Commission (EC), Joint Research Centre, Institute for Environment and Sustainability, Sustainability Assessment Unit, Via Enrico Fermi 2749, 21027 Ispra, Italy for 18% of global greenhouse gas (GHG) emissions (Weidema et al. 2008; Leip et al. 2010); because of the emissions of CO_2, CH_4 and N_2O occurring during crops production for animal feed and the animal rearing (Steinfeld et al. 2006). Livestock production has significant environmental impacts including greenhouse gas (GHG) emissions (Stanford University 2010). According to the FAO (2006), the global demand for beef and milk in 2050 is expected to rise up by 72 and 82%, respectively, compared with 2000, and thus the GHG emissions from these sectors need to be reduced considerably. Identification of the best reduction strategies requires a detailed analysis of the environmental loads produced by each food product, in particular livestock products, during the entire life cycle in order to identify the hotspots and to compare them. For this purpose, the Life Cycle Assessment (LCA), a method that analyses products along the whole production cycle, their use and waste management (Guinée et al. 2002), is well suited for this type of analysis and has been used for the determination of environmental impacts of livestock products (de Vries and de Boer 2010). The structure of this chapter differs from the others in this book because of the numerous types of supply chains in the livestock sector in terms of both species and products (meat, milk, wool, eggs). They obviously use several production technologies and different amounts of resources and so produce different environmental impacts. Consequently, the chapter is ordered according to the different supply chains starting from beef and dairy production, through sheep and goats, and ending with pigs and poultry.

5.2 Overview of Product Based Life Cycle Assessment Methods on Livestock[1]

Livestock products are often shown as amongst the most harmful for the environment. This view is, inter alia, supported by the findings of the Input-Output Environmentally-Extended Analyses conducted for the United Nations Environment Programme (UNEP 2010) and the European Commission's Joint Research Centre (Tukker et al. 2006). According to peer-reviewed environmental assessments conducted by the United Nations Food and Agriculture Organizations (FAO) at global level, the greenhouse emissions tied to livestock are also significant and effective mitigation measures should shaped and implemented (Gerber et al. 2010, 2013; MacLeod et al. 2013; Opio et al. 2013). Such studies contributed, among others, to indicate livestock as top priority for the agenda on the environment of policy makers. Also because of the numerous scientific challenges when accounting for the environmental burdens, pressures and benefits tied to livestock supply chains, several have been the efforts made to advance science on life cycle assessment in these sectors. See the following sections for more detail on the scientific literature produced so far. Building somehow on these efforts, high has been the proliferation of product-specific environmental assessment methods, mostly developed for ei-

[1] The views expressed in this chapter are those of the authors and do not necessarily represent the views of the European Commission and FAO

ther commercial purposes linked to environmental communication or in support of a suite of policy measures on eco-friendlier farming practices or low-carbon biofuel productions. This section introduces these product-specific environmental assessment methods to pave the way to the development of harmonised methods in the context of consensus building initiatives in this field such as, amongst others, the FAO-led Livestock Environmental Assessment and Performance (LEAP) Partnership (LEAP 2014a), the European Commission's Product and Organisation Environmental Footprint (EC 2014) and the European Food Sustainable Consumption and Production Round Table (Food SCP RT 2014). These initiatives involve governments, business representatives, and civil society and strive, to varying extents, for ensuring equal footing in steering the development process of the technical guides. The livestock-specific environmental assessment methods were mostly identified through a search for relevant PCRs in the repositories of the prominent programme operators of product performance-based environmental communication schemes (Subramanian 2012, GEDnet 2014) that was coupled with an internal consultation round within the European Food Sustainable Consumption Round Table in late 2013 (Food SCP RT 2013a). This search was restricted to methods that are freely available. The assessment methods found can be grouped as follows:

Group A[2]: product category rules (PCR) developed in the context of type III environmental labelling schemes established according to ISO 14025 (Boeri 2012; Brondi 2013; IERE 2006; Japanese CFP scheme 2011a, b, c; Marino et al. 2011; Palm 2010; Pernigotti 2011; Sessa 2013a, b, c).

Group B: environmental assessment methods released by and for business associations. For example, the International Dairy Federation and DairyUk have issued carbon footprinting methods (Carbon Trust 2010; IDF 2010). It must be noted that some PCRs have also been prepared, commented and endorsed by specific business associations such as, e.g., Assocarni for the PCR on meat of mammals (Boeri 2012);

Group C: product-based carbon footprinting methods developed in support of policy measures for lowering the environmental performance of farming practices both at product and at organisation level (Tuomisto et al. 2013; Bochu et al. 2013);

Group D: Sector-specific carbon footprinting methods underpinning the peer-reviewed life cycle assessments conducted by authoritative bodies such as FAO (Gerber et al. 2010; 2013; MacLeod et al. 2013; Opio et al. 2013) and the European Commission's Joint Research Centre (Leip et al. 2010);

Group E: Product-based environmental footprinting methods to calculate greenhouse gas emissions of biofuels and biogas out of livestock supply chains (EU 2009). The methods sorted out by product category under concern are presented in Table 5.7. The methods have been grouped in three major categories, namely: products from ruminant supply chains (other than dairy products), dairy products and poultry products. The names of these product categories were set for illustrative sake and are not necessarily aligned with reference international codes for prod-

[2] Further PCRs are seemingly under development in the context of The Sustainability Consortium (TSC, 2014) and the French labelling scheme laid down in the national law generally known as Grenelle de l'Environment (Cros et al. 2010, French Parliament 2010)

ucts. Given the wide range of products coming out from livestock supply chains, Table 5.7 is to be conceived as a partial overview of the methods available to date on livestock products. In addition, new methods are on the pipeline of a few labelling schemes. For instance, The Sustainability Consortium has been developing new methods on, amongst others, beef, milk, butter, cheese, yoghurt, chicken, eggs, and pork (TSC 2014). Similarly, the new PCRs in the context of the French labelling initiative are seemingly under development on dairy products, and on meat and co-products from bovine, poultry and pork supply chains (ADEME AFNOR 2014).

Despite the proliferation of environmental assessment methods in the livestock sector, it is high the methodological misalignment between technical documents applicable to the same product category. If the methods listed in Table 5.7 are analysed against the criteria set for PCR characterization purposes in the context of the PCR Guidance Development Initiative (Ingwersen and Subramanian 2013), misalignment areas can be clearly spotted. From a rapid screening, it sounds clear that, currently, there is lack of convergence on how to set functional unit, system boundaries, allocation rules and, last but not least, the scope of the methods in terms of environmental issues covered. Data quality requirements are also another major issue. Nevertheless, we decided to leave the discussion on this point out the scope of this chapter for practical reasons. The in-depth analysis of all datasets would have, in fact, required a dedicated project. Functional units look different across methods from the different groups presented in Table 5.1, across PCRs of different food categories and even across PCRs of the same product category. Of course, the application context of the methods falling into the different groups matters and seems sufficient per se to justify such misalignment. For example, the approach adopted to set up the function unit in the context of product-based B2B communication (see Group A for PCRs) is different from the sector-specific environmental reporting methods (see Group B and D). Going through PCRs, we found out that divergences on functional unit exist. For example, the functional unit for chicken meat is equal to "one pound of meat at the processing plant exit gate" according to IERE (2006), it shall be expressed as 1 kg of poultry meat and the required packaging according to the International EPD System PCR on meat of poultry (Palm 2010), and it should be expressed as "per unit weight (100 g of contents amount)" according to the PCR on chicken developed in the context of the Japanese pilot program on the carbon footprint of products (Japanese CFP scheme 2011b). The methods screened also diverge in terms of system boundary. Beyond the differences in terms of coverage of processes related to e.g. supply of capital goods (e.g. machineries, buildings, greenhouses, etc.) and labour (transport of farmers and other workers), the definition of system boundary varies across methods because of lack of harmonization on how to define co-products, by-products and waste streams. For example, manure is a co-product according to e.g. the FAO report on the global assessment of GHG emissions and mitigation opportunities (Gerber et al. 2013), and according to the European Commission's Product Environmental Footprint (PEF) Guide (EC 2013a). In contrast, manure is not a co-product according to a number of other references such as e.g. Boeri (2012), IERE (2006), and Sessa (2013a, b, c). As no GHG emissions from the farming stage are associated to biogas and biodiesel produced

Table 5.1 Environmental assessment methods per product category and groups

Product category	PCRs (Group A)	Business association guidelines(Group B)	Carbon footprint calculator (Group C)	Reference studies (Group D)	Laws (Group E)
Products from ruminant supply chains (other than dairy products)	Meat of mammals (Boeri 2012), Meat (IERE, 2006) Finished bovine leather (Pernigotti 2011) Leather footwear (Brondi 2013)		Cattle, pigs, sheep, goats, and other small ruminants (Bochu et al. 2013)	Pig sector (MacLeod et al. 2013) Ruminant sectors (Opio et al. 2013) Large ruminants (Leip et al. 2010)	Biogas and biofuels from manure and tallow (EU 2009)
Dairy products	Raw milk (Sessa 2013a) Processed liquid milk and cream (Sessa 2013b) Youghurt, butter, and cheese (Sessa 2013c)	Dairy products (Carbon Trust 2010) Dairy (IDF 2010)		Dairy sector (Gerber et al. 2010)	Biogas and biofuels from manure and tallow (EU 2009)
Poultry products	Meat (IERE, 2006) Meat of poultry (Japanese CFP scheme 2011b; Palm 2010) Poultry eggs (Japanese CFP scheme 2011a; Marino et al. 2011) Down and feather (Japanese CFP scheme 2011c)		Poultry (Bochu et al. 2013)	Chicken sector (MacLeod et al. 2013)	

from manure and tallow in the so-called Renewable Energy Directive (EU 2009), we deduce that manure and tallow are not co-products either according to such law. How to distribute environmental burdens, pressures and benefits among livestock co-products remains one of the most debated and unresolved issues. Assessment results drastically change, depending on the allocation approach adopted. This issue is particularly evident in livestock-related LCA methods where diverging approaches for dealing with process multifunctionality exist. For example, Gerber et al. (2013) have followed the following approach for their FAO report: among edible products (e.g. meat and eggs; and beef and milk), the allocation is based on protein content; between edible and non-edible products (e.g. milk, meat and fibre), the allocation is based on economic value of outputs; no emissions are allocated to the by-products from the slaughtering stage (e.g. offal, skins, blood). In the European Commission report conducted on livestock sector contribution to GHG emissions in EU (Leip 2010), allocation of emissions between multiple products throughout the supply chain is generally performed according to the nitrogen content of the products. The only exception was the allocation of CH_4 emissions from enteric fermentation and manure management of dairy cattle, which is allocated to milk and beef on the basis of the energy requirement for lactation and pregnancy). In the context of the International EPD System PCR on finished bovine leather (Pernigotti 2011), allocation at the slaughtering stage should be conducted among raw hide, comestible goods and scraps according to physical allocation (mass). In the PCR on meat of mammals (Boeri 2012), the allocation between meat, milk and leather should be conducted according to an estimate of their economic value. According to IERE (2006), all impacts should be allocated to meat. Allocation according to energy content is recommended by the Renewable Energy Directive (EU 2009) for co-products from the production of fuels. As said, tallow and manure are considered as sources of biofuels in such context. Last but not least, methods were found diverse in terms of environmental impact categories covered. Several are the methods dealing with GHG emissions only and coming up with figures on the carbon footprint of: the livestock sector as a whole (Gerber et al. 2013; Leip 2010), dairy sector (IDF 2010; Gerber et al. 2010), dairy products (Carbon Trust 2010), pig and chicken supply chains (MacLeod et al. 2013), ruminant supply chains (Opio et al. 2013). With the exception of PCRs developed in the context of the Japanese pilot project on the carbon footprint of products, which by definition covered GHG emissions only, all other PCRs listed in Table 5.1 adopt a multi-criteria perspective in the sense that a range of impact categories are covered. Nevertheless, impact assessment models recommended are often different, especially between different schemes (cfr. EIRE 2006 with Marino et al. 2011). This is a major issue that heavily affect the interpretation of assessment results. To support decision making processes and avoid that environmental information is deliberately disclosed in a misleading way, the European Commission and the European Food Sustainable Consumption and Production Round Table have recently recommended a list of assessment models and characterisation factors to be used in the context of environmental communication (both B2B and B2C). See the PEF Guide (EC 2013b) and ENVIFOOD Protocol (Food SCP RT 2013b) for more detail. Unlike these initiatives, the LEAP Partnership has been focussing on GHG emissions and other few impact categories when developing LCA guidelines on feed, poultry, small and large ruminants (LEAP 2014). This narrower scope is justified by the seemingly more consolidated science behind GHG emission accounting. Nevertheless, LEAP will not limit itself to GHG emissions. At present, efforts are on-going to explore how best set up a common framework to assess not only negative, but also positive impacts of livestock on biodiversity. Consensus will also be sought on issues such as e.g. use of water and of nutrients (LEAP 2014).

5.3 Beef Cattle

In the following, a description of the main aspects of this sector at the international and European levels is presented. Then, 34 international LCA studies on beef cattle production published in peer-reviewed journals, scientific reports or international conference proceedings are analysed. The study selection covers all the LCA applications to beef production systems published in the last 10 years. Methodological problems connected with the application of LCA in the beef cattle production sector are analysed in detail, starting with a critical comparative analysis of the LCA case studies. Finally, hotspots for the implementation of the LCA methodology in the beef production sector are identified in order to manage the methodological problems presented above. (Table 5.2)

5.3.1 The Beef Cattle Sector: Main Aspects

Over the years, the world market for beef has suffered a decline in terms of number of farm animals, production and consumption. The largest losses are found among the developed countries, mainly because of the economic downturn. However, trends in some emerging countries, particularly Brazil, India and Argentina, have seen improvements in terms of consumption and production, as they are managing to meet their domestic demand and overcome the shortcomings of the United States and the European Union, whose countries have increased import volumes. In 2010, 56 million t of meat were produced worldwide, according to USDA data (FAO 2006). The United States appear to be the major producer with nearly 12 million t of meat produced, followed by Brazil with 9 million tons and the EU−27 with about 8 million t. In the European context, France has by far the EU's largest cattle herd, with more than 19 million animals, followed by Germany (about 12.7 million) and Britain (10.3 million.). Italy, Ireland, Spain and Poland are each home to around 6 million cattle. Cattle farming is a significant component of European livestock sector, and many LCA studies of milk and meat production have been performed in recent years (Basset-Mens 2008). In this regard, Weidema et al. (2008) estimated that the 24 % of the environmental impacts of overall European consumption were attributable to milk and meat. The breeding of cattle has always accompanied the

Table 5.2 List of references included in the literature review and their main characteristics

Reference	Fu	Method	Main boundaries	Geographical areas	Time boundaries
Basarab et al. (2010)	1 kg of live weight	LCA and economic analysis	Cradle to slaughterhouse gate LCI basis	Canada (Alberta)	*Not specified*
Beauchemin et al. (2010)	1 kg of meat	LCA and HOLOS model for LCI	Cradle to farm gate	Canada	8 years
Beauchemin et al. (2011)	1 kg of carcass weight	HOLOS model for LCI and LCA	Cradle to farm gate	Canada (Western)	8 years (2003–11)
Bonesmo et al. (2013)	kg of CO_2-eq/kg of live weight kg of CO_2-eq/FPCM[a]	HOLOS model adapted for Norwegian dairy and beef production system	Cradle to farm gate	Norway	1 year
Casey and Holden (2006)	1 kg of live weight (after the first rearing year)	LCA	Cradle to farm gate	Ireland	*Not specified*
(Cederberg and Stadig 2003)	1 kg of energy corrected milk	LCA	Cradle to farm gate	Sweden	*Not specified*
Cederberg et al. (2009)	1 kg of carcass weight equivalent 1 kg of meat exported to Europe	LCA	Cradle to gate	Brazil	1 year (2005)
Clarke et al. (2012)	1 kg of carcass and ha of land occupied	GBSM model and a partial LCA	Cradle to farm gate	Ireland	*Not specified*
De Vries and de Boer (2010)	1 kg of meat 1 kg of protein 1 kg of average daily intake	Review of LCAs	Cradle to international retailer	OECD countries	*Not specified*
Doreau et al. (2011)	1 kg of live weight gained	LCA	Cradle to farm gate	*Not specified*	*Not specified*
Edwards-Jones et al. (2009)	kg CO_2-eq/kg live weight	LCA	Cradle to farm gate	Wales	*Not specified*
Flysjö et al. (2012)	1 kg of ECM[b]	LCA	Cradle to farm gate	Sweden	*Not specified*
Foley et al. (2011)	kg CO_2-eq-per ha	GBSM model and a partial LCA	Cradle to farm gate	Ireland	1 year
Leip et al. (2010)	1 kg of carcass meat	CAPRI model and LCA	Cradle to farm gate	EU–27	*Not specified*

Table 5.2 (continued)

Reference	Fu	Method	Main boundaries	Geographical areas	Time boundaries
Nguyen et al. (2010)	1 kg of meat at the slaughterhouse	LCA	Cradle to farm gate	Europe	*Not specified*
Nguyen et al. (2012a)	1 kg of carcass weight 1 ha of land occupied	LCA	Cradle to farm gate	*Not specified*	1 year
Nguyen et al. (2012b)	1 kg of carcass weight 1 ha of land occupied	LCA	Cradle to farm gate	Francia	1 year
Nijdam et al. (2012)	kg CO_2-eq/kg product	Review of LCAs	Cradle to farm gate	*Not specified*	*Not specified*
Ogino et al. (2004)	1 live animal	LCA	Cradle to farm gate	Japan	*Not specified*
Ogino et al. (2007)	1 marketed beef calf (8 months)	LCA	Cradle to farm gate	Japan	*Not specified*
Oishi et al. (2013)	1 kg of live weight	LCA and economic analysis	Cradle to farm gate	Japan	*Not specified*
Pelletier et al. (2010)	1 kg of live weight	LCA	Cradle to farm gate	United States (Upper Midwest)	*Not specified*
Peters et al. (2010a)	*n.a* [c]	Hybrid LCA On-site and I/O LCI	Cradle to farm gate	Australia	2 years
Phetteplace et al. (2001)	*n.a.*	LCA for Carbon Footprint	Cradle to farm gate	US	*Not specified*
Ridoutt et al. (2011)	kg CO_2-eq/kg of live weight	LCA	Cradle to farm gate	Australia	*Specified*
Ridoutt et al. (2012a)	1 kg of live weight	LCA	Cradle to farm gate	Australia	*Not specified*
Ridoutt et al. (2012b)	L H_2O eq/kg live weight m^2 year/kg of live weight	LCA	Cradle to farm gate	Australia	*Not specified*
Ridoutt et al. (2013)	kg CO_2-eq/kg of live weight L H_2O-eq/kg of live weight	LCA	Cradle to farm gate	Australia	*Not specified*
Roer et al. (2013)	kg CO_2-eq/kg of carcass weight	LCA	Cradle to farm gate	Norway	*Not specified*
Roy et al. (2012)	kg of CO_2-eq/kg of meat (or gr of protein or MJ of energy)	LCA	Cradle to fork	Japan	*Not specified*

Table 5.2 (continued)

Reference	Fu	Method	Main boundaries	Geographical areas	Time boundaries
Vergè et al. (2008)	kg of CO_2-eq/kg of live weight	LCA	Cradle to farm gate	Canada	20 years
Weidema et al. (2009)	Meat and dairy products consumed in EU−27 kg $CO_{2\ eq}$/kg of carcass weight	LCA and environmental impacts, monetarisation	Whole life cycle	EU−27	*Not specified*
Weis and Leip (2012)	kg CO_2-eq/kg of raw milk kg CO_2-eq/kg of eggs	LCA with CAPRI model	Cradle to farm gate+slaughtering	EU−27	20 years
Williams et al. (2006)	1 t of carcass weight 20,000 eggs 10,000 l milk	LCA	Cradle to farm gate	UK	*Not specified*
Zonderland-Thomassen et al. (2013)	L H_2O-eq/kg of live weight	LCA	Cradle to farm gate	New Zealand	1 year

[a] Fat and Protein Corrected Milk (FPCM)
[b] Energy Corrected Milk (ECM)
[c] *n.a.* not available

evolution of agriculture in Italy, accounting for a total of 5.592.700 heads (ISTAT 2010). As can be seen from the Italian census data, beef cattle account for 15 to 42% of the total number of cattle reared in the country, with 862.660 heads (ISTAT 2010). The differences between breeding techniques strongly influence the production and the economic results and, above all, the environmental impacts of the production system. Beef production systems are managed essentially by two techniques: the fattening of calves and the breeding of suckler cows. The first is characterised by the purchase of calves of different age and weight: new-born calves 10 days old and weighing 30–40 kg, calves aged 2–3 months weighing 70–120 kg, light weanlings between 8 and 10 months weighing 270–300 kg or heavy weanlings 14–16 months old and weighing 380–480 kg; these are fattened until they reach the ideal weight for slaughter, i.e. 550–650 kg at 15–18 months. The fattening of calves can be done by both extensive methods, that require loose housing which allows the animals to move freely and to develop their muscle mass, or intensive methods, such as tethered housing, where the animal is tied or penned within its location, and deprived of freedom of movement. The beef production achieved in lowland areas through fattening cattle of high genetic merit, often imported from abroad, can cause socioeconomic deterioration. Indeed, the high cost of imported weanlings adversely affect the economic result of the production system. Furthermore, the intensive farming is responsible for environmental degradation due to the excess nitrogen production. The suckler cow production system can be divided into two categories: the cow-calf line and the calf-heifer line. The most common is the cow-calf line, which requires the purchase of heifers or calves of high genetic merit that remain on the farm until the end of their career and are bred to provide calves to be sold and fattened in other stalls. This activity represents an economic opportunity, not only for the lowland areas, but also the hills and mountainous areas which are otherwise difficult to exploit and can be abandoned, causing their environmental degradation. The calf-heifer line is a more intensive breeding method involving the purchase of heifers on the market that are impregnated as soon as possible. The young cows are fattened for slaughter before or after weaning the calf. Both breeding systems can be conducted in the confined wild state. Breeding in confinement is typical of farms also engaged in crop production (especially corn), which have significant amounts of crop residues and use manure to maintain soil fertility. Wild and semi-wild breeding is practised only in the marginal areas for the purpose of environmental restoration. This production system mainly exploits forage resources through grazing.

5.3.2 Literature Review on LCA Application in Beef Cattle Sector

The results reported in the scientific publications collected in this review are as different as the studies analysed. The variability is essentially because of the difference in the production systems and methodological choices (functional unit, system boundaries, allocation method, etc.). The GHG emitted by specialised beef production systems vary from 22 to 40 kg CO_2 equivalents per kg of meat; whereas for meat from dairy cow systems values are lower, from 14 to 19 kg CO_2 equivalent per kg of meat (Sonneson et al. 2009). These results were confirmed by Nguyen et al. (2010) who report 27.3 kg CO_2 equivalents per kg of meat for suckler cow and calf-beef production systems and an average of 17.9 kg CO_2 equivalent per kg of meat for dairy calf and beef systems. This huge variation is largely because of the very wide variety in beef production systems, which range from very intensive to very extensive (Nijdam et al. 2012). Several studies aiming to identify ways to reduce the environmental impacts of ruminants have focussed exclusively on the analysis of GHG (Martin et al. 2010; Eckard et al. 2010). However, the most critical aspect remains the evaluation of how the implementation of these practices could produce a net reduction of environmental impacts, assessing, for instance, other impact categories (Beauchemin et al. 2011). In this regard, only five studies exclusively dedicated to GHG emission assessment are included in this review; others extend the analysis to other impact categories (e.g. energy use, acidification potential, eutrophication potential or land use). Among these, some studies focus on the analysis of the effects of different diet on the production of CH_4 from enteric fermentation (Doreau et al. 2011; Oishi et al. 2013). A significant factor in LCA analysis of beef production systems is the definition of the unit of product with respect to which the environmental impacts are defined (functional unit—FU). According to Nijdam et al. (2012), the most commonly used functional unit for meat is either kilogram of carcass weight or live weight. This uniformity is not evident in practice, as the values assigned by each author to the respective FU vary considerably depending on the production system analysed, the rearing species, and the traditional and local slaughtering activities. Different FUs are found in the studies which evaluate the effects of production process modifications on environmental impacts. Nguyen et al. (Nguyen et al. 2012a, b) when focussing on the effects produced by the different animal management strategies and different feed crop rotations, use as FU, respectively, 1 kg of carcass weight and 1 ha of land occupation; Doreau et al. (2011) evaluate the effects of different diets on the GHG emissions based on the unit increase in animals weight. A common characteristic of all the analysed studies is the heterogeneity in the system boundary (SB) definition. The variety and the complexity of beef meat transformation processes is a critical methodological point for LCA analysis. The life cycle is usually considered at the farm gate, confining the analysis to the rearing phase and disregarding the slaughtering and transformation processes. The descriptions of the productive phases that characterise the life cycle from cradle to farm gate are not always consistent in the examined studies. Basarab et al. (2010) and Cederberg et al. (2009) include the transport of animals to the slaughterhouse in the SB, whereas Ogino et al. (2004) consider the disposal of animal wastes (manure and slurry) to be part of the SB. In general, capital goods and internal and external transport are excluded from the SB (Oishi et al. 2013; Nguyen et al. 2012a, b; Ridoutt et al. 2011; Basarab et al. 2010; Beauchemin et al. 2010; Leip et al. 2010; Pelletier et al. 2010; Ogino et al. 2004, 2007; Casey and Holden 2006; Cederberg and Stadig 2003). The agricultural phase in beef cattle rearing is restricted to fodder and grassland production that consists of farm operations (fertilisation, pesticide use, etc.), and is responsible for high environmental impacts of the entire production system. The inclusion of crop production in beef rearing system impact assessment is a critical and debated question, for the analysis of which we refer the reader to Sect. 5.2.2. Only Basarab et al. (2010) use primary data because they focus on a specific area (Alberta, Canada) which specialises in beef production. Beef production systems are characterised by a high number of co-products and by-products. Thus, allocation is a key methodological issue in environmental impact assessment for this sector. The collected studies avoid co-product allocation, defining as their goal the assessment of environmental impacts generated only by beef production systems, which is totally different from assessment of milk production. Only Casey and Holden (2006) use, respectively, 1 kg of live weight and 1 kg of live weight gained as FU to avoid impact allocation between milk and meat. Manure, the main by-product, is included within the SB in all the studies analysed, because it is considered as organic fertiliser that returns directly (including the agricultural phase) or indirectly into the natural cycle. However, the polluting emissions produced by manure management operations are always included in the Life Cycle Inventory (LCI). The Life Cycle Impact Assessment (LCIA) varies among the studies, especially for the impact categories and the methods used for their assessment. All the analysed articles, excluding that by Weidema et al. (2009), stop at the classification and characterisation impact stages, which are obtained by different methods: IPCC 2007 (Climate Change 2007); EDIP (Hauschild and Potting 2003); CML (Guinée et al. 2002); CED (Frischknecht et.al. 2003); Impact 2000+ (Jolliet et al. 2003); and Ecoindicator 99 (Goedkoop and Spriensma 2001), depending on the impact indicators chosen for the assessment. Among these, the impact categories designed to measure the environmental impacts in terms of GHG, non-renewable energy use, eutrophication and acidification potential and land occupation are the most common ones. In general, other variables being constant (intensive, extensive, conventional or organic rearing systems), the cow calf-beef production system has greater impact than beef production systems. The phase with greatest impact is animal rearing, due to the emission of enteric CH_4 and NH_3 and N from animal excreta, the major source of environmental loads. Data availability remains a complex problem, as witnessed by the considerable time dedicated in all studies to system definition and inventory construction. LCA analyses built on primary data are not common, especially because of the beef production system complexity and the broad variability of climate conditions. Almost all the studies analysed in this review (n=34) use data collected and developed by third party organisations (national statistics agencies, non-government organisations, professional associations, etc.), derived from literature, or collected from dedicated LCA databases (LCAFood or Ecoinvent). Data uncertainty and LCIA result evaluation are almost wholly absent from the analysed studies, with the exception of Casey and Holden (2006), Weidema et al. (2008), Pelletier et al. (2010), Nguyen et al. (2010), Foley et al. (2011), Bonesmo et al. (2013) and Roer et al. (2013), all of whom report the evaluation of uncertainty and sensitivity of both input data and LCIA results.

5.3.3 Strategies to Mitigate the Impacts

Given the prospects of growth in consumption of meat and milk by 2050 (FAO 2006), debate on how to produce animal products in a sustainable way is taking place among the scientific community. Environmental performance improvement options can be classified, according to their effects, in two main areas: agricultural improvements and rearing and breeding technologies enhancing environmental performance of the fostering phase. In relation to the first aspect, agronomic production techniques may be improved by substitutions of current inputs, such as chemical pesticides and fertiliser, with lower impact inputs, such as organic ones, and by replacement of fossil energy sources with renewable; it is possible to reduce nitrate leaching, N_2O and ammonia emission, by planting catch crops during winter and reducing liquid manure pH. Land use may be reduced, thus improving growing practice: cereal yields can easily be increased by increased input of fertiliser, plant protection agents, better management and intensive cereal cultivation in low yield areas. Multiple use of such cultivation techniques produces a growth of emissions per ha but, as a result of the increased fertilisation, production increases too: the emissions per ton of cereal produced will decrease and, as many authors underline, the overall effect on cereal production may be a reduction in land use and ammonia emissions with only small changes in other emissions (Weidema et al. 2009). Animal husbandry has a strong impact in all categories. The most important pollutant in the impact categories acidification, terrestrial eutrophication, and respiratory inorganics is ammonia, mostly generated by manure production and handling. Beef fattening diets are generally well balanced and have a low N content that cannot be further reduced, especially as regards grazing animals. On the other hand, ammonia emission from liquid manure represents a problem only in beef fattening units, and can be very limited (up to 60 or 70 %). Limiting ammonia emissions provides a manure richer in ammonia N for plant fertilisation and saves on chemical fertilisers. Nitrogen leaching is responsible for aquatic eutrophication impacts that can be reduced by optimised protein feeding and by the use of manure N as a substitute for fertilisers, resulting in less leaching of N and fewer N_2O emissions.

Management of feedlot and ranch can improve environmental performance of breeding farms. In particular, Capper (2011) underlined that reducing time-to-slaughter may reduce CO_2 eq emission because the growth phase requires more energy than fattening. The methane and the dinitrogen oxide from enteric rumen fermentation in cattle contribute equally to about 90 % of GHG emissions. Methane emission is correlated with fatty acid diet contents so addition of fats to cattle feed can have a positive environmental effect in relation to global warming potential (Grainger and Beauchemin 2011). GHG emission can also be reduced by the use of liquid manure for biogas production to reduce consumption of energy from fossil sources. This has a threefold effect; according to Sommer et al. (2001), the methane emission from the manure will be reduced by 40 % or 1.1 kg methane per Mg manure, the N_2O emissions will be reduced by 14 g per Mg manure, and, at the same time, the methane produced will replace energy from fossil sources and thereby reduce the overall contribution to global warming.

5.3.4 *Other Methodological Measures and Innovative Tools for Product Environmental Assessment: Carbon, Water and Land Footprints*

5.3.4.1 Carbon Footprint

The Carbon Footprint (CF) shows the amount of greenhouse gases (GHG) emitted during a product's lifecycle (Röös et al. 2013). This environmental impact indicator is an increasingly important method for reporting the climate change impacts of food production and is fast becoming one of the key indicators of environmental sustainability. With regard to the livestock sector, several studies, focussed on the evaluation of the CF of different beef products, have been carried out. Edwards-Jones et al. (2009) evaluated the impact of beef products in the UK using primary data from three farms. Within a system that considers GHG produced from cradle to farm gate, producing 1 kg of lamb releases on average almost 3 kg CO_2 eq/kg live weight and for the production of 1 kg of beef they estimated 3.15 kg CO_2 eq/kg live weight of GHG emissions. With wider system boundaries, that included production of farm crops for animal feed, the amount of GHG emitted was almost 15 times higher for both lamb and beef (Edwards-Jones et al. 2009). Likewise, differences in the amount of GHG emitted from beef production depend on the cattle farming system used (intensive fattening, extensive pastoral, etc.). Nijdam et al. (2012) reviewed 15 LCA studies on beef production in a variety of cattle farming, finding that the production of 1 kg of extensively farmed beef results in three to four times as many greenhouse gas emissions as the equivalent amount of intensively farmed beef. According to these authors, the differences in feed transformation efficiency are higher in intensive systems; but for both systems, they found that methane from enteric fermentation and emissions from manure are, by far, the most important contributors to the CF (Nijdam et al. 2012). Few studies use empirical methods; usually, the GHG emissions of livestock production systems are calculated with the standardised IPCC approach (Tier 2). Ridoutt et al. (2011) used this approach to assess the GHG emissions from beef production in Australia, extending the GHG emissions calculation to agricultural soils after inorganic nitrogen fertiliser application and to the residue of cultivated leguminous pastures. A hybrid approach was used by Peters et al. (2010a) to perform an environmental life cycle assessment of Australian red meat production. Detailed on-site process modelling and input-output analysis were used to build a Life Cycle Inventory (LCI) and to assess the CF and the total energy consumption of three different Australian supply chains. They compared the grass-fed with the lot-fed systems, finding lower total GHG emissions for the latter; the additional effort in producing and transporting feeds was effectively offset by the increased efficiency of meat production in feedlots.

5.3.4.2 Water Footprint

Freshwater consumption is another relevant impact of agriculture and related production activities, accounting for around 70% of global freshwater withdrawals (UNESCO-WWAP 2009). Consequently, several studies in recent years have focussed on the application of a single indicator LCA-based WF in agriculture, in order to find a possible solution for reducing the pressure on freshwater resources from agriculture and food production. Ridoutt et al. (2012b) applied an LCA-based WF calculation method to the Australian beef cattle production system. Taking data directly from farms, they selected six geographically defined production systems, in order to cover a broad range of production method (pasture and feedlot finishing), product (yearling through to heavy steers), environment (high-rainfall coastal country to semi-arid inland country) and local water stress (Ridoutt et al. 2012b). All the flows from surface and groundwater into the farming system were included in the LCI. Moreover, the reduction in flows from the farming land base to surface and groundwater as a result of the operation of farm dams used for livestock watering was considered, together with the direct use of water in animal rearing and the water use associated with the production of all the inputs entering the system. To calculate the water footprint in units of L H_2O eq, they multiplied each spatially differentiated instance of water use by the locally relevant WSI and divided it by the global average WSI (0.602) (Ridoutt et al. 2012b).

5.3.4.3 Land Footprint

An innovative approach to land use in LCA analysis of beef production sectors has been proposed by Ridoutt et al. (2012, 2013). They consider it from a qualitative rather than only a quantitative approach (e.g. m^2.yr), and suggest the net primary productivity of potential biomass (NPP0, g C.m^2.yr−1) as an indicator to account for land's intrinsic productivity capacity. Comparing six beef production systems (from cradle to farm gate), they report a variability of NPP0 for kg of live weight between 86 and 176 m^2.yr-e, where 1 m^2.yr-e is 1 m^2 of land occupied on global average NPP0 (Ridoutt et al. 2012). According to the authors, this indicator, called the land use footprint, is easy to calculate from existing databases and allows us to consider the different pressure exerted globally on the land resources, depending on productivity (Ridoutt et al. 2012). In a further study, Ridoutt et al. (2013) propose a normalisation step, in order to make the different life cycle impact category indicators comparable. In particular, they perform the normalisation of the carbon footprint (Ridoutt et al. 2011), the water footprint (Ridoutt et al. 2012) and the land use footprint (Ridoutt et al. 2012, 2013) in relation to beef production systems in Australia by using the global economic system for 1995 to 2000 as reference (Ridoutt et al. 2013). Although they find no correlation between these indicators, their study is a first attempt to overcome the lack of comprehensiveness when considering indicators as stand-alone environmental indicators.

5.3.5 Comparative Analysis of Different Types of Breeding and Final and Processed Products

Using data on the composition of the entire beef production system in the EU−27 (Weidema et al. 2008), Nguyen et al. (2010) studied four systems (one suckler cow-calf system and three dairy bull systems). The results of this study show that the suckler cow-calf system has the lowest environmental efficiency because the higher quantity feed for kg of meat is followed by higher manure production. The dairy bull system with calves slaughtered at 12 months emerged as the most efficient (Nguyen et al. 2010). Furthermore, in another study analysing the most common beef production system in France, Nguyen et al. (2012a) focussed on the possible scenarios for GHG emissions reduction. The analysed scenarios included changes in grazing management and in herd and diet management, as well as a combination of all such strategies. As regards the 10 alternative scenarios, the authors also found that their combination could reduce the current impact of 13–28 % per kg of live weight (Nguyen et al. 2012b). Three beef production systems in the USA were analysed by Pelletier et al. (2010): (1) directly weaned calves in the herd, (2) weaned calves on grazing which ended up in feedlots and (3) calves finished directly on pasture. They found that the last resulted in the highest impacts in terms of cumulative energy use, ecological footprint, greenhouse gas emissions and eutrophying emission impact categories. The environmental efficiency, in terms of non-renewable energy consumption and GHG emissions, of three specialised and two mixed (crop-livestock) farms, was studied by Veysset et al. (2010), in the search for management options for income maximisation. For the assessment of the economic and environmental performances of the systems they used two models: Opt'INRA to optimise the economic input, and PLANETE to assess the environmental performances. From the economic perspective, the authors found higher efficiency in mixed crop-livestock farms, because of higher management flexibility, especially in crop-based rather than grassland-based farms. However, from the environmental perspective, also crop-based farms had restricted opportunities in non-renewable energy consumption and GHG production improvement, due to the reduced number of possible solutions found by the authors, for the three studied systems (Veysset et al. 2010).

5.3.6 Hotspots

The difference in beef production systems is determined by a series of farm characteristics, including the rearing species, the number of animals, the type of production (milk or meat), the rearing system (conventional or organic; intensive or extensive, etc.), manure management, the presence or absence of agricultural activities for feed production supporting the livestock system. All the selected studies found that animal rearing was the phase of beef production system with the greatest impact (Peters et al. 2010a); this was mainly caused by the emission of enteric CH_4,

NH_3 and of N from animal excreta to be the major cause of environmental loads (Ogino et al. 2004, 2007; Casey and Holden 2006; Cederberg et al. 2009; Beauchemin et al. 2010; Nguyen et al. 2010; de Vries and de Boer 2010; Basarab et al. 2010; Veysset et al. 2010; Foley et al. 2011; Nijdam et al. 2012; Oishi et al. 2013). Many studies therefore combine LCA analysis with models for the optimisation of farm resources (e.g. Beauchemin et al. 2010; Pelletier et al. 2010; Nguyen et al. 2010) or direct their attention to the assessment of environmental impacts depending on different diets or rearing techniques (e.g. Ogino et al. 2004, 2007; Basarab et al. 2010; Doreau et al. 2011; Nguyen et al. 2012a, b; Clarke et al. 2012; Ridoutt et al. 2012). Another relevant issue in environmental impact assessment of beef production and the whole livestock sector is land occupation. Usually, land use and land use changes are considered in LCI as the amount of land occupied by processes or by raw material production (or extraction). This is also the case for beef production, whereby the land use impact category is measured in terms of the m^2 of land required to produce a certain amount of meat in a defined period of time. Moreover, the literature also suggests that LCA coupled with other approaches provides much more comprehensive information for environmentally conscious policy-makers, producers, and consumers in selecting sustainable products and production processes (Roy et al. 2009). Thus, the integration of LCA analysis with farm economic efficiency models has assumed major importance in the livestock sector and beef production in recent years (Beauchemin et al. 2011; Oishi et al. 2011; Clarke et al. 2012), although others use different methods to assess the environmental loads generated by livestock production (Veysett et al. 2010).

5.4 Dairy Cattle

5.4.1 Literature Review on LCA Application to Milk and Dairy Products and Problematic Approaches

Several research studies about the application of LCA methodology to milk and dairy products have been published in the last 10 years. In this paragraph a critical review of the literature LCA studies regarding the evaluation of the environmental performance of milk and other dairy products is reported, in an attempt to summarise the main issues, both methodological and technical, that these studies highlight. The selection of peer-reviewed LCA articles for inclusion in the comparative analysis was based mainly on the year of publication and on the main scope of the studies; the older studies (prior to 2010) were excluded because of the large number of available articles and the fact that their results are often used as primary data or for comparisons in more recent studies. In addition, the older studies are already discussed in some LCA reviews (de Vries and de Boer 2010).

According to these selection criteria, milk is the most studied dairy product (seven studies), followed by cheese (four studies) whereas only one article about yogurt has been reviewed. Among the twelve works analysed, two considered only the carbon footprint of the product (Thoma et al. 2013; Yan et al. 2011), and the remaining ten assessed a higher number of indicators. It is evident that the older studies about milk production (Castanheira et al. 2010) assessed only raw milk production, whereas more recent ones have tried to probe more deeply. First, some studies enlarged the system boundaries up to the processing plant (Gonzales-Garcia et al. 2013) or up to the end of life of the product (Thoma et al. 2013). Furthermore, some research papers compared the environmental performance of different farming approaches, focussing on the differences between intensive and extensive systems and organic and traditional ones (O' Brien et al. 2012; Yan et al. 2011; Guerci et al. 2013). One study compared the results of a traditional LCA with those of the Environmental Product Declaration (EPD) of milk in the International EPD® System (Fantin et al. 2012).

Without exception, on-farm activities have been found to be the main environmental hotspots for milk production, followed by the production of feed, particularly concentrates; seasonal-grass based systems can have a lower impact thanks to their lower resource use and their production of fewer pollutants from concentrate feed compared with forage and shorter manure storage periods. As regards other dairy products, five studies have been analysed: one about yogurt (Gonzales-Garcia et al. 2013b), and four about cheese (Gonzales Garcia et al. 2013c, d; Kim et al. 2013; Van Middelaar et al. 2011). The two articles about cheese written by the same author (Gonzales-Garcia) adopted a similar approach, although treating different kinds of cheese. All these studies agreed that the production of milk is the main hotspot for the most common impact categories considered: global warming, eutrophication, acidification and photochemical ozone formation potentials. However, in the majority of the works, the authors tried to identify the hotspots of cheese production over which the manufacturer has direct control.

An interesting approach is proposed by Van Middelaar et al. (2011), who tried to make a combined economic and environmental evaluation of cheese by using the parameter eco-efficiency, which expresses the gross value added of a unit of environmental impact (global warming, land use or energy use).

The critical analysis of the LCA studies showed that, similarly to other products derived from livestock, the handling of multifunctionality, frequently solved by applying allocation approaches, is one of the main critical issues in the environmental assessment of cheese and milk. In fact, the environmental load of milk production has to be divided between all the outputs of the rearing process: milk, meat and skin. Furthermore, the dairy factory generally produces more than one product, which implies that the whole impact should be allocated among all of them. Another problem related to the assessment of the environmental performance of dairy products transformed in medium and large dairy plants is that they often use milk supplied by different farmers with different rearing systems. Obtaining primary data from all of them is frequently a problem, and thus it is common practice to include in the inventory analysis primary average data obtained from a representative sample of farms or information from the literature. (Table 5.3)

Table 5.3 List of references included in the literature review and their main characteristics

Reference	LCA	Other tool	Product
Castaneheira et al. (2010)	X		Milk
Fantin et al. (2012)	X		Milk
Gonzales-Garcia et al. (2013a)	X		Milk
Gonzales-Garcia et al. (2013b)	X		Yogurt
Gonzales-Garcia et al. (2013c)	X		Cheese
Gonzales-Garcia et al. (2013d)	X		Cheese
Guerci et al. (2013)	X		Milk
Kim et al. (2013)	X		Cheese
O' Brien et al. (2012)	X		Milk
Thoma et al. (2013)	X	CF	Milk
Van Middelaar et al. (2011)	X	Eco-efficiency analysis	Cheese
Yan et al. (2011)		CF	Milk

5.4.2 Methodological Problems Connected with the Application of Life Cycle Assessment for Dairy and Dairy Products: Critical Analysis of International Experiences

5.4.2.1 Goal and Scope

Most of the articles analysed had a similar purpose, i.e. to evaluate the potential environmental burdens of milk and cheese production chains. Furthermore, some authors compared different production or farming systems or included economic evaluations such as economic efficiency and the evaluation of a benchmark. The main goals of the studies about milk production analysed here were both to assess the potential environmental impact of the product (Gonzalez-Garcia et al. 2013a; Castanheira et al. 2010; Thoma et al. 2013) and to compare different agricultural or breeding management systems (Yan et al. 2013; Guerci et al. 2013; O'Brien et al. 2012).Only one study compared the results of LCA and an EPD for the same product and performed a critical analysis of the existing product category rules (Fantin et al. 2012).

The goals of Gonzalez-Garcia et al. (2013a), Castanheira et al. (2010) and Thoma et al. (2013) were similar, though Thoma et al. (2013) were focussed only on the carbon footprint of milk production. Gonzalez-Garcia et al. (2013a) aimed at evaluating the environmental performance and the energy balance of the production of UHT milk in Portugal and at identifying the hotspots in the production chain. The authors chose a Portuguese dairy factory, using best available technologies (BATs) for the assessment. Castanheira et al. (2010) aimed to identify the processes with the largest environmental impact, and considered a typical Portuguese dairy farm. The goal of Thoma et al. (2013) was to determine GHG emissions associated with consumption of 1 kg of milk by US consumers. Yan et al. (2013), Guerci et al. (2013) and O'Brien et al. (2012) applied LCA methodology to different farming systems. Yan et al. (2013) performed a carbon footprint study, the purpose of which was to compare two systems for milk production in a grass-based, rotational grazing system:

one used nitrogen fertilisers for pasture production and the other used white clover, which is an alternative to nitrogen fertilisers (applies biological nitrogen fixation). Guerci et al. (2013) aimed at assessing the environmental impacts of milk production of different farming systems (organic versus conventional, confinement systems versus pasture systems and different annual production levels) and at identifying their strengths and weaknesses. Finally, O'Brien et al. (2012) compared the environmental impacts of seasonal grass-based and confinement dairy farms, following an LCA approach. The main purpose of Fantin et al. (2012) was to compare the environmental performance of milk production with the published EPD of a similar product by following the requirements of the PCR document for milk of the International EPD® System and critically analysing it. In fact, the authors discussed the main key issues affecting the comparability of different EPDs for the same product.

As regards dairy products, Gonzalez-Garcia et al. (2013b) focussed their analysis on the assessment of environmental impacts and energy balance from the production of different types of yogurt.

The studies about LCA of cheese stemmed from different needs but the goals and scope of the studies analysed show some similarities. The main difference is that Gonzales-Garcia (2013c, d) and van Middelaar et al. (2011) focussed on the environmental performance of a specific product with the aim of quantifying its environmental impact and identifying the most impacting processes. Furthermore van Middelaar et al. (2011) assessed the ecological impact in the context of the economic efficiency through the eco-efficiency parameter. Kim et al. (2013) performed a more strategic analysis, aiming at defining a benchmark for the US cheese producers and at providing stakeholders with information about the environmental impact of cheese.

5.4.2.2 Functional Unit

In milk and dairy product LCAs, two main approaches towards functional unit definition are presented: the first considers only the mass of the product regardless of its composition and its water content; the second one takes into consideration the nutritional value of the product, normalising the mass to a certain energy or fat and protein content. It should be pointed out that the choice of a "corrected functional unit", such as fat and protein or energy content, could be an efficient approach for covering the nutritional value of dairy products as well and could allow comparison of the results of different studies.

Three studies among seven on milk production referred to 1 kg of energy corrected milk (ECM), a correction factor used by the dairy industry to determine the amount of energy contained in milk and based on fat and protein content (Gonzalez-Garcia et al. 2013a; Yan et al. 2013; Guerci 2013). Three LCAs referred to a certain amount of product (1 L, 1 kg or 1 t) (Fantin et al. 2012; Thoma et al. 2013; Castanheira et al. 2010). The comparative analysis by O' Brien et al. (2012) was based on different functional units: 1 t of fat and protein corrected milk (FPCM), 1 t of milk solids (MS), the on-farm area occupied and the total area occupied.

Concerning dairy products, Gonzalez-Garcia et al. (2013b) referred to the production of 1 t of yogurt ready for consumption. In the studies about cheese production, the most commonly adopted functional unit was 1 kg of cheese (Gonzales-Garcia et al. 2013c; d; van Middelaar et al. 2011). Only one study among the four analysed referred to 1 t of cheese on a dry weight basis (Kim et al. 2013).

5.4.2.3 System Boundaries

As regards studies on milk production, the most common approaches regarding system boundaries definition are from cradle to farm gate and from cradle to gate. The former included only farms' activities until raw milk production (Castanheira et al. 2010; O'Brien et al. 2012; Yan et al. 2013; Guerci et al. 2013). On the other hand, the latter took into account the pasteurisation and packaging processes at dairy plants, excluding the distribution and use phases (Gonzalez-Garcia et al. 2013a; Fantin et al. 2012). Moreover, a third kind of study considered a cradle-to-grave approach (Gonzalez-Garcia et al. 2013b; Thoma et al. 2013).
The production of capital goods (machinery and buildings) and road infrastructures were excluded from all these studies as well as land use and soil quality changes caused by cultivation-related activities. Regarding the studies on LCA of cheese, different approaches towards system boundaries definition were adopted. In particular, the production of milk was common to all the studies analysed, whereas the other cheese production phases considered vary. Gonzales-Garcia et al. (2013a, b) analysed the process from farm to cheese manufacturing plant gate. Kim et al. (2013) considered all the processes of the life cycle, from cradle to grave, and finally van Middelaar et al. (2011) evaluated the phases between milk production and sale.

5.4.2.4 Availability and Quality of Data

As far as data quality is concerned, the studies on milk production can be divided into two main groups: studies using primary data for farms or dairies (Fantin et al. 2012; Guerci et al. 2013; Yan et al. 2013; Gonzalez-Garcia et al. 2013b); studies using secondary data for farms (Castanheira et al. 2010; O' Brien et al. 2012; Gonzalez-Garcia et al. 2013a, b; Thoma et al. 2013).

Castanheira et al. (2010) and O'Brien et al. (2012) did not use primary data, but only secondary ones from previous studies. Moreover, in Gonzalez-Garcia et al. (2013a, b) data regarding the foreground processes for the production of raw milk were obtained from Castanheira et al. (2010) The foreground data for dairy factories consisted of average annual data obtained by on-site measurements. Thoma et al. (2012) used data collected from several sources such as the USDA's National Agricultural Statistical Service and Economic Research Service, peer-reviewed literature and other technical literature and an extensive nationwide survey of dairy farm operations. On the other hand, Fantin et al. (2012) used primary data for both farm activities and dairy processing and packaging. Gonzalez-Garcia et al. (2013b) used primary data from the dairy factory regarding transportation to wholesale and retail stages, whereas they use literature data for the use phase. The authors of two studies collected primary data from a large sample of farms: Yan et al. (2013) used data obtained from experimental systems in 16 Irish farms; Guerci et al. (2013) collected data from 12 dairy farms, five from Denmark (two of which were organic), two from Germany which differed in their summer feeding systems (confinement vs. pasture), and five from Italy (all of which used confinement feeding). All studies used literature data and the Ecoinvent database for background data.

A critical methodological issue in LCAs of dairy products is often the calculation of methane and nitrogen emissions because of the management and agronomic use of chemical and organic fertilisers, such as manure and slurry. Methane emissions from enteric fermentation and manure management and emissions of nitrous oxide, nitrogen oxides and ammonia from manure management were generally calculated according to IPCC 2006 and EMEP/EEA Corinair 2009 (Castanheira et al. 2010; Gonzalez-Garcia et al. 2013a, b; Fantin et al. 2012; Guerci et al. 2013). On the other hand, O'Brien et al. (2012) excluded the emissions of manure in pastures. When considered, phosphorus emissions were often calculated in accordance with Nemeck and Kagi suggestions (Fantin et al. 2012; Guerci et al. 2013).

Nitrogen and carbon dioxide emissions from livestock respiration were not taken into account in any study. Carbon dioxide sequestration by crops was accounted for in Guerci et al. (2013) but it was not considered by O'Brien et al. (2012), Gonzales-Garcia et al. (2013c) or Fantin et al. (2012).

Data about the production of milk employed in the cheese manufacturing plants were not always available for LCA analysis. Among the four studies considered, two (Gonzales-Garcia et al. 2013c; van Middelaar et al. 2011) used primary data for the production of milk, whereas the remaining two (Gonzales-Garcia et al. 2013d; Kim et al. 2013) used secondary data from scientific literature (Castanheira et al. 2010; Thoma et al. 2012). However, in the two cases where primary data were used, milk was provided to the cheese manufacturer by different farms, but only a few, representative of the average situation, were included in the evaluation. Data related to the cheese production process derived from a primary source in all the studies included in this literature review. Similarly to milk production, background data were generally taken from the literature or databases, particularly the Ecoinvent database.

5.4.2.5 Allocation Methods

As regards milk production, different approaches were applied to solve multifunctionality problems: no authors used system expansion or substitution, whereas several studies applied allocation, mainly on an economic or biological basis. Only one study (Fantin et al. 2012) adopted a conservative approach and allocated all impacts to milk production.

Yan et al. (2013) applied economic allocation for concentrate feed production and economic allocation between milk and meat. In Gonzalez-Garcia et al. (2013a), economic allocation was applied in the case of dairy farms in order to partition the environmental burdens between meat and milk, which are based on historical market prices in Portugal. As regards the dairy factory studied by Gonzalez-Garcia et al. (2013a), different types of UHT milk were produced: simple milk and cocoa milk, as well as butter and cream. Nevertheless, the authors considered the whole system as a black box and applied mass allocation for the assessment of the co-products (milk, cream and butter). Guerci et al. (2013) and Thoma et al. (2013) used biological allocation, based on the feed energy required to produce the amount of milk and meat at farm level. Moreover, Thoma et al. (2013) used economic allocation for feed crop processing, and mass balance of milk solids (fat and protein contents) for the allocation between milk and cream.

In the production of yogurt two main co-products have been identified (Gonzalez-Garcia et al. 2013b): yogurt and animal fodder, to which impacts were allocated following a mass-based partitioning approach.

In the production of cheese two main multi-output processes have been identified: the production of milk that also implies the production of meat, manure, calves and skin, and the manufacture of cheese, generally accompanied by the production of whey and other co-products, such as cream. In three studies reviewed (Gonzales-Garcia et al. 2013c, d; van Middelaar et al. 2011) the allocation of the impacts of the farm is done on an economic basis. The allocation factor for milk is explicated in the two studies (Gonzales-Garcia et al. 2013a; van Middelaar et al. 2011) where primary data are used to model the milk production, which is respectively equal to 92 and 87 %.

The allocation of impacts caused by cheese manufacturing plants is managed in different ways depending on the study. That which produces San Simon da Costa cheese (Gonzales-Garcia et al. 2013c) did not have specific equipment for whey processing, which was sent to the wastewater treatment plant. Therefore in this case cheese was the only output of the system and no allocation was needed. The authors also analysed an alternative scenario with whey valorisation in which they allocated the impact to the whey and cheese according to their fat content. Gonzales-Garcia et al. (2013b) analysed the effect of different allocation approaches and found that mass allocation improved the impact of cheese compared with the economical one, because the economic value of whey per unit of mass is lower than that of cheese. Gonzales-Garcia et al. (2013d) and van Middelaar et al. (2011) subdivided the impact of cheese production according to the economic value of co-products. Kim et al. (2013) adopted another approach: the impact of milk was subdivided between the co-products on the basis of their fat and protein content, whereas the environmental load of all the other materials, such as steam and electricity, was allocated following the economic approach.

5.4.2.6 Life Cycle Impact Assessment (LCIA)

The studies analysed applied different impact assessment methods, which included different characterisation factors and environmental indicators. However, the studies reviewed display a certain level of coherence in their choice of impact assessment method. The majority of the LCAs reviewed considered more than one impact category, whereas two studies among twelve assessed only the potential impact on climate change (Yan et al. 2011; Thoma et al. 2013). The potential impact on global warming was without exception evaluated in line with the IPCC guidelines (IPCC 2006).

The characterisation method most commonly adopted was the CML 2001 (Gonzales-Garcia et al. 2013a, b, c, d; Castanheira et al. 2010; O'Brien et al. 2012). These authors focussed their analysis on the following impact categories: abiotic depletion (ADP), acidification (AP), eutrophication (EP), global warming (GWP), ozone layer depletion (ODP), photochemical oxidant formation (POFP). Only two studies considered the impact on land use (Gonzales-Garcia et al. 2013b; Guerci et al. 2013). The toxicological impact categories and the cumulative energy demand (CED) are evaluated by Gonzalez-Garcia et al. (2013a, b). It must be noted that water depletion was not considered in any of the studies analysed. On the other hand, both Fantin et al. (2012) and Guerci et al. (2013) selected the impact assessment methods recommended by the International EPD system. Two studies evaluated biodiversity (Guerci et al. 2013; Kim et al. 2013). Finally, Kim et al. (2013) utilised the ReCiPe method (Goedkoop et al. 2009) and the Usetox (Rosenbaum et al. 2008) whereas van Middelaar et al. (2011) considered the model proposed by Thomassen et al. (2009). Two studies related the environmental impacts to eco-efficiency analysis by means of the gross value added of the product (Van Middelaar et al. 2011) or combined them into a single score expressed in monetary units (Guerci et al. 2013).

5.4.2.7 Critical Analysis

The studies analysed often performed sensitivity analyses to evaluate improving actions in the dairy production chain, discussed the effect of allocation approaches on total LCA results or compared the results with literature on LCA studies of the same product. Gonzalez-Garcia et al. (2013b) discussed, via sensitivity analysis, some improvement actions, which could contribute to the reduction of the overall environmental performance of yogurt production. As regards raw milk production, optimised farm management can lead to the reduction of the potential environmental impact. The main improvements that can be applied to farm activities are:

- increase the consumption of grass silage instead of maize silage;
- lower the use of concentrates or use concentrates with a lower environmental impact (e.g. domestic or regionally produced rapeseed meal instead of imported soybean meal);
- reduce the use of high protein concentrate meals in order to lower nitrogen losses;
- increase the length of the grazing season in order to reduce the storage of manure and store manure under aerobic conditions and target and reduce N fertiliser application (Gonzalez-Garcia et al. 2013b; O' Brien et al. 2012).

As regards the dairy factory, the minimisation of milk losses (which involves the increase of total yogurt production and the reduction of the co-product dairy fodder at the same time), the reduction in the total energy requirements in the dairy factory, the use of gas fuelled boilers instead of oil fuelled ones, reductions in travel distances and energy consumption in both retail and consumption phases would lead to environmental improvements. On the other hand, the consideration of dairy fodder as an avoided product does not allow reductions in any impact categories (Gonzalez-Garcia et al. 2013b, d; van Middelaar et al. 2011). The recovery of whey in cheese production plants increases the total environmental impacts of the processes analysed, except for the impact category EP, thanks to the reduction of phosphate emissions. However, it should be recognised that the system analysed releases two value-added outputs and the impact allocated to cheese according to the fat content is lower than in the base case scenario (Gonzales-Garcia et al. 2013c). Increasing the cheese ageing period would lead to a worsening of the results (Kim et al. 2013). Furthermore normalised results highlight that cheese production mainly affects the categories aquatic eutrophication, aquatic ecotoxicity and terrestrial acidification, and it is possible to lower these impacts through energy conservation and water conservation/treatment activities (Kim et al. 2013). Gonzalez-Garcia et al. (2013a), Fantin et al. (2012) and Castanheira et al. (2010) compared their results with other literature studies on milk production. They stated that their results fall within the range of literature values. Moreover, they found that the main flows affecting the results are the same, although their contributions to the total results are different because of the different assumptions made and models used in the studies. However, Gonzales-Garcia et al. (2013a) identified some differences from other studies, such as the allocation approach, data sources, characterisation factors, farm management practices and enteric fermentation emission factors, which do not allow a comprehensive and detailed comparison with LCA literature results. Consequently, when the results of different studies on the same product are compared, these aspects should be taken into account, and a sensitivity analysis which considers the assumptions and uncertainty of the results should be performed (Gonzalez-Garcia et al. 2013a).

Yan et al. (2013) found that the difference in carbon footprint between the systems investigated is in agreement with other similar studies (the other three studies show that white clover reduces the carbon footprint of milk) and discussed the main differences (carbon sequestration, stocking density). The authors also performed a ratio sensitivity analysis which examined the effect of emission factors on the comparison between the two systems. The analysis revealed that to reverse the ranking of white clover and nitrogen fertilisers systems, changes to emission factors and assumptions had to be much greater than the uncertainty ranges found in the literature.

As regards an allocation approach, Gonzalez-Garcia et al. (2013a) discussed the effect of different allocation methods among milk, cream and butter on the total life cycle results: in addition to a mass allocation approach, the authors performed a sensitivity analysis in which economic and protein-based allocations were applied to the system. The results showed that economic allocation improved the environmental performance of milk production by 34 %, whereas protein-based allocation worsened the results by up to 5 %. Gonzales Garcia et al. (2013b) analysed the effect of different allocation approaches and found that mass allocation improved the impact of cheese more than the economical one, because the economic value of whey per unit of mass is lower than that of cheese.

Guerci et al. (2013) performed two correlation analyses: the first on the impact categories and the second between impact categories and main parameters of dairy farms. The former showed strong and positive relations between GWP, acidification, eutrophication and energy use, whereas land use was negatively related to the four categories. The latter found that feed efficiency affected several impact categories (significant negative correlation with global warming, acidification and eutrophication). This supports the theory that better animal efficiency (in terms of feed conversion rate) is one of the ways of reducing the environmental impact in milk production. A positive relation was observed between GWP, acidification, energy use, biodiversity and the amount of grassland of the farmed area, whereby the farms with the largest amount of grassland had cows grazing during the summer season. Overall, the results of the study showed that the improvement of greenhouse gas emissions would lead to an improvement in the environmental performance of the dairy farm.

Van Middelaar et al. (2011) assessed the eco-efficiency of the processes of the supply chain, expressing the gross value added per unit of environmental impact (GWP, land use and energy use). and found that. the least eco-efficient product production process is concentrate production.

5.4.2.8 Environmental Hotspots

On-farm activities were found to be the main environmental hotspots for milk and other dairy productions regardless of the impact category considered (Gonzalez-Garcia et al. 2013a, b, c, d; Fantin et al. 2012; Guerci et al. 2013; O' Brien et al. 2012; Castanheira et al. 2010; Van Middelaar et al. 2011; Kim et al. 2013). Larger contributions from the farm subsystem are made by enteric fermentation, the production of animal feed, airborne and waterborne emissions from farm activities as well as manure management and spreading (Gonzalez-Garcia et al. 2013a, d; Fantin et al. 2012; Guerci et al. 2013). As regards the proportion of the different compounds and their environmental impacts, the global warming and the photochemical ozone formation potentials of milk production are primarily influenced by methane emissions because of enteric fermentation and manure management and secondarily by feed production (Castanheira et al. 2010; Fantin et al. 2012). Particularly, enteric methane, manure deposition, fertiliser spreading, fertiliser production, electricity production, indirect nitrous oxide emissions, slurry storage, concentrate production, and slurry spreading account for 95 % of the total GHG emissions (Yan et al. 2013). The nitrogen that volatilises in the form of ammonia from manure and fertilisers significantly affects the acidification potential and, to a lesser extent, the eutrophication potential, which is mainly influenced by nitrate emissions (Castanheira et al. 2010; Fantin et al. 2012).

Concerning the different dairy cow feeds, concentrates have the higher environmental load (Castanheira et al. 2010; O' Brien et al. 2012). Furthermore the CF of milk is 11–23 % lower for white clover systems compared with nitrogen fertiliser systems because of the fact that methane, carbon dioxide and nitrous oxide emissions are significantly higher for the latter (Yan et al. 2013). The most extensive use of land is for organic farms, supporting the theory that such farms generally need more land to produce feed because of their lower crop yields The farms that had the lowest impact on biodiversity losses were organic (Guerci et al. 2013). In addition, a simplified sensitivity analysis performed by Guerci et al. (2013) showed that, accounting for the emissions from direct land use change would increase the impact of conventional farms, whereas it would remain the same for organic farms. The same authors also found that grasslands have an important role in GWP mitigation and in reducing biodiversity losses, especially on organic and pasture-based farms. These effects are probably because of a greater capacity for carbon sequestration. In addition, farms with more grassland are more self-sufficient in feed so they avoid the heavy impact of commercial feed production and transport on total energy consumption. The influence of grassland on lowering acidification could be because of the lower fertiliser input for this type of crop. Correlation analysis showed that land occupation is significantly reduced when the farming intensity increased (stocking rate, N surplus and use of fertiliser) and when crop production on the farmland increased (Guerci et al. 2013).

As regards dairy plants, energy consumption (electricity, fossil fuels), packaging production, transport-related activities and on-site emissions are the main contributors to the environmental impacts of this phase (Gonzalez-Garcia et al. 2013a; Fantin et al. 2012).

Concerning other dairy products, the production of powdered and concentrated milk needed for yogurt production is the main hotspot for the dairy factory phase, mainly because of the high energy consumption required for their production processes. Moreover, it was found that the production of packaging materials and energy requirements contributes significantly to the yogurt environmental profile. Finally, the distribution phase, consumption at the household and final disposal showed a low contribution (Gonzalez-Garcia et al. 2013c). Although the production of milk is the main environmental concern of cheese production, Gonzales-Garcia et al. (2013c, d) and Kim et al. (2013) focussed on the environmental impact of cheese manufacturing plants. The main contributions to GWP by the manufacturing plant are related to the combustion of fossil fuel both for energy production and for transport (Gonzales-Garcia et al. 2013c, d; Kim et al. 2013). Furthermore, boilers fuelled with oil and wastewater treatment plants contribute significantly to ADP, AP, ODP, POCP and EP (Gonzalez-Garcia et al. 2013c; Kim et al. 2013). Finally, the smoking process performed with birch wood has an important influence on AP, POFP and GWP mainly because of the wood supply chain and combustion (Gonzales-Garcia et al. 2013c).

5.5 Sheep and Goat

5.5.1 Comparative Analysis of Life Cycle Thinking Approaches in the Sheep and Goat Sector

The literature review of LCT approaches in the sheep and goat sector was performed through consultation of scientific databases and search engines, including Scopus, Web of Knowledge, Google Scholar and Google, as well as LCA conferences and the websites of Ecosystem Assessment, FAO, and IPCC. Selected studies, published from 2010 onwards, refer to various countries. Following a homogeneous framework adopted for all sectors addressed in this book, the review tries to show the prospects and constraints of LCT as a tool to assess the environmental impact of sheep and goat production. There is less research, globally, on LCT involving the sheep and goat sector compared to other livestock sectors, possibly because of the great diversity of situations (including species, products, and intensity of land used), or because of its secondary economic importance and low political weight, as de Rancourta et al. (2006) argue about Mediterranean and other European areas. The literature on the environmental assessment of sheep and goat production appears to be focussed on the differences in the livestock systems, and there are few, if any, studies considering the wider systemic perspectives of both, that of sustainability and that of two-way relationships.

5.5.1.1 Life Cycle Assessment Applications: State of the Art

Very few studies have addressed the environmental impact of sheep and goat sector using LCA methodology. These studies that directly or indirectly refer to the wider term "sheep and goat", including not only meat production, but also sheep and goats in general (for milk, wool and dairy products). Generally, LCA analysis of the sheep and goat sectors presents a comparison of several species and their products, often including cow, sheep and goat, and of the breeding methods used.

In the following, firstly we briefly explain the main objectives and results of the studies reviewed, than we make a detailed and comparative analysis involving each LCA step.

As argued above, few studies were found that specifically applied LCA methodology to the sheep and goat sector. The study of Head et al. (2011) showed that sheep have a high impact on biodiversity, greenhouse gases and health. Despite significantly greater land use for goat livestock, production of goat's milk had a slightly lower impact than that of cow's milk, because dairy goats require less feed. Comparing conventional and organic lamb farming, Head et al. (2011) demonstrated that conventional lamb has almost twice the impact than that of organic lamb because the former eats a greater percentage of soy-based concentrate, with negative implications for biodiversity. Organic lamb had a slightly higher negative impact on climate change and human health, than conventional lamb, because manure was used instead of chemical fertiliser on the wheat straw.

Two studies by Kanyarushoki et al. (2008, 2010) shown that per 1000 kg of, goat milk had a greater negative impact than cow milk. Moreover, goat farms had greater impact per hectare of land occupied, except on climate change (Kanyarushoki 2010). The authors estimated the emissions, non-renewable energy and land occupation of several farms through the EDEN model, which is a Microsoft® Excel-based tool (van der Werf et al. 2009). Impacts were compared using two functional units (FU): (a) per 1 t of fat and protein corrected milk (FPCM) sold; and (b) on-farm plus estimated off-farm hectares utilised. The authors decided to avoid allocation between animal and crop products and separated the farms into two parts: production of crop products not used for animal production, and all other farm processes. In the final step, economic data was used to determine the impacts of milk and animal production.

Michael (2011) applied the standardised LCA methodology to identify and evaluate the carbon footprint, water and energy efficiency of five animal product industries. The results were significantly influenced by whether the animal species was a ruminant or non-ruminant, and whether competitive feed conversion ratios were achieved. The non-ruminant animal species were highly efficient in terms of emissions compared to the ruminants, provided that the enterprises were well managed and feed conversion rates were high. Ruminants (dairy sheep and dairy goats) had a significant burden of enteric emissions and methane.

Koch et al. (2013) applied the AGRIBALYSE® programme launched by the French Environment and Energy Management Agency (ADEME) to create a Life Cycle Inventory (LCI) database of French agricultural products. AGRIBALYSE® was built with two aims: (i) to create an LCI database to provide data for the environmental labelling of food products; and (ii) to share data to enable the agricultural and food industries to assess the production chain and reduce environmental impacts. AGRIBALYSE® provided 136 LCI data sets for arable, horticultural and livestock products. The data for the production systems and direct emission was processed using Excel®, while the indirect flows were added using SimaPro® to obtain the LCI and LCIA data sets.

After this general presentation of LCA studies involving sheep and goat production, the next section (§ 5.5.2) makes a comparison between them, following LCA steps and trying to highlight the prospects and constraints of this methodology.

5.5.1.2 Other Life Cycle-Based Methodologies and Tools: The Carbon Footprint

While the literature review does not present studies combining LCA with other methodologies, such as Social LCA and Life Cycle Costing, numerous Life Cycle Thinking (LCT) approaches of the Footprint indicator Family have been used for the sheep and goat sector. Carbon footprint (CF) is one of the most common, followed by water footprint (WF) indicator. The literature review include these two indicators because of their strong similarities with and complementarities to LCA assessment (see, for example, the EC-JRC PEF Guide, 2013; Boulay et al. 2013; Fang et al. 2014) and of the international initiatives developed worldwide (presented in § 1.2.4.2 and 1.2.4.3). The review does not intend to completely describe the state of the art of the CF literature for the sheep and goat sector. We selected studies from 2010 and strictly related to LCT approach to give some insight into the main aspects relevant to LCA methodology perspectives.

Biswas et al. (2010) compared the emissions (CO_2, N_2O and CH_4) performance of three different Australian products (sheep's wool, sheep's meat and wheat) in three adjacent plots (mixed pasture, wheat and sub-clover). The system boundaries adopted a cradle to farm gate perspective and was divided into two main stages, pre-farm and on-farm. An economic allocation method was used to calculate the input and output of co-products. The input/output data of the LCI was linked to the relevant libraries in SimaPro 7. They reported that the life cycle greenhouse gas (GHG) emissions of 1 kg of sheep's wool were approximately three times higher than the GHG emissions of the sheep's meat production. on the on-farm stage contributed the most significant portion of total emissions. CH_4 emissions from enteric methane production and from the decomposition of manure accounted for a significant portion (83–90 %) of the total emissions from sub-clover and mixed pasture production. A sensitivity analysis was carried out showing that the GHG emissions were very sensitive to the fluctuation of prices of sheep meat and wool (respectively +/−3.5 and +/−14 %).

Many authors calculated the environmental impact of goat/sheep production through the joint application of the LCA approach and specific models, including:

- the Cranfield model (Williams et al. 2006);
- the EDEN model (Kanyarushoki et al. 2008; 2010);
- the Capri model (Weiss and Leip 2012);
- the Global Livestock Environmental Accounting model (GLEAM) (Opio et al. 2013).

Willams et al. (2006) proposed first the Cranfield model, which was used as a reference point by the other studies presented below. Williams et al. (2012) developed the systems model for the stratified UK sheep industry to provide the activity data input for the life cycle assessment of the Cranfield model. This includes the bio-physical performance of the lowland, upland and hill sheep flocks. The LCA analysis of the production of lamb meat took into account the different sizes of the breeds and consequent feed requirements, different types of land and consequent yields of grass (and management requirements), and different rates of lamb growth and ewe productivity. The FU was a 1000 kg edible lamb carcass at the national level; the system boundary was the farm gate. Enteric methane was calculated using the IPCC (2006) Tier 2 formula, and the results were expressed as LCIs using the characterisation factor of the IPCC for GWP and of the CML for other impacts (i.e., eutrophication potential, acidification, abiotic resource use, type of land). The baseline results were compared with alternative scenarios that considered changes in sheep management, changes in genetic potential and management quality (including animal health), and in the emission factor for enteric methane. The Cranfield model was also used in Phase One of the English Beef and Sheep Production Roadmap set by a steering group of industry organisations led by EBLEX (the organisation for sheep and beef producers in England). In the second phase of the Roadmap a different model was employed—the E-CO2 system—using real data at farm level. This model used Carbon Trust, IPCC 2006 and PAS 2050 methodology (BSI 2008) to calculate the GWP of beef and sheep production.

A GHG footprint study for exported New Zealand lamb (Ledgard et al. 2010) assessed the full life cycle CF of lamb from farms, through to cooking and eating the meat, and the disposal of waste and sewage. Emissions (CH_4, N_2O, CO_2, refrigerant) referred to a 100 g portion of raw, purchased meat as the functional unit. This study used a biophysical allocation for different animal types on farms, based on the amount of feed they consumed. An economic allocation was used for lamb meat, mutton, wool and, at the meat processing stage, meat and non-meat products. At the farm stage a private data set covering nearly 500 farms throughout New Zealand was sampled, to be statistically representative of the sheep farming sector, and stratified to cover the wide range of different farm types (from extensive high country through to more intensive rolling land). 100-year global warming potential (GWP100) conversion factors were used to convert methane and nitrous oxide emissions. The total footprint was divided into 80 % for the on-farm stage, 3 % for meat processing, 5 % for (oceanic shipping) transportation and 12 % for the consumer phase (excluding consumer transport).

Weiss and Leip (2012) carried out another interesting study to estimate GHG fluxes for all emission sources of the agricultural sector. Estimates of GHG (CH_4, N_2O and CO_2) fluxes referred to the main European livestock products (meat, milk and eggs) according to a cradle to gate attributional life-cycle assessment, including emissions from land use and land use change. Calculations were made using the CAPRI modelling system, considering on-farm and off-farm fluxes and emissions from land use changes. The quantification of methane emissions from enteric fermentation and manure management followed a Tier 1 approach for sheep and goats. Allocation of emission fluxes to multiple outputs was based on the nitrogen content in the products. As far as CH_4 emissions from dairy cattle (enteric fermentation and manure management), the energy requirement for lactation and pregnancy was used to allocate emissions from milk and young animals. For most animal products, except sheep and goat meat and milk, emissions from foregone carbon sequestration dominated enhanced carbon sequestration in managed grasslands leading to net emissions. Emission intensities differed considerably between the EU-27 countries for all products examined, due to many factors (productivity, dependency of imported feed products, and share of pasture in the animal feed diet). The comprehensive approach of this paper, and its peculiarities compared to other studies, were reviewed in Bellarby et al. (2013).

Gac et al. (2012) studied the carbon footprint (CH_4, N_2O, CO_2) of French and New Zealand lamb production from cradle to farm gate for the year 2008, comparing two contrasting systems: in-shed lamb *vs.* grass lamb. Each system was analysed using a methodology developed to fit its own country, namely GES'TIM (Gac et al. 2010) for France and the Overseer® model (Ledgard et al. 2010) for New Zealand. The impact on climate change was assessed by using the GWP100 proposed by IPCC (2006). A common mass allocation was firstly used to allocate impacts to either meat or wool. The differences in the average CF in the two countries underline the importance of country specificities of both environmental context and the socio-economic characteristics of local livestock systems. In fact, the higher carbon footprint of French lamb was due to the use of external feed input and the fact that sheep are housed in-shed for part of the year, with emissions from manure management. Conversely, in New Zealand, where productivity is often higher due to warmer climatic conditions, the animals stay outside all year eating perennial pastures and therefore there are no gaseous emissions linked to external food production and manure management. Carbon sequestration in pastoral soils can potentially have a significant effect on reducing the carbon footprint at farm level. This paper presents an interesting sensitivity analysis that showed how results were highly dependent on methodological choices. Firstly, the effects of allocation method were tested: economic *vs.* mass allocation. There was a small difference between countries when mass allocation was used, and a much larger difference using economic allocation; this was because in New Zealand wool has an economic value for carpet making, whereas in France it has little economic value. Another sensitivity analysis was performed by comparing the same methodologies across both countries. Audsley and Wilkinson (2012), using the Cranfield system model, explored options for reducing UK GHG emissions from crop and livestock production systems considering cradle to farm gate boundaries. Among livestock production, sheep systems included hill, upland and lowland, pure and crossbred flocks. Emissions were expressed as GWP100 in tonnes CO_2-eq per unit of product. For each system, emissions of nitrous oxide were calculated using the IPCC Tier 1 methodology (IPCC 2006). GHG emissions were always higher for ruminants due to the methane emitted during rumination. Differences between upland and lowland sheep were small in terms of GHG emissions/kg of product at the farm gate. The best alternative system in terms of reduced emissions compared to the combined typical systems was identified for each livestock sector using the Cranfield model. The potential reductions in GHG emissions ranged from 7 % for dairy beef and poultry meat to 21 % for sheep meat.

Chatterton et al. (2012) developed an integrated livestock-ecosystems linear programming model to assess the economic and environmental impacts of the livestock sector in the UK. For this, the Cranfield Model was combined with a grassland productivity model and a soil erosion model to assess the environmental consequences of the livestock sector. A model was also developed to calculate soil erosion. The output of the LCA model were linked within the linear programming framework. The objective function to be maximised was the sum of the various ecosystem services (Provisioning + Regulating + Cultural), which were converted to a common monetary valuation system. The results show the importance of the use of a systems-based LCA approach in identifying the trade-offs between the cultural benefits of extensive systems and the potential efficiencies of more intensive systems.

The study of Eady et al. (2012) was interesting in that it used the whole suite of approaches recommended in the ISO guidelines to model co-production at the farm level, in an attempt to best represent the mixed farming system. Studying a single case study farm in Western Australia, the authors compared the CF of products with and without quantifying the benefits of mixed farming system, and compared different methods of modelling co-products. The mixed farm being studied produced distinct products (Merino wool, sheep meat and grains) that were modelled from cradle to gate using system expansion. Co-production from the sheep activity was modelled using allocation, comparing biophysical and economic relationships. As in the other studies previously discussed, the authors concluded that when compared to biophysical allocation, economic allocation shifted the environmental burden to the higher value co-products and away from the products with high resource use.

Brock et al. (2013) determined the emission profile and carbon footprint of wool production in south Wales. GHG emissions were estimated at the pre-farm and on-farm stages of production, the second being the most relevant. This study is interesting in that it tested how the emissions profile varied according to calculation method and assumptions As in other studies, the total emissions were apportioned to wool and co-products, based on economic allocation. This study also showed that the calculated emissions for wool production changed substantially, under an economic allocation method, by changing the farm emphasis from wool to meat production (41 % decrease) and by changing wool price (29 % variability). Other sensitivity analyses referred to changes in the fibre diameter (23 % variability) and fleece weight (11 % variability). The paper excluded carbon sequestration.

The implications of land occupation for CO_2 was addressed by Schmidinger and Stehfest (2012) who calculated the missing potential carbon sink of producing or not producing a certain livestock product. The applied methodology related land occupation data from LCA studies to the potential carbon sink as calculated by the IMAGE model and its process-based spatially explicit carbon cycle model. The total GHG effect of a product was calculated as the sum of the emissions along the product chain according to conventional LCA (not including direct emissions from land-use change) plus the CO_2 emission or missed potential carbon uptake due to land-use occupation in terms of kg CO_2-eq/kg product. The authors accounted for regional differences (world region), heterogeneity in land-use, and different time horizons (30-year, 50-year and 100-year time horizon). Calculations showed that the CO_2 consequences of land occupation were in the same order of magnitude as the other process-related greenhouse gas emissions of the LCA, and depended largely on the production system. The highest CO_2 implications of land occupation were calculated for beef, sheep, and goat.

Ripoll-Bosch et al. (2013) explored whether accounting for the multifunctionality of sheep farming affected the CF of lamb meat. Three farming systems (the pasture-based system, the mixed sheep-cereal system, and the industrial system, or zero-grazing) in Spain were considered representative. The study's main data sources include the FAO and national statistics. The authors computed from cradle to farm gate because post-farm gate processes were assumed to be equal for each system, and, therefore, were excluded from the analysis. The CF assessment followed the attributional approach. They quantified emissions (CO_2, N_2O and CH_4) using a model processed in MS Excel that consisted of four main modules: (i) herd structure and performance (but no herd dynamics considered); (ii) feed production (assessed both, whether on farm or off-farm production); (iii) animal feeding; and (iv) manure management. Calculations of emission in the model were based on a Tier 2 level. The GWP values used to convert methane and nitrous oxide into CO_2-eq were taken from IPCC (2007). The highest GHGs emissions involved the pasture based livestock system. When accounting for multifunctionality, the lowest GHGs emission were for the pasture-based system and the highest for the zero-grazing system.

Liang et al. (2013) studied GHG emissions from the livestock sector (swine, cow, beef, goat and poultry) in Beijing based on average data between 2007 and

2009. They covered the structure and relative proportions of diverse livestock, and adjusted related coefficients to the local situation. In this study, the assessments of total GHG emissions (only CH_4 and N_2O) was computed, together with the relative proportion in different processes (enteric fermentation, inside barn and waste management).

The Global Livestock Environmental Assessment Model (GLEAM), used in Opio et al. (2013), is a process-based static model that simulates the functioning of livestock production systems. It consists of five main modules: herd module, manure module, feed basket module, system module and allocation module, and two additional modules for the calculation of direct and indirect on-farm energy and post-farm gate emissions. The authors presented a life cycle analysis of the GHG emissions arising from ruminant supply chains. The average emission intensity for products from ruminants were estimated in terms of kg CO_2-eq/kg fat and protein corrected milk for milk, and in terms of kg CO_2-eq/kg carcass weight for meat.

In the Italian literature, a growing interest in this topic can be seen among a group of researchers inside the Animal Science and Production Association (ASPA). From these studies, we selected two papers on the GHG emissions of the Italian sheep sector. Atzori et al. (2013a) studied the differences in primary and secondary CO_2-eq emissions among four simulated scenarios of dairy sheep production in Sardinia, all able to produce the same amount of milk per year. Using the Tier 3 approach of the IPCC (2006), an Excel® spreadsheet simulated different processes: animal categories, land use, soil management, biomass available (pasture or hay), and purchased feeds. A specific sub-model was built to estimate farm CO_2-eq emissions, including methane from enteric fermentation, methane and nitrous oxide from manure management, CO_2-eq from fertilisers and fuel and from purchased feeds. Animal requirements, dry matter intake and nitrogen excretion were estimated based on locally developed equations; enteric emissions were based on IPCC sheep coefficients, whereas emissions from manure management were based on dairy cattle IPCC tables. Simulated scenarios considered four farms with high and medium-low production levels, with or without pasture, with different percentages of on-farm or purchased feed or forage. Results suggest that a reduction of emissions takes place with high production levels and on-farm feed production.

Atzori et al. (2013b) aimed to assess the GHG emission of the Italian sheep sector by accounting for CH_4 from enteric fermentation, CH_4 and N_2O from manure management, both expressed as CO_2-eq using the Tier 3 approach. A meta-modelling approach was applied within each animal category, to estimate: diets and metabolizable energy requirements for maintenance, activity, cold stress and production; emitted methane as a percentage of metabolizable energy intake; nitrogen excretion; and emission factors for CH_4 and N_2O from manure. Their results for enteric fermentation emissions were higher than those proposed by the IPCC Tier 1 guidelines.

The study of Jones et al. (2014) estimated the cradle to farm gate CF of 64 sheep farms across England and Wales for a single year using empirical farm level production data, in terms of kg CO_2-eq/kg live weight finished lamb. Default IPCC Tier 1 emission factors and data from the literature were used for reporting direct and indirect emissions (CH_4, N_2O and CO_2). Variation in the CFs relating to both system type and management was assessed. A non-parametric test was used to make comparisons between the footprints of lowland, upland and hill farms; between farms categorised by breeding ewe flock size; and then between farms categorised by area. Multiple linear regression models and dominance analysis indicated the four farm management variables with the highest impact on the size of the carbon footprint of finished lamb (head/ewe; lamb growth rate; the percentage of ewe and replacement ewe lamb flock not mated; and concentrate use). Shared inputs, such as fertilisers, were allocated based on total grazing livestock units. Emissions were shared between categories of sheep products (finished lambs, live lambs, culls sold for meat, breeding sheep and wool) using economic allocation.

5.5.1.3 Other Life Cycle-Based Methodologies and Tools: The Water Footprint

To obtain further insight about other life cycle-base methodologies and footprint indicators we chose to study the water footprint. This choice partly reflects the same arguments as discussed for the carbon footprint review, and is partly motivated by the relevance of water impact in the sheep and goat sector. Following the same criteria as CF, we present a selection of literature and do not intend to give a complete picture of the state of art in the sector, but only to suggest some useful points for future improvement in the development of complementary LCT approaches.

Several approaches to estimating water use and its impacts have been developed, each differing in the types of water included, whether the upstream or downstream processes were considered, and the characterisation of environmental impacts. Among these approaches, we only report some of those studies below that refer to a life cycle approach. In many cases, authors calculated the water impact of sheep and goat production through the joint application of LCA and specific models, such as the MEDLI model (Peters et al. 2010b) and the OVERSEER® nutrient budget model (Zonderland-Thomassen et al. 2012). The study of Peters et al. (2010b) aimed to account for water use in southern Australian red meat production, considering three supply systems, among which was a sheep-meat supplier. The functional unit of this LCA was defined as the delivery of 1 kg of hot standard carcass weight (HSCW) meat to the meat processing works product gate for wholesale distribution. The water input and output was allocated to red meat production in accordance with the relative mass of the red meat and its by-products. The authors used a hydrological model based on MEDLI, a model for analysing effluent reuse systems, and a climate file. One critical point in the WF estimation was whether, and which, environmental consequences result from water being an input to the system. Construction of the life cycle inventory, and characteristics of the water source, such as whether (1) it is renewable, (2) extraction exceeds the renewal rate, and (3) whether the extracted water is returned to the original watercourse in full, must be understood in order to determine whether water use is sustainable. The quality of water output and the time reference are aspects relevant in the WF estimation that could be difficult to manage in an LCA framework. As suggested in Peters et al. (2010b), if the frame of reference is a particular year, then changes to foreground production systems that occur from year to year and that threaten biodiversity are overlooked. Chatterton et al. (2012), in a study for EBLEX, the organisation for beef and lamb levy payers in England, combined the Cranfield LCA model with the WaSim[3] water simulation model to establish a water footprint for English beef and sheep production. Their assessment took into account all input and output of water linked to the production of beef and sheep meat—from hill, upland and lowland ewes—to calculate water use per kilogram of meat. The total footprint was accounted for almost entirely by green water (97 %), required for feed crop and grass production. The grey water (only nitrate leaching) accounted for the remaining 3 %. Hill systems had a much higher use of blue water because grass yields were significantly lower and thus green water footprints were much greater per ton of grass required.

Mekonnen and Hoekstra (2012) estimated the WF of animal products and compared it with the WF of some crops. Different production systems (grazing, mixed and industrial) and feed composition per animal type and country (China, India, Netherland, USA) were considered. They assessed the WFs of growing feed crops using a grid-based dynamic water balance model that took into account local climate, soil conditions and data on irrigation at a high spatial resolution. They considered sheep and goat, as meat animals, together with beef cattle, chicken and pigs. Sheep and goats were the least impactful animal category. The paper is very interesting for the relevance it gave to the different production systems (grazing, miked, industrial production) in calculating the total, blue and grey WFs. The study is also interesting for the methodological problems it raised: uncertainties due to a lack of data; assumptions made and combination of different data sources. Some aspects ignored in the paper (the indirect water footprints of materials used in feed production and animal raising; the potential pollution by fertilisers other than nitrogen or by pesticides or other agro-chemicals; the grey water footprint from animal wastes) are particularly relevant for industrial production systems, resulting in an underestimation of their WF.

The WF of pastoral farming systems in New Zealand (NZ) was the topic of a study by Zonderland-Thomassen et al. (2012). Survey data from representative sheep and beef farm systems was used to deal with variation in production systems. The cradle-to-farm gate life cycle required for the production of milk, beef, and sheep meat was analysed. Economic allocation was applied when dividing the WF between milk and meat. Biophysical allocation based on feed intake was used when dividing the WF between beef cattle and sheep, while economic allocation was used when dividing the WF for sheep between meat and wool. A WF approach compliant with LCA principles was used to assess the stress-weighted WF. The eutrophication potential was also assessed. Water losses associated with evapotranspiration from irrigated pasture, as well as nitrate leaching and phosphate runoff were

[3] The Water balance-Simulation Model (WaSiM) has been developed by Schulla (Schulla, J., 1997; Schulla, J., Jasper, K., 2007) to evaluate the influence of climate change on water balance.

computed using the hydrological sub-model in the OVERSEER® nutrient budget model (Wheeler et al. 2003).

5.5.2 The Implementation of Life Cycle Assessment in the Sheep and Goat Sector: Methodological Problems

Following the same approach as used in other sectors, we analysed the LCA ISO 14044 specific requirements in the domain of sheep and goat LCA studies, suggesting some points for future reflection and improvement.

5.5.2.1 Goal and Scope

All the selected papers claimed to assess the environmental impact of different goat/sheep products, in some cases both meat and milk, and in other cases only one product. According to the different goals of the papers, some considered only a single species (sheep or goats), while many studies carried out comparative environmental assessment of several vegetable and/or livestock products. Table 5.4 lists the selected articles, specifying the methodology used for the analysis—LCA, CF, WF—and the product investigated. As shown in the table, there are studies that considered only one product, others that looked at many co-products of the same livestock, and others that compared several products among different livestock. There were case studies and sectorial analyses. Even with this variety of goals and scopes, no relevant methodological problems appeared in the definition of the first LCA step.

5.5.2.2 Functional Unit

The selected LCAs studies commonly defined the FU as the mass of the product leaving the farm gate, but with different specific criteria. Many examples of FU could be drawn from the papers (see Table 5.4) even when they referred to the same product. Different FUs do not permit comparison of results from different LCA studies because, as is well known, the FU allows the comparison of alternative systems of products with a similar function.

As argued later on, few studies test the robustness of their results against different FUs with a sensitivity analysis.

The choice of a proper FU is one aspect that deserves more attention in defining homogeneous standards, but it does not require further methodological advances. The literature review of the sheep and goat sector leads to conclusions in accordance with the guidelines proposed in the previous paragraph concerning the choice of the functional unit. This choice would require a standardisation in relation to the objectives of the study and the phases included in the system boundaries. Qualitative

Table 5.4 Articles reporting on the implementation of LCT tools in the sheep/goat sector

Reference	FU	System Boundaries	Products	Methodology
Williams et al. (2006)	1 t of sheep meat and 10 m^3 of goat milk,	Farm gate	Sheep meat and goat milk	Cranfield LCA model
Kanyarushoki et al. (2008)	1000 kg of fat-and protein corrected milk (FPCM), and hectares of land utilised	Cradle-to-farm gate and from farm to the retailer entrance gate	Goat and cow milk, dairy products	LCA
Kanyarushoki et al. (2010)	1000 kg of FPCM, and hectares of land utilised	Cradle-to-farm gate and from farm gate to the retailer entrance gate	Goat and cow milk, dairy products	LCA
Ledgard et al. (2010)	100 gm portion of lamb meat	Full life cycle of meat, from farm to consumption and consumer waste stages	Lamb meat	CF
Peters et al. (2010b)	1 kg of (HSCW) sheep meat	Cradle-to-farm gate and processing for wholesale distribution	Sheep meat	WF
Head et al. 2011	1 kg of product	From farm to the supermarket	Goat milk, lamb meat, sheep meat, and goat dairy products	LCA
Michael (2011)	1 kg of milk adjusted for fat and protein content	Farm gate	Sheep and goat milk	LCA
O'Mara (2011)	The gross energy content of the commodities	Cradle-to-farm gate	Goat/sheep meat and milk	CF
Ripoll-Bosh et al. 2011	1 kg of meat lamb	Cradle-to-farm gate	Lamb meat	CF
Audsley et al. (2012)	kg of product fresh weight, MJ edible energy, kg edible protein	Farm system	Sheep meat	Cranfield LCA model
Chatterton et al. (2012)	Ton of sheep meat	Livestock sector	Sheep meat	Cranfield LCA model
Eady et al. (2012)	Grain: 1 t, Wool: 1 kg of greasy wool, sheep: one animal	Cradle-to-farm gate	Wool, sheep meat and grains	CF
Gac et al. (2012)	1 kg of total sheep live	Cradle-to-farm gate	Lamb meat	CF

Table 5.4 (continued)

Reference	FU	System Boundaries	Products	Methodology
Mekonnen and Hoekstra (2012)	Ton of product	Cradle-to-farm gate	Sheep/goat meat	WF
Schmidinger and Stehfest (2012)	kg product, m^2/year	Product chain+carbon uptake	Beef, milk, pork, poultry, crops, sheep and goat	CF (LCA+IMAGE)
Zonderland-Thomassen et al. (2012)	kg live weight and kg fat and protein corrected milk (FPCM)	Cradle-to-farm gate	Sheep meat and milk	WF
Weiss and Leip (2012)	1 kg of carcass meat and 1 kg of raw milk	Farm gate, including slaughtering	Sheep/goat meat and milk	CF
Williams et al. (2012)	1000 kg carcass lamb	Farm gate	Lamb meat	Cranfield LCA model
Atzori et al. (2013a)	Litre of milk and no. of ewes	Cradle-to-farm gate	Sheep	CF
Atzori et al. (2013b)	1 kg of sheep meat, 1 kg of sheep milk, the number of livestock, kg of 4% fat milk	Cradle-to-farm gate	Meat and sheep milk	CF
Brock et al. (2013)	1 kg of greasy wool	Cradle-to-farm gate	19 µ wool production	CF
Koch et al. (2013)	1 kg or 1 L of product	Cradle-to-farm gate	Sheep/goat milk and lambs	LCA
Liang et al. (2013)	The average number of major livestock	Livestock husbandry systems, waste treatment systems, and agricultural use of livestock waste/manure	Livestock	CF
Opio et al. (2013)	kg of carcass weight and kg of FPCM	Cradle-to-farm gate and farm gate to retail	Goat/sheep meat and milk	CF
Ripoll-Bosch et al. (2013)	1 kg of lamb live weight	Cradle-to-farm gate	Lamb meat	CF
Jones et al. (2014)	1 kg of life weight lamb	Cradle-to-farm gate	Sheep meat	CF

Legend: *CF* Carbon Footprint/GHG, *LCA* Life Cycle Assessment; *WF* Water Footprint indicators should be used when the assessment of environmental load is related to the final products, such as by correcting the amount of milk to an energy corrected milk (ECM) basis, or by specifying the amount of meat in terms of animal parts. As in some of the selected papers, it is suggested that selecting multiple functional units or assessing the variability of results against different FU might be more accurate in a sensitivity analysis.

5.5.2.3 System Boundaries

The reviewed studies encompass all the processes in goat/sheep production: from production and the application of fertiliser, pesticides and herbicide for forage to goat/sheep cheese or meat processing. In particular, the production of input, such as fertiliser, pesticides (Biswas et al. 2010; Koch et al. 2013; Ripoll-Bosch et al. 2011, 2013; Williams et al. 2006), herbicide (Biswas et al. 2010), and their application (Head et al. 2011; Jones et al. 2014; Koch et al. 2013; Weiss and Leip 2012; Williams et al. 2006, 2012) were included in some system boundaries. The production of feed (O'Mara 2011; Koch et al. 2013; Opio et al. 2013; Ripoll-Bosch et al. 2011, 2013; Williams et al. 2006, 2012) and forage were inside the system boundaries of some studies (Jones et al. 2014; Kanyarushoki et al. 2008; 2010; Ripoll-Bosch et al. 2011; Williams et al. 2006). Mekonnen and Hoekstra (2012) considered amount of feed consumed per animal category, per production system and per country; while Atzori et al. (2013a) used purchased feeds, animal categories, land use, soil management and biomass available (pasture or hay). Some system boundaries included livestock husbandry systems and agricultural use of livestock waste/manure (Head et al. 2011; Jones et al. 2014; Liang et al. 2013; Williams et al. 2006). Some studies (Kanyarushoki et al. 2008, 2010; Opio et al. 2013) included the entire chain for goat/sheep milk and goat/sheep meat, from farm gate to retail entrance gate. Weiss and Leip (2012) and Mekonnen and Hoekstra (2012) included farm gates and slaughter, others (Gac et al. 2012; Koch et al. 2013; Williams et al. 2012) only the farm gates. Ledgard et al. (2010) calculated the GHG emissions across the full life cycle of meat, from farm to consumption and consumer waste stages. Peters et al. (2010b) assessed the environmental impact of the delivery of meat to the meat processing works' product gate for wholesale distribution; Head et al. (2011) included the entire food chain for 98 different animal products from farm to the supermarket. Authors usually clarified the processes excluded in the definition of the system boundaries of their studies:

- the post-farm dairy chain of goat/sheep sector (Atzori et al. 2013a, 2013b; Audsley and Wilkinson 2012; Biswas et al. 2010; Gac et al. 2012; Koch et al. 2013; Liang et al. 2013; O'Mara 2011; Ripoll-Bosch et al. 2011, 2013; Jones et al. 2014; Zonderland-Thomassen et al. 2012; Williams et al. 2006, 2012);
- the production of medicines (Gac et al. 2012; Head et al. 2011; Opio et al. 2013; Ripoll-Bosch et al. 2013; Zonderland-Thomassen et al. 2012) and their use (Kanyarushoki et al. 2008);

- machinery and buildings (Gac et al. 2012; Ripoll-Bosch et al. 2013; Head et al. 2011; Jones et al. 2014; Kanyarushoki et al. 2008, 2010; Zonderland-Thomassen et al. 2012);
- in many papers, the impacts associated with the land use change, biomass burning, biological fixation, emission from non-N fertilisers and lime;
- emissions from processing, transport, packaging, retail, consumption (Opio et al. 2013; Weiss and Leip 2012), consumer transport (Ledgard et al. 2010) and waste from the products (Head et al. 2011; Opio et al. 2013; Weiss and Leip 2012).

There were generally different reasons for the exclusions. Exclusions are motivated by the low entity of some impacts, as well as by the high degree of uncertainty in the data (Head et al. 2011), or by limitations in the availability of emission data (Jones et al. 2014; Opio et al. 2013). Secondly, exclusions may have been motivated by lack of methodology or consensus on the quantification approach (Opio et al. 2013). Finally, exclusions were also made due to a lack of appropriate characterisation factors (Kanyarushoki et al. 2008, 2010).

Among the selected studies, some considered different types of breeding and breeding systems. Williams et al. (2012) calculated the environmental impact of lamb meat taking into account the different sizes of the breeds and consequent feed requirements, different types of land and consequent yields of grass, and different rates of lamb growth and ewe productivity. Ripoll-Bosch et al. (2011, 2013) and Gac et al. (2012) studied low-mid and highly intensive productive systems, while other studies showed the environmental impact of the conventional and organic systems (Head et al. 2011; Peters et al. 2010b; Williams et al. 2006). Other scholars (Audsley and Wilkinson 2012; Head et al. 2011; Kanyarushoki et al. 2008, 2010; Mekonnen and Hoekstra 2012; Opio et al. 2013; Peters et al. 2010b; Williams et al. 2006) compared the environmental impact of ruminants and small ruminants; others (Michael 2011) compared non-ruminant animal species and ruminants. The time boundaries were specified in all studies reviewed, although due to a lack of availability of data, sources usually referred to different years. Some studies made reference to 1 year, others studies referred to 1 year but used average data (Kanyarushoki et al. 2008; 2010; Liang et al. 2013; Mekonnen and Hoekstra 2012; Zonderland-Thomassen et al. 2012), some used data from two (Peters et al. 2010b) or more years (Koch et al. 2013).

Concerning the system boundaries definition, a major problem in the livestock environmental assessments occurs when farms have surfaces designated for fodder production. In this situation, a holistic integrated approach and system expansion are needed to assess the environmental impact of both vegetable and livestock production cycles, and this approach increases the complexity of the analysis. The consideration of vegetable the production cycle to support animal nutrition requires a global assessment which considers the different use of farm land and related issues (different crops, rotation), as well as the effects of land use changes. This global assessment, even if more suitable from a conceptual perspective, makes things more difficult and increases the amount of data, the complexity of calculations, the assumptions required, and the uncertainty of results. Analysis of the literature suggests that greater attention must be devoted to the specification of geographical and time boundaries of the studies, especially considering the relevance of spatial and temporal dimensions in livestock management and environmental impact. Finally, future developments in system boundaries were linked to availability and quality of data, dealt with in the following paragraph, because the lack of data is often a reason to omit some processes from the system boundaries.

5.5.2.4 Availability and Quality of Data

Availability and quality of data is one of the most critical issues when applying LCA approaches. The literature review and the comparative analysis reveal that there is a need for further development towards more complete and reliable data.

As illustrated below, the selected papers adopted different approaches and assumptions with reference to the data used in the analysis.

Some authors used both data on farm activities and data from databases. The databases used were, for example the Ecoinvent (Michael 2011; Opio et al. 2013; Ripoll-Bosch et al. 2013; Head et al. 2011; Kanyarushoki et al. 2008, 2010; Williams et al. 2006) and the SimaPro (Biswas et al. 2010; Koch et al. 2013; Michael 2011;Williams et al. 2006).

Many papers took data from the literature. The list of data sourced from the literature is very long and varies between studies. Some took data related to the animal husbandry system (Atzori et al. 2013a, b; Biswas et al. 2010; Head et al. 2011; Liang et al. 2013; Mekonnen and Hoekstra 2012; O'Mara 2011; Opio et al. 2013 Williams et al. 2006, 2012), manure management (Opio et al. 2013) and enteric fermentation (Atzori et al. 2013a; Gac et al. 2012; Head et al. 2011; Jones et al. 2014; Liang et al. 2013; O'Mara 2011; Weiss and Leip 2012; Williams et al. 2006, 2012). Data from other literature was used for the emission of N_2O (Atzori et al. 2013b; Head et al. 2011; Liang et al. 2013, O'Mara 2011; Ripoll-Bosch et al. 2011, 2013; Williams et al. 2006, 2012), carbon dioxide CO_2 (Atzori et al. 2013a; Audsley and Wilkinson 2012; Head et al. 2011; Ledgard et al. 2010; Ripoll-Bosch et al. 2011, 2013; Weiss and Leip 2012), and the emission factor for carbon/ solid storage (Gac et al. 2012; Liang et al. 2013). Other papers used literature data on GHG emissions from the production (Head et al. 2011; O'Mara 2011; Ripoll-Bosch et al. 2013) and application of pesticides and herbicides (Gac et al. 2012; Head et al. 2011; Ripoll-Bosch et al. 2013), and fertiliser (Gac et al. 2012; O'Mara 2011; Head et al. 2011; Ripoll-Bosch et al. 2013). Some authors used data on CH_4 emissions (Atzori et al. 2013a, b; Audsley and Wilkinson 2012; Biswas et al. 2010; Head et al. 2011; Jones et al. 2014; Liang et al. 2013; O'Mara 2011; Ripoll-Bosch et al. 2013), on the deforestation (Opio et al. 2013), on crops (Ripoll-Bosch et al. 2013; Head et al. 2011; Weiss and Leip 2012; Williams et al. 2006, 2012), and feed production (Mekonnen and Hoekstra 2012; O'Mara 2011; Opio et al. 2013; Ripoll-Bosch et al. 2013). Data on slaughter (Head et al. 2011; Weiss and Leip 2012) and land use (Head et al. 2011; O' Mara 2011; Weiss and Leip 2012) also comes from the literature. Other authors (Chatterton et al. 2012) did not specify in detail data sources. Due to the variability of farming practices, soils and climate, it was often difficult to construct a realistic "national average" production system. For this reason, Koch et al. (2013) created several LCI datasets for the same product, for different farming practices or regions, and made different data quality controls. Due to the lack of some data, some studies dealt with the problem of uncertainty, especially of GHG emissions. The uncertainty of GHG emissions in the agricultural sector is due to numerous complex factors, such as a high variability in emission factors, especially in N_2O emissions from agricultural soils (Weiss and Leip 2012). Liang et al. (2013) adjusted GHG coefficients related to a China-specific situation; while the CAPRI database (Weiss and Leip 2012) applies an internal procedure to correct data automatically, filling data gaps or removing data errors, such as statistical outliers or implausible breaks in a time series. In Williams et al. (2006), the measurements of pollutants were all associated with errors and the authors reduced uncertainty in results by aggregating components. In Opio et al. (2013) uncertainties were associated with the variables used in the calculation of emission factors, in estimates of activity data (e.g. animal populations and herd parameters) and assumptions made. The analyses of uncertainty were based on the Monte Carlo (MC) simulation approach, which enables an investigation into how input uncertainty propagates through the lifecycle emissions model.

In the study of Mekonnen and Hoekstra (2012), there were several uncertainties in the quantification of the water footprint of animals and animal products, due to a lack of data, so that many assumptions were made: for example, when using aggregated data taken from official statistics or by combining different data sources from statistics and literature there is an assumption of similarity, with data that is not country-specific and/or not system-specific, that may reduce the credibility and comparability of results. The above review reported that data problems arise in both the agricultural and the following phases of the product chain (for example, transport, manufacturing, and packaging, which are often excluded from the system boundaries). The obvious suggestion to develop the database for the future goes hand in hand with the suggestion to include a sensitivity check of data quality in the studies.

5.5.2.5 Allocation Methods

Allocation describes how "input" and "output" are shared between the product studied and co-products. Co-product handling is a crucial issue because it could have a significant effect on the final LCA results (Flysjö et al. 2011). Allocation can be complex because of multiple output from processes, and of multiple use of output. For example, with reference to sheep and goats, you can consider multiple joint productions and co-products such as milk, meat production, and wool. The choice of allocation method, as well as possibly affecting results, should be evaluated together with the scope of the assessment and the functional unit used. As discussed later, economic allocation is the most frequent approach because it reflects the value of the products to society and the driving forces for production; it is related to the economic value of the co-products, taking into consideration the relative incidence of single joint production compared to total revenue of farms. With price fluctuations and spatial variability, the economic allocation could be different in time and space. Moreover, the relative importance of production can change when it is expressed in livestock units or area, so making comparisons very difficult. The problem of allocation is linked to the type of LCA: a physical approach, as proposed by ISO 14044, is preferred in consequential LCA, while an economic approach could be suitable in attributional LCA (Weiss and Leip 2012). In the LCA studies reviewed, product allocation is often based on the economic values of co-products (Biswas et al. 2010; Gac et al. 2012; Head et al. 2011; Jones et al. 2014; Kanyarushoki et al. 2008, 2010; Ledgard et al. 2010; Opio et al. 2013; Ripoll-Bosch et al. 2013; Zonderland-Thomassen et al. 2012; Williams et al. 2006). In Weiss and Leip (2012) allocation was based on the nitrogen content in the products, and other considered biophysical allocation was based on the amount of feed consumed (Zonderland-Thomassen et al. 2012); on the protein content (Opio et al. 2013), on the metabolic energy required to produce each co-product (Koch et al. 2013), or on the relative mass (Peters et al. 2010b). Sometimes biophysical allocation was used together with economic allocation (Gac et al. 2012; Koch et al. 2013; Ledgard et al. 2010; Michael 2011; Opio et al. 2013; Zonderland-Thomassen et al. 2012). Even with the variety of approaches used in the reviewed papers, the allocation procedure does not pose methodological problems. It is suggested that studies consider the opportunity to adopt mixed allocation rules and, most of all, to test the variability of results in a sensitivity analysis.

5.5.2.6 Life Cycle Impact Assessment (LCIA)

The LCA studies presented different stages of LCIA. Some used several impact categories to show results, as reported in Table 5.5. Other authors used a single impact category to analyse the environmental performance of sheep and goats in terms of CF or WF (see Table 5.4). In many studies, the authors described the evaluation method used. The IPCC 2006 method was the most used by authors, mainly addressing GHG emissions. Kanyarushoki et al. (2008) used CML 2001 and Cumulative Energy Demand. Williams et al. (2006) followed the IPCC 2001 method using timescales of 20, 100 and 500 years, and the CML method. Some scholars used other methods, such as CML and the IPCC, 2007 method (Williams et al.2012), IPCC (2007) (Gac et al. 2012; O'Mara 2011) and the ReCiPe (hierarchic) method (Head et al. 2011). Koch et al. (2013) used different calculation methods for each category impact; for example, they used IPCC, 2006 to evaluate Greenhouse gas emissions, the Recipe method to assess water quality and CML2002 for resource depletion.

The most commonly considered environmental impact categories are listed in Table 5.5. Firstly, it is important to highlight the global warming potential, which was the aim of many papers in the review. Eutrophication and acidification potential (Head et al. 2011; Kanyarushoki et al. 2008, 2010; Williams et al. 2006, 2012) and finally, land occupation (Head et al. 2011; Kanyarushoki et al. 2008; 2010;

Table 5.5 Impact categories considered in the reviewed studies

	Impact categories and Indicators																
References	EP	GWP	ODP	AP	POCP	HTP	FAETP	MAETP	TETP	ARU	CC	PF	IR	LO	LT	LU	Non-renewable energy use
Williams et al. (2006)	✓	✓	–	✓	–	–	–	–	–	✓	–	–	–	–	–	✓	✓
Kanyarushoki et al. (2008)	✓	–	–	✓	–	–	–	–	✓	–	✓	–	–	✓	–	–	✓
Kanyarushoki et al. (2010)	✓	–	–	✓	–	–	–	–	✓	–	✓	–	–	✓	–	–	✓
Head et al. (2011)	✓	–	✓	✓	✓	✓	✓	✓	✓	–	✓	✓	✓	✓	✓	–	–
Michael (2011)[a]	✓	✓	✓	✓	–	–	–	–	–	–	–	–	–	–	–	–	✓
Chatterton et al. (2012)[b]	✓	✓	–	–	–	–	–	–	–	–	–	–	–	–	–	–	–
Williams et al. (2012)	✓	✓	–	✓	–	–	–	–	–	✓	–	–	–	✓	–	–	✓
Koch et al. (2013)[c]	✓	✓	✓	✓	–	✓	–	–	–	–	✓	–	✓	✓	✓	✓	–

Legend: *EP* eutrophication potential, *GWP* global warming potential, *ODP* ozone layer depletion potential, *AP* acidification potential, *POCP* photochemical oxidation, *HTP* human toxicity potential, *FAETP* fresh water ecotoxicity, *MAETP* marine ecotoxicity, *TETP* terrestrial ecotoxicity, *ARU* abiotic resources used, *CC* climate change, *PF* particulate matter formation, *IR* ionising radiation, *LO, LT, LU* land occupation, transformation and use, *NREU* non-renewable energy use

[a] Michael (2011) also considered water use (litres)

[b] Chatterton et al. (2012) also evaluated soil erosion, pesticides, N leaching, ammonia, faecal contamination, *Chryptosporidium* and total area used

[c] More impact categories than those specified in the Table are reported in the paper

Williams et al. 2012) were less considered. Chatterton et al. (2012) evaluated the impact of the livestock sector in terms of soil erosion, pesticides, eutrophication, N leaching, greenhouse gas emissions, ammonia, faecal contamination, *Chryptosporidium* and total area used. The impact categories considered by Michael (2011) were water use (litres), energy use (MJ), global warming potential (CO_2 equivalent), ozone depletion potential (CFC−11 equivalent), acidification potential (SO_2 equivalent) and eutrophication potential (PO_4).

Previous considerations lead to the argument that LCIA is an issue on which methodological problems occur, asking for a future advances. Three aspects deserve the most attention: land use and land use change, water assessment, and carbon storage, impact categories that are particularly important in the environmental assessment of livestock sectors.

5.5.2.7 Interpretation and Tools Supporting the Interpretation Analysis

The ISO standard distinguishes some elements that should be considered in the interpretation phase: (i) identification of the significant issues based on the results of the LCI and LCIA phases; (ii) evaluation that considers completeness, sensitivity and consistency checks; (iii) conclusions, limitations, and recommendations.

1. Identification of the significant issues based on the results of the LCI and LCIA phases. All reviewed studies reported information on the interpretation phase, and it was possible to identify the significant environmental issues. Most studies assessed the whole system's impacts, others showed the most impactful steps, impact categories, impactful substances or materials.
2. Evaluation that considers completeness, sensitivity and consistency checks. As far as completeness, the reviewed studies quite often declared exclusions and recognised their limitations, sometimes considered in a sensitivity analysis or through consistency checks. Firstly, the complexity of LCA methodology applied to the agricultural sector would ask for a methodological innovation to integrate the multifunctionality of agriculture in the LCA analysis. Multifunctionality recognises that agriculture also contributes non-tradable goods, such as environmental and landscape services (Kanyarushoki et al. 2010; Ripoll-Bosch et al. 2013). For this reason, Ripoll-Bosch et al. (2013) considered the cultural ecosystem services provided as co-products: beyond the primary function of producing animal products, the sheep farming systems in Spain that they studied also provide other services or public goods, such as landscape conservation, cultural heritage, preservation of biodiversity, or fire prevention. Secondly, because of system heterogeneity, the use of mixed data (at farm level and at national/international scale; from field and from database) and the methodological assumptions, have effects on the results. Even if methodologically accurate, the nature of estimated results is sometimes recognised but not supported with measures to appreciate the difference between real and potential impacts. According to the research goals, different aspects are tested through a sensitivity analysis. Some authors presented

different scenarios to investigate how varying the results affected sheep management (Atzori et al. 2013a; Chatterton et al. 2012; Jones et al. 2014; Ripoll-Bosch et al. 2011, 2013; Williams et al. 2012), the functional unit (Kanyarushoki et al. 2008, 2010), or the allocation rules (Biswas et al. 2010; Gac et al. 2012; Michael 2011; Zonderland-Thomassen et al. 2012). Kanyarushoki et al. (2008, 2010) compared cow and goat specialised dairy farms in two French regions, and investigated how varying the results affected the functional unit. In terms of hectares of land occupied, goat farms had a higher impact (Kanyarushoki et al. 2008, 2010), except in climate change (Kanyarushoki et al. 2010). As far as a sensitivity analysis related to the allocation rule, some studies considered the effect of price fluctuation (Biswas et al. 2010). Others changed the allocation method to show differences in results between economic and mass-based allocation, in some cases concluding that their results are highly dependent on this methodological choice (Gac et al. 2012), in other cases showing that few differences emerged (Michael 2011; Zonderland-Thomassen et al. 2012).

3. Conclusions, limitations, and recommendations. The reviewed studies quite often recognised their limitations, as discussed in the previous point, but only sometimes performed a sensitivity analysis or estimated errors. The papers often made proposals regarding recommendations and mitigation strategies, even if not all mitigation strategies were site-specific, as more appropriate (Nicholson et al. 2001; Gerber et al. 2010; Blake et al. 2004). Michael (2011) made methodological recommendations, suggesting that the LCA methodology should be improved to enable appropriate recognition and to focus on products with special properties (e.g. lactose content) other than fats and protein in milk from dairy sheep and dairy goats. According to Liang et al. (2013) relevant strategies that should be considered to reduce GHG emissions from the livestock sector are related to improving rearing technologies, breeding, and developing a large-scale biogas industry. In O'Mara (2011), the mitigation potential for enteric CH_4 emissions was considered as three issues: improved feeding practices, use of specific agents and diet additives, and management changes and improved animal breeding. Ledgard et al. (2010) proposed to create tools for emission mitigation, such as the minimisation of enteric fermentation methane through breeding or vaccines, and the reduction of nitrous oxide emissions through soil additives and nitrogen management practices. Lipson et al. (2011) proposed interventions both to reduce methane production and to improve the efficiency of water used by goats. Firstly, they suggested managing grazing to reduce methane production by encouraging goats to consume younger, more easily digestible forage. Secondly, they suggested improving the efficiency of water used by goats both by using water-efficient feed crops that can increase the productive efficiency of livestock water use, and fodder trees and forage legumes that also reduce erosion and improve transpiration, soil structure and soil fertility. Other mitigation strategies proposed in the paper refer to genetic selection, animal breeding and vaccination, to increase feed conversion or to reduce enteric methane emissions. Authors, citing some literature, noted that strategies to mitigate the environmental impact of livestock production may come with some risks. For

example, increased dietary reliance on crop residues in order to increase the water use efficiency of ruminant livestock may be simultaneously counterproductive to the goal of reducing greenhouse gas emissions because ruminant consumption of residual crop material increases enteric methane production during digestion. Whatever the strategies, it is necessary to evaluate emissions from livestock on temporal and spatial scales, to identify problems and trends, and to prevent environmental degradation (Liang et al. 2013). As shown, the selected papers usually reported mitigation strategies but it was not very common to find a sensitivity analysis to check for invariability of the results toward changes in strategies or hypothesis. The proposed mitigation refers to the results of other studies or suggests future research developments. Because of the sensitivity of LCA results, a stronger international standardisation of procedures and methodological advances are necessary.

5.5.2.8 Comparative Analysis of the Different Types of Breeding, Final and Processed Products

According to the specific goal, scope and system boundaries adopted in the selected papers, many different types of comparison result:

- different types of breeding and breeding system, between ruminants and small ruminants (Audsley and Wilkinson 2012; Head et al. 2011; Kanyarushoki et al. 2008, 2010; Mekonnen and Hoekstra 2012; Opio et al. 2013; Peters et al. 2010b; Williams et al. 2006);
- between ruminant and non-ruminant animal species (Michael 2011);
- between conventional and organic systems (Head et al. 2011; Peters et al. 2010b; Williams et al. 2006);
- between low-mid and highly intensive productive systems (Ripoll-Bosch et al. 2011, 2013; Gac et al. 2012);
- between different sizes of breeds and consequent feed requirements, different types of land and consequent yields of grass (and management requirements), different rates of lamb growth and ewe productivity (Williams et al. 2012).

5.5.2.9 Critical Review

In order to assess the scientific and technical validity of the study and improve its credibility, a critical review (CR) should be carried out by an external independent panel of experts, following international (the ISO 14040-series), or other national, product specific or case-specific standards. The literature review provides evidence that a CR of experts is not generally applied, both as a simple peer review of the final report (apart from the journal review process), or as a more integrated quality assurance process. This is not specific to the scientific literature about the sheep and goat sector, but is general to all the field applications, although it plays an important role in the quality assurance of LCA studies. All the reported studies appear to include the phases required by the specific LCA standard. Some difficulties could be described involving the completeness of information given by papers about the content of LCA steps, which might be useful to a CR. These difficulties mostly concern the critical review of the inventory analysis (adequacy of data and its validation, calculation and sensitivity analysis) and the critical review of the interpretation phase a (data quality assessment and a sensitivity analysis). The process of critical review is one aspect that needs a solution for the credibility of the future development of LCA and LCT approaches (Klöpffer, 2013; Weidema et al. 2013). The ongoing revision of ISO TS 14071 could become the basis for improving the review process of life cycle based standards.

5.6 Pig Production

Nine Life Cycle Assessment (LCA) studies were selected from peer-reviewed scientific journals and scientific reports. All these studies were aimed at assessing the environmental loads of pig production and at highlighting the hotspots in the production chain. The studies published in the last 10 years were selected (Table 5.6).

The studies refer principally to northern Europe and to the production of small pigs slaughtered to obtain fresh meat whereas no information is reported for larger pigs slaughtered for meat suitable for derived edible products that represent the main goal of the pig production chain in southern Europe. Principally, the studies selected evaluated environmental loads related to the pig production chain until the farm gate (Cederberg and Flysjo 2004; Basset-Mens and van der Werf 2005). Some of them accounted for subsequent stages, such as plant processing (Reckmann et al. 2013), and retail (Dalgaard et al. 2007). Some studies were a cradle to grave contribution wherein meat consumption and waste disposal were included in the cycle (Kingstone et al. 2009; Thoma et al. 2011). Some studies compared organic versus non-organic production (Cederberg and Flysjo 2004; Basset-Mens and van der Werf 2005; Williams et al. 2006) or pork meat with other sources of protein such as tofu (Håkansson et al. 2005). No studies were combined with evaluation of economic impacts (e.g. life cycle costing, net present value, etc.).

5.6.1 Goal and Scope

Although specific goals and scope differed among studies, they all aimed principally to investigate the environmental performance of different pig production systems in present or future scenarios. The environmental loads of different pig farms system such as indoor vs. outdoor, slatted floor vs. bedding and compound vs. liquid-fed were assessed (Kingstone et al. 2009). The greenhouse gas emissions (GHG) associated with pork meat production until consumer was evaluated for the USA (Thoma et al. 2011). They assessed only the global warming potential (GWP) because it is indicative of opportunities for improved energy efficiencies

Table 5.6 Summary of the reviewed studies on LCA pig production

Reference	FU	Main boundaries	Geographical areas	Time boundaries
Basset-Mens and van der Werf (2005)	1 kg of live weight; 1 ha	Cradle to farm gate	France	1996–2001
Cederberg and Flysjö (2004)	1 kg of bone and fat free meat; 1 ha	Cradle to farm gate	Sweden	*10–20 years future scenarios*
Dalgaard et al. (2007)	1 kg of carcass weight	Cradle to delivered final retail destination	Danish	*1995 and 2015*
Håkansson et al. (2005)	20 g of complete proteins	Cradle to final consumption including waste disposal	Sweden	*Not specified*
Kingstone et al. (2009)	1 kg of slaughter pork consumed	Cradle to final consumption including waste disposal	British	2007
Reckmann et al. (2013)	1 kg of slaughter weight	Cradle to slaughter gate	Germany	2010/11
Stephen (2011)	1 kg of live weight	Cradle to farm gate	British	*Not specified*
Thoma et al. (2011)	4 oz boneless meat	Cradle to final consumption including waste disposal	USA	2008–09
Williams et al. (2006)	1 kg of carcass weight	Cradle to farm gate	British	*Not specified*

or conservation and moreover the authors believed that this impact could act as a useful baseline level of GHG which would be beneficial if voluntary carbon trading markets become viable in the future. The environmental impacts of producing grower-finisher pigs (12–105 kg) were evaluated with different diet scenarios (Stephen 2011). The diet scenarios were: conventional soya-based diet, homegrown bean-based diet, homegrown pea-based diet and homegrown lupin-based diet. A Danish study compared the environmental impact of Danish pork in 2005 with that produced in 1995 (base scenarios) and considered different scenarios for the year 2015. Additionally, the same study also compared the Danish results with those of pork produced in Great Britain and the Netherlands. It evaluated the effect on the environmental profile of pork production of the improvement in the number of weaning piglets for sow and in finishing feed conversion rate (Dalgaard et al. 2007). Different future pig production systems were evaluated (Cederberg and Flysjo 2004) when different aspects of sustainability were prioritised. Production in the future scenario was focussed on animal welfare, environmental care and high quality products at low prices. The LCA was used to assess how the management of different pig production systems impacted on the environment (Basset-Mens and van der Werf 2005). The systems compared were conventional good agricultural practice, a French quality label scenario and a French organic scenario. A German study assessed the environmental impacts of pork production, highlighting the hotspots in the pork cycle as well as the performance of a sensitivity analysis to determine whether different methodological and input parameters impacted on results (Reckmann et al. 2013). Pork meat and Tofu were compared when LCA was aimed to assess the environmental load of different sources of protein (Håkansson et al. 2005).

5.6.2 Functional Unit

The functional unit (FU) was not homogeneous among the studies because it was analysed at different stages of the pig production chain. The FU was 1 kg of live weight evaluated at the farm gate (Cederberg and Flysjo 2004; Basset-Mens and van der Werf 2005). Other studies evaluated the FU at the gate of the slaughterhouse (Reckmann et al. 2013). The FU was 1 kg of bone and fat-free meat when the aim of the study was to evaluate the final function at consumer of edible parts of pig meat. This was because the amount of meat consumed does not reflect the slaughter meat since the consumer does not eat chop bones and usually not the fat (Thoma et al. 2011). In the LCA used to measure the environmental load of British pork consumption the FU was set as 1 kg of pork product as consumed by the consumers (Kingstone et al. 2009). No studies were found that had chosen the nutritional property of meat such as protein or energy content as the FU. One hectare of arable land was used in studies that compared different production systems, organic versus conventional, and it reflected the function of non-market goods such as environmental services on local scale (Cederberg and Flysjo 2004; Basset-Mens and van der Werf 2005).

5.6.3 System Boundaries

System boundaries varied with the scope of the researches. Thus, for each of the reviewed LCAs the system boundaries began and finished at different stages of the production chain. Crop and feed production, pig housing (including enteric fermentation and manure management) were considered when the functional unit was 1 kg of live weight at the farm gate (Basset-Mens and van der Werf 2005; Williams et al. 2006) or 1 kg of slaughter weight estimated at the abattoirs when slaughter operations were included (Reckmann et al. 2013) or 1 kg of slaughter weight evaluated at the distribution depots when transport from slaughterhouse to retail was also considered (Dalgaard et al. 2007). In cradle-to-grave studies the contribution of feed and pork production, delivery to processor, processing, packaging (included production of raw materials and ultimate disposal), distribution, retail, consumption and waste disposal were considered (Kingstone et al. 2009; Thoma et al. 2011). The data for the crop and feed production were related to the amount of feed needed to meet pigs' dietary requirements in the different stages of stock production. Different feed compositions were considered among the studies and the main components were soy and sunflower meal, wheat, barley, fish meal and corn (Basset-Mens and van der Werf 2005; Delgaard et al. 2007; Reckmann et al. 2013). The animal production phase comprised breeding, weaning and fattening phases. Indoor or outdoor pig housing systems were compared for non-organic sow and weaning stages, whereas fattening stock were always modelled a as entirely housed. In organic systems, the phases of breeding, weaning and finishing were modelled as an outdoors system (Williams et al. 2006). Different slaughter weights were considered as this aspect affects the time that pigs remain at the piggery and influences the environmental burden associated with feed consumption, manure production and related operations (Stephen 2011). A British study accounted for manufacturing, maintenance and housing of capital goods such as vehicles, building and machinery (Williams et al. 2006). Generally, veterinary input for insemination or consumption of medicals was excluded.

5.6.4 Availability and Quality of Data

Data were provided from different sources between studies. Primary data for inventories of crop and feed production were provided by feed factories and used in a German study including the use of fertilisers, fossil fuels and other resources. (Reckmann et al. 2013). In a Danish study (Dalgaard et al. 2007) the LCA data on barley, heat and electricity were from the LCA food database (www.LCAfood.dk), data on soy meal imported from Argentina were derived by previous study. The total amount of feed consumed by the pig during its life was calculated on the basis of data from BPEX (British Pig Executive). The content of the feed was based on recipes of feed mixtures from a Danish feed company. Data on energy use and disposal of animal by-products at the abattoir were derived from the Green Accounts from Horsens slaughterhouse (2007) and from the processor of animal by-products (DAKA 2007). In a US study the raw data were provided by industry experts and standard pork industry handbooks. Regionally specific data for feed crops were taken from the farm extension service and from the National Agricultural Statistical. Additional input data for fuels and electricity consumption for crop production were obtained from the technical literature, state agricultural extension services, the US Department of Energy, the USDA, and other academic institutions. Transport emissions from producer to processor and from processor to distributor were calculated from information provided by industrial sources (Thoma et al. 2011). In a French study (Basset-Mens and van der Werf 2005) production and delivery data of inputs for crop production were derived in accordance with Nemecek and Heil (2001). The BUWAL 250 database (BUWAL 1996) was used to assess road and sea transport loads. Data associated with building construction were from Kanyarushoki (2001). Ammonia emissions from field were estimated according to ECETOC (1994). NH_3, N_2O and CH_4 emissions were treated according to IPCC (1996) and UNECE (1999). In a British study (Williams et al. 2006) data were obtained from disparate sources. Many data on farm management, productivity and typical inputs were taken from standard texts. Values for fertiliser use and manure composition also came from Defra's RB209 (MAFF 2000) and Surveys of British Fertiliser Practice (Defra 2001–05). Data on pesticide use came from annual pesticide surveys. Gaseous emissions of ammonia, nitrous oxide and methane came mainly from the UK's national inventories, which also supplied some activity data (proportion of manure spread on arable and grassland). Other production data came from the expertise of the project team, the scientific literature and the Ecoinvent LCA database.

5.6.5 Allocation of Burdens to Co-products

The division or extension of the processes is referred to in the ISO standardised guidelines (ISO 14044) as a priority in order to avoid allocation. When a system under study produces more useful outputs and it is impossible to divide it into subsystems the environmental loads must be allocated correctly to the co-products. The allocation may be done on the basis of economic, physical or functional properties (price, mass, or protein contained). Total loads in the studies analysed were assigned mainly to products and co-products in proportion to total revenue or weight. Some crops and animals produce more than one product, for example oil and meal from soybean or milk and meat in dairy farms. As reported previously, the studies analysed here analysed only fresh pig meat and not derivates or co-products, so all burdens were allocated solely to the functional unit. When co-products were generated in feed production or slaughter operations the economic allocation was applied in five studies. The protein source used in pig feed management (soy and rapeseed meal) with the co-product generated by the extraction (soy and rapeseed oil) was allocated by economic value. Moreover, a sensitive analysis was performed to verify the outcomes when all environmental loads were charged on soy and rapeseed meal or when mass allocation was used between oil and meal (Cederberg and Flysjo 2004). In Reckmann et al. (2013), the economic allocations for soy meal and soy oil were 66.3 and 33.7 % respectively. For finishers and sows, where culled sows represent the co-products of pig production, resource use and emissions were allocated with an economic value of 6–7 % and 93–94 % for sows and finishers, respectively (Basset-Mens and van der Werf 2005). Pig systems produce prime meat from finishers, but culling breeding stock (sows and boars) also produces meat. This meat is mainly consumed as processed foods and its quality is generally considered lower and reflected in its lower price, typically less than 25 % of the price of prime meat. On this basis the burdens for the lower economic value of secondary meat were allocated and indicated a reduction of the potential output of prime meat of less than 5 % (Williams et al. 2006). At the processing gate slaughter meat and waste such as offal and blood are also produced. It was reported that these waste products were used as pet food or fertiliser, digested anaerobically or sent to landfill. The allocation of these co-products is generally very difficult because the precise type (quality) and amount are generally not known. The allocation ratio of these co-products was calculated using economic US census data relating to other species. An allocation ratio was used that assigned 89 and

11 % of the greenhouse gas burden to the meat processing and rendering operations respectively (Thoma et al. 2011). Moreover, the same authors allocated the GHG emissions related to retail processes such as energy use and refrigerant loss as a function of pork meat mass sold. Finally, for the consumer phase they allocated by pork meat mass emissions related to the transport from retail to home, refrigeration, cooking, food loss and waste disposal.

5.6.6 *Life Cycle Impact Assessment* (*LCIA*)

The environmental impacts analysed differed among the studies. Global warming potential (GWP_{100}) evaluated on the basis of a 100-year time horizon, eutrophication potential (EP), and acidification potential (AP) were the commonest ones analysed. Some studies also evaluated other impact categories such as photochemical ozone formation (Cederberg and Flysjo 2004), terrestrial ecotoxicity (Basset-Mens and van der Werf 2005), and ozone depletion and photochemical smog (Delgaard et al. 2007). Moreover, in some cases the consumption of resources such as land use, primary energy use, abiotic resource use and pesticide use were also counted (Basset-Mens and van der Werf 2005; Hakansson et al. 2005; Williams et al. 2006; Kingstone et al. 2009). Only one study assessed a single impact (GWP) and the GWP equivalents were adopted from IPCC 2007 (Thoma et al. 2011). The EDIP method (Wenzel et al. 1997. Version. 2.03) was used for the LCIA, whereas the characterisation factors for methane and nitrous oxide were chosen according to IPCC 2001 (Dalgaard et al. 2007). In a French study (Basset-Mens and van der Werf 2005) the characterisation factors for the impact category were adopted in agreement with Guinée et al. (2002). Eco-indicator'99 (H), a model that displays 11 impact categories, was used for the impact assessment but only GWP, land and fossil fuel use were considered in the LCA (Hakansson et al. 2005). The main GWP sources, expressed as kg of carbon dioxide equivalent, were nitrous oxide and methane, both from feed production and pig housing, and carbon dioxide from fossil fuel. For the EP characterisation the major chemical compounds included were nitrate and phosphate leaching in the water and ammonia emission in the air. The EP was quantified mainly as a phosphate equivalent but it was also expressed as a nitrate equivalent (Delgaard et al. 2007) or kg of O_2-eqv./FU (Cederberg and Flysjo 2004). Ammonia (NH_3) and sulphur dioxide (SO_2) emitted respectively from the agricultural phase (field and housing) and fossil fuel combustion were considered in the assessment of acidification potential. The AP was reported as sulphur dioxide equivalent and in one case as mol H^+/g (Cederberg and Flysjo 2004). It must be noted that, when reported, the characterisation factors adopted for the different impact categories differed among studies. Methane was accorded 21 (Basset-Mens and van der Werf 2005), 23 (Dalgaard et al. 2007) or 25 (Thoma et al. 2011) kg CO_2 equivalent for GWP_{100}. Nitrous oxide was treated as 296 (Dalgaard et al. 2007), 298 (Thoma et al. 2011) or 310 (Basset-Mens and van der Werf 2005) kg CO_2 equivalent for GWP_{100}. For eutrophication potential 1 kg NH_3 and 1 kg NO_3 were treated as 0.44 and 0.43 kg PO_4 equivalents (Williams et al. 2006) or 0.35 and 0.1 kg PO_4

equivalents (Basset-Mens and van der Werf 2005). Water depletion was not considered in any of the studies taken into consideration.

5.6.7 The Interpretative Analysis

Farm operations, including feed production and animal housing (animal and manure management), were the biggest hotspots in the pork chain for most of the impacts analysed. The GWP was mainly affected by emission of nitrous oxide from crops, by nitrous oxide and methane emitted from manure management and by methane emitted from enteric fermentation. The contributions of the different GHG to the total GWP were 38, 33 and 29 % for CO_2, N_2O and CH_4 respectively. The feeding stage was the greatest source of CO_2 emissions (82 %), whereas pig housing and slaughterhouse stages accounted for 5 and 13 % respectively. N_2O emissions were mainly related to feed production (95 %), and the majority of CH_4 (93 %) was related to the pig housing stage. The feeding, housing and slaughterhouse stages accounted for 63, 30 and 7 % to total GWP, respectively. Within the pig housing operations, the finishing stocks were indicated as the main contributors to GWP (Reckmann et al. 2013). When the LCA was a cradle-to-grave assessment the slaughtering and retail operations were minor sources of GHG emissions. The consumption phase that included transport, refrigerator, cooking and waste disposal was reported to account 20 (Kingstone et al. 2009) and 30 % (Thoma et al. 2011) of total GWP.

The eutrophication potential was related for the 52 % to feed production, 40 % to pigs housing and 8 % to slaughterhouse processes (Reckmann et al. 2013). A British study reported that indoor pork production on slatted floors was associated with 96 % of eutrophication potential, whereas all the processes from the farm gate to consumption and final waste disposal accounted for only 4 %. AP was 0.18 kg SO_2 equivalent for the farming phase compared with an overall impact per kg of meat produced of 0.19 kg SO_2 equivalent (Kingstone et al. 2009). In a Danish study it was reported that the greatest contribution to eutrophication was from nitrate leaching (62 %) followed by ammonia emission (32 %). Nitrate and ammonia were related to crop operations, pig housing and manure application (Dalgaard et al. 2007). The main contribution to AP derived from animal housing (76 %) and feed production (23 %). AP was related mainly to the ammonia emitted from manure/slurry in the housing, during storage and after field fertiliser application. The pigs housing and feed production accounted for 87 and 13 % of the total ammonia emitted, respectively. Total ammonia was related to the 93 % of the total acidification potential. With regard to stock, the fattening stage was identified as the greatest source of AP compared with the sow and weaning stages (Reckmann et al. 2013). High feed efficiency (less feed/FU) and greater grain yield were associated with a lower release of nitrifying substances (Cederberg and Flysjo 2004). The EP analysed for different scenarios in French pork production showed that for kg of pork it was higher in the organic scenario; on the other hand, it was lower in the organic scenarios for hectares. Acidification potential was highest both for kg of pig and for hectares in the good agricultural practice scenario (Basset-Mens and van der Werf 2005).

5.6.8 Critical Review

All the studies reviewed herein agree that, compared with the use of data provided by a bibliography, the improvement in the availability of direct data relating to the different stages of production, in particular for the main feed used, could improve assessments. Results reported for GWP of pork production in the USA evaluated at the farm gate were within the range reported in the literature. The slaughter operations, packaging and transport made a lower contribution. On the other hand, retail refrigeration, transport to home and cooking operations were significant contributors to the overall carbon footprint. The authors reported that the two main factors affecting the carbon footprint were the change in manure management from deep pit to anaerobic lagoon, and the allocation method (economic) for feed co-products. Finally, they indicated that greater sustainability could be achieved for GHG reductions associated with technologies that capture or convert the methane from anaerobic lagoons (Thoma et al. 2011). Hotspot and marginal improvement were discussed principally for eutrophication and acidification potential because the emissions of nitrate to water and ammonia to air have a direct and immediate impact on a regional scale (Basset-Mens and van der Werf 2005). The authors indicated that eutrophication potential could be reduced with optimal nitrogen use and the introduction of catch crops in the rotation adopted for feed production. Ammonia could be reduced by nutritional and manure management strategies. Reduction of the protein intake or improvement of their utilisation by the animals could reduce the amount of nitrogen in manure. Emissions of NH_3 from animal housing could be reduced by better control of microclimatic conditions (lower air velocity and temperature), improved frequency of manure removal and covering of slurry stores. Moreover, fast and effective incorporation into the soil could minimise the NH_3 emissions during manure application (Basset-Mens and van der Werf 2005). In British pork production improvement in environmental sustainability could be achieved mainly in the pig farm phase. The farming methods of loose housing and outdoor breeding make a significantly higher contribution to eutrophication and acidification than pigs raised indoors on slatted flooring. Moreover, the impacts could be reduced with a greater feed efficiency, a high number of pigs per litter and correct manure management. Environmental improvement from farm gate to grave could be achieved with high energy efficiency at the abattoirs (energy and heat) and at home with the use of AA rated appliances (refrigerator and cooker) and at retail with optimal refrigeration (Kingstone et al. 2009). Greater protein content in feed was associated with higher emissions or nitrous oxide, ammonia and nitrate from housing, manure storage and manure field application. A sensitive analysis indicated that when protein content in the fattening feed mixtures was decreased from 18 to 16 % the global warming, eutrophication and acidification potential decreased by 2, 5 and 7 % respectively (Dalgaard et al. 2007). The same authors highlighted that the concept of food miles is often used incorrectly. Indeed, the overall GWP for kg of British pork was no lower than the kg of Danish pork, which included the transport of meat from the Danish slaughterhouse to Great Britain (transport accounted for less than 1 % of total GWP). On the other hand, because of the higher productivity efficiencies in Denmark, the eutrophication and acidification potentials were lower than those for the British pork. Therefore, the idea of choosing "home products" as more environmentally friendly in view of the shorter distance from farm to retail is not justifiable if feed efficiency, high production and correct manure management are not encouraged. A Swedish study reported a greater feed consumption (+ 14 %) in the "animal welfare" scenario compared with the "environment" scenario. The higher feed consumption for kg of pork produced was because of lower piglets/sow production and the higher feed intake related to animal movements in outdoor systems. Therefore, these data suggest that production systems which guarantee a better degree of animal welfare may lead to greater feed consumption and lower efficiency of resources utilisation. Moreover, the impossibility of using synthetic amino acid in organic systems implies a higher concentration of protein in the ration and a greater amount of nitrogen in the excreta as a consequence. Finally, the authors suggest the introduction of ammonia filters in the pig house ventilation system to reduce ammonia emissions would have a significant impact on eutrophication and acidification potential (Cederberg and Flysjo 2004).

5.7 Poultry

In the following, a description of the main aspects of this sector at the international and European levels is presented. Then 11 international LCA studies on poultry production published in peer-reviewed journals, scientific reports or international conference proceedings are analysed (Table 5.7). All the LCA applications to poultry production systems published in the last 6 years excluded the work of Williams et al. (2006) that we found critical were selected for this review. Methodological problems connected with the application of LCA in this sector are examined, starting with a critical comparative analysis of the LCA case studies. Finally, the environmental hotspots are identified in order to develop possible solutions to the problems presented.

5.7.1 The Poultry Sector: Main Aspects

In 2010 poultry production worldwide reached about 79 million t, with an increase in production of 5.8 % over the previous year. According to FAO data, the main producers are the USA, China, the European Union and Brazil, which together deliver as much as 77 % of total production. The production of poultry meat in the EU 27 in 2010 increased by 4.3 % to just over 12 million t. The main producing countries are: Germany, the UK, Italy, France and Spain, all of whom saw production increase substantially compared with previous years. The main production in the European Union is represented by chicken meat and the proportion further increased (+ 3.7 %)

Table 5.7 List of references included in the literature review and their main characteristics

Reference	FU	Method	Main boundaries	Geographical areas	Time boundaries
Wiedemann et al. (2010)	1 kg of carcass weight 1 kg of live weight	LCA	Cradle to meat industry gate	Australia	1 year
Pardo et al. (2012)	0.6 kg of sliced chicken breast fillet	LCA	Cradle to grave	Spain	*Not specified*
Bengtsson and Seddon (2013)	1 t of roast chicken	LCA	Cradle to retailer	Australia	1 year
Leinonen et al. (2012)	1 t of carcass weight	LCA	Cradle to farm gate	UK	*Not specified*
Davis and Sonesson (2008)	1 meal (chicken- based) per person	LCA	Cradle to grave	Sweden	*Not specified*
Pelletier (2008)	1 t of live weight kg of CO_2 eq per kg of product	LCA	Cradle to farm gate	USA	*Not specified*
MacLeod et al. (2013)	(CW or eggs) and kg of CO_2 eq per kg of protein	LCA	Cradle to retail gate	World	*Not specified*
Cederberg et al. (2009)	1 kg of carcass weight	LCA for CF	Cradle to farm gate	Sweden	*15 years*
Katajajuuri et al. (2008)	1 t of honey marinated sliced broiler chicken fillet	LCA	Cradle to slaughterhouse gate	Finland	*Not specified*
Pelletier et al. (2013)	For example, kg CO_2-eq per 1000 tonne of eggs	LCA for CF	Cradle to slaughterhouse gate	USA	*1 year*
Williams et al. (2006)	1 kg of carcass weight	LCA	Cradle to farm gate	UK	*Not specified*

in 2010, comprising about 80 % of the total poultry meat. The turkey production increased by 6 %, whereas the duck production remained stable (+0.6 %). At the national level the slaughter poultry (including wild game) recorded a growth of 1.8 % over the previous year, with 548.191 million animals slaughtered, according to the ISTAT data for 2010. The chickens are the most prevalent, accounting for 89.66 % of the total, followed by turkey with 5.16 % and wild game with 3.67 %. The Veneto region in Italy slaughtered as much as 42.9 % of total poultry meat, followed by Emilia Romagna with 19.6 % and Lombardy with11.7 %.

5.7.2 Literature Review on LCA Application to Poultry Production

5.7.2.1 Goal and Scope

The poultry supply chain is recognised by several authors and researchers as the most environmentally efficient among the different meat production systems. For this reason, only a few studies have focussed on the assessment of environmental loads generated up to the farm gate (Pelletier 2008; Cederberg et al. 2009; Leinonen et al. 2012) and most target the assessment of the whole supply chain (or different post-farm activities) in order to identify the hotspots and strategies to improve the environmental sustainability of poultry production and consumption in post-farm processes (Williams et al. 2006; Katajajuuri et al. 2008; Davis and Sonesson 2008; Wiedemann et al. 2010; Pardo et al. 2012; Bengtsson and Seddon 2013). The environmental impact assessment of different chicken production systems was the goal of recent studies by Boggia et al. (2010), who compared the conventional with the free-range system; the first two also included the organic production system. One differs from the other two production systems regarding the use of organic feed (Leinonen et al. 2012). In particular, Boggia et al. (2010) compared the conventional boiler production system with two organic production systems called "organic" and "organic plus", that differ for the restrictive requirements in terms of animal welfare considered in the second system.

5.7.2.2 Functional Unit

The functional unit (FU) was not homogeneous among the studies because it was analysed at different stages of the chicken production chain. The most common FU used in the reviewed studies was either 1 t of carcass weight (Leinonen et al. 2012; Wiedemann et al. 2010) or 1 kg live weight (Pelletier 2008; Wiedemann et al. 2010) or 1 kg of product at a number of different endpoints in the supply chain (Pardo et al. 2012; Bengtsson and Seddon 2013). Only a few studies used different FUs related to downstream processes. Davis and Sonesonn (2008) estimated the environmental impacts of two different chicken-based meals in order to identify the opportunities for environmental load reduction in the consumption phase. Similarly, Pardo et al. (2012) and Katajajuuri et al. (2008) used FUs related to further processed chicken meat, respectively 0.6 kg of sliced chicken breast fillet packaged in modified atmosphere and 1 kg of carcass weight of marinated breast fillet, in order to verify the environmental sustainability of post-farm processes. MacLeod et al. (2013) assessed the impacts in kg of CO_2-eq per kg of protein content in order to make the results comparable among different livestock products.

5.7.2.3 System Boundaries

Usually the studies related to the environmental impact estimation of poultry productions consider the cradle to farm gate system boundaries (SB). This, according to Bengtsson and Seddon (2013), is because of the lack of sufficiently detailed information in the cradle to retail or consumer supply chains. Hence, the reviewed studies were divided between those that analysed the supply chain only to the farm gate (Leinonen et al. 2012; Cederberg et al. 2009, Pelletier 2008), those that included slaughter processes (Katajajuuri et al. 2008; Williams et al. 2006), and those that included the other downstream processes (packaging, distribution, transport to retailer, etc.) to the use and disposal phases (Pardo et al. 2012; Davis and Sonesson 2008; Wiedemann et al. 2012; Bengtsson and Seddon 2013). Although all the studies include the upstream processes from the raw material extractions, the studies including post-farm processes usually simplify the input/output flows related to the agricultural phase. Wiedemann et al. (2010) did not include soil carbon fluxes in their analysis or the use of irrigation in wheat production; Davis and Sonesson (2008) did not consider the pesticides used in feed crop production because of missing data. This was also the case for capital goods and other emission sources (cleaning materials, waste treatments, etc.) that are usually not included in environmental impact assessment of the entire supply chain.

5.7.2.4 Availability and Quality of Data

Data were provided by three main sources in these studies, national inventory studies, simulated supply chain studies and industry data studies, depending on the scope of the studies. Cederberg et al. (2009), Pelletier (2008),Williams et al. (2006) and Wiedemann et al. (2010), performed national inventory studies, aiming to provide results that were representative of the country in question. All these studies collected data from a wide range of sources: national statistics, literature, direct involvement of industry and commercial stakeholders. Bengtsson and Seddon (2013) used data collected directly from an Australian industry that covered all the phases of the supply chain of chicken production. The same approach was used by Katajajuuri et al. (2008), who investigated a simulated Finnish chicken supply chain using data from the literature and commercial facilities. Leinonen et al. (2012) used an approach that applied a structural and mechanistic model to assess energy, materials, animal performance, crop production and nutrient input/output flows for the UK broiler industry. Davis and Sonesson' study (2008) was based only on previous LCA studies and explored the effects of improving sustainability measures in the post-farm phases of integrated supply chains. Most of the studies used IPCC methods for estimating GHG emissions arising from several sources within the poultry production chain: the use of energy and the handling of manure and wastes (MacLeod et al. 2013; Wiedemann et al. 2010; Leinonen et al. 2012; Cederberg et al. 2009; Williams et al. 2006). Only two studies used different sources for assessing the emissions arising from the poultry supply chain (Pelletier 2008; Davis and Sonesson 2008) The first used data from previous poultry GHG and the latter used a model for Swedish supply chains' input-output assessment (System Analysis for Food and Transport—SAFT). Only a few studies collected data from different chicken rearing systems. Conventional, free-range and organic farming were compared by Wiedemann et al. (2010) and Leinonen et al. (2012); Bengston and Seddon (2013) and Boggia et al. (2010) compared only conventional and free-range systems. The only relevant difference between the three systems is related to the higher feed consumption and manure production of free-range systems (Leinonen et al. 2012); a further difference in the organic system is the use of organic feed for animal rearing.

5.7.2.5 Allocation of Burdens to Co-products

Co-products were identified at three points along the poultry supply chains. The first allocation point was the production of meat from breeding hens and eggs in the breeding system. The second allocation point was the production of litter and meat chickens at the grow-out farm. The third allocation point was at the slaughterhause, between primary and secondary products. The allocation methodologies used in the reviewed studies varied according to their specific goal and scope. The most frequently used method was economic allocation, which resulted in a much larger share of the impacts being allocated to meat production than other less valuable co-products (Bengtsson and Seddon 2013; Davis and Sonesson 2008). The reason for using the economic allocation was that it reflects the objective of the industry that optimises the products and co-products to achieve the highest economic return (Bengtsson and Seddon 2013). Katajajuuri et al. (2008) used meat mass to allocate the environmental impacts between co-products and Pelletier (2008) and Pelletier et al. (2013) used a gross energy content allocation criterion (mass-adjusted gross chemical content) in order to reflect the real biological flows and the associated environmental impacts. For the second allocation point described above, almost all the studies used the system expansion methodology for litter (manure and bedding) nutrients, accounting for manufacture and application of synthetic fertilisers and including all the emissions arising from the use of the fertilisers (e.g. Pelletier 2008; Wiedemann et al. 2010). The lack of data on the quantities of manure transferred from livestock production into arable production systems in Sweden forced Cedeberg et al. (2009) to avoid allocation by distributing all the resources used and related emissions from manure to chicken primary products. Davis and Sonesson (2008), who focussed on the analysis of chicken meal consumption habits, used system expansion to allocate the environmental impacts between chicken meal production and wastes produced after the consumption phase. They used system expansion replacing oil and coal (50/50) with wastes produced at the end of the chicken-based meals chain in heat production.

5.7.2.6 Life Cycle Impact Assessment (LCIA)

The most common areas of environmental impacts contribution analysed in the reviewed studies were energy use, GHG emission, ozone depletion, water use and those impact categories closely related to the feed production phase (MacLeod et al. 2013; Leinonen et al. 2012; Wiedemann et al. 2010; Pelletier et al. 2013; Pardo et al. 2012; Pelletier 2008). The methods used for assessing the environmental impacts of the chicken supply chain were different. For energy use impact assessment the CED (cumulative energy demand) was the most common among the reviewed studies (Wiedemann et al. 2010; Pardo et al. 2012; Pelletier et al. 2013; Pelletier 2008). The impacts related to GHG emissions were quantified by an IPCC (2006) Tier 2-type approach (Cederberg et al. 2009; Pelletier et al. 2013; MacLeod et al. 2013) or assessment of their contribution to climate changes with the GWP as indicator (Leinonen et al. 2012; Wiedemann et al. 2012). Some studies used complex impact assessment methodology: CML2- Baseline 2000 or Recipe 2008 (Pardo et al. 2012, Pelletier 2008) or different impact categories (Bengtsson and Seddon 2013; Davis and Sonesson 2008; Prudêncio da Silva et al. 2008; Katajajuuri et al. 2008). On a global scale, MacLeod et al. (2013) found that chicken supply chains produce 58 million t of eggs and 72 million t CW (Carcass Weight) annually and related GHG emission of 606 million t CO_2-eq; average emission intensity ranges from 5.4 kg CO_2-eq/kg CW for meat and 3.7 kg CO_2-eq/kg eggs. A similarly impressive result was obtained by Katajajuuri el al. (2008), who analysed the environmental impacts of 1 t of marinated and sliced broiler fillet in Finland and found a GWP of 3,635 kg of CO_2-eq Very low values were found by Boggia et al. (2010), however, who identified GHG emissions that varied between 0.66 and 0.70 kg of CO_2-eq per kg/kg of broiler meat. The organic rearing system was found to have the greatest impact in terms of GHG emissions. Indeed, Wiedemann et al. (2010) found that this system produced 2.86 kg of CO_2-eq/kg of CW, more than the 2.38 and 1.89 CO_2-eq/kg of CW found for the free-range and conventional systems respectively. However, organic systems resulted in a lower environmental impact in terms of energy use (12.8 MJ/kg CW) than free-range and conventional systems (respectively 16.8 MJ/kg CW and 20.4 MJ/kg CW). Different results were reported by Leinonen et al. (2012), who found the organic system had the greatest impact in terms of EP (eutrophication potential), primary energy use and land occupation than the conventional system, that recorded lower impacts for GWP (global warming potential) and AP (acidification potential). The differences between the two studies are because of the different FUs. Indeed, the latter used 1 t of CW as FU, implying a smaller number of chicks but a longer productive cycle than that of the conventional system. The longer cycle meant a higher amount of feed needed in organic system with feed production that resulted in all the analysed studies as the most impactful phase in the poultry supply chain.

5.7.2.7 Interpretation

The most relevant areas of environmental impact in the poultry supply chain were, for all the analysed studies, feed production followed by grow-out housing, meat processing and breeding. These results can be ascribed to fuel, energy and fertiliser use in feed production and to manure production and management in the grow-out phase (Katajajuuri et al. 2008; Davis and Sonesson 2008; Wiedemann et al. 2010; MacLeod et al. 2013; Pelletier et al. 2013; Bengtsson and Seddon 2013). The contributions from feed production ranged from ~45 to 82.4 % (Katajajuuri et al. 2008; Pelletier 2008). When different impact categories were considered, feed production was found to be responsible for ozone depletion, acidification potential and eutrophication potential (Pelletier et al. 2008). In particular, Davis and Sonesson (2008) found that almost 90 % of eutrophying emissions originated from feed production and that ammonia (NH_4) emitted from farm activities was the main culprit responsible for acidification. When the overall environmental impacts of the feed production phase were considered, nitrous oxide was found to be the dominant emission source, ranging from 49 to 59 % (Wiedemann et al. 2010). Cederberg et al. (2009) found that carbon dioxide contributed 39–47 % of the overall emissions throughout the supply chain. Analysing the poultry supply chain from cradle to meat processing gate, Bengtsson and Seddon (2013) identified the grow-out phase as the most relevant in terms of global warming and non-renewable energy depletion; the meat processing phase contributed only 10–20 % of the overall environmental impact for more than 10 impact categories analysed, mainly energy consumption, water use and wastewater production. Among the different meat production systems, poultry meat production appears to be the most environmentally efficient because of several factors (identified by Pelletier et al. 2008) in a protein energy return on investment[4] (EROI) equal to 17.7 % compared with reported values for other livestock production systems such as beef (2.5 %), sheep (1.8 %) and pigs (7.1 %) (Pimentel 2004). The most relevant were identified by Williams et al. (2006) in very high feed conversion and high daily gain.

5.7.3 Critical Analysis

The reviewed studies agree that the use of primary data for the different production phases, in particular those related to feed production, could reduce the uncertainty

[4] EROI is a dimensionless index used to compare the relative efficiency of energy use per unit protein produced by different food systems.

of impact assessment. At the same time, almost all of the studies identify the chicken meat and egg production chains as the most efficient of the livestock production systems from the environmental perspective. The agricultural phase is recognised as the most relevant in terms of environmental emission generation, not only for conventional production systems but also for free-range and organic systems. Results reported for free-range and organic grow-out systems are almost the same as for the conventional ones, as regards the longer chick production cycle, the higher mortality rate and consequently the higher daily feed intake and manure production. Moreover, chicken production (meat and eggs) is defined as having the highest feed conversion ratio (FCR) and thus constitutes more sustainable livestock production. These data, considering the high proportion of environmental impacts allocated to the agricultural phase, suggest that the adoption of low impact agricultural practices by feed ingredient suppliers (which, for example, reduces the use of energy-intensive synthetic fertilisers and emissions resulting from their application) can significantly affect the environmental performance of chicken products. MacLeod et al. (2013), who analysed GHG emissions from the poultry sector on a global scale, found room for environmental impact reduction in several areas: reduction of land use changes (LUC) and improving efficiency in fertilisation management and in energy use both on-farm and off-farm. Davis and Sonesson (2008), analysing two meal types based on chicken in Sweden, found a promising emissions reduction strategy in terms of replacing oil and coal in heat energy production with the wastes generated at the end of life of the supply chain. The same authors recommend a shift in consumption habits, i.e. increasing the consumption of poultry-derived proteins because they have a lower environmental impact than other animal protein sources. This proposal seems difficult to achieve, and furthermore, as suggested by the same authors, requires a balance between the increase of the impacts because of increased demand for poultry protein and the reduction of the production of other livestock sectors.

Conclusions and Lessons Learned

This environmental impact assessment of the livestock sector presents some critical issues that may occur regardless of the methodologies used, but that in the case of LCA or LCA-based methodologies make the evaluation extremely complex. The complexity of approaches, data requirements, and model specifications has become so high that some standardisation is necessary to make things more credible and comparable. The variability in the results of all the livestock supply chains is caused by the difference in the production systems and methodological choices (functional unit, system boundaries, allocation method, etc.). For instance, in beef production systems a very wide variety, ranging from very intensive to very extensive (Nijdam et al. 2012), was observed; in dairy farms, which generally produce more than one product, the whole impact of dairy activities should be shared and allocated between all of them and the environmental performance of their processed products also depends on the use of milk supplied by different farms with different rearing systems. Of special interest in LCA analysis of the livestock production systems was the definition of the functional unit (FU), or, rather, the unit respecting which the environmental impacts are defined. The choice of a "corrected functional unit", such as fat and protein or energy, could be an efficient approach which takes into account the nutritional value of livestock products and allows the comparison of the results of different studies. Livestock products differ in terms of production techniques and economic values, protein content and live weight (Nguyen et al. 2012a). Thus, the use of more complex FUs is mandatory for studies focussed on the evaluation of the environmental and economic impacts of the whole beef supply chain (Weidema et al. 2009) or comparison of different livestock sectors (beef, pigs, chickens, sheep and goats, etc.) (de Vries and de Boer 2010). The choice of FU is critical, as pointed out in Sect. 1.4, for studies addressing environmental impacts and load allocation of milk and meat in beef production (Cederberg and Stadig 2003). A common characteristic of all the analysed studies is the heterogeneity in system boundaries' (SB) definition. Besides the variety and the complexity of livestock transformation processes, a relevant critical methodological point for LCA analysis, the inclusion of crop production (fodder especially) in rearing systems' impact assessment is a critical and debated question. Meat and milk production systems are characterised, moreover, by a high number of co-products and by-products, let alone the production of both meat and milk. Almost all the studies reviewed consider the cradle to farm gate life cycle and exclude capital goods from the analysis. However, the environmental impact from capital goods has been included in some recent publications which found that capital goods contribute significantly to the total impact of agricultural production systems (Blengini and Busto 2009; Frischknecht et al. 2007). As regards different methods of impact assessment and classification there are several approaches (often IPCC 2007; EDIP; CML; CED; Impact 2000+ and Eco-indicator 99) that are choosen according to the goal and scope of the studies and their effectiveness in showing results. The phase with the greatest impact, in all the studies, is animal rearing; enteric CH_4, NH_3 and N from animal excreta are the major culprits responsible for environmental loads. The land use impact category is particularly relevant for beef and dairy production, which has the highest impact compared with other meat production systems (pigs and chickens). Data availability remains a long-standing problem and is hard to solve, as witnessed by scientific studies dedicated to system definition and inventory construction. A more complete picture of the environmental impacts (some of which have not been adequately addressed so far) and of the phases along the whole chain should be included as improvements for future research. A strong interaction between research experts and economic organisations (e.g. farmer's associations) could make the LCA methodology useful in the decision-making process connected to the definition of an environmental chain strategy. This interaction is useful for many reasons: to support LCA data requirements, improving and expanding databases; to support the standardisation process and levels; to stress the main gaps in current knowledge on which future research and developments should be focussed.

From a methodological perspective, there are many studies oriented to the evaluation of environmental impacts; only a few of them (Weidema et al. 2009; Basarab et al. 2010; Van Middelaar et al. 2011; Oishi et al. 2013) combine LCA with the evaluation of economic impacts (e.g. life cycle costing, net present value, value added, etc.). A few studies combine LCA with farm simulation models (Beauchemin et al. 2010; Leip et al. 2010; Clarke et al. 2012), since they can give useful information for the improvement of rearing systems and related environmental impacts.

Only Weidema et al. (2009), using a hybrid methodology combining macroeconomic data (input/output tables) with the emissions generated by the analysed productive processes (Suh et al. 2004), evaluate the whole life cycle from cradle to fork. They include the transformation, marketing and use phases in the SB.

In LCA analysis of the livestock sector, economic allocation is the most frequently used method although the ISO standards recommend physical or biological criteria (carcass weight, protein content, etc.) in preference (Yan et al. 2011). Indeed, De Vries and de Boer (2010) and Nguyen et al. (2012a) used these allocation criteria to identify studies to include in their review of the environmental impacts of different livestock productions and to assess the environmental impacts of four beef farming systems. Moreover, economic allocation methodology does not account for the environmental benefit produced by the milk system with the reduction of biological methane and ammonia emissions (Cederberg and Stadig 2003). Thus, the application of system expansion is preferable. When a system expansion is applied, for example to dairy and beef production systems, it is assumed that the meat from both the culled dairy cow and the raised dairy calf replaces beef meat produced in a cow-calf system. The choice of meat and also the production system used to obtain this by-product is crucial because the amount of environmental impact from beef production depends on it (Flysjö et al. 2012). However, these allocation methods could be avoided, according to Weis and Leip (2012), who suggest allocating input and output flows of the processes for raising and fattening young animals for meat, and dividing the activities of dairy and suckling cows for milk production into the raising of young animals during pregnancy (which is allocated for meat) and the production of milk. The management of by-products, in LCA analysis, is another critical point. For livestock production, this is the case of manure because of its dual simultaneous effect (Garnett 2009): manure increases the nutrients in the soil and also the soil's carbon sequestration potential (FAO 2001). It was estimated that, globally, almost 22 % of the total nitrogen and 38 % of the total phosphate used for agriculture productions derive from animal excreta, of which half come from beef production (UNEP, n.d.). At the same time, manure, according to the report "Livestock's Long Shadow—Environmental Issues and Options" (FAO 2006), is responsible for N_2O and CH_4 emissions, contributing 5 % to global GHG emissions. In the majority of the studies analysed, manure management is considered as a means for the production of organic fertiliser that, depending on its relative nitrogen content,[5] is used as a substitute for chemical fertiliser (e.g. Casey

[5] Manure composition and thus its rate of chemical nitrogen fertiliser substitution are assessed by different methodologies related to the analysis of the physiological mechanisms of animals and diet composition (Ogino et al. 2004; Pelletier et al. 2010; Basarab et al. 2010; Leip et al. 2010).

and Holden 2006; Beauchemin et al. 2010; Nguyen et al. 2010; Basarab et al. 2010; Beauchemin et al. 2011). This methodological approach allows the authors to assess the environmental impacts allocated to livestock production better without leaving out the manure use impacts (linked to acidification, eutrophication and GHG emissions); but Garnett (2009) considers it incomplete, because as a natural source of nitrogen, manure reduces the need for chemical fertiliser production and transport. According to Weidema et al. (2008) and Leip et al. (2010), manure must be considered in any system expansion approach aspiring to perform good LCA analysis. This means considering the impacts of manure fertilisation on the entire chemical N fertiliser supply chain, defining the level of substitution of chemical nitrogen fertilisers by manure (it varies from 20 to 60 % depending whether manure is spread during grazing or collected from stables) (Nguyen et al. 2010). Many LCA studies on livestock production consider the land use impact categories only in terms of the m^2 required annually for livestock production (Cederberg and Stadig 2003; Cederberg et al. 2009; Doreau et al. 2011; Ridoutt et al. 2011; Flysjö et al. 2012). This category is particularly relevant for beef production, which has higher impact than certain other meat production systems (pigs or chickens). In particular, as regards meat and dairy production, to obtain 1 kg of beef meat some 27–49 m^2 of land are needed (de Vries and de Boer 2010). The high values of land required for beef production are related to two factors: the low efficiency of feed conversion rate and the lowest number of annual progeny compared with pigs and chickens. Within the beef production system, the impact on the land use category is higher for suckler-cows than the system in which the herds are a co-product of milk (de Vries and de Boer 2010). However, despite the importance of including GHG emissions because of land use changes (Flysjö et al. 2012), there is no consensus on how to include those emissions in environmental impact estimates. Three methods which include land use and land use changes in LCA of milk production were analysed by Flysjö et al. (2012), who clearly showed how GHG emission estimates differed depending on the methodology used. Many LCA analyses deal with this problem by identifying the required land to produce a specific amount of output in a given period of time (Lindeijer 2000). This type of information is useful for evaluating land use efficiency, but, according to Nguyen et al. (2010), there are other issues that need to be considered: “the opportunity cost of land” (Garnett 2009, pp. 493), or rather the cost of the land if it was used for other purposes, and the potential land use change that derives from an increase in demand for land and land products. The opportunity cost of land use has been estimated in terms of emitted CO_2 as between 2.8 and 2.2 kg CO_2/m^2 year depending on whether it was converted to crop production or grassland (Nguyen et al. 2010) In another study, Nguyen et al. (2012b), using this indicator, found a potential impact reduction on GW between 20 and 48 % if grasslands were converted to forests rather than to annual crops. Roer et al. (2013) considering four sub-processes in the life cycle of bovine meat and milk production (concentrate, forage, cattle rearing and others) found that the GWP of meat production varies from 17.7 to 18.4 CO_2-eq/kg of carcass weight. This is also the only impact category which depends on cattle rearing and accounts for 45–48 % of the total GWP of the whole life cycle.

The methodological innovations emerging from this review seek to limit this huge variability by focusing on a combination of models from representative livestock farms and related emissions assessment with the LCA analysis. A thorough review of these methodological approaches can be found in Crosson et al. (2011), who summarise the GHG emissions per kg of product in 35 whole farm modelling studies (from 31 published papers) of beef and dairy cattle production systems. Beauchemin et al. (2010, 2011) use the HOLOS model, which is based on IPCC methodology, to assess CH_4, N_2O and CO_2 emissions at the farm level. This model accounts for all the emissions linked to the beef production supply chain (fertiliser and herbicide production and transport, feed production, etc.). Modelled results were used in LCA to assess the environmental impacts of beef production in Canada. Bonesmo et al. (2013) used the HOLOS model adapted to the Norwegian situation (HOLOSNor) to assess the GHG emission intensity of Norwegian dairy and beef production systems. They found that the main culprits responsible for GHG emissions per kg of carcass weight were, in order of relevance: soil's nitrous oxide emissions, indirect energy use, soil C loss and enteric methane which was not significantly correlated to the variation in total GHG emissions per kg of carcass weight (Bonesmo et al. 2013). The same approach was used in a GGELS project (Leip et al. 2010), in which the CAPRI model was used to define the six livestock systems of EU−27 (Loudjani et al. 2010). The same method has been used to assess the GHG emissions and removals in the whole EU livestock sector (at regional scale) including: methane, nitrous oxide, carbon dioxide and also land use and land use changes (Weiss and Leip 2012) and CAPRI model results were used as input and output of the LCI. Oishi et al. (2013) evaluated the economic and environmental impacts resulting from changes in the age of animals at slaughter and in diet in the cow-calf system (race Japanese Black) in Japan, using as indicators the actualised net income and the environmental impacts from an LCA analysis. The input for economic evaluation was based on the continuous coupling system that was found (Oishi et al. 2011) to be the most economic and efficient one. The LCA analysis was built on the system presented by Ogino et al. (2007); a cradle to farm gate system with 1 kg of total live weight as FU. Then, the LCA results were normalised and the relative contribution of each category to the environmental impact of the whole system aggregated into a single dimensionless indicator, as suggested by Brentrup et al. (2004), for comparison with the economic indicator. Finally, a multi-criteria analysis was used to aggregate global warming (GWP), acidification potential (AP) and eutrophication potential (EP) impact categories following the approach suggested by Hermann et al. (2007). The results of this complex study showed that increasing culling parity to an economically efficient level can reduce the total environmental impact; changes in diet have no effect on environmental and economic impact (Oishi et al. 2011). Capper (2011) also pointed out that reducing time-to-slaughter can represent an option for decreasing CO_2-eq emissions per unit of beef because of the lifetime dilution of maintenance energy costs.

The combination of a bioeconomic model for livestock management with partial LCA (Carbon Footprint, Ecological Footprint, etc.) in order to assess the environmental impacts of livestock production systems represents an innovative approach to the environmental impact assessment for this sector. One example is the use of GBSM (Grange Beef System Model) with a partial LCA analysis in order to assess the GHG emissions of beef production systems The integration of farm management models with LCA analysis has also been suggested (e.g. Beauchemin et al. 2011; Oishi et al. 2011; Foley et al. 2011; Clarke et al. 2012), and some studies use other impact assessment methodologies to quantify the environmental loads produced by livestock production. All these attempts, which are in line with Place and Mitloehner (2012), represent efforts to account for the complex biogeochemical processes that occur within the rumen of cattle fed on different diets and also to account for varying management strategies such as age-to-slaughter, which can meaningfully alter the environmental load per unit of beef.

References

ADEME-AFNOR. (2014). Affichage Environnemental des produits de grande consummation. http://affichage-environnemental.afnor.org/. Accessed 7 April 2014.

Atzori, A. S., Rassu, S. P. G., & Pulina, G. (2013a, June). *Partial carbon footprint of dairy sheep farms: simulated results from four different scenarios*. Abstract presented at the 20th Association for Science and Animal Production Congress, Italy.

Atzori, A. S., Mele, M., Cappucci, A., & Pulina, G. (2013b June). *Emission of greenhouses gas in Italy: An inventory for the sheep sector using the Tier 3 of IPCC Guidelines*. Presentation presented at the 20th Association for Science and Animal Production Congress, Italy.

Audsley, E., & Wilkinson, M. (2012, October). *Using a model-based lca to explore options for reducing national greenhouse gas emissions from crop and livestock production systems*. Paper presented at the Proceedings of the 8th International Conference on Life Cycle Assessment in the Agri-Food Sector, France.

Bartl, K., Gómez, C. A., & Nemecek, T. (2011). Life cycle assessment of milk produced in two smallholder dairy systems in the highlands and the coast of Peru. *Journal of Cleaner Production, 19*(13), 1494–1505.

Basarab, J., Bauer, J., Chase, D., Chaudhary, N., Dobson, C., Goddard, T., Kolk, J., Laate, E.A., Li, X., Nolan, S., Stadlwieser, R., & Strankman, P. (Eds.). (2010). *Evaluating environmental and economic impact for beef production in Alberta using life cycle analysis—Final report*. Walterloo: Conestoga-Rovers & Associates.

Basset-Mens, C. (2008, November). *Estimating the carbon footprint of raw milk at the farm gate: Methodological review and recommendations*. Paper presented at the 6th International Conference on Life Cycle Assessment in the Agri-Food Sector, Zurich, Switzerland.

Basset-Mens, C., & van der Werf, H. M. G. (2005). Scenario-based environ- mental assessment of farming systems: The case of pig production in France. *Agricultural Ecosystem and Environment, 105*, 127–144.

Beauchemin, K. A., Henry Janzen, H., Little, S. M., McAllister, T. A., & McGinn, S. M. (2010). Life cycle assessment of greenhouse gas emissions from beef production in western Canada: A case study. *Agricultural Systems, 103*(6), 371–379.

Beauchemin, K. A., Janzen, H. H., Little, S. M., McAllister, T. A., & McGinn, S. M. (2011). Mitigation of greenhouse gas emissions from beef production in western Canada—Evaluation using farm-based life cycle assessment. *Animal Feed Science and Technology, 166*, 663–677.

Bellarby, J., Tirado, R., Leip, A., Weiss, F., Lesschen, J. P., & Smith, P. (2013). Livestock greenhouse gas emissions and mitigation potential in Europe. *Global Change Biology, 19*, 3–18.

Bengtsson, J., & Seddon, J. (2013). Cradle to retailer or quick service restaurant gate life cycle assessment of chicken products in Australia. *Journal of Cleaner Production, 41*, 291–300.

Biswas, W. K., Graham, J., Kelly, K., & John M. B. (2010). Global warming contributions from wheat, sheep meat and wool production in Victoria, Australia e a life cycle assessment. *Journal of Cleaner Production, 18*, 1386–1392.
Blake, R. W., & Nicholson, C. F. (2004). Livestock, land use change, and environmental outcomes in the developing world. In E. Owen, T. Smith, M. A. Steele, S. Anderson, A. J. Duncan & M. Herrero (Eds.), *Responding to the livestock revolution: The role of globalisation and implications for poverty alleviation* (pp. 133–153). Nottingham University Press.
Blengini, G. A., & Busto, M. (2009). The life cycle of rice: LCA of alternative agri-food chain management systems in Vercelli (Italy). *Journal of environmental management, 90*(3), 1512–1522.
Bochu J. L., Metayer N., Tuomisto, H., Haastrup, P., Constantin, A. L., Pelletier, N., De Camillis, C., & Leip, A. (2013). Development of Carbon Calculator to promote low carbon farming practices—User guidance manual for the Carbon Calculator. Deliverable to the European Commission, Joint Research Centre, Institute for Environment and Sustainability, Italy. http://mars.jrc.ec.europa.eu/mars/Projects/LC-Farming. Accessed 7 April 2014.
Boeri, F. (2012). Product Category Rules (PCR): Meat of mammals. PCR 2012:11, Version 2.0, Prepared by Life Cycle Engineering, COOP Italia, Assocarni, Italy for: The International EPD System, Stockholm, Sweden. www.environdec.com. Accessed 7 April 2014.
Boggia, A., Paolotti, L., & Castellini, C. (2010). Environmental impact evaluation of conventional, organic and organic-plus poultry production systems using life cycle assessment. *World's Poultry Science Journal, 66*(01), 95–114.
Bonesmo, H., Beauchemin, K. A., Harstad, O. M., & Skjelvåg, A. O. (2013). Greenhouse gas emission intensities of grass silage based dairy and beef production: A systems analysis of Norwegian farms. *Livestock Science, 152*(2), 239–252.
Boulay, A. M., Hoekstra, A. Y., & Vionnet, S. (2013). Complementarities of water-focused life cycle assessment and water footprint assessment. *Environmental Science & Technology, 47*, 11926–11927.
Brentrup, F., Küsters, J., Lammel, J., Barraclough, P., & Kuhlmann, H. (2004). Environmental impact assessment of agricultural production systems using the life cycle assessment (LCA) methodology II. The application to N fertilizer use in winter wheat production systems. *European Journal of. Agronomy, 20*, 265–279.
Brock, P. M., Graham, P., Madden, P., & Alcock, D. J. (2013). Greenhouse gas emissions profile for 1 kg of wool produced in the Yass Region, New South Wales: A Life Cycle Assessment approach. *Animal Production Science, 53*, 495–508.
Brondi, C. (2013). Product Category Rules (PCR): Leather footwear. PCR 2013:15, Version 1.01, prepared by Synesis Consortium, Italy for: The International EPD System, Stockholm, Sweden. www.environdec.com. Accessed 7 April 2014.
BSI. (2008). *Guide to PAS 2050. How to assess the carbon footprint of goods and services*. British Standards, London, UK.
BUWAL (1996). Ökoinventare für Verpackungen. Schriftenreihe Umwelt Nr. 250/1+2, Bundesamt für Umwelt, Wald und Landschaft, Bern, Switzerland. (In german).
Capper, J. L. (2011). The environmental impact of beef production in the United States: 1977 compared with 2007. *Journal of Animal Science, 89*, 4249–4261.
Carbon [A-Za-z_-'&;]{3,20} (2010, September). Guidelines for the Carbon Footprinting of Dairy Products in the UK. Publication produced with the sponsorship of and endorsed by DairyUK and of DairyCo., Carbon Trust Footprinting Company Limited, London, UK.http://www.dairyco.org.uk/resources-library/research-development/environment/carbon-footprinting-dairy-products-in-the-uk/#.U0bKFPmSzA0. Accessed 7 April 2014.
Casey, J. W., & Holden, N. M. (2006). Quantification of GHG emissions from sucker-beef production in Ireland. *Agricultural Systems, 90*(1), 79–98.
Castanheira, É. G., Dias, A. C., Arroja, L., & Amaro, R. (2010). The environmental performance of milk production on a typical Portuguese dairy farm. *Agricultural Systems, 103*, 498–507.
Cederberg, C., & Stadig, M. (2003). System expansion and allocation in life cycle assessment of milk and beef production. *The International Journal of Life Cycle Assessment, 8*(6), 350–356.

Cederberg, C. & Flysjö, A. (2004). *Environmental assessment of future pig farming systems*. Quantification of three scenarios from the FOOD21 synthesis work, SIK report 723, 1–54.

Cederberg, C., Meyer, D., & Flysjö, A. (Eds) (2009). *Life cycle inventory of greenhouse gas emissions and use of land and energy in Brazilian beef production*. SIK-The Swedish institute for food and biotechnology.

Chatterton, J., Audsley, E., Graves, A., Morris, J., & Williams, A. (2012, October). *Using systems-based lca to investigate the environmental and economic impacts and benefits of the livestock sector in the UK*. Paper presented at the Proceedings of the 8th International Conference on Life Cycle Assessment in the Agri-Food Sector, France.

Clarke, A. M., Crosson, P., Brennan, P. (2012, October). *An assessment of greenhouse gas emissions and economics of grass based suckler beef production systems*. Paper presented at the Proceedings of the 8th International Conference on Life Cycle Assessment in the Agri-Food Sector, France.

Climate Change. (2007). IPCC fourth assessment report. The Physical Science Basis. http://www.ipcc.ch/ipccreports/ar4-wg1.htm.

Crosson, P., Shalloo, L., O'Brien, D., Lanigan, G. J., Foley, P. A., Boland, T. M., & Kenny, D. A. (2011). A review of whole farm systems models of greenhouse gas emissions from beef and dairy cattle production systems. *Animal Feed Science and Technology*, *166*, 29–45.

Daesoo, K., Greg T., Nutter, D., Milani, F., Ulrich, R., & Norris, G. (2013). Life cycle assessment of cheese and whey production in the USA. *International Journal of Life Cycle Assessment*, *18*(5), 1019–1035

DAKA (2007). Grønt regnskab. 2005/2006. Daka a.m.b.a. www.daka.dk.

Dalgaard, R., Halberg, N., & Hermansen, J. E. (2007). *Danish pork production. An environmental assessment*. DJF Animal Science No 82.

Davis, J., & Sonesson, U. (2008). Life cycle assessment of integrated food chains—A Swedish case study of two chicken meals. *The International Journal of Life Cycle Assessment*, *13*(7), 574–584.

de Rancourta, M., Foisb, N., Lavínc, M. P., Tchakériand, E., & Vallerande, F. (2006). Mediterranean sheep and goats production: An uncertain future. *Small Ruminant Research*, *62*(3), 167–179.

Devendra, C. (2010). Concluding synthesis and the future for sustainable goat production. *Small Ruminant Research*, *89*, 125–130.

De Vries, M., & De Boer, I. J. M. (2010). Comparing environmental impacts for livestock products: A review of life cycle assessments. *Livestock Science*, *128*(1), 1–11.

Doreau, M., Van Der Werf, H. M. G., Micol, D., Dubroeucq, H., Agabriel, J., Rochette, Y., & Martin, C. (2011). Enteric methane production and greenhouse gases balance of diets differing in concentrate in the fattening phase of a beef production system. *Journal of Animal Science*, *89*(8), 2518–2528.

Dyrmundsson, O. R. (2006). Sustainability of sheep and goat production in North European countries. From the Arctic to the Alps. *Small Ruminant Research*, *62*, 151–157.

Eady, S., Carre, A., & Grant. T. (2012). Life cycle assessment modelling of complex agricultural systems with multiple food and fibre co-products. *Journal of Cleaner Production*, *28*, 143–149.

EC. (2013a). Product Environmental Footprint (PEF) Guide. Annex II to the Commission Recommendation of 9 April 2013 on the use of common methods to measure and communicate the life cycle environmental performance of products and organisations. Official Journal of the European Union, 2013/179/EU.

EC. (2013b). Product Environmental Footprint (PEF) Guide. DG-JRC Ref. Ares (2012) 873782–17/07/2012.

EC. (2014). European Commission's Single Market for Green Products Initiative. http://ec.europa.eu/environment/eussd/smgp/index.htm. Accessed 7 April 2014.

Eckard, R. J., Grainger, C., & de Klein, C. A. M. (2010). Options for the abatement of methane and nitrous oxide from ruminant production: A review. *Livestock Science*, *130*, 47–56.

ECETOC (1994). Ammonia emissions to air in western Europe. Technical Report No. 62. European Chemical Industry Ecology and Toxicology Centre, Brussels, Belgium.

Edwards-Jones, G., Plassmann, K., & Harris, I. M. (2009). Carbon footprinting of lamb and beef production systems: insights from an empirical analysis of farms in Wales, UK. *The Journal of Agricultural Science, 147*(06), 707–719.

EU. (2009). Directive 2009/28/EC of the European Parliament and of the Council on the promotion of the use of energy from renewable sources and amending and subsequently repealing Directives 2001/77/EC and 2003/30/EC.

European Union. (2013). *Agriculture, forestry and fishery statistics*. Luxembourg: Eurostat pocketbooks.

Fang, K., Heijungs, R., & de Snoo, G. R. (2014). Theoretical exploration for the combination of the ecological, energy, carbon, and water footprints: Overview of a footprint family. *Ecological Indicators, 36*, 508–518.

Fantin, V., Buttol, P., Pergreffi, R., Masoni, P. (2012). Life cycle assessment of Italian high quality milk production. A comparison with an EPD study. *Journal of Cleaner Production, 28*, 150–159.

Flysjö, A., Cederberg C., Henriksson M., & Ledgard S. (2011). How does co-product handling affect the carbon footprint of milk? Case study of milk production in New Zealand and Sweden. *International Journal of Life Cycle Assessment, 16*, 420–430.

Flysjö, A., Cederberg, C., Henriksson, M., & Ledgard, S. (2012). The interaction between milk and beef production and emissions from land use change–Critical considerations in life cycle assessment and carbon footprint studies of milk. *Journal of Cleaner Production, 28*, 134–142.

Foley, P. A., Crosson, P., Lovett, D. K., Boland, T. M., O'Mara, F. P., & Kenny, D. A. (2011). Whole-farm systems modelling of greenhouse gas emissions from pastoral suckler beef cow production systems. *Agriculture, Ecosystems & Environment, 142*(3), 222–230.

Food and Agriculture Organization of the United Nations FAO. (2001). *Soil carbon sequestration for improved land management*. World Soil Resources Reports 96. FAO, Rome, Italy

Food and Agriculture Organization of the United Nations FAO. (2006). *Livestock's Long Shadow—Environmental Issues and Options*. Rome: FAO.

Food and Agriculture Organization of the United Nations FAO (2009). *The state of food and agriculture 2009: Livestock in the balance*. Rome: FAO.

Food and Agriculture Organization of the United Nations FAO. (2012). Milk availability trends in production and demand and medium-term outlook. http://www.fao.org/docrep/015/an450e/an450e00.pdf. Accessed 20 March 2014.

Food and Agriculture Organization of the United Nations FAO. (2013). *Statistical yearbook. World Food and Agriculture*. Rome, Italy.

Food SCP RT. (2013a) *Production product category rules: Needs, the role of the protocol, and round table governance*. Working paper by the European Food Sustainable Consumption and Production Round Table (SCP RT) Working Group 1, Brussels, Belgium.

Food SCP RT. (2013b). *ENVIFOOD protocol, environmental assessment of food and drink protocol.* European food sustainable consumption and production round table (SCP RT), Working Group 1, Brussels, Belgium. http://www.food-scp.eu/. Accessed 7 April 2014.

Food SCP RT (2014). European food sustainable consumption and production round table. http://www.food-scp.eu/. Accessed 7 April 2014.

French Parliament. (2010). Loi No 2010–788 du 12 juillet 2010 portant engagement national pour l'environnement. French law No 2010–788, 12 July 2010.

Frischknecht R., Jungbluth N., et al. (2003). *Implementation of life cycle impact assessment methods*. Final report ecoinvent 2000, Swiss Centre for LCI. Duebendorf, CH. www.ecoinvent.ch.

Frischknecht, R., Althaus, H. J., Bauer, C., Doka, G., Heck, T., Jungbluth, N., Kellenberger, D., & Nemecek, T. (2007). The environmental relevance of capital goods in life cycle assessments of products and services. *International Journal of Life Cycle Assessment, 12*(1), 7–17.

Gac, A., Cariolle, M., Deltour, L., Dollé, J. B., Espagnol, S., Flénet, F., Guingand, N., Lagadec, S., Le Gall, A., Lellahi, A., Malaval, C., Ponchant, P., & Tailleur, A. (2010). GES'TIM—Guide méthodologique pour l'estimation des impacts des activités agricoles sur l'effet de serre.

Réalisé dans le cadre du projet "Gaz Ã Effet de Serre et Stockage de Carbone en exploitations agricoles" (CASDAR 6147).

Gac, A., Ledgard, S., Lorinquer, E., Boyes, M., & Le Gall A. (2012, October). *Carbon footprint of sheep farms in France and New Zealand and methodology analysis*. Paper presented at the Proceedings of the 8th International Conference on Life Cycle Assessment in the Agri-Food Sector, France.

Garnett, T. 2009. Livestock-related greenhouse gas emissions: Impacts and options for policy makers. *Environmental Science & Policy, 12*(4), 491–503.

GEDnet. (2014). Global environmental declarations network. http://gednet.org/.

Gerber, P., Vellinga, T., Opio, C., Henderson, B., & Steinfeld, H. (2010). *Greenhouse gas emissions from the dairy sector: A life cycle assessment*. Animal Production and Health Division, FAO, Rome, Italy.

Gerber, P. J., Steinfeld, H., Henderson, B., Mottet, A., Opio, C., Dijkman, J., Falcucci, A., & Tempio, G. (2013). *Tackling climate change through livestock—A global assessment of emissions and mitigation opportunities*. Rome: FAO.

Goedkoop, M., & Spriensma, R. (2001). The eco-indicator99: A damage oriented method for life cycle impact assessment: Methodology report. http://irs.ub.rug.nl/dbi/4581696db734f.

Goedkoop, M., Heijungs, R., Huijbregts, M., De Schryver, A., Struijs, J., & van Zelm, R. (2009). ReCiPe 2008. A life cycle impact assessment method which comprises harmonised category indicators at the midpoint and the endpoint level, 1st Edition. http://www.pre-sustainability.com/download/misc/ReCiPe_main_report_final_27-02-2009_web.pdf. Accessed 10 March 2014.

González-García, S., Castanheira, E., Dias, A. C., & Arroja, L. (2013a). Using life cycle assessment methodology to assess UHT milk production in Portugal. *Science of the Total Environment, 442*, 225–234.

González-García, S., Castanheira, E., Dias, A. C., & Arroja, L. (2013b). Environmental performance of a Portuguese mature cheese-making dairy mill. *Journal of Cleaner Production, 41*, 65–73.

González-García, S., Castanheira, E., Dias, A. C., & Arroja, L. (2013c). Environmental life cycle assessment of a dairy product: The yoghurt. *International Journal of Life Cycle Assessment, 18*, 796–811.

González-García, S., Hospido, A., Moreira, M. T., Feijoo, G., & Arroja, L. (2013d). Environmental Life Cycle Assessment of a Galician cheese: San Simon da Costa. *Journal of Cleaner Production, 52*, 253–262.

Grainger, C., & Beauchemin, K. A. (2011). Can enteric methane emissions from ruminants be lowered without lowering their production? *Animal Feed Science and Technology, 166*, 308–320.

Guerci M., Trydeman Knudsen, M., Bava, L., Zucali, M., Schönbach, P., & Kristensen, T. (2013). Parameters affecting the environmental impact of a range of dairy farming systems in Denmark, Germany and Italy. *Journal of Cleaner Production, 54*, 133–141.

Guinée, J. B., Gorrée, M., Heijungs, R., Huppes, G., Kleijn, R., & De Koning, A. (2002). Handbook on life cycle assessment. Operational guide to the ISO standards, 1–708. Institute for Environmental Sciences, Leiden, The Netherlands.

Haenlein, G. F. W. (2004). Goat milk in human nutrition. *Small Ruminant Research 51*(2), 155–163.

Håkansson, S., Gavrilita, P., & Bengoa, X. (2005). Comparative life cycle assessment. pork vs tofu. Life Cycle Assessment, 1N1800, Stockholm. 1–17.

Hauschild, M., & Potting, J. (2003). Spatial differentiation in life cycle impact assessment—The EDIP2003 methodology. Institute for Product Development Technical University of Denmark

Head, M., Sevenster, M., & Croezen, H. (2011). Life cycle impacts of protein- rich Foods for Superwijzer. http://www.ce.nl/?go=home.downloadPub&id=1252&file=2329_finalreportM-HSD.pdf. Accessed 12 January 2014.

Hermann, B. G., Kroeze, C., Jawjit, W. (2007). Assessing environmental performance by combining life cycle assessment, multi-criteria analysis and environmental performance indicators. *J. Cleaner Prod. 15*, 1787–1796.

IDF. (2010). *A carbon footprint approach for dairy. The IDF guide to standard lifecycle assessments methodology for the dairy sector*. Bulletin of the International Dairy Federation, 445/2010, Brussels, Belgium.

IERE. (2006). Product Category Rules (PCR): Meat. Earth sure meat environmental product declarations. Institute for Environmental Research and Education (IERE), Washington, USA. http://iere.org/programs/earthsure/product-category-rules-pcrs/. Accessed 7 April 2014.

Ingwersen, W., & Subramanian, V. (2013). Guidance for Product Category Rule Development. Version 1.0. Product Category Rule Guidance Development Initiative. http://www.pcrguidance.org. Accessed 7 April 2014.

IPCC. (2006). IPCC guidelines for national greenhouse Inventories. http://www.ipccnggip. iges. or.jp/public/2006gl/index.html. Accessed 29 Sept 2013.

IPCC (2007). IPCC guidelines for national greenhouse Inventories. From http://www.ipcc.ch/publications_and_data/publications_and_data_reports.shtml#1.

Iribarren, D., Hospido, A., Moreira, M. T., & Feijoo G. (2011). Benchmarking environmental and operational parameters through eco-efficiency criteria for dairy farms. *Science of the Total Environment, 409*, 1786–1798.

Istat. (2010). 6 Censimento dell'Agricoltura 2010, Istitutio Naziionale di Statistica, Roma, Italy.

Japanese CFP scheme. (2011a). Product Category Rules (PCR): Market poultry eggs. PCR ID: PA-CN–01. The carbon footprint of products calculation and labeling pilot program. Japanese carbon footprint of products scheme. http://www.cms-cfp-japan.jp/english/pcr/pcrs.html. Accessed 7 April 2014.

Japanese CFP scheme (2011b). Product Category Rules (PCR): Chickens. PCR ID: PA-CP–01. The carbon footprint of products calculation and labeling pilot program. Japanese carbon footprint of products scheme. http://www.cms-cfp-japan.jp/english/pcr/pcrs.html. Accessed 7 April 2014.

Japanese CFP scheme. (2011c). Product Category Rules (PCR): Down and feather. PCR ID: PA-CR–01. The carbon footprint of products calculation and labeling pilot program. japanese carbon footprint of products scheme. http://www.cms-cfp-japan.jp/english/pcr/pcrs.html. Accessed 7 April 2014.

Jolliet, O., Margni, M., Charles, R., Humbert, S., Payet, J., Rebitzer, G., and Rosenbaum, R. (2003). IMPACT 2002+: A New Life Cycle Impact Assessment Methodology. *International Journal of Life Cycle Assessment, 8*(6) 324–330.

Jones, A. K., Jones, D. L., & Cross P. (2014). The carbon footprint of lamb: sources of variation and opportunities for mitigation. *Agricultural Systems, 123*, 97–107.

Kanyarushoki, C. (2001). Evaluation de l'impact environnemental des bâtiments porcins. Application de la méthode d'analyse du cycle de vie. Mémoire de fin d'études DESS. Université Rennes 1, INRA, Rennes, France.

Kanyarushoki, C., Fuchs, F., & van der Werf, H. M. G. (2008, November). *Environmental evaluation of cow and goat milk chains in France*. Paper presented at the 6th International Conference on Life Cycle Assessment in the Agri-Food Sector, Switzerland.

Kanyarushoki, C., van der Werf, H. M. G., & Fuchs F. (2010, September). *Life cycle assessment of cow and goat milk chains in France*. Paper presented at the of LCA food 2010, Italy.

Katajajuuri, J. M., Grönroos, J., & Usva, K. (2008, November). *Environmental impacts and related options for improving the chicken meat supply chain.* Paper presented at the 6th International Conference on Life Cycle Assessment in the Agri-Food Sector, Switzerland.

Kim, D., Thoma, G., Nutter, D., Milani, F., Ulrich, R., & Norris, G. (2013). Life cycle assessment of cheese and whey production in the USA. *The International Journal of Life Cycle Assessment, 18*(5), 1019–1035.

Klöpffer, W. (2013, August). *The critical review according to iso 14040 + 44-how and why it came about.* Paper presented at the 6th International Conference on Life Cycle Management, Sweden.

Kingstone, C., Meyhoff Fry, J., & Aumonier, S. (2009). *Scoping life cycle assessment of pork production.* Environmental resource management, final report, 1–33.

Koch, P., & Salou, T. (2013). AGRIBALYSE®: Rapport Méthodologique—Version 1.0. Ed ADEME. Angers. France. p 384.

LEAP (2014a). Environmental performance of animal feeds supply chains: Guidelines for quantification. Livestock Environmental Assessment and Performance Partnership. Rome: 33 FAO. http://www.fao.org/partnerships/leap/en/. Accessed 7 April 2014.

LEAP (2014b). Greenhouse gas emissions and fossil energy demand from poultry supply chains: Guidelines for quantification. Livestock Environmental Assessment and Performance Partnership. Rome: FAO. http://www.fao.org/partnerships/leap/en/. Accessed 7 April 2014.

LEAP (2014c). Greenhouse gas emissions and fossil energy demand from small ruminant supply chains: Guidelines for quantification. Livestock Environmental Assessment and Performance Partnership. Rome: FAO. http://www.fao.org/partnerships/leap/en/. Accessed 7 April 2014.

Ledgard, S. F., Lieffering, M., McDevitt, J., Boyes, M., & Kemp, R. (2010). A greenhouse gas footprint study for exported New Zealand lamb. Report prepared for the Meat Industry Association, Balance Agri-Nutrients, Landcorp and MAF. http://www.fedfarm.org.nz/files/2010-Lamb-CarbonFootprint.pdf. Accessed 17 March 2014.

Leinonen, I., Williams, A. G., Wiseman, J., Guy, J., & Kyriazakis, I. (2012). Predicting the environmental impacts of chicken systems in the United Kingdom through a life cycle assessment: Broiler production systems. *Poultry Science*, *91*(1), 8–25.

Leip, A., Weiss, F., Wassenaar, T., Perez, I., Fellmann, T., Loudjani, P., Tubiello, F., Grandgirard, D., Monni, S., & Biala, K. (2010). *Evaluation of the livestock sector's contribution to the EU greenhouse gas emissions (GGELS)–final report*. European Commission, Joint Research Centre.

Liang, L., Lal, R., Duc, Z., Wua, W., & Menga, F. (2013). Estimation of nitrous oxide and methane emission from livestock of urban agriculture in Beijing. *Agriculture, Ecosystems and Environment*, *170*, 28–35.

Lindeijer, E. (2000). Review of land use methodologies. *Journal of Cleaner Production*, *8*(4), 273–281.

Lipson J., Reynolds T., & Anderson L. (2011). Environmental Implications of Livestock Series: Goats EPAR Brief No.56 http://evans.washington.edu/files/EVANS_UW_Request_156_Environmental_Impacts_of_Livestock_Goats_0.pdf. Accessed Jan 2014.

Loudjani, P., Wassenaar, T., & Grandgirard, D. (2010). Typology of livestock production systems in Europe. In: Leip, A., Weiss, F., Wassenaar, T., Perez, I., Fellmann, T., Loudjani, P., Tubiello, F., Grandgirard, D., Monni, S., & Biala, K. (2010). *Evaluation of the livestock sector's contribution to the EU greenhouse gas emissions (GGELS)–final report*. European Commission, Joint Research Centre; pp. 61–70.

MacLeod, M., Gerber, P., Mottet, A., Tempio, G., Falcucci, A., Opio, C., Vellinga, T., Henderson, B., & Steinfeld, H. (2013). *Greenhouse gas emissions from pig and chicken supply chains—A global life cycle assessment*. Rome: FAO.

MAFF, 2000. Fertiliser Recommendations for Agricultural and Horticultural Crops (RB209) (7th edition). The Stationery Office, London.

Marino, M., Sessa, F., Niccolucci, V., & Bastianoni, S., (2011). Product Category Rules (PCR): fresh hen eggs in shell. PCR 2011:15, Version 1.0, prepared by Life Cycle Engineering and Ecodynamics Group, Siena University, Italy for: The International EPD System, Stockholm, Sweden. www.environdec.com. Accessed 7 April 2014.

Martin, C., Morgavi, D. P., & Doreau, M. (2010). Methane mitigation in ruminants: From the rumen microbes to the farm scale. *Animal*, *4*, 351–365.

Mekonnen, M. M., & Hoekstra, A. Y. (2012). A global assessment of the water footprint of farm animal products. *Ecosystems*, *15*, 401–415.

Michael, D. (2011). *Carbon reduction benchmarks & strategies. New animal products*. Rural Industries Research and Development Corporation. ISBN 978-1-74254-247-8.

Nemecek, T., Heil, A. (2001). SALCA—Swiss agricultural life cycle assessment database, version 12, December 2001. FAL, Swiss Federal Research Station for Agroecology and Agriculture, Zuerich, Switzerland.

Nguyen, T. L. T., Hermansen, J. E., & Mogensen, L. (2010). Environmental consequences of different beef production systems in the EU. *Journal of Cleaner Production*, *18*(8), 756–766.

Nguyen, T. T. H., van der Werf, H. M. G., Eugène, M., Veysset, P., Devun, J., Chesneau, G., & Doreau, M. (2012a). Effects of type of ration and allocation methods on the environmental impacts of beef-production systems. *Livestock Science, 145*(1), 239–251.

Nguyen, T. T. H., Doreau, M., Eugène, M., Corson, M. S., Garcia-Launay, F., Chesneau, G., & van der Werf, H. M. G. (2012b, October). *Effect of farming practices for greenhouse gas mitigation and subsequent alternative land use on environmental impacts of beef cattle production systems*. Paper presented at the 8th Intenational Conference on Life Cycle Assessment in the Agri-Food Sector, France.

Nicholson, C., Blake, R., Reid, R., & Schelhas, J. (2001). Environmental impacts of livestock in the developing world. *Environment, 43*, 2–17.

Nijdam, D., Rood, T., & Westhoek, H. (2012). The price of protein: Review of land use and carbon footprints from life cycle assessments of animal food products and their substitutes. *Food Policy, 37*(6), 760–770.

O'Brien, D., Shalloo, L., Patton, J., Buckley, F., Grainger, C., & Wallace, M. (2012). A life cycle assessment of seasonal grass-based and confinement dairy farms. *Agricultural Systems, 10*, 33–46.

Ogino, A., Kaku, K., Osada, T., & Shimada, K. (2004). Environmental impacts of the Japanese beef-fattening system with different feeding lengths as evaluated by a life-cycle assessment method. *Journal of Animal Science, 82*(7), 2115–2122.

Ogino, A., Orito, H., Shimada, K., & Hirooka, H. (2007). Evaluating environmental impacts of the Japanese beef cow–calf system by the life cycle assessment method. *Journal of Animal Science, 78*(4), 424–432.

Oishi, K., Ibi, T., Kahi, A. K., & Hirooka, H. (2011). Optimal culling strategy in relation to biological and economic efficiency and annualized net revenue in the Japanese Black cow–calf production system. *Journal of Agricultural Science, 149*, 783–799.

Oishi, K., Kato, Y., Ogino, A., & Hirooka, H. (2013). Economic and environmental impacts of changes in culling parity of cows and diet composition in Japanese beef cow–calf production systems. *Agricultural Systems, 115*, 95–103.

O'Mara, F. P. (2011). The significance of livestock as a contributor to global greenhouse gas emissions today and in the near future. *Animal Feed Science and Technology, 166*, 7–15.

Opio, C., Gerber, P., Mottet, A., Falcucci, A., Tempio, G., MacLeod, M., Vellinga, T., Henderson, B., & Steinfeld, H. (2013). *Greenhouse gas emissions from ruminant supply chains—A global life cycle assessment*. Rome: FAO.

Palm, D. (2010). Product Category Rules (PCR): meat of poultry, fresh or chilled; Meat of poultry, frozen. PCR 2010:13, Version 1.0, prepared by IVL—Swedish Environmental Research institute, Sweden for: The International EPD System, Stockholm, Sweden. www.environdec.com. Accessed 7 April 2014.

Pardo, G., Ciruelos, A., Lopez, N., Gonzàlez, L., Saioa, R., & Zufia J. (2012, October). *Environmental improvement of a chicken product through life cycle assessment methodology*. Paper presented at the 8th Intenational Conference on Life Cycle Assessment in the Agri-Food Sector, France.

Pelletier, N. (2008). Environmental performance in the US broiler poultry sector: Life cycle energy use and greenhouse gas, ozone depleting, acidifying and eutrophying emissions. *Agricultural Systems, 98*(2), 67–73.

Pelletier, N., Pirog, R., & Rasmussen, R. (2010). Comparative life cycle environmental impacts of three beef production strategies in the Upper Midwestern United States. *Agricultural Systems, 103*(6), 380–389.

Pernigotti, D. (2011). Product Category Rules (PCR): finished bovine leather. PCR 2011:13, Version 1.0, prepared by the Giada Agency of Provincia di Vicenza, Italy (contractor: Aequilibria di Pernigotti Daniele) for: The International EPD System, Stockholm, Sweden. www.environdec.com. Accessed 7 April 2014.

Peters, G. M., Rowley, H. V., Wiedemann, S., Tucker, R., Short, M. D., & Schulz, M. (2010a). Red meat production in Australia: Life cycle assessment and comparison with overseas studies. *Environmental Science & Technology, 44*, 1327–1332.

Peters, G. M., Wiedemann, S. G., Rowley, H. V., & Tucker R.W. (2010b). Accounting for water use in Australian red meat production. *International Journal of Life Cycle Assessment, 15*, 311–320.

Phetteplace, H. W., Johnson, D. E., & Seidl, A. F. (2001). Greenhouse gas emissions from simulated beef and dairy livestock systems in the United States. *Nutrient Cycling in Agroecosystems, 60*(1–3), 99–102.

Pimentel, D. (2004). Livestock production and energy use. *Encyclopedia of Energy, 3*, 671–676.

Place, S. E., & Mitloehner, F. M. (2012). Beef production in balance: Considerations for life cycle analyses. *Meat Science, 92*(3), 179–181.

Reckmann, K., Traulsen, I., & Krieter, J. (2013). Life cycle assessment of pork production: A data inventory for the case of Germany. *Livestock Science, 157*, 586–596.

Ridoutt, B. G., Sanguansri, P., & Harper, G. S. (2011). Comparing carbon and water footprints for beef cattle production in Southern Australia. *Sustainability, 3*(12), 2443–2455.

Ridoutt, B. G., Page, G., Opie, K., Huang, J., & Belotti, W. (2012a, October). *Assessing carbon, water and land use footprints for beef cattle production in Southern Australia*. Paper presented at the 8th Intenational Conference on Life Cycle Assessment in the Agri-Food Sector, France.

Ridoutt, B. G., Sanguansri, P., Freer, M., & Harper, G. S. (2012b). Water footprint of livestock: Comparison of six geographically defined beef production systems. *The International Journal of Life Cycle Assessment, 17*(2), 165–175.

Ridoutt, B. G., Page, G., Opie, K., Huang, J., & Bellotti, W. (2013). Carbon, water and land use footprints of beef cattle production systems in southern Australia. *Journal of Cleaner Production, 73*, 24–30.

Ripoll-Bosch, R., de Boer, I. J. M., Bernués, A. & Vellinga T. V. (2011). Greenhouse gas emissions throughout the life cycle of Spanish lamb-meat: a comparison of three production systems. In: A. Bernués, et al. (Eds.), Economic, social and environmental sustainability in sheep and goat production systems. CIHEAM/FAO/CITA-DGA Options Méditerranéennes: Série A. Séminaires Méditerranéens; 100, 125–130.

Ripoll-Bosch R., Díez-Unquera B., Ruiz R., Villalba D., Molina E., Joy M., Olaizola A., & Bernués A. (2011). An integrated sustainability assessment of mediterranean sheep farms with different degrees of intensification. *Agricultural Systems, 105*, 46–56.

Ripoll-Bosch, R., de Boer, I. J. M., Bernués, A., & Vellinga, T. V. (2013). Accounting for multifunctionality of sheep farming in the carbon footprint of lamb: A comparison of three contrasting Mediterranean systems. *Agricultural Systems, 116*, 60–68.

Roer, A. G., Johansen, A., Bakken, A. K., Daugstad, K., Fystro, G., & Strømman, A. H. (2013). Environmental impacts of combined milk and meat production in Norway according to a life cycle assessment with expanded system boundaries. *Livestock Science, 155*(2), 384–396.

Röös, E., Sundberg, C., Tidåker, P., Strid, I., & Hansson, P. A. (2013). Can carbon footprint serve as an indicator of the environmental impact of meat production? *Ecological Indicators, 24*, 573–581.

Rosenbaum, R. K., Bachmann, T. M., Gold, L. S., Huijbregts, M. A., Jolliet, O., Juraske, R., Koehler, A., Larsen, H. F., MacLeod, M., Margni, M., McKone, T. E., Payet, J., Schuhmacher, M., Meent, D. V., & Hauschild, M. Z. (2008). USEtox—The UNEP-SETAC toxicity model: recommended characterisation factors for human toxicity and freshwater ecotoxicity in life cycle impact assessment. *The International Journal of Life Cycle Assessment, 13*(7), 532–546.

Roy, P., Nei, D., Orikasa, T., Xu, Q., Okadome, H., Nakamura, N., & Shiina, T. (2009). A review of life cycle assessment (LCA) on some food products. *Journal of Food Engineering, 90*(1), 1–10.

Roy, P., Orikasa, T., Thammawong, M., Nakamura, N., Xu, Q., & Shiina, T. (2012). Life cycle of meats: an opportunity to abate the greenhouse gas emission from meat industry in Japan. *Journal of Environmental Management, 93*(1), 218–224.

Schmidinger, K., & Stehfest, E. (2012). Including CO2 implications of land occupation in LCAs—Method and example for livestock products. *International Journal of Life Cycle Assessessment, 17*, 962–972.

Schulla, J. (1997). Flussgebietsmodellierung und Wasserhaushalts-Simulation mit WaSiM-Modellbeschreibung. Eidgenössische technische Hochschule Zürich, Zürich, pp. 189 http://www.wasim.ch/downloads/doku/wasim/schulla_1997.pdf. Accessed 10 April 2014.

Schulla, J., Jasper, K. (2007). Model Description WaSiM-ETH. http://www.wasim.ch/downloads/doku/wasim/wasim_2007_en.pdf. Accessed 11 April 2014.

Sessa, F. (2013a). Product Category Rules (PCR): Raw milk. PCR 2013:16, Version 1.01, prepared by Granarolo S.p.A., Life Cycle Engineering, ENEA, Distretto Latte Lombardo, Politecnico di Milano, Universita' degli studi di Milano, Caseificio dell'Amiata, ICS and Latteria Montello, Italy for: The International EPD System, Stockholm, Sweden. www.environdec.com. Accessed 7 April 2014.

Sessa, F. (2013b). Product Category Rules (PCR): Processed liquid milk and cream. PCR 2013:17, Version 1.01, prepared by Granarolo S.p.A., Life Cycle Engineering, ENEA, Distretto Latte Lombardo, Politecnico di Milano, Universita' degli studi di Milano, Caseificio dell'Amiata, ICS and Latteria Montello, Italy for: The International EPD System, Stockholm, Sweden. www.environdec.com. Accessed 7 April 2014.

Sessa, F. (2013c). Product Category Rules (PCR): yoghurt, butter, and cheese. PCR 2013:18, Version 1.01, prepared by Granarolo S.p.A., Life Cycle Engineering, ENEA, Distretto Latte Lombardo, Politecnico di Milano, Universita' degli studi di Milano, Caseificio dell'Amiata, ICS and Latteria Montello, Italy for: The International EPD System, Stockholm, Sweden. www.environdec.com. Accessed 7 April 2014.

Sommer S. G., Møller H. B., & Petersen S. O. (2001). *Reduktion af drivhusgasemission fra gylle og organisk affald ved biogasbehandling*. [In Danish, summary in English]. Foulum: Danish Institute of Agricultural Sciences. (DJF Rapport no. 31).

Sonesson, U., Cederberg, C., & Berglund, M. (2009). *Greenhouse gas emissions in beef production—Decision support for climate certification*. Report 2009:4

Stanford University. (2010). Environmental and social impact of the 'livestock revolution'. ScienceDaily. www.sciencedaily.com/releases/2010/03/100316101703.htm. Accessed 1 April 2014.

Steinfeld, H, Gerber, P, Wassenaar, T, Castel, V, Rosales, M., & de Haan, C. (2006). *Livestock's long shadow: Environmental issues and options*. Rome: FAO.

Stephen, K. L. (2011). *Life cycle assessment of UK pig production systems: The impact of dietary protein source*. Masters of Philosophy. The University of Edinburgh. 1–134.

Subramanian, V. (2012). *Integration of product category rules from north america into the global PCR database*. Washington, DC: PRé North America Inc.

Subramanian, V., Ingwersen, W., Hensler, C., & Collie H. (2012). Comparing product category rules from different programs: learned outcomes towards global alignment. *The International Journal of Life Cycle Assessment, 17*(7), 892–903.

Suh, S., Lenzen, M., Treloar, G. J., Hondo, H., Horvath, A., Huppes, G., Joillet, O., Klann, U., Krewit, W., Moriguchi, Y., Munksgaard, J., & Norris, G. (2004). System boundary selection in life-cycle inventories using hybrid approaches. *Environmental Science & Technology, 38*(3), 657–664.

Thoma, G., Martin, R. E., Nutter, D., Ulrich, R., Maxewell C., & Frank G. (2011). *National life cycle carbon footprint study for production of US swine*. 1–64.

Thoma, G., Jennie, P., Darin, N., Shonnard, D., Ulrich, R., Matlock, M., Dae Soo, K., Neiderman, Z., Kemper, N., East, C., & Adom, F. (2013). Greenhouse gas emissions from milk production and consumption in the United States: A cradle-to-grave life cycle assessment (2008). *International Dairy Journal 31*, S3–S14.

Thomassen, M. A., Dolman, M. A., van Calker, K. J., & de Boer, I. J. M. (2009). Relating life cycle assessment indicators to gross value added for Dutch dairy farms. *Ecological Economics, 68*, 2278–2284

Tukker, A., Huppes, G., Guinée, J. B., Heijungs, R., Koning, A., van Oers, L., Suh, S., Geerken, T., van Holderbeke, M., Jansen, B., & Nielsen, P. (2006). Environmental impact of products (EIPRO)—analysis of the cycle environmental impacts related to the total final consumption of the EU–25. European Commission, Joint Research Centre, Institute for Prospective Technology Studies., Seville, Spain.

Tuomisto, H. L., Angileri, V., De Camillis, C., Loudjani, P., Pelletier, N., Nisini, L., & Haastrup, P. (2013). Final technical report: Certification of low carbon farming practices. JRC Technical Reports. European Commission, Joint Research Centre. Luxembourg. http://mars.jrc.ec.europa.eu/mars/Projects/LC-Farming. Accessed 7 April 2014.

TSC. (2014). The sustainability consortium—food, beverage & agriculture sector working group.http://www.sustainabilityconsortium.org/fba/. Accessed 7 April 2014.

UNECE (1999). Control options/techniques for preventing and abating emissions of reduced nitrogen compounds. EB.AIR/WG.5/1999/8. United Nations Economic and Social Council, Economic Commission for Europe, Geneva.

UNEP. (2010). Assessing the environmental impacts of consumption and production: Priority products and materials, a report of the working group on the environmental impacts of products and materials to the international panel for sustainable resource management. (Hertwich, E., van der Voet, E., Suh, S., Tukker, A., Huijbregts M., Kazmierczyk, P., Lenzen, M., McNeely, J. & Moriguchi, Y).

UNEP undated. Animal waste management in livestock farms: Practical environmental policies and reviews. http://www.agrifoodforum net/home.asp. Accessed 20 January 2014.

UNESCO-WWAP (2009) *The United Nations world water development report 3: Water in a changing world*. The United Nations Educational, Scientific and Cultural Organization, Paris.

van der Werf, H. M. G., Kanyarushoki, C., & Corson, M. S. (2009). An operational method for the evaluation of resource use and environmental impacts of dairy farms by life cycle assessment. *Journal of Environmental Management*, *90*, 3643–3652.

van Middelaar, C. E., Berentsen, P. B. M., Dolman, M. A., & de Boer, I. J. M. (2011). Eco-efficiency in the production chain of Dutch semi-hard cheese. *Livestock Science*, *139*, 91–99

Vergé, X. P. C., Dyer, J. A., Desjardins, R. L., & Worth, D. (2008). Greenhouse gas emissions from the Canadian beef industry. *Agricultural Systems*, *98*(2), 126–134.

Veysset, P., Lherm, M., & Bébin, D. (2010). Energy consumption, greenhouse gas emissions and economic performance assessments in French Charolais suckler cattle farms: Model-based analysis and forecasts. *Agricultural Systems*, *3*(1), 41–50.

Weidema, B. P., Wesnæs, M., Hermansen, J., Kristensen, T., & Halberg, N. (2009). *Environmental improvement potentials of meat and dairy products*. JRC Scientific and Technical Reports.

Weiss, F., & Leip, A. (2010). Methodology for quantification of greenhouse gas and ammonia emissions from livestock sector the EU–27. In: A. Leip, F. Weiss, T. Wassenaar, I. Perez, T. Fellmann, P. Loudjani, & K. Biala (2010). Evaluation of the livestock sector's contribution to the EU greenhouse gas emissions (GGELS)–final report. European Commission, Joint Research Centre, pp. 71–139.

Weiss, F., & Leip, A. (2012). Greenhouse gas emissions from the EU livestock sector: A life cycle assessment carried out with the CAPRI model. *Agriculture, Ecosystems and Environment, 149,* 124–134.

Wenzel, H., Hauschild, M. & Alting, L. (1997). Environmental assessments of products—vol. 1: Methodology, tools and case studies in product development. London. Chapman & Hall.

Wheeler, D. M., Ledgard, S. F., de Klein, C. A. M., Monaghan, R. M., Carey, P. L., McDowell, R. W., & Johns, K. L. (2003). *OVERSEER® Nutrient Budgets—moving towards on-farm resource accounting*. Paper presented at the 65 New Zealand Grassland Association Conference, New Zealand.

Wiedemann, S., McGahan, E., Grist, S., & Grant, T. (2010). Life cycle assessment of two Australian pork supply chains. In proceedings of the 7th international conference on life cycle assessment in the agri-food sector (LCA Food 2010) (Vol. 2, pp. 227–232). Università degli Studi di Bari.

Williams, A. G., Audsley, E., & Sandars, D. L. (2006). *Determining the environmental burdens and resource use in the production of agricultural and horticultural commodities*. Main Report. Defra Research Project IS0205. Bedford: Cranfield University and Defra. www.silsoe.cranfield.ac.uk and www.defra.gov.uk.

Williams, A., Audsley, E., & Sandars, D. (2012, October). *A systems-LCA model of the stratified UK sheep industry.* Paper presented at the 8th International Conference on Life Cycle Assessment in the Agri-Food Sector, France.
World Bank. (2009) *Minding the stock: Bringing public policy to bear on livestock sector development*. Report no. 44010-GLB. Washington, DC.
Yan, M. J., Humphreys, J., & Holden, N. M. (2011). Review an evaluation of life cycle assessment of European milk production. *Journal of Environmental Management*, *92*, 372–379.
Yan M. J., Humprheys J., Holden N. M. (2013). The carbon footprint of pasture-based mill production: Can white clover make a difference? *Journal of Dairy Science 96*, 857–865.
Zonderland-Thomassen, M. A., Lieffering, M., & Ledgard, S. F. (2013). Water footprint of beef cattle and sheep produced in New Zealand: Water scarcity and eutrophication impacts. *Journal of Cleaner Production, 73,* 253–262.

Chapter 6
Life Cycle Assessment in the Fruit Sector

Alessandro K. Cerutti, Gabriele L. Beccaro, Simona Bosco, Anna I. De Luca, Giacomo Falcone, Angela Fiore, Nathalie Iofrida, Agata Lo Giudice and Alfio Strano

Abstract Fruit products are generally considered to be some of the less environmentally damaging foods in occidental diets. In fact studies investigating the carbon footprint of different food choices have reported that fruit is the category with the least environmental impact. However, these studies use data from environmental

A. K. Cerutti (✉) · G. L. Beccaro
Department of Agriculture, Forestry and Food Science, University of Turin,
Via Leonardo da Vinci 44, 10095 Grugliasco, Turin, Italy
e-mail: alessandrokim.cerutti@gmail.com

G. L. Beccaro
e-mail: gabriele.beccaro@unito.it

S. Bosco
Institute of Life Sciences, Scuola Superiore Sant'Anna, via S. Cecilia, 3 56127 Pisa, Italy
e-mail: s.bosco@sssup.it

A. I. De Luca · G. Falcone · N. Iofrida · A. Strano
Department of Agriculture, Mediterranean University of Reggio Calabria,
Feo di Vito, 89122 Reggio Calabria, Italy
e-mail: anna.deluca@unirc.it

G. Falcone
e-mail: giacomo.falcone@unirc.it

N. Iofrida
e-mail: nathalie.iofrida@unirc.it

A. Strano
e-mail: astrano@unirc.it

A. Fiore
Department of Science, International Doctoral School in Bioecosystems and Biotechnology
University of Basilicata, Viale dell'Ateneo Lucano 10, 85100 Potenza, Italy
e-mail: angela.fiore@unibas.it

A. Lo Giudice
Ionian Department of Law, Economics and Environment, University of Bari,
Via Lago Maggiore angolo via Ancona, 74121, Taranto, Italy
e-mail: agata.logoiudice@uniba.it

B. Notarnicola et al. (eds.), *Life Cycle Assessment in the Agri-food Sector*,
DOI 10.1007/978-3-319-11940-3_6

assessments of generic fruit production, which take no account of specific issues within orchard systems and fruit supply chains. Indeed, modern food production is very diverse, with high levels of specialisation and complexity. These features inevitably affect methodologies in the application of LCA to food products and agro-systems. It is therefore important to study what has already been done regarding standardisation of application protocols in order to make appropriate comparisons between products. In the present chapter, a review of LCA application in fruit systems is presented: papers from international journals, national journals, and conference proceedings have been reviewed. In particular, it can be assumed that mainstream research on the LCA applied to fruit production systems began around 2005; most of the papers were published in 2010 and 2012 in conjunction with international conferences on LCA in the agri-food sector. The review covers all the main criteria for conducting an LCA in fruit production systems. Specific issues considered are: aims and scopes, system boundaries, product considered, functional unit, data origin, life cycle-based methodology adopted, and environmental impact assessment method used. Furthermore this chapter investigates two aspects that are rarely considered in LCA studies of fruit systems: the role of nurseries in determine environmental impacts and the carbon storage properties of orchards.

Keywords Orchard modelling · Perennial crops · Carbon storage · Nursery · Sustainable fruit growing

6.1 Introduction

6.1.1 Overview

Fruit products are generally considered to have a lower environmental impact potential than most foods in western diets. For example, Carlsson-Kanyama et al. (2003) quantified the energy consumption of different diets and reported an average of 5 MJ per kg of in-season fruit (26 MJ per kg of out-of-season fruit), 15 MJ per kg of vegetables, 17 MJ per kg of bread and flour products, 33 MJ per kg of dairy products, 37 MJ per kg of meat, and 75 MJ per kg of fish products. On the other hand, compared with other food products, fruit production is considered to be an intensive agricultural system in terms of inputs of pesticides and fertilisers as well as investments in capital and material (e.g. Mouron et al. 2006a). Indeed, the embodied energy of orchard infrastructures, such as hail nets and irrigation pipes, is higher than in other cropping systems.

Studies examining the carbon footprint of different food choices have reported that fruit is the food category with the lowest environmental impact potential (e.g. Wallén et al. 2004; Berners-Lee et al. 2012). However, these studies use data from environmental assessments of generic fruit production which do not take into account specific issues within orchard systems and fruit supply chains. Indeed, different results may be obtained in relation to the production system (e.g. conventional or organic), the production site (specific soil and climate conditions affecting yield and agronomic performances), or the retailing system (long-term cold storage may dramatically influence the environmental performance of the product). Recently, Mouron et al. (2012) demonstrated that the same apple cultivation in five European regions may have completely different protection requirements, leading to very different environmental impacts.

High levels of specialisation, diversification, and the complexity of orchard systems inevitably affect the method involved in applying life cycle assessment (LCA) to food products and agro-ecosystems (Notarnicola et al. 2012a). It is therefore important to study the work that has already been done regarding the standardisation of methods in order to make appropriate comparisons between products.

The main aim of this chapter is to describe a reference framework for choosing the best settings for LCA applications in fruit production systems. Therefore Sect. 6.1 continues with an overview of fruit production in the international context and its environmental burdens. Sect. 6.2 introduces the application of the LCA methodology to fruit production systems and describes the contexts where this approach is used. Sect. 6.3 describes state-of-the-art practice in LCA in fruit production in the light of an in-depth literature review. In particular, in this section, both general and critical aspects are highlighted, such as the selection of the functional unit (FU) and ways of modelling the orchard, and also the selection of the impact categories and support tools for pesticide distribution or nutrient balances. Finally, in Sect. 6.4 lessons learned are presented and discussions on methods for realising the most reliable LCA applications in fruit production systems are proposed.

6.1.2 The fruit industry in Europe

In 2011, world production of fruit was 637,575,625 t, mostly concentrated in Asia (51 %) and America (23 %) (FAOSTAT 2013). In Europe, 71,626,657 t of fruit were produced, corresponding to around 11 % of the fruit produced in the world, with significant contributions by Italy (24.23 % of the fruit produced in Europe), Spain (21.79 %) and France (12.28 %). The main European fruit productions are reported in Table 6.1.

The important role played by the Asian markets is even more evident from analysis of production trends over the past 10 years: whereas America, Europe, Africa and Oceania record fairly constant fruit production, in Asia it has increased by about 55 %, making China and India the highest producers of fruit in the world, responsible for about 20 and 14 % of world production respectively.

6.1.3 Main Environmental Burdens Related to Fruit Production

In general, fruit production is considered an agricultural sector with low environmental impact in comparison with the herbaceous crops sector (Granatstein and Kupferman 2006) and other food sectors (Carlsson-Kanyama et al. 2003; Cuadra and Bjorklund 2007; Frey and Barrett 2007; Garnett 2006). Some authors underline that fruit production requires less bioproductive land compared with animal and

Table 6.1 European fruit production (in t of product and ha of cultivation) in 2011 compared with the world and Europe (data elaborated from FAOSTAT 2013)

	Production (t of fruit)		Cultivation (ha of orchards)	
	World	Europe	World	Europe
Major fruits				
Apple	75,484,671	15,196,149	4,745,442	1,048,874
Pear	23,952,157	3,352,615	1,626,270	190,649
Apricot	3,900,828	935,184	488,344	106,310
Cherry	2,241,424	871,475	379,946	193,323
Peach and nectarine	21,510,180	4,329,917	1,571,880	284,149
Plum	10,999,162	2,747,782	2,488,685	527,069
Lemon and lime	15,183,758	1,343,221	958,646	75,693
Orange	69,461,782	6,274,573	3,912,780	309,866
Mandarin and clementine	26,030,014	3,210,889	2,245,666	171,708
Kiwifruit	1,490,000	693,769	94,125	38,515
Berry fruits				
Blueberries	368,804	45,217	83,529	14,471
Currants	664,275	652,178	114,733	112,864
Raspberries	637,765	479,708	111,391	93,924
Strawberries	4,308,179	1,435,279	242,371	158,611
Nut trees				
Almond	1,942,242	357,375	1,651,560	656,626
Chestnut	2,023,019	121,380	540,732	85,019
Hazelnut	742,993	163,139	619,843	99,280
Walnut	3,418,502	381,511	965,552	157,966

some horticultural products (Gerbens-Leenes et al. 2002); others argue that perennial habitats (potentially) host natural pest predators and therefore benefit the food-webs (Simon et al. 2010).

In order to evaluate the environmental burdens of fruit systems it is necessary to identify which processes are involved, which resources are being used, and what kind of emissions they result in. The field operations associated with the greatest impacts can be grouped into six main groups: pest management, irrigation, fertilisation, soil management, weather damage prevention and tree management.

1. Pest and disease management

Pest and disease management involves a number of operations aimed at mitigating pest damage in fruit production. The main goal of the operation is to keep as much fruits as possible suitable for the market. Pest management can vary considerably depending on the production protocol. In conventional farming, the main approach to pest management is spraying with pesticides to eradicate harmful organisms. It is well known that synthetic pesticides have several limitations and serious harmful effects on the environment and human health (Holb 2009). Furthermore, the complete eradication of orchard pests is considered impossible without seriously compromising the environment; their increasing resistance to pesticides is also a problem (Suckling et al. 1999). In integrated and organic farming, different strategies are applied but the usual approach is to consider a pest as a natural organism with its own life cycle and natural enemies. By taking action in specific periods of the life cycle (e.g. mating disruption) or supporting natural pest enemies, it is possible to achieve improvements in fruit quality with less resource consumption (e.g. Reganold et al. 2001; Mila i Canals et al. 2006; Suckling et al. 1999). However, most of the non-chemical control measures are not widely used because of their high labour costs and/or time limitations during the season (e.g. Holb 2009; Suckling et al. 1999).

Orchards are among the most intensively sprayed agricultural systems, for the purposes of avoiding visible fruit damage and satisfying international commercial quality standards (Simon et al. 2010). Furthermore, pests and disease that are host-tree permanent may remain in the orchard for many years and require continuous control. The main environmental risks related to the use of conventional pesticides are the negative effects on the animal and plant communities exposed to them both in the orchard and in other ecosystems, both terrestrial and aquatic, to which pesticides are lost (Gliessman 2007).

2. Fertilisation

In orchards, fertilisation is required in order to supply the nutrients needed by the trees. Fertilisers are the result of industrial synthesis processes or mining or they can be by-products such as manures or plant residuals. The most important elements are nitrogen, phosphorus, and potassium. Nitrogen plays an important role in the vegetative development of trees and thus in tree management strategies (Nesme et al. 2006). As a consequence, N-fertilisers result in more pruning time and associated impacts. Deciduous fruit trees have low N demands compared with open field crops (Sanchez et al. 1995) and loss of nutrient surplus from orchards happens, especially during the winter and early spring when trees are not actively taking up N. Thus, both the amount and timing of fertiliser application are relevant in nutrient management (McDonald 2007) and in environmental assessment of orchards (Page 2009). Fertilisers are also a source of air pollution in terms of emissions of ammonia and nitrous oxide, and phosphate and nitrate are the main emissions affecting ground and surface water. Traditionally, fertilisers reach the plant through direct application to the soil, but nowadays several alternative techniques are used, especially fertigation, whereby fertilisers are mixed in the irrigation water or given directly to trees through foliar application.

Although some authors point out that fertilising with manure is a common practice for fruit production (Amiri and Fallahi 2009; Wei et al. 2009), only a few studies can be found where the use of different manures is compared in fruit orchards (Cerutti et al. 2011c). Although manures can improve soil characteristics and consequently plant growth and yield (especially in poor soils), they are ineffective in nutrient-rich soils or in combination with fertigation and high irrigation.

3. Irrigation

Numerous studies have demonstrated that water has a major influence on fruit growth and quality. If precipitation does not fulfil the demands of the culture, an irrigation system is needed to reach market standards in terms of fruit quality. There is a range of different irrigation systems, each associated with different environmental impacts (Mila i Canals and Polo 2003), mostly because of differing energy consumption. The energy consumption in the irrigation systems is strongly related to the climatic conditions in the area, the water source (surface water or groundwater), the water consumption of the culture, and the irrigation type. Usually, the higher the water requirement, the higher the energy consumed for irrigation.

As regards fertilisation, the best environmental practices can be achieved with the reutilisation of wastewater from other systems. Recent studies have demonstrated the possibility of applying treated municipal wastewater in orchards where the risk of transmitting disease is minimal (Palese et al. 2009).

4. Soil management and weed control

Soil quality is considered a key factor in human wealth because it is linked to several aspects of socio-ecosystems, e.g. food production, water quality, energy demand, and waste disposal (Lal 2009). It is well known that bad agricultural practices, such as over-fertilisation, excessive use of pesticides and irrigation, removal of crop residues and others, dramatically decrease soil quality (Lal 2009). Therefore, soil management and the effects on soil quality are important aspects to consider in evaluation of the sustainability of fruit production systems.

Soil plays an important role in the orchard system, as it does in terms of the quality of the fruit produced (e.g. McDonald 2007; Glover et al. 2000) and the environmental impacts associated with it (Mila i Canals and Polo 2003). Nutrients, water, and organic matter meet in the soil and careful management can improve fertiliser use efficiency and the need for application of pesticides and thus affect both the commercial and the environmental aspects of the production.

The goals of good orchard floor management are to improve soil quality, control erosion and weeds, and reduce surface runoff and leaching of pesticides and fertilisers (Glover et al. 2000).

An important agricultural technique used to prevent soil erosion and maintain a good soil structure for water infiltration is the use of cover crops (with either pure stands or stands where the legumes are mixed with other crops) under fruit trees. The purpose of the cover crops is to improve soil fertility through nitrogen fixation and recovery of nutrients from deep soil layers, enhance biological control of pests by providing a reservoir for pest predators, and modify the microclimate of the orchard (Gliessman 2007).

Different orchard floor management techniques are associated with different environmental impacts related to the consumption of resources (principally fuel, land, and water) and production of pesticides and fertilisers.

From an agricultural perspective the debate about the use of chemical or mechanical weeding is still open (McDonald 2007). The use of residual herbicides has been shown to be beneficial to tree growth and yield, but they leave the soil surface without a protective cover for much of the year, which can have a range of undesirable effects such as induction of soil compaction and reduction in water holding and infiltration capacity. On the other hand, mechanical weeding degrades the soil structure, decreases nutritional reserves, and can harm tree growth by injuring shallow-growing fine roots. Furthermore, mechanical weeding requires consumption of diesel, which can offset the environmental benefits of avoiding chemical products.

At present, the most environmentally friendly technique for controlling weeds is the use of post-emergent herbicides such as glyphosate for foliar action because they allow for regrowth of weeds during the winter so the soil is not permanently bare (Merwin et al. 1995). The use of post-emergent herbicides allows the production of biomass that returns to the soil and helps to build organic matter and foster the biological activity of the soil. Although various studies have investigated the different effects of approaches to weed control in orchards, proper environmental assessments are still missing.

Mulching is attracting more and more attention as an environmentally friendly method of weed control, but not so much is known about how organic mulches affect biological activity and nutrient availability in perennial cropping systems (Forge et al. 2003).

5. Tree management and harvest

Reaching equilibrium between growth and fruiting is one of the main objectives of the fruit grower. There are several different practices aimed at maximising the efficiency of fruit load, increasing fruit size, guaranteeing homogeneous colour, and hindering biennial bearing. These operations are divided into three main categories: branch pruning, thinning, and tree training. All these techniques are usually used for tree management in one season.

Branch pruning is usually carried out with hand-operated pruners, loppers and saws, usually from ladders. Commonly used machine-operated equipment includes compressed air-operated pruners and motorised hydraulic ladders (hydra-ladders) when trees are tall. Thinning can be performed by hand, usually on hydra-ladders, or with chemical thinning agents, usually used only in conventional or integrated fruit production.

Some authors (Granatstein and Kupferman 2006) suggest that the most environmentally friendly harvest can be achieved with an orchard manual management regime, in which all agronomic procedures are aimed at restricting plant height in order to avoid the use of machinery. They argue that such an orchard management regime is beneficial for three reasons: it is economical because of faster returns, higher quality fruit, and lower labour costs for maintenance; it is kinder to the environment because of optimal pest management by means of integrated methods; and it is socially responsible because fewer ladders mean fewer worker injuries.

6. Weather damage prevention

Weather damage prevention in orchards may be very different in different parts of the world, but it is almost always considered an important aspect of successful fruit production. Weather damage, although often related to brief extreme weather conditions, may strongly reduce the yield by destroying flowers, lowering the commercial quality of the fruits, or even harming the trees. The most studied extreme weather conditions that can inflict damage on the crop are hail and frost. The first problem is mainly resolved by the use of hail protection nets that are photo-neutral in order not to reduce light interception. These protection nets are the result of advanced technology and can generate environmental impacts in all their life cycle stages. Frost damage may be avoided by several different techniques, the selection of which depends on the frequency of frost events, water availability, and the economic importance of the plantation.

6.2 Overview of Life Cycle Thinking Methodologies and Approaches in the Fruit Sector

6.2.1 Environmental Assessment of Fresh Fruits and Fruit Products: State of the Art of International Practices

As highlighted in the previous chapters, the application of the LCA methodology in the agri-food sector has been widespread since the 1980s. In fact, the life cycle approach is a useful tool for finding new and alternative methods of production which can reduce environmental impacts, thus increasing food products' sustainability. Despite this, its applications in the fruit sector are still not very common, and many critical issues still need to be resolved. As a consequence, environmental assessment studies began only at the end of the 1990s and the scientific core of LCA's application in the fruit sector is very recent (see Sect. 6.3.2).

Nowadays LCA applied in the fruit sector is used as the scientific base in several international practices such as labels and declarations regarding the environmental sustainability of fruit products and supply chains. Among several environmental declaration schemes, the most important are the EU-ENVIFOOD Protocol and the International EPD® System.

The ENVIFOOD protocol is described in detail in Chap. 1. In this framework, fresh fruit products belong to group 1 and they are expected to be studied in the full life cycle including the use phase if relevant in the PCR.

The application of the International EPD® System to the fruit sector is represented by two main product category rules (PCR) of fresh fruit, "Fruits and nuts, except kiwifruit" (S-P–00369) and "kiwifruit" (S-P–00310) and three PCR on prepared products, namely "Fruit juices" (CPC 2143), "Jams, fruit jellies and marmalades" (CPC 21494), and "Other prepared and preserved fruit and nuts", the last one under development at the time of writing. These documents attempt to merge theoretical aspects to provide a scientifically sound assessment of the impacts and practical aspects of collecting data and managing assessment. Indeed, the amount of work required to collect high-quality data has been recognised as a major obstacle in the production of PCRs by small and medium-sized businesses (SMEs) (Zackrisson et al. 2008).

General PCRs for fresh fruits recognise that standard sampling is quite unlikely to render a representative yield in kg of product per hectare or yield factor per square metre of cropland required to produce 1 kg of product (Fruits and nuts—2012:07 p. 10, Sect. 7.4). Therefore three options are given: (1) adopting a typical yield factor (m2/kg) previously agreed between the interested parties in the area under evaluation and based on agronomic parameters and historical data for the area; (2) sample inflows/outflows from orchards of the same fruit in a group of farms to obtain an average yield factor; or (3) considering every production period as a unique batch in EPD terms, in which case the period of validity of the EPD will cover only a single production period.

Furthermore, three scientific studies have been published on the application of EPD® schemes to the fruit sector (Ingwersen 2010; Cellura et al. 2012a; Svanes and Aronsson 2013), mainly to test or discuss some of the methodological choices for fruits and nuts. Ingwersen (2010) presents a new method to assess the range of environmental products in the fresh product industry; it is designed for use in EPD® schemes for pineapple production in Costa Rica, and compares conventional and organic management. Monte Carlo simulation was used to create a range of comparability, so "Best" and "Worst" performances represent 0.5 and 99.5 % of the range of environmental performance respectively.

Taking into account the integration of EPD® schemes with other methodologies, Cellura et al. (2012a) couple organisation-specific tools, such as Environmental Management Systems (EMS) (European Parliament 2009) and Environmental Product Declarations (IEC 2008), with LCA methodology and environmental planning over an extended region, with special regard to vegetable cultivation. In particular, the study extends the approach of product-oriented management systems (POEMS) (Ardente et al. 2006) (specifically designed for single firms) to the ESCM of an entire agricultural district, creating a new approach called district-oriented management systems (DOEMS).

The EPD® scheme was used as a reference methodology by Svanes and Aronsson (2013) to elaborate the CF of a Cavendish banana supply chain. They used both ISO 14067 (Draft International Standard, DIS) (ISO 2012) and, for some methodological choices, the PCR for fruits and nuts of the International EPD® System (2009).

6.2.2 *Other Life Cycle Methodologies and Tools for Product Environment Assessment*

6.2.2.1 Simplified Life Cycle Assessment (S-LCA)

Simplified-life cycle assessment (S-LCA) is an important way to spread the application of LCA among different stakeholders other than LCA experts, such as decision-makers and SMEs. Organisations that aim at identifying their product life cycle environment impacts could use a standard LCA method, but the expensive access to the large amount of life cycle data, the lack of in-house expertise, and the costs associated with its implementation make it highly necessary to define simplified LCA approaches (Arzoumanidis et al. 2013). Thus, strategies to obtain simplification can be achieved in many ways, at the level of both LCI and LCIA.

Considering that the principles applied to the LCI phase of LCAs of food products are often not very clear, Mourad et al. (2007) proposed a simplified method for creating a useful dataset for different stakeholders in the agricultural chain with limited LCA skills (farmers, environmental managers, and decision-makers) as a reference for perennial crop evaluation. Such an S-LCA has been applied in one paper on the environmental profile of orange production in Brazil (Coltro et al. 2009). The method was based on well-accepted universal principles of stoichiometry applied to grain or fruit growth, whereby minimum estimations were introduced in the inventory mass balance; the elementary composition of the agricultural produce and photosynthesis principles were also taken into account.

6.2.2.2 Footprint Labels (Carbon Footprint, Water Footprint, Ecological Footprint)

The use of a single indicator rather than a complete LCA solves the complexity of LCA results, but raises the prospect of burden shifting. This approach can unfairly promote products that do not necessarily have a better overall environmental performance or environmental footprint (Weidema et al. 2008). However, there is a growing interest on the part of non-governmental organisations and retail chains with regard to this indicator thanks to its ease of understanding and communication of the impact of everyday products (Finkbeiner 2009). Concerns about Greenhouse gas (GHG) emissions and their effect on global warming have resulted in the calculation of the Carbon footprint (CF) of many areas of human activity. Calculating the CF of food products presents specific difficulties for Life cycle inventory (LCI) data collection, since the processes involved cannot be easily standardised (Cowell and Clift 1997; Haas et al. 2000; Mourad et al. 2007). Also, there is an ongoing debate on biogenic carbon (see Sect. 6.3.7). Indeed, agricultural activities constitute a significant source of CO_2, CH_4, and N_2O emissions, but the implementation of sustainable management practices within agricultural systems has been shown to reduce these emissions, which can be attributed to the natural capacity of agricultural wood biomass and soils to absorb and store CO_2 (Janssens et al. 2003). Many guidelines have focused on the GHG life cycle of goods and services. They include the publicly available specification PAS2050, developed by the British Standards Institute and the Carbon Trust (BSI 2008), the French Bilan Carbone (ADEME 2010), and the GHG Protocol drawn up by the World Resources Institute and the World Business Council for Sustainable Development (WBCSD/WRI 2009). Moreover, two specific ISO standards on product carbon footprints, ISO 14067 (ISO 2013), and on water footprints (ISO 2013) are in preparation. The application of carbon footprint analysis in the fruit sector is represented by five studies: pear (Liu et al. 2010), apple, kiwifruit (McLaren et al. 2010; Blanke 2013), mandarin orange, strawberry and peach (Yoshikawa et al. 2008) and one prepared and preserved product, orange juice (Spreen et al. 2010).

Water footprint analysis has been applied to fruit products as a single indicator in just one paper (Ingwersen 2012). Water use was estimated as a stress-weighted water footprint with the equation adapted from Ridoutt and Pfister (2010) and water consumed on the farm was estimated with the FAO CROPWAT model (FAO 2009) (Ingwersen 2012). In Stoessel et al. (2012) the irrigation inventory for imported crops was calculated according to Pfister et al. (2011). It is worth noting that estimation of the direct and indirect use of water has been included in other ten studies, with particular regard to water used for irrigation.

The application of the Ecological footprint analysis (EFA) along with an LCA was carried out only by Cerutti et al. (2010) for nectarines, in order to compare different environmental assessment methods. The comparison of two different assessment methods applied to the same productive process resulted in a strong accordance in terms of a single score for LCA and EFA values. EFA has been recognised by the European Commission's Joint Research Centre as a scientific reference methodology for the Environmental Footprint, a harmonised framework for sustainability assessment of a product, expected to be in line with ISO standards on life cycle assessment and recognised scientific methodologies (European Commission's Joint Research Centre 2011).

6.3 Methodological Problems Connected with the Application of Life Cycle Assessment in the Fruit Sector: Critical Analysis of International Experiences

6.3.1 Literature Review

The aim of the review is to highlight the state of the art of LCA applications to fruit sectors and to compare the different methods applied.

The review was conducted in several phases. In-depth bibliographic research was performed initially, mainly through the use of scientific literature research engines and databases of peer-reviewed literature, ranging from official published literature (i.e. scientific journals, conference proceedings and books, identified by ISBN, ISSN or DOI numbers) to the so-called "grey literature" (e.g. theses and dissertations).

In the first step, 55 studies were collected and then four criteria were applied to refine the literature sample. In particular, references were excluded from the analysis if:

1. the agricultural phase of the fruit systems was not considered in the system boundaries of the assessment, because of the intrinsic specificities of the agricultural phase for fruit supply chains;
2. the studies did not perform a full LCA, including all steps described in ISO 14040 and series;
3. the studies did not concern case studies but review themselves;

4. the studies were focused on olive or grape orchards, as these two production systems have already been investigated in Chap. 2 and 3.

Following these criteria, 41 papers were selected, of which 21 were published in international journals and 14 were contributions to international conferences. Other papers (6) comprised proceedings of Italian conferences, papers in Italian journals or doctoral dissertations.

Afterwards, a synthesis matrix (following Petti et al. 2010 and Cerutti et al. 2011a) was arranged in order to classify all papers according to 40 parameters, ascribable to eight themes as shown in Table 6.2: Reference, Method, Boundaries, Data, Functional Units, Agricultural assessment, Life Cycle Interpretation (LCI), Results. Further details of some of the most important parameters characterising LCA studies are given in dedicated paragraphs, and generic information is provided in the present section.

Referring to the "Goal and scope" parameter (Table 6.2) and following the classification proposed by Cerutti et al. (2011a), the aims of the studies reviewed can be grouped as follows:

- Evaluations of environmental performances of products, processes, or supply chains through the calculation of impacts. In some cases, hotspots are highlighted and environmental burdens are quantified; in other cases, particular attention is paid to specific categories such as energy consumption and GWP.
- Methodological suggestions. These papers are aimed at proposing guidelines to improve the development of LCA applications, suitable management options at process level, integrations of statistical tools.

Table 6.2 Parameters used to construct the synthesis matrix

Issue	Parameters considered
Reference	Authors, title, year, literature typology, product, farming method
Method	Study typology, methodology, goal and scope, LCIA method, LCIA phases, LCIA categories, land use change assessment, water consumption assessment
Boundaries	Main boundaries, boundary specification, boundary exclusions, geographical boundaries, time boundaries
Data	Primary data, databases, bibliographic sources, data quality checks
Functional units	Functional units, allocations, system expansions
Agricultural assessment	Pesticide diffusion models, pest management, nutrient balance models, nutrients, field emission accounting, carbon storage accounting, nursery impacts (potential)
Life Cycle Interpretation (LCI)	System with most impact, LC phase with most impact, Dominant impact category, substance or material with most impact, other interpretations, sensitivity analysis
Results	Main numerical results

- Comparisons. This kind of study highlights differences between products and/or cultivars, or in cases of single products between different farming systems, such as organic, integrated, and conventional farming or open field and protected crops; in other cases, the environmental burden of each phase along a life cycle or a supply chain is assessed. The main purpose of comparisons is to improve environmental performances by helping the decision-making process of responsible actors.

With regard to the "Methodology" parameter, 26 papers apply the LCA only, and others integrate it with other methodologies, such as principal component analysis (Soler-Rovira and Soler-Rovira 2008; Mouron et al. 2006a), economic analysis (Mouron et al. 2006b; Pergola et al. 2011, 2013; Strano et al. 2013; Khoshnevisan et al. 2013), CF (McLaren et al. 2010; Spreen et al. 2010), EFA (Cerutti et al. 2010), and energy analysis (Pergola et al. 2011, 2013).

6.3.2 *Overview of the Review and General Aspects*

Papers were published in the period 2005–2013, with peaks in 2010 and 2013; in particular, six papers were published during the period 2005–2007, 17 papers from 2008 to 2010, and the remaining 18 from 2011 to 2013, showing an increasing interest in these last years in sustainability concerns regarding fruit production. Figure 6.1 shows the distribution of different work types across years.

Among the papers analysed, 31 of them concern LCA applications to specific products, 4 are both applicative and comparative contributions, 3 are purely methodological, and the others are applicative and methodological (2) and applicative, comparative, and methodological (1).

The full list of papers with main characteristics is presented in Table 6.3.

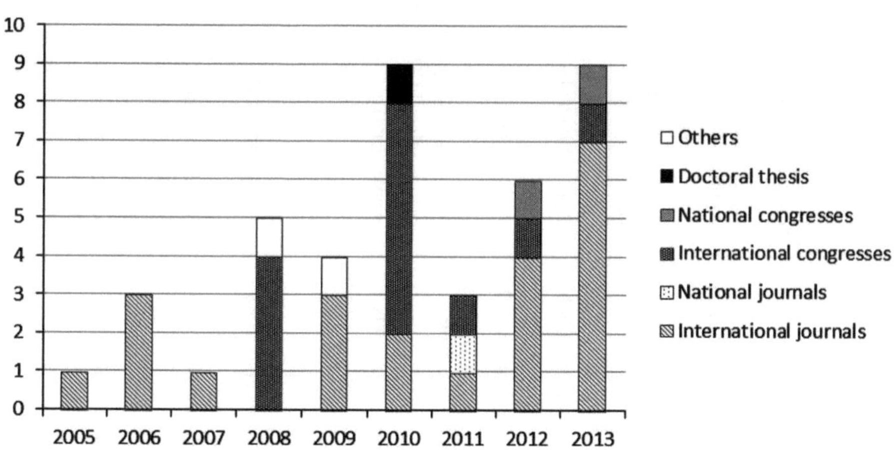

Fig. 6.1 Distribution of different work types across years

Table 6.3 List of references included in the literature review with main characteristics

Reference	Product and FU	Method	Main boundaries	Geographical areas	Time boundaries
Sanjuan et al. 2005	Oranges (1 kg)	LCA	Cradle to farm gate	Spain (Comunidad Valenciana)	1 year (2000)
Milà i Canals et al. 2006	Apple (1 t)	LCA	Cradle to farm gate	New Zealand (Central Otago, Hawke's Bay)	One growing season (1999–2000)
Mouron et al. 2006a	Apple (1 ha and total receipt)	LCA, PCA, statistical risk assessment	Cradle to farm gate	Switzerland (Eastern and Central)	4 years (1997–2000)
Mouron et al. 2006b	Apple (1 ha and total receipt)	Income analysis, LCA	Cradle to farm gate	Switzerland (Eastern and Central)	4 years (1997–2000)
Milà i Canals et al. 2007	Apple (1 kg)	LCA	Cradle to international retailer	National and international scale	One growing season (1999–2000)
Coltro et al. 2008	Frozen concentrated orange juice (FCOJ) (1 t)	LCA	Cradle to gate LCI basis	Brazil (Sao Paolo areas)	One growing season (2002–2003)
Fakouri et al. 2008	Apple (1 kg)	LCA	Cradle to grave	Sweden	*Not specified*
Soler-Rovira et al. 2008	Apple (1 kg and 1 ha of orchard)	Principal component analysis (PCA) and LCA	Cradle to international retailer	36 selected countries	1998–2008
Williams et al. 2008	Strawberry (1 t)	LCA	Cradle to international retailer	Spain, United Kingdom	*Not specified*
Yoshikawa et al. 2008	Mandarin orange, strawberry, apple, grape, peach	LCA, I-O analysis, process analysis	Gate to grave	Japan	*Not specified*
Beccali et al. 2009	Oranges and lemons (essential oil, natural juice, concentrated juice)	LCA	Cradle to gate (industry gate)	Italy (Sicily)	One growing season (2005)

Table 6.3 (continued)

Reference	Product and FU	Method	Main boundaries	Geographical areas	Time boundaries
Blanke et al. 20092009	Apple (1 kg)	LCA	Cradle to grave (including end of life)	Germany, New Zealand, South Africa	*Not specified*
Coltro et al. 2009	Oranges for frozen concentrated orange juice (FCOJ) (1 t)	LCI	Cradle to gate	Brazil (Sao Paolo areas)	One growing season (2002–2003)
Ribal et al. 2009	Oranges—navel ancient varieties (1 kg)	LCA, Data Envelopment Analysis (DEA)	Cradle to gate	Spain (Comunidad Valenciana)	*Not specified*
Beccali et al. 2010	Oranges and lemons (essential oil, natural juice, concentrated juice)	LCA	Cradle to gate (industry gate)	Italy (Sicily)	One growing season (*not specified*)
Cerutti et al. 2010	Nectarine (1 t)	LCA, EFA	Cradle to gate	Italy (Piemonte)	Orchard lifespan (20 years)
Clasadonte et al. 2010	Peach (1 kg)	LCA	Cradle to market	Italy (Sicily)	Orchard lifespan (15 years)
Clasadonte et al. 2010	Oranges (1 t)	LCA	Cradle to national retailer	Italy (Sicily)	Orchard lifespan (15 years)
Ingwersen 2010	Pineapple (one serving–USDA standard)	LCA	Cradle to international retailer	Costa Rica, Gainesville.	*Not specified*
Knudsen et al. 2010	Orange juice (1 l)	LCA	Cradle to international retail	Brazil, Germany, Denmark	*Not specified*
Liu et al. 2010	Pear (1 t)	LCA (consequential approach)	Cradle to retailer	China (Beijing suburb and Liaoning Province)	One growing season (2007–2008)

Table 6.3 (continued)

Reference	Product and FU	Method	Main boundaries	Geographical areas	Time boundaries
McLaren et al. 2010	Apple, kiwifruit (1 kg)	Carbon footprint (PAS 2050), LCA	Cradle to consumption	New Zealand, UK	One growing season (2006–2007) for apple; two growing seasons (2003–2005) for kiwifruit
Spreen et al. 2010	Orange juice (1 acre of orchard)	LCA	Cradle to gate	Florida	Orchard lifespan (30 years)
Cambria and Pierangeli, 2011	one-year and two-year walnut tree seedlings (100 seedlings)	LCA	Cradle to gate	Italy (Basilicata)	Orchard lifespan (20 years)
Cerutti et al. 2011b	Apple (1 kg)	LCA	Cradle to consumption	Italy (Piemonte)	Orchard lifespan (20 years)
Pergola et al. 2011	Table grape (1 kg and 1 ha of orchard)	LCA, LCC, Energy analysis	Cradle to gate	Italia (Puglia)	One growing season (2009–2010)
Dwivedi, et al. 2012	NFC orange juice (1.893 l)	LCA	Cradle to Grave	Florida	Orchard lifespan (30 years)
Ingwersen 2012	Pineapple (one serving—USDA standard)	LCA	Cradle to international retailer	Costa Rica	One growing season (*not specified*)
Lo Giudice and Mbohwa 2012	Citrus fruit (1 ton oranges)	LCI	Cradle to national market	Italy (Sicily)	One growing season (2003–2004)
Mouron et al. 2012	Apple (1 ha of orchard)	Sustain OS methodology	Cradle to gate	Switzerland, Germany, The Netherlands, France, Spain	(*Not specified*)
Pirilli et al. 2012	Clementines (1 t and 1 ha of orchard)	LCA	Gate to gate	Italy (Calabria)	One growing season (*not specified*)

Table 6.3 (continued)

Reference	Product and FU	Method	Main boundaries	Geographical areas	Time boundaries
Stoessel et al. 2012	Apple, avocado, banana, citrus, grape, kiwifruit, pear, pineapple (1 kg)	LCA	Cradle to international market	Switzerland	*(Not specified)*
Alaphilippe et al. 2013	Apple (mass, surfarce)	LCA	Cradle to gate	France (southern regions)	4 years
Blanke 2013	Apple (1 hectare of orchard and 1 kg of apple)	PAS 2050–1 (hort)	Cradle to gate	*not defined*	*Not defined*
Cerutti et al. 2013	Ancient apple cultivars (1 t, 1 ha of orchard; 1000 € income)	LCA	Cradle to gate	Italy (Piemonte)	Orchard lifespan (20 years)
Girgenti et al. 2013	Raspberries, blueberries (125-g flow pack)	LCA	Cradle to grave	Italy (Piemonte)	Orchard lifespan (11 years—Raspberries, 17 years blueberries)
Lo Giudice et al. 2013	Oranges (1 t)	LCA	Cradle to national market	Italy (Sicily)	Orchard lifespan (50 years)
Khoshnevisan et al. 2013	Strawberry (1 ton of GH and OF strawberries)	LCA, economic analysis, artificial neural networks	Cradle to farm gate	Guilan province, Iran	1 crop year 2011–2012
Pergola et al. 2013	Lemon and orange (1 kg, 1 ha of orchard)	LCA, LCC, energy analisys	Cradle to grave	Italy (Sicily)	Orchard lifespan (50 years)
Strano et al. 2013	Clementine (1 ha of orchard)	LCA+LCC	Cradle to gate	Italy (Calabria)	Orchard lifespan (40 years)
Svanes and Aronsson 2013	Banana (1 kg)	Carbon footprint (ISO-DIS 14067, PAS 2050–1)	Cradle to grave	Costa Rica, Norway	1 year (2008)

6.3.2.1 Fruit Production Systems Investigated

Regarding the typologies of product investigated, citrus and apples were the fruits that particularly captured the attention of researchers. Moreover, 27 papers (61.3 %) concern the production of a single fresh fruit, whereas 13 papers analyse more than one product: for example, multiple fruits or vegetables, Seven papers assess processed fruit products, such as juices or essential oils, and one study concerns the LCA of 1 and 2-year walnut tree seedling production (Cambria and Pierangeli 2011).

Furthermore, all the studies on oranges describe integrated production, with one study concerning just the LCI analysis phase. In all of these studies, Spanish and Italian realities are taken into account and the production phase is highlighted as having the main environmental burdens because of the huge water consumption during the irrigation phase.

Regarding the studies on apple production systems, they have all a comparative aim. In particular, Mila i Canals et al. (2006) evaluate the environmental impacts of three commercial apple orchards compared with two reference systems in New Zealand (conventional and integrated productions). Results show that more than 50 % of the environmental impacts are owed to energy-related emissions, and that production of pesticides and agricultural machinery are significant in terms of the overall energy consumption of the orchard. In particular, pesticide production represented 10–20 % of energy consumption, and machinery production accounted for 7–12 % of energy consumption. Another paper (Fakouri et al. 2008) compares conventional and organic production, focussing on three possible transportation scenarios. In this study, conventional production and transportation by diesel truck scenario resulted in the phases with the greatest impact. Furthermore, Cerutti et al. (2011) take into consideration the environmental assessment of different apple supply chains in Northern Italy. The aims were: to identify the environmental hotspots in production system performance; to describe management strategies to increase environmental performance; to compare the environmental burden of different food products; and to evaluate the environmental properties of the supply chain. The environmental assessment was conducted from a producer/retailer point of view, comparing different transport strategies from the same area of production. The scenario with a long-distance retail system was the one with the greatest impact, and GWP the category with the greatest impact. Another paper on apple production systems in Northern Italy is that by Cerutti et al. (2013) in which the environmental impact of three ancient apple cultivars is compared with that of the commercial cultivar Golden Delicious. This cultivar had the best environmental efficiency in terms of impact per mass of product; nevertheless ancient cultivars have lower impact per hectare of orchard as they are low-input/low-output systems. The last paper in the review of apple production systems (Alaphilippe et al. 2013) compares nine orchard systems: conventional, organic and low-input orchards were planted with three apple cultivars differing in their disease susceptibility and the Golden Delicious conventional system was used as the reference. Low-input systems generally displayed the highest environmental performances, but the overall environmental performance of a production system was not dramatically improved when conventional practices were replaced by low-input or organic practices.

Peach and nectarine systems are studied in two papers (Clasadonte et al. 2010a; Cerutti et al. 2010). The first aims to: profile the environmental burden of this production; compare the environmental burden of different cultivars; and compare different farming practices in terms of early and late ripening. In this study, the irrigation process was the system which had the greatest impact, and electricity in the field phase was the category which had the greatest impact. The second study simply uses a nectarine orchard as a case study for testing differences in environmental impact assessment using LCA or EFA.

Two papers focus on pineapples in particular: Ingwersen (2010) conducted an LCA that could be used as a background document for developing a PCR (EPD® scheme) with the intention of making the results comparable with all fruit products: human toxicity and ecotoxicity were the main impact categories. The other paper by Ingwersen (2012) focused on the LCA of fresh pineapple from the farm to retail shelf in the USA. The farming stage was found to be the one with the greatest impact; packaging was significant because of the packing material, and refrigeration was the primary contributor to impact during distribution, although small in terms of overall impact.

Small fruits (berries) are studied in just one study (Girgenti et al. 2013). In particular, the primary energy consumption and GWP of producing a 125-gramme flow pack of blueberry and raspberry are calculated. Because of the small fruit mass per unit of sale, packaging played a fundamental role in determining the environmental impact of the functional unit.

Finally, the paper on walnut tree seedlings (Cambria and Pierangeli 2011) is one of the few examining the nursery phase, which is usually excluded from the system boundaries because data are not easily accessible (see Sect. 6.3.6). In particular, the paper aimed to assess the environmental impacts of walnut tree seedlings in an Italian nursery destined for the high-quality timber supply chain. Plastic materials were found to be responsible for most of the environmental impact.

6.3.2.2 Papers on Fruit Processing

Processed fruit products are also investigated with LCA methodology. In particular, Beccali et al. (2009) discuss the use of an LCA approach to investigate the production of citrus-based products: essential oil, natural juice, and concentrated juice from oranges and lemons. The main aims of this work are: to assess mass and energy inputs and outputs in the production stages from citrus cultivation to transport of the final products, including indirect environmental impact related to energy source generation and water and raw material production; and to evaluate the environmental impact of the examined products, in order to identify the most significant issues and to suggest suitable options to reduce the environmental impact of the production system. Fertilisers and diesel showed the highest environmental impact. Citrus-based fruits were studied in another paper by the same research group (Beccali et al. 2010), which represents a step forward compared with the paper of 2009. In this case, starting from the results obtained in the former study the aim of the paper was to carry out an improvement analysis of the assessed production system, proposing sustainable scenarios for saving water and energy to reduce environmental burdens.

Frozen concentrated orange juice (FCOJ) was analysed in another paper by Coltro et al. (2008) but from a different point of view: in this study, an assessment of the energy use at several steps of the life cycle of this product was carried out for qualifying and quantifying the main environmental aspects of the examined orange juice. The results showed that since 50 % of the total energy used by the system is from renewable resources, the GWP of FCOJ is about 70 % of the total energy use. Furthermore, the major energy use could be attributed to the orange cultivation phase (71 %), followed by FCOJ production system (23 %) with a small contribution by the transport stage (6 %). The greater contribution of the orange cultivation was related to the amount of orange used for FCOJ production (about 10 t of orange: 1 t of FCOJ). About 86 % of the energy use in the FCOJ production stage is non-renewable. The reason for this is the energy required for concentrating the orange juice and keeping it frozen.

6.3.2.3 Papers on Fruit Supply Chains

Several LCA studies have been developed with the aim of comparing different food products from different countries. Williams et al. (2008) carry out a comparative LCA between the production of strawberries following different production protocols (conventional and integrated cultivation, open field, and greenhouse) in Spain and in the UK. This approach was applied to achieve several goals: profiling the environmental burden of the fruit product; evaluating the environmental properties of a supply chain; and investigating the environmental benefits of local vs. foreign production. The agricultural phase (for UK strawberries) and transport (for Spanish strawberries) were the phases with the greatest impact.

Finally, an environmental assessment of organic orange juice imported from Brazil to Denmark was performed by Knudsen et al. (2010). Several supply chains were considered: small-scale organic, small-scale conventional, and large-scale organic. The aims were: identifying the environmental hotspots in the product chain of organic orange juice originating from small-scale farms in Brazil and imported to Denmark; comparing the environmental impact at the farm gate of the organic orange production with a comparable conventional and a large-scale organic orange production in the same region in São Paulo, Brazil. The organic small-scale orange plantations showed the lowest value for the different categories (except for land use, which was slightly higher), followed by either the conventional small-scale and the organic large-scale with the highest value for GWP and eutrophication, or the organic large-scale and the conventional small-scale with the highest value for non-renewable energy use and acidification. The supply chain of Brazilian orange juice was also studied by Coltro et al. (2009) but in this case only an LCI analysis was performed, intended to characterise Brazilian orange production in terms of farm size, cultivated varieties, watering system, and tillage practices.

In studies focused on the supply chain which use life cycle-based methods energy consumption is usually considered as a proxy for environmental impact assessment.

For example Blanke and Burdick (2009) assessed the energy balance (as part of an LCA) for home-grown apples compared with imported apples from South Africa and New Zealand. The objectives of the study were to: assess the sustainability of the products and supply chain; improve resource conservation and management; study energy consumption of fruit storage versus shipment; study influence of higher yields (in New Zealand) compared with shorter distance (in South Africa); and to identify hotspots of excessive energy consumption in fruit storage versus ship transportation or food chain subsequently. The transportation of apples from South Africa by sea cargo had the greatest environmental impact. A similar study was performed by Mila i Canals et al. (2007) to see whether or not "food miles" can be considered as a relevant indicator for the environmental impacts associated with foods. In particular this study focussed on the primary energy use (PEU) of a comparative evaluation of domestic (European) vs. imported apples (from Southern America and New Zealand): the results showed similarities in the total PEU ranges for EU and New Zealand apples during the European spring and summer. Conversely, in autumn and winter PEU values are generally higher for apples imported than European apples consumed in Europe. However, this last consideration may not be true if apples for consumption in one EU country are imported from another EU country, because energy use for road transportation has a significant influence on the result.

6.3.3 Functional Units and Allocation Strategies

Twenty studies use a mass unit, such as the kilogram or the ton of product, depending on the boundaries considered and the characteristics of products assessed (processed or consumed). Other studies take into consideration different FUs such as: a litre (Ingwersen 2010, 2012; Knudsen et al. 2011), a serving of the product (Ingwersen 2010) or a surface unit such as the acre (Spreen et al. 2010) or the hectare (Mouron et al. 2006a, b, 2012; Soler-Rovira and Soler-Rovira 2008; Pirilli et al. 2012; Pergola et al. 2011, 2013; Strano et al. 2013; Alaphilippe et al. 2013; Blanke 2013; Knudsen et al. 2011).

Furthermore, some studies base their assessments on the purchasable item unit, such as a pack (Dwivedi et al. 2012; Girgenti et al. 2013), a plant (Fakouri 2008) or a number of seedlings (Cambria and Pierangeli 2011).

In many cases, the same FU is used to assess and compare many different products (Beccali et al. 2009, 2010; McLaren et al. 2010; Milá i Canals et al. 2006; Stoessel et al. 2012). In other cases more FUs are applied to assess the same production system with different purposes, such as: energy consumption per kg of product (Blanke and Burdick 2005); combinations of mass assessments and surface-based assessments and/or economic ones (Knudsen et al. 2011; Cerutti et al. 2013; Pergola et al. 2013; Alaphilippe et al. 2013; Girgenti et al. 2013; Blanke 2013); management performance influence on environmental impacts through the integration of statistical tools (Mouron et al. 2006a, 2006b).

Concerning the allocation step in LCA, it is important to underline that the life cycle phases of a food product may consist in multifunctional processes, with a range of co-products and by-products: this makes the allocation a complex and sensitive step. In this sense, ISO (2006a, b) suggest avoiding allocations if possible, for example by dividing the processes or expanding system limits (Liu et al. 2010).

Among the studies reviewed, 25 papers do not apply allocation criteria or do not explicitly specify if a method is used, whereas other papers use mass allocation (9), combined mass and energy allocations (4) and flow allocations (2). A few other papers use economic, energy, or mass, economic, and land-based allocation criteria.

6.3.4 System Boundaries and Orchard Modelling

The boundaries of a studied fruit system have to be set in accordance with the aim of the study. First it is necessary to define the "main boundaries", which are related to the different stages of the fruits' productive and logistic chain; consumption and end of life of fruit residues after consumption can be included too. In Fig. 6.2 the whole life cycle of fruits is represented, divided into upstream processes, agricultural core processes, and downstream processes. The upstream processes are the ones occurring from the production of raw materials (the cradle) to the transport of all productive inputs to the field (the farm gate). A non-exhaustive list of productive inputs comprises: fertilisers, pesticides, organic amendments, fuels, agricultural machinery, electricity, water, propping/covering structure of the orchard (wood, plastics, metal), greenhouse structure, bins, working tools. The agricultural core processes comprise all the annual operations performed for the management of the crop which occur within the boundaries of the farm (from gate to gate): for example, irrigation, fertilisation, pest management, weed management, pruning, thinning, harvesting. This phase accounts for all the direct emissions to soil, water, and atmosphere occurring in consequence of the application of fertiliser and pesticides, such as N_2O emissions related to nitrogen fertilisation (usually these data are estimated with models, because inventory data are not often available). Also CO_2 uptake from

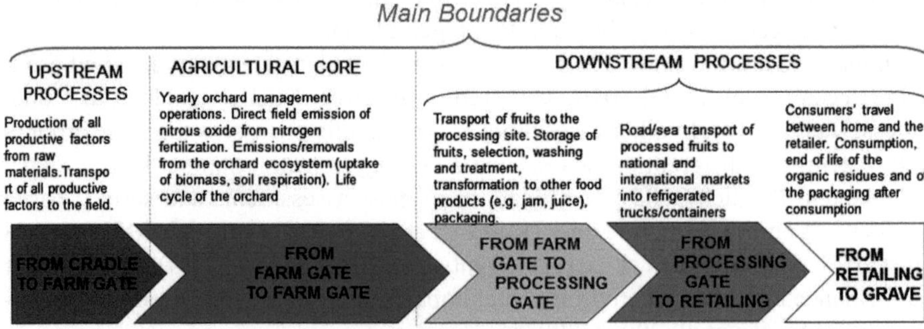

Fig. 6.2 Main boundaries of a LCA study about fruit production and process involved in each sub-phase of the fruit life cycle: from cradle to grave

biomass in photosynthetic processes and biogenic CO_2 emissions from soil can be included within this phase if a model or direct measurement of flows is used.

The downstream processes can be divided into three phases:

- From farm gate to processing gate, comprising transport of fruit to the processing site, storage, selection, washing and conservation treatments, transformation to other food products (e.g. jams, fruit juices), packaging.
- From the processing gate to retailing, including all transport and storage activities occurring at national and international level, by road or sea, in trucks or containers (which may be refrigerated), directed to wholesale retailers, fresh markets, distribution centres, and final retailers.
- From retailing to grave, meaning consumer's travel between home and retailer, consumption as fresh or processed product (refrigeration required), waste treatment of fruit's organic residues and packaging after consumption.

The upstream phase and the agricultural core phase have nearly always been included in the main boundaries of the reviewed studies, because they are closely related to production of fruits, and because they often account for an important part of the overall environmental impact, especially that owed to fertiliser production and machine utilisation. Another activity which significantly affects the environmental sustainability of fruit production is the production of capital goods of the farm such as the agricultural machinery, the propping/covering structure of the orchard (cement, zinc-aluminium poles, metal wire for the support of irrigation pipes, cement blocks, plastic hail nets), the buildings; these phases have been included within the boundaries in 47 % of the reviewed studies.

Soil to a depth of 1 m has to be considered as part of the system and included within the boundaries of the study (Mila i Canals 2003); therefore all differences in soil should be considered as impacts (organic matter, heavy metals). These differences can be related to land use change, land management change, or the usual management regime. Only 10 % of the reviewed studies have included the soil within the system boundaries. (See Sect. 6.3.7 on accounting for carbon emission and fixation in orchard ecosystem.)

Production of organic soil amendments as compost and manure is, usually, not included within the upstream processes, because they are waste pertaining to the life cycle of different products or processes (meat, milk, other agricultural products, forest management). The storage and management phases, as well as the use phase of organic amendments, can be included within the main boundaries, and system expansion can also be considered such as the quantity of mineral fertilisers avoided if organic amendments are used.

Table 6.4 reports the percentage of inclusion of the three phases of the downstream processes in the main boundaries of the reviewed studies; the phase of fruit consumption and end-of-life treatment of fruit residues and packaging is rarely examined because, in most cases, they have a minor impact in comparison with the entire life cycle. The storage, processing, transportation, and retailing phases are considered in more than half of the cases, in terms of examining the sustainability of the logistic chain (distribution at local or international level).

Table 6.4 Percentage of reviewed studies which include downstream processes in main boundaries

	% of inclusion in reviewed studies
From farm gate to processing gate	56
From processing gate to retailing	50
From retailing to grave	18

If the examined fruits are related to perennial crops, in addition to decisions about main boundaries decisions about "time boundaries" have to be taken as well, which means modelling the orchard life cycle (Fig. 6.3). It can last from 15 to 50 years, and can be subdivided into several stages relevant from a LCA perspective and variable in duration depending upon the species (Mila i Canals 2003).

1. Nursery: production of seedlings (upstream process, 2±3 years)
2. Establishment stage: plantation of trees and installation of the support/cover structure (1±3 years)
3. Stage of young trees: annual orchard management operations are carried out, low harvested yield because of the smaller size of trees (3±5 years)
4. Stage of adult trees: all annual operations are carried out, high harvested yield (10±30 years)
5. Stage of old trees: all annual operations are carried out, harvested yield falls because of the trees' age (0±5 years)
6. Destruction stage: the trees are removed and usually burned for production of domestic heat or in open air, and the field is prepared for future crops (usually 1 year)

It is fairer to consider the whole life cycle of the orchard within the time boundaries of the study and to allocate the environmental impact of the low-yield stage among the

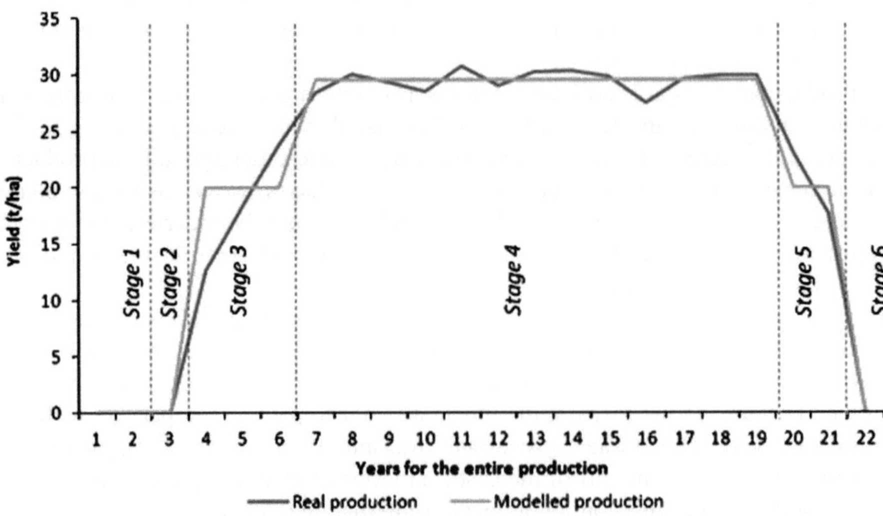

Fig. 6.3 Time boundaries of the orchard life cycle (agricultural core phase). Real production line refers to an apple orchard in Cuneo province, northern Italy (Data from Cerutti et al. 2011b)

Table 6.5 Percentage of reviewed studies about perennial fruit crops which include different phases of orchard life cycle within the time boundaries

	% of inclusion in reviewed studies
Nursery stage	18
Establishment stage	29
Low production stage—young trees	18
High production stage	89
Low production stage—old trees	18
Destruction stage	18

fruits leaving the system throughout all the years. However, the difficulty of data recovery regarding the low-yield stage, nursery, and establishment and destruction stages, means LCA analysts generally consider only the stage of adult trees in most cases. Table 6.5 summarises the review of the selected studies regarding time boundaries.

For annual crops (e.g. strawberries), the time boundaries matters are not so relevant; the nursery phase or seed production phase can be considered in an LCA study.

6.3.5 *Availability and Quality of Data*

Generally, as highlighted by Sonnemann et al. (2011), in LCA studies a combination of data sources is used to gather generic, specific, or average data from two main types of sources: primary sources (e.g. direct interviews) or secondary ones (e.g. scientific literature or databases, some of which are included in several LCA software). According to this classification, in 28 of the reviewed papers, the authors make use of primary sources for data collection, mainly through questionnaires and interviews of farmers or organisations; 29 studies use an LCA database (55.2% of cases the Ecoinvent or one of its versions, alone or in combination with others), and the rest of the studies do not specify which database is used, if any. Concerning data-gathering from scientific literature, 28 papers (68.3%) refer to a variable range of bibliographic sources, from one (Beccali et al. 2010) to about 12 (Fakouri et al. 2008).

To ensure relevant, reliable, and easily accessible data (von Bahr and Steen 2004), a data quality check is recommended (ISO 2002). However, only 17% of the authors conducted some form of data control, mostly in qualitative terms; only one paper (Beccali et al. 2009) confirmed that the data complied with quality requirements according to the European Platform on LCA (European Commission 2006).

6.3.6 *The Role of the Nursery*

To increase knowledge about the environmental impacts of agricultural production with a detailed LCA, all the life cycle phases should be taken into account. In particular, the nursery plays an important role because it usually represents the first link in fruit supply chains and a fundamental input to the plantation (Beccaro et al. 2014).

Furthermore, according to Nicese and Lazzerini (2013), in comparison with traditional agriculture, the plant nursery industry is generally characterised by increased land, technology, and resource use. It is beyond question that plant nurseries need high levels of inputs (raw materials, energy, structures) and thus significantly contribute to environmental impacts (Nicese and Lazzerini 2013).

Otherwise, since the potential contribution of nursery activities on total impacts depends on a multitude of conditions, according to Bessou et al. (2013) on perennial crops, it is appropriate to include the nursery stage, even if its relative contribution could be negligible. However, although many authors stress the importance of considering the nursery in environmental impact assessment (Milá i Canals and Clemente Polo 2003; Cerutti et al. 2013), the lack of data could represent a difficulty (Cerutti et al. 2013).

As evidence of this last issue, within scientific literature on LCA in the fruit sector, only six articles (out of 41 papers) specifically considered the role of the nursery in their applications. Four of them implemented the nursery phase thanks to the availability of primary data (Cambria and Pierangeli 2011; Cerutti et al. 2013; Girgenti et al. 2013), and the others used secondary data (Cerutti et al. 2011b; Stoessel et al. 2012).

In particular, Cambria and Pierangeli (2011) focused exclusively on the nursery phase, i.e. with system boundaries "from cradle to (nursery) gate", of walnut tree (*Juglans regia* L.) seedlings in the south of Italy, comparing young plants of different ages and adopting 100 seedlings as a functional unit.

Girgenti et al. (2013) focus on the nursery as the "pre-farm phase" of blueberry (1 year) and raspberry (2 years) production in northern Italy to evaluate the relative GHG emissions. They made some suggestions for improving the phase regarding the reduction of substratum quantity that means, in turn, less energy consumption by production and transportation.

Cerutti et al. (2010) considered the nursery stage for nectarines (2 years) as the average occurring of processes and resources consumption needed to obtain rootstocks, scions, and young plants. Nursery inputs, compared with the orchard stages, mean less fossil fuel consumption but more fertilisers and chemical products.

For apple production, Cerutti et al. (2011) analysed the nursery as one of the six stages considered, with the purpose of quantifying the main environmental impacts of the supply chain in Piedmont (Italy) and evaluating the relative impact of production and retail (with particular attention to the impact of transportation). In a further LCA application to apple varieties, Cerutti et al. (2013) counted the stock resources (plastics, steel, piling wood and plants) as average nursery processes and as resources needed to obtain rootstocks, scions and young plants for plants per hectare of the given orchard design.

In their study of 34 assorted fruits and vegetables of a large Swiss retailer, Stoessel et al. (2012) considered the seedling phase elaborating average data from secondary sources.

Finally, Cellura et al. (2012a, b), although referring to protected crops, can be considered as good base from which to investigate nursery processes because some agronomic operations and technologies are similar to those used in the early growing phase of perennial crops (land preparation for seeding, mulching, pavilions, or tunnels, etc.). In particular, Cellura et al. (2012a) assessed the ecoprofile of each product and the effect of each life cycle step in the total environmental impacts. The results showed that zucchinis' life cycle involved the highest impacts, except for waste generation, mostly because of their life cycle. Tunnel and pavilion greenhouses were characterised by comparable ecoprofiles, although slightly higher environmental impacts were associated with the former. Production of greenhouses and packaging resulted in phases with the greatest impact. Another study (Cellura et al. 2012b) applied LCA for evaluating the energy consumption and environmental burdens associated with the production of the same protected crops. In this case, LCA was used as a "decision tool" for addressing local policies for sustainable production and consumption patterns. Furthermore, the ecoprofiles of these products were estimated to identify supply chain elements with the highest impact in terms of global energy requirements, GHG emissions, eutrophication, water consumption, and waste production. Also, in this case, zucchinis in pavilions showed the main impacts. Pavilion and tunnel greenhouses displayed comparable ecoprofiles, although pavilion greenhouses were associated with slightly higher environmental impacts, especially with respect to waste production; production of greenhouses and packaging resulted in phases with the greatest impact.

6.3.7 Methods of Carbon Storage Accounting

The issue of the carbon cycle is generally omitted from LCA analyses applied to perennial cropping systems (Bessou et al. 2013) such as some fruit orchards.

The accounting of carbon storage in a fruit orchard ecosystem can be divided into two types:

- carbon temporarily stored in the above-ground and below-ground tree biomass for 13–30 years (life cycle of the orchard);
- the medium-to long-term soil carbon stock change, related to the balance between inputs of organic matter in soil (senescent leaves, pruning material, thinned fruits, soil grass cover, dead roots, compost, and manure addition), and outputs in the form of CO_2 emissions, because of organic matter degradation processes.

The carbon stock variations of tree structure and soil can in turn be attributed to two anthropogenic decisions:

- land use change (LUC), if the land was not used previously for agriculture or if different crops were grown;
- land use management change, if agricultural practices changed during the same tree crop management (e.g. tillage, irrigation, pruning management, soil grass cover,…).

At present, the IPPC provides guidelines mainly for LUC and few for land use management change. Even emissions and removals linked to land use change are almost absent from LCA studies on fruit products. Categorical estimates are used in the IPCC 2006 guidelines tier one approach for estimating the C stock changes in typical situations. These are based on four categories of land use, three categories of tillage and four categories of input with regard to crop residues and manure. Alternatives to categorical estimates are measurements and modelling, which increase the accuracy of results (Knudsen 2010).

Only four of the reviewed studies take into account at least one of the abovementioned carbon stock variations (details in Table 6.6), probably because at present there is no international consensus on an appropriate methodology because of the complexity and variability of this phenomenon in agricultural eco-systems, which is dependent on a number of variables (pedoclimatic conditions, orchard management regime, crop species response).

The possible influence on CO_2 field emissions of the key characteristics of perennial crops in general, such as their spatial structure (rows and inter-rows) or the use of irrigation, notably in southern regions, is in most cases disregarded (Bessou et al. 2013).

According to ISO 14067:2013, GHG emissions and removals arising from fossil and biogenic carbon sources and sinks should be included in the CFP and documented separately in the CFP study report. The CO_2 emissions arising from biogenic carbon sources are because of burning and degradation of biomass and microbial activity in soil, whereas CO_2 removals are to the result of photosynthetic processes.

Other sources and sinks which should be included in the CFP study, according to 14067, are: GHG emissions and removals occurring as a result of direct land use change, assessed in accordance with the relevant sections of the IPCC Guidelines for National GHG Inventories; soil carbon change should be included if not already calculated as part of direct land use change.

The PAS 2050 methodology includes change in the carbon content of soil because of direct land use change, but excludes change in existing agricultural systems (McLaren et al. 2010). In accordance with PAS 2050–1:2012, the carbon stock change should be linearly amortised over a period of 20 years.

6.3.8 *Water Management Assessment*

Because of the problems of scarcity and depletion of water resources, the optimisation of their use in agricultural processes is crucial in terms of quantity and quality, and also from an LCA perspective (Milá i Canals and Clemente Polo 2003).

In most papers on fruit production, water management assessment is seen in terms of irrigation impact and water requirements. Generally, simple accounting of irrigation water in fruit production calculates water volume (m^3) per functional unit (ha or kg of product) (Williams et al. 2008; Beccali et al. 2009; Ribal 2009; Lo Giudice et al. 2012, 2013; Pirilli et al. 2012; Soler-Rovira and Soler-Rovira 2008; Strano et al. 2013).

Ingwersen (2010, 2012) uses specific models to assess the water footprint. Assessing an LCA application to Costa Rican pineapple production, the author estimates water use by considering a stress-weighted water footprint, as suggested by

Table 6.6 Overview of the reviewed studies which include some elements of carbon storage in the LCA analysis

Reference	Description of carbon storage accounting
Coltro et al. 2009	CO_2 fixation by orange trees has not been taken into account, but CO_2 fixation by the fruit was considered according to the methodology described by Mourad et al. (2007): it calculates the CO_2 uptake for the photosynthesis of fruits, starting from a balanced basic photosynthetic reaction and from the carbohydrate content of the fruit. This methodology can be extended to the whole plant, but needs deep knowledge of the elementary composition of the plant
McLaren et al. 2010	Soil carbon content change measurement after 20 years (LUC) and distribution of the impact across those 20 years, as suggested by PAS 2050. As an alternative to measurement it is possible to use available data related to similar pedoclimatic conditions, orchard management practices, similar carbon content in soil at the orchard establishment time, similar previous land use
Knudsen 2010	Changes in the soil organic carbon (C) depending on orchard management practices were estimated with the simple tier 1 methodology in the IPCC 2006 guidelines. Results are given from both a 20-year and a 100-year perspective. The IPCC estimation method covers a period of 20 years, as hereafter the soil is assumed to have reached a new 'steady state' C content
Ingwersen 2012	Carbon footprint from land transformation was estimated only when conversion from primary or secondary forest was reported. In this case, carbon loss was estimated by identifying the historical Holdridge life zone (Holdridge 1967) of the farm occupied and summing the carbon in living biomass (Helmer and Brown 2000) and the estimated soil carbon and dividing this carbon loss over 20 years. If primary forest is cleared to create a pineapple farm it would result in a carbon footprint of approximately ten times larger than the carbon footprint without land use change. Sites on previously cultivated land are clearly preferable

Ridoutt and Pfister (2010) and Pfister et al. (2009), i.e. adapting the unitless water stress index (WSI) and using the FAO CROPWAT model (FAO 2009), parameterised with site-specific climatic and soil data and plant-specific parameters. In Stoessel et al. (2012), the irrigation inventory for imported crops is calculated according to Pfister et al. (2011).

6.3.9 Modelling the End of Life

Among the reviewed studies, only four papers provide specifications on the product's end of life. In Clasadonte et al. (2010b), as well as in Lo Giudice et al. (2013) and Lo Giudice and Mbohwa (2012), an LCA implementation is carried out which considers the end of life of a ton of oranges (cv. Tarocco) produced in Sicily. In particular, the authors consider the disposal of the non-edible part (the peel) and the separate waste collection of the municipal solid waste and its recovery as compost. They do not take into account the end-user consumption as they assume its impacts are negligible. Lo Giudice et al. (2013) find that the potential impacts of oranges' end of life (i.e. peel composting) correspond to about 2.56 % (of the total damage) contributing to each damage category as follows (accounted through IMPACT 2002+): resources (0.759 %), human health (5.01 %), climate change (2.14 %) and ecosystem quality (2.25 %). In terms of impact categories, the "composting of the non-edible orange part" represents about 0.76 % of "non-renewable energy", 6.5 % of "respiratory inorganics" and 2.1 % of "global warming".

Yoshikawa et al. (2008) consider the production, distribution, and consumption of ten different fruits and 14 vegetables in Japan; in particular, within the system boundaries and for use and end-of-life phases, the authors evaluate cooking in the household and management of solid waste, using a hybrid LCA method (combination of I-O analysis and process analysis).

The emission reduction potential is estimated for some hypothetical scenarios related to: optimisation of "local consumption" transport distance; "consumption in season" by replacing 20 % of greenhouse crop consumption with garden farming crops; "food loss reduction" by reducing 20 % of food loss in households; "food recycling" of all food waste (50 % by composting, 50 % for energy recovery by methane fermentation).

The end-of-life stage in terms of disposal of the packaging material is considered in Girgenti et al. (2013) for an LCA application to blueberry and raspberry production in northern Italy. In particular, the authors assume a hypothetical end-of-life scenario concerning the incineration of 20 % of plastic materials and the disposal of the remainder in a refuse tip. The authors suggest possible solutions related to packaging which include the replacement of plastic materials with biodegradable or other low-impact materials.

6.3.10 Life Cycle Impact Assessment (LCIA)

Reviewing the 41 papers, it is clear that all the authors conducted an LCIA phase, except Coltro et al. (2009) and Lo Giudice and Mbohwa (2012), who focused their studies on LCI of citrus fruits.

In 24 of the papers analysed (59 %), LCIA included only the mandatory phases of classification and characterisation (e.g. Sanjuan et al. 2005; Milá i Canals et al. 2006; Spreen et al. 2010; Blanke 2013), and midpoint methods have been applied. Among the latter, the method most often applied is CML 2 (six studies), as a unique method (Ribal et al. 2009; Pergola et al. 2011; Cambria and Pierangeli 2011), or combined with other impact assessment methods (Sanjuan et al. 2005; Beccali et al. 2009). As regards the other studies, the SALCA method (Swiss Agricultural Life Cycle Assessment) was applied by Mouron et al. (2006a, b) and Alaphilippe et al. (2013); Milá i Canals et al. (2006) and Knudsen et al. (2011) applied the EDIP method; Stoessel et Al. (2012) used the RECIPE midpoint method.

Many studies analyse only the impact of GHG emissions with the methodologies suggested by PAS 2050 (McLaren 2010; Ingwersen 2012; Blanke 2013) and ISO Norm 14067 (ISO 2012; Svanes and Aronsson 2013) or calculate the GWP from IPCC emission factors (Fakouri et al. 2008; Liu et al. 2010; Spreen et al. 2010;

Dwivedi et al. 2012; Girgenti et al. 2013). Others develop a classification and characterisation phase using different indicators but without specifying a methodology (Williams et al. 2008; Yoshikawa et al. 2008; Coltro et al. 2008).

Some authors complete their LCIA with optional phases like normalisation (7 %), grouping (9.3 %), and weighting (23.3 %). Specifically, Pergola et al. (2013) and Khoshnevisan et al. (2013) normalised impact results by applying the CML 2 method; four authors grouped impact categories (Milá i Canals et al. 2007; Soler-Rovira and Soler-Rovira 2008; Blanke and Burdick 2009; Ingwersen 2010).

Milá i Canals et al. (2007) and Blanke and Burdick (2005) did not specify which LCIA method they used, whereas Soler-Rovira and Soler-Rovira (2008) applied the CML 2 method but the normalised values of their life cycle analysis of each impact category were added per crop production (LCA crop indicator) and per transport (LCA transport indicator), and the sum of these two served as an overall potential environmental impact indicator (LCA total).

Ingwersen (2010) used many indicators from different impact assessment methods: the carbon footprint according to IPCC GWP 100, the virtual water/stress-weighted water footprint suggested by Riddout and Pfister (2010), pesticide toxicity in terms of USETox, energy use through NR cumulative energy demand, and eutrophication, acidification, and smog formation from TRACI (US EPA).

Among the studies that developed the weighting phase, three used the Eco-Indicator 99 method (Cerutti et al. 2010; Pirilli et al. 2012; Strano et al. 2013), three applied the IMPACT 2002+ method (Clasadonte et al. 2010a, 2010b; Lo Giudice et al. 2013), two applied the EDIP method (Cerutti et al. 2010, 2013). Employing a different approach, Mouron et al. (2012) used growing system typologies and geographical references as weighting criteria; moreover, they used different indicators, i.e. terrestrial and aquatic ecotoxicity potential and human toxicity potential were calculated according to Guinée (2002); demand for non-renewable energy resources was estimated according to Hischier et al. (2009); global warming potential over 100 years (GWP 100a) was calculated as described in IPCC (2006) and the EDIP97 method was applied to assess the eutrophication potential (EP).

As regards the impact categories considered in the papers reviewed, the most analysed is GWP 100a, followed by acidification potential (AP), EP, ozone layer depletion (ODP) and photochemical oxidation (POCP). Figure 6.4 shows how often these impact categories were used by the authors.

Despite the great importance of land use in evaluation of environmental impacts of fruit production systems, only four studies looked at land use changes.

6.3.11 Interpretation and Tools Supporting the Interpretation Analysis

The interpretation phase is the last methodological step of the four LCA phases proposed by ISO standards (14040:2006; 14044:2006).

Most of the studies reviewed discuss the interpretation phase in results or conclusions (e.g. Stoessel et al. 2012; Dwivedi et al. 2012; Cerutti et al. 2013; Lo Giudice et al. 2013; Strano et al. 2013).

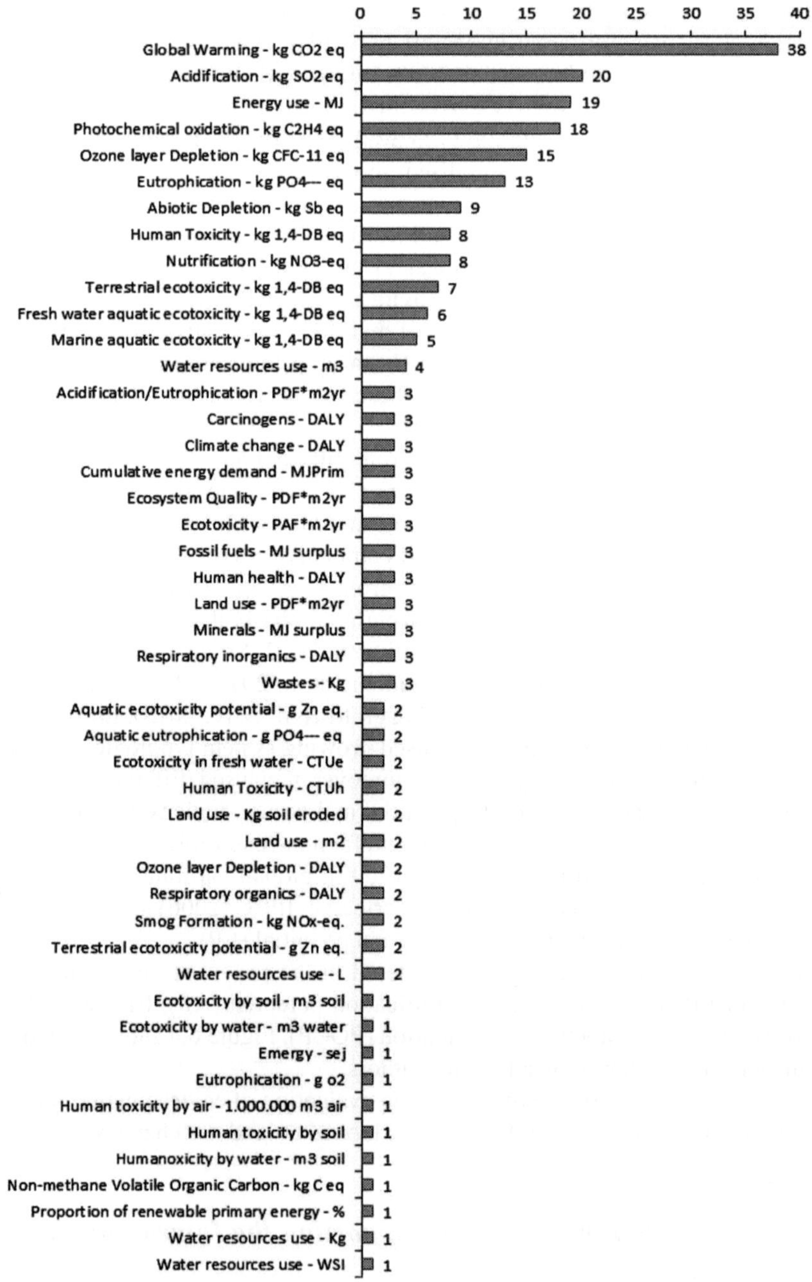

Fig. 6.4 Breakdown of impact categories used in LCA fruit studies

Furthermore, two indicators which are very important in the agricultural domain, i.e. land use and water consumption, have been ignored by all except Soler-Rovira and Soler-Rovira (2008) and Beccali et al. (2010).

As regards productive process inputs, the greatest attention has been paid to fertilisers (e.g. Sanjuan et al. 2005; Beccali et al. 2009; Liu et al. 2010; Dwivedi et al. 2012) and in particular to emissions by nitrogen fertilisation, a key practice in both agronomic and environmental terms (Pirilli et al. 2012; Cerutti et al. 2013; Khoshnevisan et al. 2013).

Other key concerns in agricultural practices are the use of chemical pesticides in phytoiatric treatments (Milá i Canals et al. 2006; Spreen et al. 2010; Dwivedi et al. 2012; Mouron et al. 2012), fuel consumption during machine utilisation (Milá i Canals et al. 2006, 2007; Fakouri et al. 2008) and materials used to build protective structures for crops and irrigation systems, such as plastics and metals (Cambria and Pierangeli 2011; Girgenti et al. 2013).

As far as the results are concerned, the most critical phase in the orchard life cycle is growth (e.g. Yoshikawa et al. 2008; Beccali et al. 2010; Knudsen et al. 2011; Lo Giudice et al. 2013), especially because of fertilisation practices (e.g. Sanjuan et al. 2005; Ribal et al. 2009; Pergola et al. 2013).

In many studies the great importance of product distribution in terms of logistics emerged, especially in the case of long distances (e.g. Milá i Canals et al. 2007; Blanke and Burdick 2005; McLaren et al. 2010; Svanes and Aronsson 2013). In studies that considered tree life in the greenhouse (Cambria and Pierangeli 2011), the construction of protection structures represented the phase with the greatest impact.

Some studies used specific tools to support interpretation of their results. Milá i Canals et al. (2006) conducted an uncertainty analysis on inventory data to estimate the margin of error of each result; they worked from literature data and expert judgments and then applied the tier one equation to estimate error propagation (Van der Sluijs et al. 2004) and establish a confidence margin for each orchard and impact category; this procedure allowed them to determine if indicators from the same impact category showed significant differences (Milá i Canals 2003).

Beccali et al. (2010) conducted a sensitivity analysis for each scenario, adopting a linear model for different elements (eco-profile of electricity, transport of citrus products, cultivation of citrus, allocation rules, scrap reuse), that allowed them to show a range of eco-profiles of the products examined.

Ingwersen (2010, 2012) applied the Monte Carlo analysis to determine the variability of each life cycle phase. The author conducted 1,000 interactions for each impact category with the SimaPro 7.2 model.

Liu et al. (2010) used a sensitivity analysis for GHG emissions and fossil fuel consumption, evidencing the high uncertainty of characterisation results, mostly influenced by specific assumptions of that study.

Knudsen et al. (2011) conducted a sensitivity analysis in relation to GWP by comparing seven growing scenarios and four processing scenarios. Cerutti et al. (2013) carried out a variance analysis with an ANOVA model (SPSS 18.0 statistical software); the authors applied it to weighting results in order to evaluate the statistical significance of the final environmental ranking.

6.3.12 Comparative Analysis

Comparative analysis is often applied by LCA practitioners and its scope can vary considerably. This kind of assessment must be conducted according to a set of essential methodological rules, such as the use of the same FU and equivalent methodological assumptions on the following issues: performances, system boundaries, data quality, allocation method, methods for evaluation of inflows and outflows, and impact assessment methods (ISO 2006b).

Several studies in the agri-food domain undertook comparisons, generally for the purpose of identifying productive systems with less impact and more eco-efficient unitary processes, and also to evaluate the importance of specific locations for both production and consumption of products and to measure potential impacts of different transformation processes.

In 28 reviewed papers (more than 68 % of the total studies analysed), the comparative analysis deals with potential environmental impacts, generally based on a comparison of growing techniques or products.

As regards growing techniques, comparisons of organic, integrated, and conventional growing systems gained the greatest attention (e.g. Milá i Canals et al. 2006; Fakouri et al. 2008; Ribal et al. 2009; Liu et al. 2010; Knudsen et al. 2011; Strano et al. 2013; Alaphilippe et al. 2013). In these cases, it emerged that conventional systems always impact more than organic/integrated ones, regardless of the kind of FU applied: surface units such as 1 ha (e.g. Pergola et al. 2013; Strano et al. 2013) or product units (e.g. Milá i Canals et al. 2006; Fakouri et al. 2008). In Liu et al. (2010), in which the application concerns a ton of pears, the organic scenario was the worst, because of the significant contribution of agricultural machinery and corresponding fossil fuel in contexts that facilitated mechanisation (flat terrain and larger farms).Some studies compare products originating from places different from where they were consumed and products originating from the same place; others compare different transport modalities (Milá i Canals et al. 2007; Blanke and Burdick 2005; Cerutti et al. 2011b). The latter highlight the importance of the logistic phase in the life cycle of a product and all of them show that the best solution is always the shortest itinerary; however, if the local product is stocked for a long period, then the environmental advantage compared with a fresh product is noticeably reduced, meaning that seasonal fruit consumption has to be preferred as it is more sustainable.

Several comparative studies include a focus on installations such as protection structures. Williams et al. (2008), Yoshikawa et al. (2008) and Khoshnevisan et al. (2013) compare protected cultivation with the open field, and find that the former has the highest environmental impact per FU. Fakouri et al. (2008) compare roofing systems (single pane glasshouse and double layer polyethylene film), identifying the double layer polyethylene film as the more efficient.

As regards comparisons of processed fruit productions, Spreen et al. (2010) and Dwivedi et al. (2012) analyse two scenarios of orange juice production, comparing replacement of old orange trees with the case without replanting; applying a FU of 1 acre, Spreen et al. (2010) find that the replanting scenario has greater impact, whereas Dwivedi et al. (2012), in relation to the impacts produced by an orange juice pack, find exactly the opposite, probably because of the reduced production of oranges. Beccali et al. (2009, 2010) compare many products derived from citrus processing (essential oils, natural juice, and concentrated lemon and orange juice) applying as FU 1 kg of final product (i.e. six FUs) and find that the production of essential oils is always the one with the greatest impact, and natural juice production is the most sustainable.

Clasadonte et al. (2010) and Cerutti et al. (2013) conduct comparative analyses between different cultivars. Clasadonte et al. (2010) compare peach cultivars (average and late ripening), in relation to 1 kg of product, and find that average ripening cultivars have greater impact. Cerutti et al. (2013) compare three ancient apple cultivars of Piedmont with the Golden Delicious cultivar, applying three different FUs (1 t of fruit; 1 ha of orchard; 1,000 € income earned by the grower), highlighting the following results: according to the land base FU (1 ha), the cultivar Golden Delicious has the greatest impact of all categories; according to the value-based FU (1,000 € of income), the Golden Delicious cultivar performs less well in all categories except for nutrient enrichment potential; results of the mass-based FU (1 t) vary, but it is possible to affirm that once more, in most impact categories, the Golden Delicious cultivar gives the best performance.

Some studies compare different unit processes (Sanjuan et al. 2005; Mouron et al 2012; Blanke 2013) in order to determine the best solution. Sanjuan et al. (2005) analyse different irrigation systems of orange orchards (groundwater and superficial water by gravity or drip irrigation) combined with different tillage techniques (tillage and no-tillage), finding that the scenario with the greatest impact is a combination of groundwater, drip irrigation, and no-tillage. Mouron et al. (2012) compare four pest management systems to ascertain the environmental impacts of the integrated pest management of an apple orchard: baseline system—non integrated pest management (IPM); advanced system 1–good IPM practices; advanced system 2–best IPM practices for pioneers; innovative system, with ecotoxicity reduced to minimum, but not yet commercially applied.

Blanke (2013) analyses different solutions for fruit thinning in apple orchards, and identifies the manual technique as the most environmentally friendly one.

Cambria and Pierangeli (2011) analyse two kinds of seedling production, 1-year and 2-year seedlings: by applying an FU of 100 seedlings they find that the latter is the one with the greater impact.

6.4 Implementation of the Life Cycle Assessment Methodology in the Fruit Sector: Lessons Learned

6.4.1 Modelling of the Orchard

The first task that has to be considered in an LCA study on fruit production systems is modelling the orchard. The main options for estimating yields highlighted in the literature review (pre-set estimations or calculation of an average—see Sect. 6.3.4) do not take into consideration the fact that major diseases or dramatic adverse climate conditions usually affect an entire production area at the same time, influencing the yield factor for the whole region (Sansavini et al. 2012). Furthermore, recent research on olive orchards in southern Italy (Notarnicola et al. 2013) has shown that statistically significant differences may occur in orchard management practices and farm performances at a regional level. As a consequence, using a local average of the yield factor might be a good way of including the variability of orchard inflows/outflows in small areas, but not the variability at a regional level or within the timescale.

A possible way of avoiding this problem is to use the annual average of orchard inflows and outflows collected over a period of years. Pirilli et al. (2012) suggested that 3 years might be sufficient but, because of the alternation of production (biennial bearing) in most of the perennial crops in Europe, an even number of years should be adopted. A 4-year time interval may be considered as a minimum requirement for data, but the optimum period of data collection should be based on a crop-specific literature review. Indirect field data may also be used to cover any missing years in the sampling. For instance, in some countries farmers are asked by their regional authorities to keep field logbooks in which they record the main inflows and outflows of their orchards. These data can be used to provide a historical weighting of the annual yield factor.

Even when full sets of field and historical data can be collected, recommendations may be needed for modelling the orchard system in LCA tools. In particular, common errors may occur when the environmental impacts of a 1-year process (e.g. orchard establishment) are balanced against multi-annual processes (e.g. fertilisation) and both process types are referred to a single (annual) FU. In order to avoid such calculation errors, a modelling procedure is suggested. In particular, six subsystems (hereafter called plans, in line with operational terminology) have been created and connected as follows (Fig. 6.5).

Plan 1: nursery. All processes and input materials used in the nursery stage can be accounted for by planting grafted plants in the orchard as the reference flow. Indeed, this process represents the connection between the nursery plan and the following parts of the orchard system.

Plan 2: establishment. All the processes that occur in the preparation of 1 ha of orchard must be included. The grafted plants are connected to the previous plan through the input of plants per hectare. Plan 2 has to lead to 1 ha of ready-to-produce orchard for connection to the next plan.

Plan 3: low production years (first part). The plan must include one sub-plan for each year of low production. Including one process for each year would correctly balance the weight of other processes that occur just once in the whole lifetime of the orchard (such as its establishment). Each of these sub-plans has to be connected through the reference flow of 1 ha of orchard and has to include an open output with the mass of fruit produced for that year. Each sub-plan considers the specific inflows and outflows of the reference year, i.e. the specific farming inputs and fruit yield. Data for these years may be obtained from field workbooks or may be modelled according to the fruit species and all the agricultural factors.

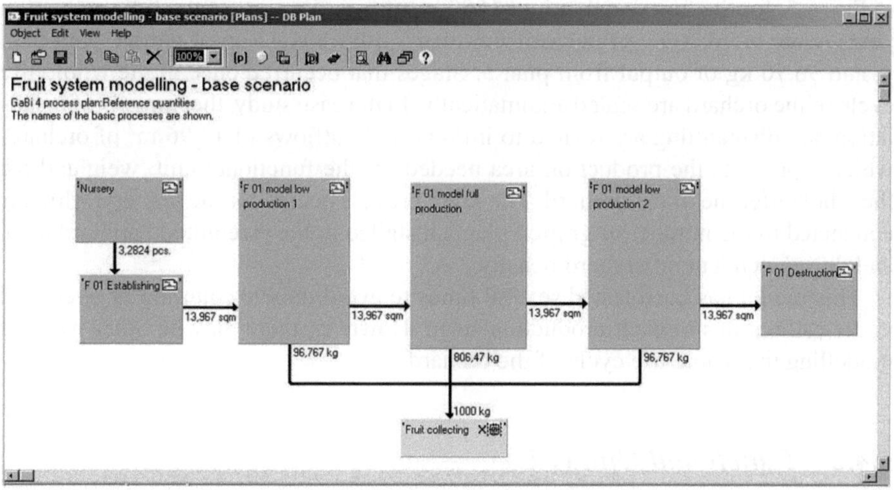

Fig. 6.5 Graphical representation of the orchard model with Gabi 4.0

Plan 4: full production years. This plan is connected to the previous one through the reference flow of orchard hectares. This plan must include one sub-plan for each year of full production which includes the specific inflows and outflows of the reference year, i.e. the specific farming inputs and the fruit yield. Inflows and outflows should be obtained from historical data or, for all years, can be considered to be the average of the data directly acquired from at least four full production years. In this case, too, each sub-plan has to be connected through the reference flow of orchard hectares, leaving the output of orchard hectares in the last year free to be connected to the next plan. In each year, the specific sub-plan for output of fruit produced has to be left open.

Plan 5: low production years (second part). This plan follows the same rules as plan 4 according to the inflows and outflows of the second tail of the model describing the orchard's senescence. Specific data for these years are very rare, but information may be obtained directly from farm managers. It is not uncommon for the orchard to be removed from production at the first signs of lower production; in this case the low production stage can be avoided and plan 4 may be directly connected to plan 6.

Plan 6: dismantling. This plan follows the same rules as the establishment plan (2), with the exception of an input of orchard hectares in the first process for connection to the previous plan and the end of an output of orchard hectares in the final process because no further connections are required.

Once the six plans are completed and connected, one last process has to be added. This is a fictitious process called "fruit collecting" (Fig. 6.5) which is needed to connect the fruit outflows from the three production plans (3–5) to a single output of fruit mass that can be fixed as the functional unit that best fits the case study (e.g. the 1000 kg of fruit in Fig. 6.5). All the inflows are automatically scaled to the weight of harvested material at each stage. For example, in the case study reported in Fig. 6.5, for the functional unit of 1000 kg of fruit, impacts of the full production years relate to 96.76 kg of output from plan 3, plus 806.47 kg of output from plan 4, and 96.76 kg of output from plan 5. Stages that occurred once in the whole life cycle of the orchard are scaled automatically. In the case study, the impacts of installation and dismantling are related to inflows and outflows of 13.96 m^2 of orchard, which represents the production area needed for the functional unit, weighted for the whole lifetime of the orchard. The same process occurs for the nursery, which is connected to the number of grafted plants installed in the "weighted" orchard area, and thus dependent on orchard density.

This model has been tested several times to avoid double counting or over- and under-estimations of each production stage. However, there may be other ways of modelling the whole life cycle of the orchard.

6.4.2 *Functional Unit* (*FU*)

From the literature review of fruit LCA it clearly emerges that the selection of the FU significantly influences the results of the assessment and its interpretation. Therefore, every assumption must be clearly explained and justified. As highlighted in Sect. 6.3.3, according to ISO (2006a), the FU represents the quantified performance of a product system to be used as a reference unit, and as the LCA is a comparative approach all the assessment process is structured around it. Furthermore, it is important to consider the following:

- the products assessed;
- the aim of the assessment;
- the boundaries (physical, temporal, geographical);
- to whom the results are addressed: the service delivered by the system in a study may vary depending on the intended audience (Cowell 1998)

According to Notarnicola et al. (2012), disregarding these issues may lead to bias and, above all, to the incorrect transposition of the reality of the product system to the identification of its functions. Nevertheless, as often occurs in agricultural systems, some products may serve more than one function at once: in this case, an FU representing more functions is needed or more FUs must be chosen (Notarnicola et al. 2012; Reap et al. 2008).

Given the results of the literature review, it is possible to comment on the selection of the FUs in relation to the typology of assessment in the fruit sector and the stakeholder to whom the assessment is addressed (Table 6.7).

If the assessment concerns a single product without any comparison, it is possible to choose the most suitable FU among a wide variety. However, in order to ensure maximal transferability of analysis results, such selection should be related to the stakeholder (farmers, consumers, policy-makers) to whom the analysis is addressed.

Indeed, if the study is directed at farmers, it can be interesting to show results by mass–(e.g. the kg or the ton) or volume–(e.g. litres) *based* FUs, but also energy–or

Table 6.7 FUs relevant to the audience and coherent with the typology of LCA study

Addressed actors Typology of the study	Farmers	Consumers	Policy-makers, territorial planners, local communities
Assessment of a single product	Mass, volume, energy, economic values	Mass of consumed product, volume, serving	Surface, mass, energy, nutrient contents
Comparing different growing practices for the same product	Mass, energy, economic values	Mass of consumed product, volume, serving, nutrient content	Mass, surface, energy, economic values
Comparing different fruit products	Surface, energy, combined measurement units	Volume of processed or consumed product, RDI, serving, nutrient content	Surface, energy, economic values

economic value-*based* FUs to highlight the phase with the greatest impact and improve the eco-profiles of production process. On the other hand, consumers may have more interest in FUs based on the mass, the serving, or the volume of the consumed product.

Moreover, among the papers reviewed, Fakouri et al. (2008) considered 1 kg of product consumed; in a similar way, Svanes et al. (2013), assessing the carbon footprint of 1 kg of Cavendish bananas, distinguished between those consumed and those sold, and gave results also per nutrient density unit as defined by Smedman et al. (2010). Ingwersen (2012) based his assessment on a serving of the product normalised against a recommended diet. The author legitimated his choice by affirming that mass-based FUs are not the most appropriate FU for food (Schau and Fet 2008), "because consumers typically do not purchase foods based on mass content, and mass content varies based on water content" (Ingwersen 2012, p. 153); however, it is important to remember that the relevance of the FU may vary according to the goal of the study and the addressees. If the LCA study concerns the assessment of a single product and it is aimed at bringing relevant information to policy-makers, territorial planners or local communities, surface-, mass-, energy-, nutrient content-*based* FUs could be the most suitable FUs.

In the case of comparative analysis, e.g. different growth systems such as organic, integrated, and conventional ones, farmers are usually more interested in mass– or energy-*based* comparisons, because productivity and costs are the main concerns in farm management. This is especially true when open field growing is compared with protected growing systems (e.g., greenhouses, tunnels). If this kind of study is addressed to consumers, the same considerations as for single-product assessments can be valid. When results are intended to aid politicians, planners, or communities in decision-making processes, it can be useful to interpret them through surface-, energy–or economic value-*based* FUs as well as mass-*based* ones.

The selection of appropriate FUs in the case of comparisons between different fruits must take into account a wide range of qualitative differences intrinsic to products:

- in some typologies of fruit not all parts are completely edible;
- considering the different yields of fruit crops, surface-*based* FUs may be unsuitable;
- compared with local fruits, imported fruits include the transport and the storage (energy consumption), as well as greenhouse production versus open field cultivations: results are not always obvious, as demonstrated by Blanke and Burdick (2005), but FUs must overcome input differences, e.g. through combined measurement units (for example, energy per kg, t, km or day);
- in the case of volume-*based* FUs, water content is another critical discriminant; for example, the yield in terms of juice varies from one fruit species to another; likewise, the water content influences the nutrients per volume unit.

These preliminary reflections suggest that, when comparing different fruit species, farmers could find more interesting surface-*based*, energy-*based* FUs or, better still, combined measurement units. From the point of view of consumers, volumes of processed or consumed products or RDI (recommended dietary intake), serving could be suitable FUs; public or local actors could find surface-, energy- or economic value-*based* FUs more useful.

An interesting reflection by Macombe and Loeillet (2013) concerns the possibility of extrapolating information by using FUs, as the relations between effects and the quantities represented by FU are usually considered proportional in environmental LCA. This could, for example, allow producers who deal with thousands of tons of a product to infer information from a study whose results are expressed in FUs with different measure units. However, in nature the relation between effects and quantities of FUs is best represented by a sigmoid than by a linear function, and proportionality appears only in a certain range of values: further researches should clearly identify the proportional relations between FUs and impacts (Macombe and Loeillet 2013).

In conclusion, selection of an FU is neither easy nor obvious, and there is no definitive rule. However, it may be sensible, albeit time-consuming, to apply more than a single FU to the same systems under study (Cerutti et al. 2013).

6.4.3 Quality of Data

With regard to the connection of data to a specific geographical area, as suggested by Milá i Canals and Clemente Polo (2003) in the context of fruit-growing systems, it is often useful to collect detailed site-dependent data for field operations, because farmers' practices may vary. This is linked to the need for primary data directly provided by the farmers, also by identifying appropriate procedures of data collection (e.g. direct interviews, standardised questionnaires, etc.). Furthermore, participative activities can be useful for making input data more reasonable, through the approval of stakeholders involved (Ingwersen 2012).

The time coverage of data collection should permit to overtake the problem of parameter variability (e.g. farmers' practices, climatic conditions, and soil characteristics) that some authors solve by using pluri-annual datasets (see Sect. 6.4.1) of the inflows and outflows of the considered system.

To verify the dependence of results on quality data, systematic procedures can be used as suggested by the ISO norms (2006a, 2006b). Mila i Canals et al. (2006) apply a gravity analysis, exploring the inventory data to estimate the error margin for each of the LCA results. Beccali et al. (2010) implement a sensitivity analysis (ISO 2006b) in order to judge the validity of collected data and to determine how changes in data and methodological choices affect results.

6.4.4 Accounting of Carbon Storage

As mentioned in Sect. 6.3.7 the accounting of carbon storage in soil-plant ecosystems is rarely considered in LCA studies of fruit products because of the high variability of the phenomenon in relation to pedoclimatic conditions, crop species behaviour, and field management practices (external input, human intervention). Carbon storage accounting is more relevant for perennial fruit crops than for annual ones such as strawberries.

The impact category for carbon storage is the "avoided" global warming potential, i.e. kg CO_2 equivalent removed from the atmosphere because it is fixed in tree organs through photosynthesis and in soil by incorporation of organic material such as senescent leaves, pruning residues, thinned fruits, compost, manure, and cover crops, which partly decompose and partly remain in the soil, increasing its carbon content since it reaches a new equilibrium. The steady state is reached faster in warm regions than in cold ones.

As regards carbon stored in tree biomass, ISO 14067:2013 indicates that the uptake of CO2 in the biomass and the equivalent amount of CO2 emissions from the biomass at the point of complete oxidation result in zero net CO2 emissions, as when trees are burned at the end of life of the orchard. Although, according to PAS 2050 and ILCD handbook methodology, it is possible to account for "temporary storage" of carbon in tree biomass, it means calculating credit for delayed CO2 emissions arising at the end of life of the orchard because of tree biomass disposal; the factor for calculate the emission credit depends on the period of temporary carbon storage (fewer than 25 years, fewer or more than 100 years).

For the estimation of changes in soil carbon stock, the most widely accepted methodologies available at present are:

- categorical estimates based on Eq. 2.25 of vol. 4 (AFOLU) of IPPC 2006 guidelines (tier 1 methodology). First, a default reference soil organic C stock (expressed in tonnes of carbon per hectare in the first 30 cm of soil depth) has to be determined by selecting the particular climate for regions and the soil category from respectively the nine and six alternatives given in Table 2.3 of AFOLU 2006 IPPC guidelines; this value is multiplied by three stock change factors from Table 5.5 of the guidelines, based on temperature and moisture regime, on the category of land use, grade of tillage, and type of crop residue management and manure input, in order to obtain the soil organic carbon content before and after change;

- simulation by a model in three steps (Knudsen 2010):

1. Determination of the baseline soil carbon content before orchard establishment through available soil analysis or the most recent literature data.
2. Estimate of change in soil carbon content equilibrium through one of the available models for simulation of the decay of C added to soil (Roth C, ICBM, Daisy, Century, DNDC, CANDY): typical data needed to run these models are amount of organic material per hectare annually added to soil (manure, compost, pruning residues, senescent leaves, thinned fruits, dead roots, inter-row soil cover), climate data (e.g. rainfalls, average temperatures, wind) soil characteristic data (e.g. texture, moisture, evaporation), agricultural practice data (e.g. fertilisation, irrigation, tillage, soil cover control)
3. Calculation of CO_2 removed from the atmosphere as the amount of carbon stored in soil after 20 or 100 years, obtained from the simulation described above, multiplied per 44/12 which is the ratio of the molecular weight of CO_2 to that of carbon.

Available measurements of soil organic carbon content in different stages of orchard life cycle can be used to control data obtained by simulation. Running the aforementioned simulation models is often time-consuming because of input data collection and management and requires specific knowledge in the field of agro-ecosystems dynamics.

ISO 14067 suggests calculating biogenic carbon emissions as well, which in this case are the CO_2 emissions arising from decomposition of organic carbon in soil, and to declare it separately. It is possible to estimate these emissions by calculating the difference between carbon input in soil and carbon left in soil at the end of the considered period, multiplied again per 44/12. Part of the biogenic CO_2 emitted will be absorbed by oceans, according to the Bern Cycle Model (Knudsen 2010).

For carbon storage accounting it is important to know the history of the field: What was the land use before the establishment of the orchard? Has a land use change occurred? Was there a change in orchard management practice during its life cycle? This information is useful for understanding the "additionality" of carbon storage related to fruit production in comparison with the baseline scenario of "no fruit production"; the concept of additionality comes from projects generating carbon credits and can help the analyst to decide which carbon storage elements to include in the LCA study and which methodology to use.

6.4.5 Life Cycle Impact Assessment (LCIA)

Impact assessment assuredly represents a critical phase in LCA analyses, because the choice of assessment methods and which elements should be considered (compulsory or discretionary ones) strongly influence the quality of results (Cerutti et al.

2011). This is particularly true in the case of agricultural systems that are strictly connected to ecosystems, and therefore specific investigations are needed.

The literature review highlighted some heterogeneity among the analyses of fruit-growing systems, probably because of the intrinsic characteristics of each agricultural process. This heterogeneity increases the variability and uncertainty of results during the LCIA; this is one of the main reasons for the use of midpoint approaches and the consideration only of compulsory assessment phases (classification and characterisation). Methods with endpoint approaches are often applied to large-scale studies, e.g. territorial studies, especially comparative ones; this is probably because of the possibility of immediate results and their easy disseminate among a wider public.

It also emerged that the most examined elements in fruit-growing sectors are often linked to climate change, water and soil eutrophication, potential acidification, and energy consumption. However, in LCA studies of permanent orchards, characterised by a very long life cycle, two important aspects should be investigated in greater depth: land use and water consumption.

Many factors justify their importance: land use, for example, entails many different issues (linked to quantitative and qualitative aspects) that should be analysed at the same time.

In contrast, using midpoint methods, many researchers conduct their analyses merely from a quantitative point of view (occupied surfaces), overlooking the effects of growing systems from a qualitative perspective (Notarnicola et al. 2012).

The use of pesticides, herbicides, and other plant protection products modifies ecological equilibriums, reduces biocoenosis, and induces instability phenomena in agro-ecosystems, where anthropic interventions are required to maintain productive efficiency. In this context, land use analysis should be reinforced by specific qualitative field surveys via endpoint methods, in order to investigate, for ex-ample, effects in terms of human toxicity and ecotoxicity.

Another fundamental aspect is the application of dispersion models for pesticides and fertilisers to understand what happens when products are dispersed, especially in the soil, and to determine and account for emissions in the inventory phase. Thus, it is possible to highlight the effects of these emissions in terms of biodiversity reduction and understand them because of the accumulation of pesticides in ecosystems. These analyses are supported by specific models (e.g. FocusPearl and PestLCI), but at times are too complex to use because of the difficulty of gathering specific data; moreover, the uncertainty of results depends on the high specificity and heterogeneity of pedoclimatic conditions in orchards (Cerutti et al. 2012).

Another aspect which needs attention during LCIA concerns water consumption. Orchards are high water-demanding systems, and therefore frequent irrigations are required, especially in climatic environments characterised by lack of precipitation. According to Notarnicola (2012), there is a lack of data on water use assessment in commercial databases. The main reason for this is is the wide variety of water use typologies, which makes not only quantification in the inventory phase difficult but also the definition of a characterisation factor. ISO 14046 (in press) should furnish useful tools for overcoming this problem.

6.4.6 Interpretation

In most of the studies reviewed, the interpretation phase consists in the identification of key aspects highlighted by LCA analysis. It may be useful, if not essential in fruit production sectors, to explain in detail which and how specific elements of the inventory contribute to environmental impacts, which are the most relevant impact categories, or the life cycle phase with the greatest impact.

In this context, uncertainty analyses are of fundamental importance because they permit us to define the completeness and consistency of results. This is especially true in fruit LCA studies, because of the intrinsic variability of agricultural systems because of their biological nature and the great variety of agricultural practices linked to specific farmers' knowledge and culture.

In terms of conducting an uncertainty analysis, it is possible to follow a procedural approach based on comparison of inventory data and results of impact assessment phase with other sources of information, such as literature references, confrontation with experts, or referencing data supplier (Heijungs and Guinée 2012). Another way is to use a numerical approach, i.e. statistical analysis: this is the most rigorous and robust way to identify eventual errors during goal and scope definition, LCI and LCIA phases. In this context, it would be sensible to conduct additional analyses from a deterministic and probabilistic perspective (e.g. Monte Carlo analysis) to validate the results obtained (Heijungs and Guinee 2012).

Indeed, as fruit-growing systems are highly heterogeneous (thanks to biotic and abiotic factors), many of the papers reviewed referred to assumptions and specific elaborations (e.g. average values of growing cycles) during goal and scope definition and LCI phases, to reduce heterogeneity and degree of uncertainty.

It would be useful to conduct at least a sensitivity analysis and a detailed description of all limitations inherent in each case study. Such an approach should produce more valid results and improve the transparency of analysis.

Finally, from the papers reviewed, it emerged that LCA is widely recognised as a useful decision-making tool, and results are often addressed to decision-makers. According to Horne et al. (2009), the interpretation phase should not be undertaken lightly, because of the growing responsibility of LCA practitioners as stewards of insights and knowledge to be used for policies.

6.4.7 Comparative Assessments

As reported in Sect. 6.3.12, one of the most common applications of the LCA methodology in the fruit sector is to comparative assessments. In particular, the review highlighted that most of the studies were conducted with the aim of developing comparative evaluations of potential environmental impacts (and improvement opportunities) of different growing techniques or different fruit products under different conditions. Other comparative applications can be linked, for example, to products deriving from fruit processing (for example, fruit juice).

Drawing general guidelines for developing comparative assessments in the fruit sector is a hard task because the scope of these studies can be very variable and so

Table 6.8 LCA guidelines (system boundaries, functional unit, and reference period) for comparative assessments

Aim of comparison	System boundaries	Functional unit	Reference period
Different growing practices	From cradle to gate or from cradle to grave	Mass- based (kg or tons) or surface-based (ha)	Seasonal period or orchard lifetime
Different cultivars	From cradle to farm gate	Mass-based (kg or tons) or surface-based (ha) or economic value-based	Orchard lifetime
Products deriving from fruit processing	From cradle to gate or from cradle to grave	Mass-based (kg or tons)	Seasonal period or orchard lifetime

can the parameters to be taken into account. Nevertheless, depending on the aim of the study, some general recommendations, at least about system boundaries, functional unit, and reference period data can be made (Table 6.8).

Furthermore, an LCA application not considered in the review but of great interest could be comparative assessments of different kinds of fruits from a sustainability perspective. As stated in Sect. 1.6 on dietary issues, in Mediterranean-type diets, fruits and vegetables are at the edge (the lowest environmental impact) of the environment-food pyramid (taking into account CF, WF, and EF) and contribute 2% to GWP, EP, and PCOP, with average values (per kg) of CF, WF, and EF, respectively, of 70 gCO_2eq/kg, 600 l, 3 global m^2/kg. Several issues have to be taken into account in such comparative assessments and giving specific guidelines is not so easy. Nevertheless, as far as the FU definition is concerned it should be possible to take into account, apart from mass-based FUs, the qualitative aspects (e.g. nutritional or chemical aspects) of the products, as suggested for comparison of different types of wine or oil (Notarnicola et al. 2012).

6.5 Conclusions

The fruit industry is certainly a relevant activity in the European agricultural economy, creating wealth and employment along the production chain, with pome fruit, kiwifruit, table grape and peach as the main contributors. In general, fruit production is considered as an agricultural sector with low environmental impacts in comparison with the herbaceous crop sector and other food sectors; nevertheless it is necessary to take into account more specific issues.

6.5.1 General Conclusions

Orchards are among the most intensively sprayed agricultural systems, in order to avoid visible fruit damage and to satisfy international commercial quality standards. Moreover, water consumption is a relevant concern in fruit production systems and energy consumption by irrigation systems could be important. Among other inputs, the use of residual herbicides has been proven to be beneficial to tree growth and yield, but they leave the soil surface without a protective cover for much of the year, which can have a range of undesirable effects such as soil compaction and reduction in water holding and infiltration capacity. Fertilisation also plays a major role in the environmental impacts of such production systems, even though deciduous fruit trees have low N demands compared with open field crops and lower nutrient loss by leaching. Another process of the system that usually leads to environmental impacts is the soil management as the soil plays a major role in terms of the quality of the fruit produced. Furthermore, careful management of soil is important to prevent soil erosion and nutrient leaching, and sustain good soil fertility and structure for water infiltration.

In general, the main results of the review can be summarised as follows.

- Among the papers analysed, 31 of them describe applications to specific products, four are both applicative and comparative contributions, three are purely methodological and the others are applicative and methodological (2) and applicative, comparative, and methodological (1).
- As regards the typologies of product investigated, citrus and apples are the fruits that have captured the attention of researchers, and only two papers were on pineapple and strawberry and just one respectively on peaches, nectarine, blueberries, kiwifruit and walnut trees.
- Processed fruit products have also been investigated, particularly the production of citrus-based products.
- Sometimes LCA studies are developed with the aim of comparing different food products from different countries, taking into account the environmental impact in the countries where fruit is cultivated (Spain, New Zealand, South Africa, Brazil), and the countries where the fruit is sold (the UK, Denmark or Europe in general).
- Depending on the boundaries considered and the characteristics of the products assessed, half of the studies use a mass unit, such as the kilogram or the ton of product, and some studies use a litre, a serving of the product, or a surface unit such as the acre or the hectare. Furthermore, some studies base assessments on the purchasable item unit, such as a pack, a plant, or a number of seedlings.
- Among the studies reviewed, 25 papers either do not apply allocation criteria or do not explicitly specify if a method is used, whereas other papers use mass allocation (nine), combined mass and energy allocations (four) and flow allocations (two). A few papers use economic, energy, mass, or land-based allocation criteria.

6.5.2 Orchard Models and Data Availability

One of the most crucial aspects is that most of the LCA studies reviewed assess perennial systems in the same way as annual crops, t considering only one growing season within the time boundaries of the system. This gives an inadequate view of the quality of the orchard model and possibly miscalculates the real environmental impact potentials of the production system.

As a result of the assessment, the most precise orchard model should be constructed according to the six stages proposed by Mila i Canals (2003). In particular, the orchard should be subdivided into several stages of variable duration depending upon the species: nursery (2–4 years); establishing stage (one occurrence); stage of young trees (2–5 years); stage of adult trees(10–30 years); stage of old trees (0–3 years); destruction stage (one occurrence). It is fairer to consider the whole life cycle of the orchard within the time boundaries of the study and to allocate the environmental impact of the low-yield stage among the fruits leaving the system throughout all years.

The nursery phase plays an important role, representing a first and fundamental input to the plantation. In fact, nurseries need high levels of inputs (raw materials, energy, structures) and contribute significantly to environmental impacts. There is a lack of studies which take into account the nursery stage; indeed, within scientific literature on LCA in the fruit sector, only six articles (out of 41 papers) specifically consider the role of the nursery in their applications.

The integration of the product's end of life in LCA applications makes the evaluation of environmental impacts more complete, and provides useful suggestions for improving the sustainability of the entire production process. Among the studies reviewed, only four papers provide specifications on the product's end of life, one of them considering the disposal of the non-edible part and the separate waste collection of the municipal solid waste and its recovery as compost.

The proposed model, including nursery and end-of-life specification, can be defined as the optimum theoretical model. Although its application is important for achieving holistic results, it is necessary to take into account the availability of data. In fact, data gathering is a pivotal step in LCA as the availability and quality of data strongly influence the reliability and usefulness of assessment results. In 31 of the reviewed papers the authors use primary sources for data collection, mainly through questionnaires and interviews of farmers or organisations. To ensure relevance, reliability and accessibility of data, a data quality check by LCA practitioners is strongly advised (ISO 2002); however, only 16% of the authors had conducted some form of control on data, mostly in qualitative terms.

Furthermore, all authors conducted the LCIA phase, except for two who used LCI. In most cases (59% of papers analysed), LCIA included only the mandatory phases of classification and characterisation and midpoint methods were applied. Many studies analyse only the impact of GHG emissions through the methodologies suggested by PAS and ISO 14067 or calculate the GWP from IPCC emission factors. Only a few authors completed their LCIA with optional phases like normalisation and weighting, which are subjective but fundamental to comparative studies.

In terms of the impact categories considered in the papers reviewed, that most analysed is GWP 100a, followed by acidification potential (AP), EP, ozone layer depletion (ODP) and photochemical oxidation (POCP). Despite their importance in evaluation of the environmental impacts of fruit production systems, only four studies treated the categories of land use, land use changes, and water consumption.

6.5.3 Efficiency and Sustainability

Several different practices aim to maximise efficiency of fruit load, increase fruit size, guarantee homogeneous colour, and prevent biennial bearing. Some authors (Granatstein and Kupferman 2006) suggest that the most environmentally friendly orchard harvest can be achieved with a manual management regime, in which all agronomic procedures are aimed at restricting plant height in order to avoid use of machinery. This has resulted in wider discussion about the difference between environmental efficiency and sustainability, and perennial plantations are a clear example of that (Cerutti et al. 2014). In fact, orchard systems with lower external inputs (such as machinery use and fertilisers) actually emit less but generally produce less fruit because of the physiology of plants. Therefore, considering exclusively the environmental impacts per unit mass of product it is possible to evaluate the eco-efficiency of the production system but not its sustainability, because efficiency does not necessarily lead to sustainability (Wackernagel and Rees 1997; van der Werf et al. 2007). As a consequence, use of a mass-based FU high input/high output system will always result in better environmental performance even if the actual impact per orchard is greater than low input/low output systems.

As this issue is still in debate, the most reasonable way to address it is to consider multiple FUs at the same time, particularly a mass-based and a land-bases FU (see Sect. 6.4.2). The use of such a functional unit reflects the perspective addressed by the particular study: the former is used in product-orientated expression of the agricultural production and the latter in land-orientated expression (Hayashi 2013).

6.5.4 Further Studies

Further studies should focus on:

1. including the multi-functionality of orchard systems in the environmental assessment, particularly in relation to the fact that orchards might have several functions other than fruit production, such as preserving genetic heritage (Donno et al. 2012) and traditional landscapes (Biasi et al. 2010). Furthermore, trees are often grown in association with other horticultural crops, especially in tropical areas, and the use of specific allocation methods or system expansion approaches should be discussed and validated;
2. modelling the role of orchards as sinks for CO_2 sequestration. Indeed, orchards, if properly managed, can have great potential for the absorption and net storage of CO_2 (Nardino et al. 2013; Palese et al. 2013) and could significantly affect results in the GWP category (Bosco et al. 2013). A discussion about how to account for carbon sequestration and temporary storage in an LCA was recently presented by Brandão et al. (2013), but specific models for orchard systems were not included;

3. consolidating results from harmonisation initiatives. As highlighted in Sect. 6.2.1, different initiatives have suggested alternative settings for LCA applications in fruit production systems. In particular, it might be interesting to have case studies in which results are validated and compared as per EPD® recommendations, either from the EnviFood protocol or other references.

References

ADEME. (2010). La méthode Bilan Carbone®. Agence de L'Environment et de la Maitrise de l'Energie. www.ademe.fr. Accessed 28 Dec 2013.

Alaphilippe, A., Simon, S., Brun, L., Hayer, F., & Gaillard, G. (2013). Life cycle analysis reveals higher agroecological benefits of organic and low-input apple production. *Agronomy for Sustainable Development, 33,* 581–592.

Amiri, M. E., & Fallahi, E. (2009). Impact of animal manure on soil chemistry, mineral nutrients, yield, and fruit quality in 'Golden Delicious' apple. *Journal of plant nutrition, 32*(4), 610–617.

Ardente, F., Beccali, G., Cellura, M., & Marvuglia, A. (2006). POEMS: A case study of an Italian wine-producing firm. *Environmental Management, 38,* 350–364.

Arzoumanidis, I., Petti, L., Raggi, A., & Zamagni, A. (2013). The Implementation of simplified LCA in agri-food SMEs. In R. Salomone, et al. (Eds.), *Product-oriented environmental management systems (POEMS)*. Ó Springer Science+Business Media Dordrecht. doi:10.1007/978–94–007–6116–2_7.

Beccali, M., Cellura, M., Iudicello, M., & Mistretta, M. (2009). Resource consumption and environmental impacts of the agrofood sector: Life cycle assessment of Italian citrus-based products. *Environmental Management, 43,* 707–724.

Beccali, M., Cellura, M., Iudicello, M., & Mistretta, M. (2010). Life cycle assessment of Italian citrus-based products. Sensitivity analysis and improvement scenarios. *Journal of Environmental Management, 91,* 1415–1428.

Beccaro, G. L., Cerutti, A. K., Vandecasteele, I., Bonvegna, L., Donno, D., & Bounous, G. (2014). Assessing environmental impacts of nursery production: methodological issues and results from a case study in Italy. Journal of Cleaner Production.

Berners-Lee, M., Hoolohan, C., Cammack, H., & Hewitt, C. N. (2012). The relative greenhouse gas impacts of realistic dietary choices. *Energy Policy, 43,* 184–190.

Bessou, C., Basset-Mens, C., Tran, T., & Benoist, A. (2013). LCA applied to perennial cropping systems: A review focused on the farm stage. *International Journal of Life Cycle Assessment, 18,* 340–361.

Biasi, R., Botti, F., Barbera, G., & Cullotta, S. (2010). The role of Mediterranean fruit tree orchards and vineyards in maintaining the traditional agricultural landscape. *ISHS Acta Horticuture, 940,* 79–88.

Blanke, M. M., & Burdick, B. (2005). Food (miles) for thought—Energy balance for locally-grown versus imported apple fruit. *International Environmental Science and Pollution Research, 12,* 125–127.

Blanke, M. M., & Burdick, B. (2009). *An energy balance (as part of an LCA) for home-grown (apple) fruit versus those imported from South Africa or New Zealand.* Joint North American LCA Conference, Boston.

Blanke, M. M. (2013). Carbon footprint of different thinning strategies in a fruit orchard. *Acta Horticulture, 998,* 125–128.

Bosco, S., Di Bene, C., Galli, M., Remorini, D., Massai, R., & Bonari, E. (2013). Soil organic matter accounting in the carbon footprint analysis of the wine chain. *International Journal of Life Cycle Assessment, 18,* 973–989.

Brandão, M., Levasseur, A., Kirschbaum, M. U. F., Weidema, B. P., Cowie, A. L., Jørgensen, S. V., Hauschild, M. Z., Pennington, D. W., & Chomkhamsri, K. (2013). Key issues and options in accounting for carbon sequestration and temporary storage in life cycle assessment and carbon footprinting. *International Journal of Life Cycle Assessment, 18,* 230–240.

BSI. (2011). *PAS 2050 specification for the assessment of the life cycle greenhouse gas emissions of goods and services*. London: British Standards Ed.

Carlsson-Kanyama, A., Ekström, M. P., & Shanahan, H. (2003). Food and life cycle energy inputs: consequences of diet and ways to increase efficiency. *Ecological economics, 44*(2), 293–307.

Cambria, D., & Pierangeli, D. (2011). A life cycle assessment case study for walnut tree (Juglans regia L.) seedlings production. *International Journal of Life Cycle Assessment, 16,* 859–868.

Cellura, M., Ardente, F., & Longo, S. (2012a). From the LCA of food products to the environmental assessment of protected crops districts: A case-study in the south of Italy. *Journal of Environmental Management, 93,* 194–208.

Cellura, M., Longo, S., & Mistretta, M. (2012b). Life Cycle Assessment (LCA) of protected crops: An Italian case study. *Journal of Cleaner Production, 28,* 56–62.

Cerutti, A. K., Bagliani, M., Beccaro, G. L., Peano, C., & Bounous, G. (2010). *Comparison of LCA and EFA for the environmental account of fruit production systems: A case study in Northern Italy*. Proceedings of LCA food 2010, 99–104, Bari, vol. II.

Cerutti, A. K., Bruun, S., Beccaro, G. L., & Bounous G. (2011a). A review of studies applying environmental impact assessment methods on fruit production systems. *Journal of Environmental Management, 92,* 2277–2286.

Cerutti, A. K., Galizia, D., Bruun, S., Mellano, G. M., Beccaro, G. L., & Bounous, G. (2011b). Assessing environmental sustainability of different apple supply chains in northern Italy. In M. Finkbeiner (Ed.), *Towards life cycle sustainability management*. Springer Science+Business Media B.V. doi:10.1007/978–94–007–1899–9_33.

Cerutti, A. K., Bagliani, M., Beccaro, G. L., Gioelli, F., Balsari, P., & Bounous, G. (2011c). Evaluation of the sustainability of swine manure fertilization in orchard through Ecological Footprint Analysis: results from a case study in Italy. *Journal of Cleaner Production, 19*(4), 318–324.

Cerutti, A. K., Beccaro, G. L., Bosco, S., Peano, C., Notarnicola, B., & Bounous, G. (2012). *State of the art of LCA application in the fruit sector*. Proceedings of 8th International Conference on LCA in the Agri-Food Sector, 403–408. St Malò.

Cerutti, A. K., Bruun, S., Donno, D., Beccaro, G. L., & Bounous G. (2013). Environmental sustainability of traditional foods: The case of ancient apple cultivars in Northern Italy. *Journal of Cleaner Production, 52,* 245–252.

Cerutti, A. K., Calvo, A., & Bruun, S. (2014). Comparison of the environmental performance of light mechanization and animal traction using a modular LCA approach. *Journal of Cleaner Production, 64,* 396–403.

Clasadonte, M. T., Matarazzo, A., & Ingrao, C. (2010a). *Life cycle assessment of Sicilian peach sector*. Proceedings of LCA food 2010, 295–300, Bari.

Clasadonte, M. T., Lo Giudice, A., & Ingrao, C. (2010b). *Life cycle assessment of the Sicilian citrus fruit field*. Proceedings of LCA food 2010, 301–306, Bari.

Coltro, L., Mourad, A. L., Germer, S. P. M., Mendonça, T. A., & Kletecke, R. M. (2008). *Energy use in the life cycle of frozen concentrated orange juice produced in Brazil*. Proceedings of the 6th international conference on life cycle assessment in the agri-food Sector, 228–233. Zurich.

Coltro, L., Mourad, A. L., Kletecke, R. M., Mendonça, T. A., & Germer, S. P. M. (2009). Assessing the environmental profile of orange production in Brazil. *International Journal of Life Cycle Assessment, 14,* 656–664.

Cowell, S. J. (1998). *Environmental life cycle assessment of agricultural systems: Integration into decision-making, centre for environmental strategy*. Guildford: University of Surrey.

Cowell, S. J., & Clift, R. (1997). Impact assessment for LCAs involving agricultural production. *International Journal of Life Cycle Assessment, 2,* 99–103.

Cuadra, M., & Björklund, J. (2007). Assessment of economic and ecological carrying capacity of agricultural crops in Nicaragua. *Ecological indicators, 7*(1), 133–149.

De Camillis, C., Bligny, J. C., Pennington, D., & Pályi, B. (2012). Outcomes of the second workshop of the food sustainable consumption and production round table working group 1: Deriving scientifically sound rules for a sector-specific environmental assessment methodology. *International Journal of Life Cycle Assessment, 17,* 511–515.

Donno, D., Beccaro, G. L., Mellano, M. G., Torello Marinoni, D., Cerutti, A. K., Canterino, S., & Bounous, G. (2012). Application of sensory, nutraceutical and genetic techniques to create a quality profile of ancient apple cultivars. *Journal of Food Quality, 35,* 169–181.

Dwivedi, P., Spreen, T., & Goodrich-Schneider, R. (2012). Global warming impact of Florida's Not-From-Concentrate (NFC) orange juice. *Agricultural Systems, 108,* 104–111.

European Commission. (2006). Documentation and quality requirements of the European platform on life cycle assessment for LCI data sets of research projects. v. 1.0.1. European Platform on LCA, Dec 22.

European Commission, Joint Research Centre, Institute for Environment and Sustainability. (2010). International Reference Life Cycle Data System (ILCD) handbook—general guide for life cycle assessment—detailed guidance. First edition March 2010. EUR 24708 EN. Luxembourg. Publications Office of the European Union. http://lct.jrc.ec.europa.eu/pdf-directory/ILCD-Handbook-General-guide-for-LCA-DETAIL-online–12March2010.pdf . Accessed on 28 Dec 2013.

European Commission, Joint Research Centre, Institute for Environment and Sustainability. (2011). "Analysis of existing environmental footprint methodologies for products and organizations: Recommendations, rationale, and alignment". http://ec.europa.eu/environment/eussd/corporate_footprint.htm. Accessed 28 Dec 2013.

Fakouri, N. B., Smejtek, L., & Denis, A. (2008). *Life cycle analysis and propositions of environmental-friendly alternatives for 3 different Swedish horticultural productions*. LCA project paper for the course: Environmental Issues in Crop production, Swedish University of Agricultural Sciences.

FAO. (2009). *CROPWAT Version 8.0. Water resources development and management service*. Rome, FAO.

FAOSTAT (2013). World Food and Agriculture Database. http://faostat.fao.org. Accessed Dec 2013.

Finkbeiner, M. (2009). Carbon footprinting-opportunities and threats. *International Journal of Life Cycle Assessment, 14,* 91–94.

Food SCP RT. (2013). *ENVIFOOD protocol, Environmental Assessment of Food and Drink Protocol, European Food Sustainable Consumption and Production Round Table (SCP RT), Working Group 1*. Brussels, Belgium.

Forge, T. A., Hogue, E., Neilsen, G., & Neilsen, D. (2003). Effects of organic mulches on soil microfauna in the root zone of apple: implications for nutrient fluxes and functional diversity of the soil food web. *Applied Soil Ecology, 22*(1), 39–54.

Frey, S., & Barrett, J. (2007). Our health, our environment: the ecological footprint of what we eat. In International Ecological Footprint Conference, *Cardiff* (pp. 8–10).

Garnett, T. (2006). *Fruit and vegetables & UK greenhouse gas emissions: exploring the relationship*. UK: Food and Climate Research Network, University of Surrey.

Gerbens-Leenes, P. W., Nonhebel, S., & Ivens, W. P. M. F. (2002). A method to determine land requirements relating to food consumption patterns. *Agriculture, ecosystems & environment, 90*(1), 47–58.

Girgenti, V., Peano, C., Bounous, M., & Baudino, C. (2013). A life cycle assessment of non-renewable energy use and greenhouse gas emissions associated with blueberry and raspberry production in northern Italy. *Science of the Total Environment, 458–460,* 414–418.

Gliessman, S. R. (2007). Agroecology: the ecology of sustainable food systems. CRC Press.

Glover, J. D., Reganold, J. P., & Andrews, P. K. (2000). Systematic method for rating soil quality of conventional, organic, and integrated apple orchards in Washington State. *Agriculture, ecosystems & environment, 80*(1), 29–45.

Granatstein, D., & Kupferman, E. (2006). Sustainable horticulture in fruit production. *In XXVII International horticultural congress-IHC2006: International symposium on sustainability through integrated and organic 767* (pp. 295–308).

Guinée, J. B. (2002). (Ed.). *Handbook on life cycle assessment—Operational guide to the ISO standards*. Kluwer Academic Publishers.

Haas, G., Wetterich F., & Geier U. (2000). Framework in agriculture on the farm level. *International Journal of Life Cycle Assessment, 5,* 345–348.

Hayashi, K. (2013). Practical recommendations for supporting agricultural decisions through life cycle assessment based on two alternative views of crop production: The example of organic conversion. *International Journal of Life Cycle Assessment, 18,* 331–339.

Heijungs, R., & Guinee, J B. (2012). An overview of the life cycle assessment method—Past, present, and future. In M. A. Curran (Ed.), *Life cycle assessment handbook: A guide for environmentally sustainable products* (pp. 15–42). Scrivener Publishing LLC & John Wiley & Sons. ISBN 978–1–118–09972–8.

Helmer, E. H., & Brown, S. (2000). Gradient analysis of biomass in Costa Rica and a first estimate of countrywide emissions of greenhouse gases from biomass burning. In C. A. S. Hall, C. L. Perez, & G. Leclerc (Eds.), *Quantifying sustainable development the future of tropical economies*. San Diego: Academic Press.

Hischier, R., Weidema, B., Althaus, H.-J., Bauer, C., Doka, G., Dones, R., Frischknecht, R., Hellweg, S., Humbert, S., Jungbluth, N., Köllner, T., Loerinick, Y., Margni, M., & Nemecek, T. (2009). *Implementation of life cycle impact assessment methods, datav2.1. Swiss centre for life cycle inventories.* Dübendorf: Ecoinvent Report.

Holb, I. J. (2009). Fungal disease management in environmentally friendly apple production–a review. In Climate Change, Intercropping, Pest Control and Beneficial Microorganisms (pp. 219–292). Springer Netherlands.

Holdridge, L. R. (1967). *Life zone ecology*. San Jose: Tropical Science Center.

Horne, R. E., Grant, T., & Verghese, K. L. (2009). Prospects for life cycle assessment development and practice in the quest for sustainable consumption. In R. E. Horne, T. Grant, & K. L. Verghese (Eds.), *Life cycle assessment. Principles, practice and prospects*. CSIRO Publishing.

Ingwersen, W. (2010). *Product category range of environmental performance for EPDs: Example of Costa Rican pineapple*. Proceedings of LCA food 2010. 337–341.Bari.

Ingwersen, W. (2012). Life cycle assessment of fresh pineapple from Costa Rica. *Journal of Cleaner Production, 35,* 152–163.

IPCC. (2006). *IPCC guidelines for national greenhouse gas inventories*. Volume 4: Agriculture, forestry and other land use.

ISO. (2002). *ISO Norm 14048:2002 Environmental management—Life cycle assessment—Data documentation format.*

ISO. (2006a). *ISO Norm 14040:2006 Life cycle assessment—Principles and framework.*

ISO. (2006b). *ISO Norm 14044:2006 Life cycle assessment—Requirements and guidelines.*

ISO. (2012). *ISO Norm 14067:2012 Carbon footprint of products-requirements and guidelines for quantification and communication.*

ISO. (2013). *ISO/DIS 14046 Environmental management—Water footprint—Principles, requirements and guidelines.*

Janssens, I. A., Freibauer, A., Ciais, P., Smith, P., Nabuurs, G. J., Folberth, G., Schlamadinger, B., Hutjes, R. W. A., Ceulemans, R., Schulze, E. D., Valentini, R., & Dolman, A. J. (2003). Europe's terrestrial bio-sphere absorbs 7 to 12% of European anthropogenic CO2 emissions. *Science, 300,* 1538–1542.

Khoshnevisan, B., Rafiee, S., & Mousazadeh, H. (2013). Environmental impact assessment of open field and greenhouse strawberry production. *European Journal of Agronomy, 50,* 29–37.

Knudsen, M. T. (2010). *A methodological approach to include soil carbon change into life cycle assessment of agricultural products*. Paper included in PhD dissertation. Aarhus University.

Knudsen, M. T., Fonseca de Almeida, G., Langer, V., Santiago de Abreu, L., & Halberg, N. (2011). Environmental assessment of organic juice imported to Denmark: A case study on oranges (Citrus sinensis) from Brazil. *Organic Agriculture, 1,* 167–185.

Lal, R. (2009). Soils and sustainable agriculture: A review. In Sustainable agriculture (pp. 15–23). Springer Netherlands.

Liu, Y., Langer, V., Høgh-Jensen, H., & Egelyng, H. (2010). Life cycle assessment of fossil energy use and greenhouse gas emissions in Chinese pear production. *Journal of Cleaner Production, 18*(14), 1423–1430.

Lo Giudice, A., & Mbohwa, C. (2012). *Life Cycle Inventory analysis of citrus fruit in Sicily*. IADIS International Conference on "Sustainability, Technology and Education 2012", pp. 101–108, Perth, Australia.

Lo Giudice, A., Mbohwa, C., Clasadonte, M. T., & Ingrao, C. (2013). Environmental assessment of the citrus fruit production in Sicily using LCA. *Italian Journal of Food Science, XXV,* 202–212.

Luo, L., van der Voet, E., Huppes, G., & Udo de Haes, H. A. (2009). Allocation issues in LCA methodology: A case study of corn stover-based fuel ethanol. *International Journal of Life Cycle Assessment, 14,* 529–539.

Macombe, C., & Loeillet, D. (2013). Social life cycle assessment, for who and why? In C. Macombe (Ed.), *Social LCAs. Socio-economic effects in value chains*. Cirad: FruiTrop Thema.

McDonald, S. J. (2007). Alternative orchard floor management practices, soil quality, nitrogen fertilizer uptake, and apple tree growth and yield. PhD Thesis. Oregon State University. p. 230.

McLaren, S. J., Hume, A., & Mitraratne, N. (2010). *Carbon management for the primary agricultural sector in New Zealand: Case studies for the pipfruit and kiwifruit industries*. Proceedings of LCA food, pp. 293–298.

Merwin, I. A., Rosenberger, D. A., Engle, C. A., Rist, D. L., & Fargione, M. (1995). Comparing mulches, herbicides, and cultivation as orchard groundcover management systems. *HortTechnology, 5*(2), 151–158.

Milá i Canals, L. (2003). Contributions to LCA methodology for agricultural systems, Site-dependency and soil degradation impact assessment. Ph. D. Thesis. Universitat Auto'noma de Barcelona, Spain. http://www.tdx.cat/handle/10803/3155. Accessed May 2010.

Milá i Canals, L., & Clemente Polo G. (2003). Life cycle assessment of fruit production. In B. Mattsson, U. Sonesson, (Eds.), *Environmentally-friendly food processing* (pp. 29–53). Woodhead Publishing Limited and CRC LLC: Cambridge.

Milá i Canals, L., Burnip, G. M., & Cowell, S. J. (2006). Evaluation of the environmental impacts of apple production using life cycle assessment (LCA): Case study in New Zealand. *Agriculture, Ecosystems & Environment, 114,* 226–238.

Milá i Canals, L., Cowell, S. J., Sim, S., & Basson, L. (2007). Comparing domestic versus imported apples: A focus on energy use. *International Environmental Science and Pollution Research, 14,* 338–344.

Mouron, P., Nemecek, T., Scholz, R. W., & Weber, O. (2006a). Management influence on environmental impacts in an apple production system on Swiss fruit farms: Combining life cycle assessment with statistical risk assessment. *Agriculture, Ecosystems & Environment, 114,* 311–322.

Mouron, P., Scholz, R. W., Nemecek T., & Weber O. (2006b). Life cycle management on Swiss fruit farms: Relating environmental and income indicators for apple-growing. *Ecological Economics, 58,* 561–578.

Mourad, A. L., Coltro, L., Oliveira, P. A. P. L. V., Kletecke, R. M., & Baddini, J. P. O. A. (2007). A simple methodology for elaborating the life cycle inventory of agricultural pro-ducts. *International Journal of Life Cycle Assessment, 12,* 408–413.

Mouron, P., Heijne, B., Naef, A., Strassemeyer, J., Hayer, F., Avilla, J., Alaphilippe, A., Höhn, H., Hernandez, J., Mack, G., Gaillard, G., Solé, J., Sauphanor, B., Patocchi, A., Samietz, J., Bravin, E., Lavigne, C., Bohanec, M., Golla, B., Scheer, C., Aubert, U., & Bigler, F. (2012). Sustainability assessment of crop protection systems: SustainOS methodology and its application for apple orchards. *Agricultural Systems, 113,* 1–15.

Nardino, M., Pernice, F., Rossi, F., Georgiadis, T., Facini, O., Motisi, A., & Drago, A. (2013). Annual and monthly carbon balance in an intensively managed mediterranean olive orchard. *Photosynthetica, 51*(1), 63–74.

Nesme, T., Bellon, S., Lescourret, F., & Habib, R. (2006). Survey-based analysis of irrigation and N fertilisation practices in apple orchards. *Agronomy for sustainable development, 26*(3), 215–225.
Nicese, F. P., & Lazzerini, G. (2013). CO2 sources and sink in ornamental plant nurseries. *Acta Horticulture, 990,* 91–98.
Notarnicola, B., Tassielli, G., & Renzulli P. A. (2012). Modelling the Agri-Food Industry with Life Cycle Assessment. In M. A. Curran (Ed.), *Life cycle assessment handbook: A guide for environmentally sustainable products* (pp. 159–184). Canada: Scrivener Publishing LLC & (Wiley).
Notarnicola, B., Tassielli, G., Renzulli, P. A. (2013). La variabilità dei dati nella LCA della produzione olivicola. In: Proceedings of the VII Congress of the Italian LCA Network, ISBN 978-88-8286-292-3, pp. 29–35.
Page G. (2009). An environmentally-based systems approach to sustainability analyses of organic fruit production systems in New Zealand. PhD Thesis. Massey University, Palmerston North, New Zealand. p. 223.
Palese, A. M., Pasquale, V., Celano, G., Figliuolo, G., Masi, S., & Xiloyannis, C. (2009). Irrigation of olive groves in Southern Italy with treated municipal wastewater: Effects on microbiological quality of soil and fruits. *Agriculture, ecosystems & environment, 129*(1), 43–51.
Palese, A. M., Pergola, M., Favia, M., Xiloyannis, C., & Celano, G. (2013). A sustainable model for the management of olive orchards located in semi-arid marginal areas: Some remarks and indications for policy makers. *Environmental Science & Policy, 27,* 81–90.
Peacock, N., De Camillis, C., Pennington, D., Aichinger, H., Parenti, A., Rennaud, J. P., Raggi, A., Brentrup, F., Sara, B., Schenker, U., Unger, N., & Ziegler, F. (2011). Towards a harmonised frame-work methodology for the environmental assessment of food and drink products. *International Journal of Life Cycle Assessment, 16,* 189–197.
Pergola, M., Favia, M., Perretti, B., Martemucci, S., Palese, A. M., & Celano, G. (2011). SostenibilitÃ dell'uva da tavola in convenzionale e in bio. *L'Informatore Agrario, 39,* 62–66.
Pergola M., D'Amico M., Celano G., Palese A. M., Scuderi A., Di Vita G., Pappalardo G., & Inglese P. (2013). Sustainability evaluation of Sicily's lemon and orange production: An energy, economic and environmental analysis. *Journal of Environmental Management, 128,* 674–682.
Petti, L., Ardente, F., Bosco, S., De Camillis, C., Masotti, P., Pattara, C., Raggi, A., & Tassielli, G. (2010). *Stato dell'arte della iife cycle assessment (LCA) nel comparto vitivinicolo.* In F. Cappellaro & S. Scalbi (Eds.), Proceeding of the IV Congress of the Italian Network of LCA, pp. 221–228.
Pfister, S., Koehler, A., & Hellweg, S. (2009). Assessing the environmental impacts of freshwater consumption in LCA. *Environmental Science & Technology, 43,* 4098–4104.
Pfister, S., Bayer, P., Koehler, A., & Hellweg, S. (2011). Environmental impacts of water use in global crop production: Hotspots and trade-offs with land use. *Environmental Science and Technology, 45,* 5761–5768.
Pirilli, M., Falcone, G., & Strano, A. (2012). *La formazione dell'inventario per l'LCA nei confronti fra processi produttivi agricoli nella clementinicoltura in Calabria.* Proceedings of the VI Congress of the Italian LCA Network, pp. 104–111.
Reap, J., Roman, F., Duncan, S., & Bras, B. (2008). A survey of unresolved problems in life cycle assessment. *The International Journal of Life Cycle Assessment, 13*(5), 374–388.
Reganold, J. P., Glover, J. D., Andrews, P. K., & Hinman, H. R. (2001). Sustainability of three apple production systems. *Nature, 410*(6831), 926–930.
Ribal, J., Sanjuan, N., Clemente, G., & Loreto Fenollosa, M. (2009). Medicion de la ecoeficienza en procesos productivols en el sector agrario. Caso de estudio sobre production de citricos. *Economia Agraria y Recursos Naturales, 9,* 125–148.
Ridoutt, B. G., & Pfister, S. (2010). A revised approach to water footprinting to make transparent the impacts of consumption and production on global freshwater scarcity. *Global Environmental Change, 20,* 113–120.
Russo, G., & Scarascia Mugnozza, G. (2005). LCA methodology applied to various typologies of greenhouses. *Acta Horticulturae, 691,* 837–843.

Sanchez, E. E., Khemira, H., Sugar, D., & Righetti, T. L. (1995). Nitrogen management in orchards. *Nitrogen fertilization in the environment*, 327–380.

Sanjuan, N., Ubeda, L., Clemente, G., Mulet, A., & Girona, F. (2005). LCA of integrated orange production in the Comunidad Valenciana (Spain). *International Journal of Agricultural Resources, Governance and Ecology, 4,* 163–177.

Sansavini, S., Costa, G., Gucci, R., Inglese, P., Ramina, A., Xiloyannis, C. (Eds.), (2012). Arboricoltura generale. Pàtron Editore Bologna, Italy.

Schau, E. M., & Fet, A. M. (2008). LCA studies of food products as background for environmental product declarations. *International Journal of Life Cycle Assessment, 13,* 255–264.

Simon, S., Bouvier, J. C., Debras, J. F., & Sauphanor, B. (2010). Biodiversity and pest management in orchard systems. A review. *Agronomy for sustainable development, 30*(1), 139–152.

Smedman, A., Lindmark-Månsson, H., Drewnowski, A., & Edman, A. K. M. (2010). Nutrient density of beverages in relation to climate impact. *Food & Nutrition Research, 54,* 1–8.

Soler-Rovira, J., & Soler-Rovira, P. (2008). *Assessment of aggregated indicators of sustainability using PCA: The case on apple trade in Spain*. Proceedings of the 6th International Conference on LCA in the Agri-Food Sector, Zurich, November 12–14, pp. 133–143.

Sonnemann, G., Vigon, B., Broadbent, C., Curran, M. A., Finkbeiner, M., Frischknecht, R., Inaba, A., Schanssema, A., Stevenson, M., Ugaya, C. M. L., Wang, H., Wolf, M-A., & Valdivia, S. (2011). Process on "global guidance for LCA databases". *International Journal Life Cycle Assessment, 16,* 95–97.

Spreen, T., Dwivedi, P., & Goodrich-Schneider, R. (2010). *Estimating the Carbon Footprint of Florida Orange Juice*. Proceedings of: The 4th International European Forum on System Dynamics and Innovation in Food Networks, pp. 95–101.

Stoessel, F., Juraske, R., Pfister, S., & Hellweg, S. (2012). Life cycle inventory and carbon and water foodprint of fruits and vegetables: Application to a Swiss retailer. *Environmental Science & Technology, 46,* 3253–3262.

Strano, A., De Luca, A. I., Falcone, G., & Gulisano, G. (2013, September). *Economic and environmental sustainability assessment of Clementine crops scenarios in Calabria*. Paper presented at the Annual SIDEA Symposium 2013, Lecce—Italy.

Suckling, D. M., Walker, J. T. S., & Wearing, C. H. (1999). Ecological impact of three pest management systems in New Zealand apple orchards. *Agriculture, ecosystems & environment, 73*(2), 129–140.

Svanes, E., & Aronsson, A. K. S. (2013). Carbon footprint of a Cavendish banana supply chain. *International Journal of Life Cycle Assessment, 18,* 1450–1464.

von Bahr, B., & Steen, B. (2004). Reducing epistemological uncertainty in life cycle inventory. *Journal of Cleaner Production, 12,* 369–388.

Van der Sluijs, J. P., Janssen, P. H. M., Petersen, A. C., Kloprogge, P., Risbey, J. S., Tuinstra, W., & Ravetz, J. R. (2004). *RIVM/MNP Guidance for uncertainty assessment and communication: Tool catalogue for uncertainty assessment*. Utrecht University: The Netherlands. http://www.nusap.net/downloads/detailedguidance.pdf. Accessed 28 Dec 2013.

van der Werf, H. M. G., Tzilivakis, J., Lewis K., & Basset-Mens, C. (2007). Environmental impacts of farm scenarios according to five assessment methods. *Agriculture, Ecosystems and Environment, 118,* 327–338.

Wallén, A., Brandt, N., & Wennersten, R. (2004). Does the Swedish consumer's choice of food influence greenhouse gas emissions?. *Environmental Science & Policy, 7*(6), 525–535.

Wackernagel, M., & Rees, W. (1997). Perceptual and structural barriers to investing in natural capital: Economics from an ecological footprint perspective. *Ecological Economics, 20,* 3–24.

WBCSD/WRI. (2009). *The greenhouse gas protocol. A corporate accounting and reporting standard.* World Resources Institute-World Business Council for Sustainable Development, Washington, DC.

Wei, X. M., Chen, B., Qu, Y. H., Lin, C., & Chen, G. Q. (2009). Emergy analysis for 'Four in One' peach production system in Beijing. *Communications in Nonlinear Science and Numerical Simulation, 14*(3), 946–958.

Weidema, B. P., Thrane, M., Christensen, P., Schmidt, J., & Løkke, S. (2008). Carbon footprint. *Journal of Industrial Ecology, 12*(1), 3–6.

Williams, A., Pell, E., Webb, J., Moorhouse, E., & Audsley, E. (2008). *Strawberry and tomato production for the UK compared between the UK and Spain.* Proceedings of the 6th International Conference on LCA in the Agri-Food Sector, pp. 254–262.

Yoshikawa, N., Amano, K., & Shimada, K. (2008). *Evaluation of environmental load on fruits and vegetables consumption and its reduction potential.* In Proceeding of: The 9th International Conference of Ecobalance, pp. 780–782.

Zackrisson, M., Rocha, C., Christiansen K., & Jarnehammar, A. (2008). Stepwise environmental product declaration: Ten SME case studies. *Journal of Cleaner Production, 16,* 1872–1886.

Conclusions

Over the last two decades the agri-food industry and all the involved stakeholders, including consumers, have been increasingly concerned about food quality and safety, as well as its sustainable production and the environmental risks and impacts linked to agri-food production processes. The "Life Cycle Thinking" approach can be one way for addressing these issues, using the Life Cycle Assessment methodology as the operative and perfect tool for finding new and alternative methods of agri-food production which can reduce environmental impacts throughout systems' life cycles.

This book represents an attempt of the "Food and Agro-industrial" Working Group of the Association of the Italian Network of LCA to highlight, in an as much as possible exhausting manner, environmental hotspots, methodological issues and best practices for the agri-food sector from a life cycle perspective with particular reference to some of the most relevant and productive agri-food supply chains within the European context, namely: olive oil, wine, cereal and derived products, livestock and derived edible products, and fruit. This book is also designed with the intent to represent a valid support tool for LCA practitioners and all the related stakeholders when developing LCAs in the agri-food sector.

As already mentioned in the preface, the book is articulated in six chapters. The first one represents an introduction and a reference basis for the following ones, providing an as exhaustive as possible overview of the key concerns, applications, and methodological issues of agri-food LCA. The relevant role and commitment of the European Commission and governments toward issues of sustainable production and consumption is highlighted together with the large number of existing eco-labelling and footprint systems. Furthermore, an overview is given of the general methodological issues arising from the development of an LCA study of a food product.

The remaining five chapters develop an as comprehensive as possible review of the state-of-the art of all the international LCA case studies, specific to each of the above mentioned agri-food supply chains. This review is the scientific basis with which the specific methodological problems are then identified together with all the hotspots of each supply chain. Each chapter also illustrates some best practices to

B. Notarnicola et al. (eds.), *Life Cycle Assessment in the Agri-food Sector*,
DOI 10.1007/978-3-319-11940-3

overcome such issues, some of which are more closely related to specific phases of each supply chain and others are common to all food sectors. In the latter case, in general terms, the agricultural stage has the highest impact during the life cycle with eutrophication, acidification and land use being the most significant categories. Also, overall, animal products, as opposed to vegetable foods, have the highest energy use. As far as the packaging phase is concerned, its impacts depend on the materials and end-of-life treatment options used, whilst transportation can have a high impact. Organic approaches can have beneficial effects on some impact categories but are at times counterbalanced by a larger energy use for fieldwork and lower yields, which in turn lead to overall greater land occupation needed for production and higher levels of eco and human toxicity.

Furthermore, in general, the main aspect that has to be underlined is that, due to the fact that the agri-food sector is characterised by so many different methodological issues, a single generally valid framework methodology cannot be designed. In fact, unlike LCA of traditional industrial products, it is more than evident that the LCA methodology applied to the agri-food sector has to take into account several key issues such as, for example, site regional-dependency, pedoclimatic conditions and data quality. What emerges is thus a need for a careful consideration of aspects regarding the choice the functional unit(s) and system boundaries, the use of as much as possible regionalised data, and the development of careful sensitivity analysis, in order to develop qualitatively representative LCAs useful for a continuous sustainability improvement of agri-food product systems.

MIX
Papier aus verantwortungsvollen Quellen
Paper from responsible sources
FSC® C105338

If you have any concerns about our products,
you can contact us on
ProductSafety@springernature.com

In case Publisher is established outside the EU,
the EU authorized representative is:
Springer Nature Customer Service Center GmbH
Europaplatz 3, 69115 Heidelberg, Germany

Printed by Libri Plureos GmbH
in Hamburg, Germany